SMART DATA PRICING

WILEY SERIES ON INFORMATION AND COMMUNICATION TECHNOLOGY

Series Editors: T. Russell Hsing and Vincent K. N. Lau

A complete list of the titles in this series appears at the end of this volume.

SMART DATA PRICING

Edited by

Soumya Sen, Carlee Joe-Wong, Sangtae Ha, and Mung Chiang

WILEY

Published by John Wiley & Sons, Inc., Hoboken, New Jersey
Published simultaneously in Canada

Library of Congress Cataloging-in-Publication Data:

Smart data pricing / edited by Soumya Sen, Carlee Joe-Wong, Sangtae Ha, and Mung Chiang
pages cm
Includes index.
ISBN 978-1-118-61166-1 (hardback)
1. Telecommunication–Pricing. 2. Resource allocation. I. Sen, Soumya, 1982-
HE7631.S57 2014
384′.043–dc23 2013051204

Printed in the United States of America

ISBN: 9781118611661

10 9 8 7 6 5 4 3 2 1

CONTENTS

FOREWORD

Smart phones, tablets, and other video and music streaming devices fuel an exploding demand for network, cloud, and content services. Providers find it difficult to increase revenue to match the investments required to address this demand. The wireless networks are getting stressed and the quality of service suffers. The experience of other industries suggests that smarter pricing mechanisms might improve the matching of resources and users and the revenue of providers, thereby increasing user welfare both in the short term and long term. Researchers are exploring this possibility and a number of recent workshops on this topic attest to the perceived urgency of developing effective approaches.

This collection of papers presents the analysis of the pricing of network services and content conducted by leading researchers from industry and academia. The topics include the following: the tension between the users' preference for simple tariffs and potential benefits of more complex schemes; the users' sensitivity to quality of service and their willingness to shift demand; economic incentives for efficient caching and infrastructure improvements; and pricing schemes for content and for cloud resources.

Researchers will welcome this timely and broad coverage of Smart Data Pricing (SDP).

JEAN WALRAND
University of California, Berkeley, CA

PREFACE

As the demand for data in both wired and wireless broadband networks continues to grow every year, Internet Service Providers (ISPs) are increasingly turning to pricing both as a congestion management tool and as a revenue generation model. This evolution in the pricing regime is evidenced by the elimination of flat-rate plans in favor of \$10/GB or higher usage based overage fees in the United States and various other countries in Asia and Europe. This rapid transition from unlimited data plans to a reign of penalty-based mechanisms, including throttling, capping, and usage-based fees, all within a span of just 4 years as witnessed in the United States is shown in Figure 1. Consequently, Smart Data Pricing (SDP) will play a major role in the future of mobile, broadband, and content. SDP refers to a departure from the traditional flat-rate or byte-counting models to considering pricing as a network management solution. Thus, SDP will impact not only end users and network operators, but will also engage content providers, policy makers, mobile advertisers, vendors, and device suppliers. SDP incorporates the following principles:

1. ***Pricing for end-user Quality of Experience (QoE) and not just linear byte-counting***: Simple policies like usage-based pricing (byte-counting) (i) force users to pay the same amount per unit of bandwidth consumed irrespective of the congestion levels on the network, and (ii) fail to account for the fact that different applications have different bandwidth requirements to attain a certain QoE for the user. SDP should try to match the price for delivering application-specific desired QoE requirements of the user to the ISP's congestion cost at the time of delivery.
2. ***Application layer control to impact physical layer resource management***: Today's smart devices, with their easy-to-use graphical user interfaces, can potentially enable consumer-specified choice for access quality. Whether done manually or in an automated mode, users' specifications of their willingness to pay for their desired QoE of different applications can be taken in as inputs at the APP layer and used to control PHY layer resource allocation and media selection (e.g., WiFi offloading versus 3G). But enabling this interaction requires consumer trials to understand how to design incentives and create interfaces that can be effective in modifying end-user behavior.

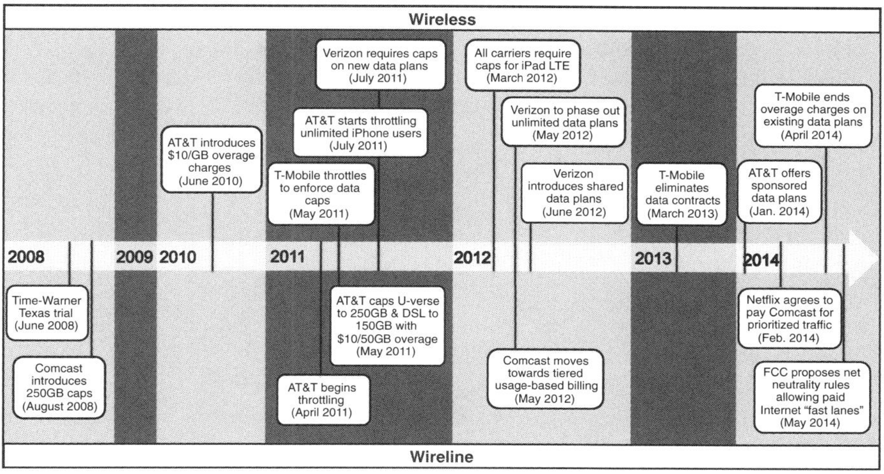

Figure 1 Timeline of the evolution in pricing plans in the United States.

3. ***Incorporating edge devices as a part of the network management system***: Instead of only managing traffic in the network core, SDP explores ways to make edge devices (e.g., smart mobile devices and customer-premise equipments like gateways) a part of the network resource allocation and management system. For example, instead of throttling traffic in the network core using the policy charging and rules function (PCRF), the edge devices (e.g., home gateways) themselves can locally regulate demand based on a user's budget, QoE requirements, and network load or available prices. Such measures to push control from the network core out to end users, while preserving the end-to-end principles of the Internet, have been gaining attention among networking research groups (for example, the M3I[1] collaboration in Europe).

SDP can refer to (a) time/location/app/congestion dependent dynamic pricing, (b) usage-based pricing with throttling/booster, (c) WiFi offloading/proactive caching, (d) two-sided pricing/reverse billing/sponsored content, (e) quota-aware content distribution, (f) shared data pricing, and any combination or extension of the above. For instance, two-sided pricing can include QoE enhancements, or it may simply refer to content providers partially subsidizing data. SDP can benefit end users, network operators, and content providers by improving users' Quality of Experience; lowering ISP congestion and CapEx/OpEx, thus increasing their revenue/profit margin and decreasing churn, and encouraging more consumption and ad revenue for content/app providers. But to realize these benefits, SDP requires pricing models that capture the interplay between technical and economic factors, as well as interfaces between network providers and content & application providers; effective user interface designs;

[1] http://www.m3i.org/

field trials; and a combination of smart ideas, systematic execution, and informed policy.

This volume of collected essays on SDP has immensely benefitted from the annual SDP Forum, which organizes workshops to bring together industry experts, academics, and regulators for in-depth discussions on the topic. SDP 2012 was held in Princeton, New Jersey, and the SDP 2013 and 2014 Workshops were was held in conjunction with IEEE INFOCOM in Turin, Italy and Toronto, Canada. The workshops have been attended by professionals from AT&T, Verizon, Comcast, NECA, Alcatel-Lucent, Cisco Systems, Qualcomm, Microsoft, ACS, and many other leading networking companies. It therefore comes with little surprise that several of the chapters in this volume have been contributed by industry researchers and showcase some cutting-edge research in this area.

The first three chapters of this book discuss SDP's feasibility in the current Internet ecosystem. The first chapter looks back on previous efforts to promote SDP and asks whether the current market climate will be more receptive. The next chapter approaches SDP's feasibility from a customer perspective, using empirical data to examine their price sensitivity. Finally, the third chapter incorporates regulatory concerns by examining network neutrality in the context of content caching.

The next three chapters address SDP's technical feasibility. The first chapter in this section develops a pricing model that accounts for the flexibility and predictability of customer demand. The second chapter focuses on wireless networks, showing how pricing can be used to make wireless resource allocation more efficient. The last chapter focuses on SDP's interface between ISPs and users, examining how the ISP can communicate prices to users through interfaces on their devices.

The next three chapters of the book shift to variants on usage-based pricing, a particular form of SDP. The first chapter examines whether usage-based pricing can in fact help ISPs by quantifying the distribution of infrastructure costs among ISP customers. The next two chapters then turn to differentiated pricing: the first of these develops a model for differentiated usage-based pricing, while the second examines the benefits of non-differentiated and differentiated pricing for ISPs and end users.

Another form of SDP, content-based pricing, is discussed in the next four chapters. The first chapter discusses a variant of usage-based or capped pricing, in which content providers subsidize the delivery of their content to end users, sponsoring users' Internet access. The second chapter shifts the focus to content delivery networks and the impact of competition on their pricing and investment, while the third chapter discusses the economics of a hybrid model in which content delivery can be offloaded to a secondary P2P network during congested times. The last chapter considers the economics of content providers, focusing on how the owners of user-generated content platforms, e.g., social networking websites, can best monetize this content.

The next four chapters discuss technical aspects of realizing economically efficient models of content delivery. The first chapter investigates the idea of opportunistic content transfer, offloading traffic to times of lower congestion with a monetary discount given during times of lower congestion. The next chapter considers a similar idea, in which sessions like content transfers can be spread over time, but with prices determined by the deadline of each session's completion. The third chapter focuses on video content, and shifts the focus away from ISPs to consider how a user might distribute a budget for consuming videos over time. Finally, the last chapter considers multicast technology and how it can alleviate network congestion.

The last two chapters of the book consider pricing in the cloud. The first chapter investigates and compares three different schemes for pricing data center resources, namely real-time instance pricing, deadline-based service level agreements, and time-dependent pricing. The last chapter proposes using combinatorial auctions to price and allocate resources in a data center while taking into account its electricity constraints.

The diversity of topics explored in these book chapters reflects SDP's broad potential impact. Indeed, SDP brings together ideas from such diverse fields as network engineering, economics, human-computer interaction, data science, and technology policy to answer fundamental questions about broadband pricing. Yet there remain significant emerging themes which this book does not cover. For instance, little rigorous analysis has been done on shared data plans, which have recently become mainstream in the U.S. Perhaps more significantly, "network neutrality" is emerging as a fundamental issue, with new regulations from the FCC and Netflix's agreement with Comcast to pay for a separate "fast lane" for its streaming traffic. And as more and more devices become connected to the Internet, pricing for the Internet of Things is becoming an important question. The emergence of these and other topics will ensure that SDP remains an exciting and relevant research topic in the years to come.

SOUMYA SEN, CARLEE JOE-WONG, SANGTAE HA, AND MUNG CHIANG

ACKNOWLEDGMENTS

We would like to thank all of the participants of the first, second, and third Smart Data Pricing Workshops, held respectively in Princeton, New Jersey on July 30 and 31, 2012; Turin, Italy on April 19, 2013; and Toronto, Canada on May 2, 2014. We are also grateful to all of the contributing authors for their time and effort, as well as our colleagues who served as reviewers for the contributions.

CONTRIBUTORS

MATTHEW ANDREWS, Bell Labs, Alcatel-Lucent, Murray Hill, NJ

RANDEEP BHATIA, Bell Labs, Alcatel-Lucent, Murray Hill, NJ

SID BHATTACHARYYA, Department of Information and Decision Sciences, University of Illinois at Chicago, Chicago, IL

JIASI CHEN, Princeton University, Princeton, NJ

MUNG CHIANG, Princeton University, Princeton, NJ

OZGUR DALKILIC, The Ohio State University, Columbus, OH

UMAMAHESWARI DEVI, IBM Research, Bangalore, India

RON DIBELKA, National Exchange Carrier Association, Inc., Whippany, NJ

HESHAM EL-GAMAL, The Ohio State University, Columbus, OH

ATILLA ERYILMAZ, The Ohio State University, Columbus, OH

SERGE FDIDA, UPMC, Paris, France

VIJAY GABALE, IBM Research, Bangalore, India

LIXIN GAO, University of Massachusetts, Amherst, MA

AMITABHA GHOSH, UtopiaCompression Corporation, Los Angeles, CA

VICTOR GLASS, National Exchange Carrier Association, Inc. Whippany, NJ

BHAWNA GUPTA, Bell Labs, Alcatel-Lucent, Murray Hill, NJ

LÁSZLÓ GYARMATI, Telefonica Research, Barcelona, Spain

SANGTAE HA, University of Colorado, Boulder, CO

JIANWEI HUANG, The Chinese University of Hong Kong, Hong Kong, China

CARLEE JOE-WONG, Princeton University, Princeton, NJ

SHIVKUMAR KALYANRAMAN, IBM Research, Bangalore, India

GEORGE KESIDIS, The Pennsylvania State University, University Park, State College, PA

FATIH KOCAK, The Pennsylvania State University, University Park, State College, PA

RAVI KOKKU, IBM Research, Bangalore, India

ATANU LAHIRI, University of Washington, Seattle, WA

TIAN LAN, George Washington University, Washington, DC

NIKOLAOS LAOUTARIS, Telefonica Research, Barcelona, Spain

SHUQIN LI, Alcatel-Lucent Shanghai

BENJAMIN LUBIN, Boston University Boston, MA; Harvard University, Cambridge, MA

DOUG LUNDQUIST, Department of Information and Decision Sciences, University of Illinois at Chicago, Chicago, IL

ANDREW ODLYZKO, University of Minnesota, Minneapolis, MN

ARIS M. OUKSEL, Department of Information and Decision Sciences, University of Illinois at Chicago, Chicago, IL

ULAS OZEN, Ozyegin University, Istanbul, Turkey

DAVID C. PARKES, Harvard University, Cambridge, MA

MARTIN I. REIMAN, Alcatel-Lucent Bell Labs

SHAOLEI REN, Florida International University, Miami, FL

MIHAELA VAN DER SCHAAR, University of California, Los Angeles, Los Angeles, CA

SOUMYA SEN, Carlson School of Management, University of Minnesota, Minneapolis, MN

MICHAEL SIRIVIANOS, Cyprus University of Technology, Limassol, Cyprus

YANG SONG, University of Massachusetts, Amherst, MA

RADE STANOJEVIC, Telefonica Research, Barcelona, Spain

STELA STEFANOVA, National Exchange Carrier Association, Inc., Whippany, NJ

JOHN TADROUS, The Ohio State University, Columbus, OH

CHEE WEI TAN, City University of Hong Kong, Hong Kong, China

ARUN VENKATARAMANI, University of Massachusetts, Amherst, MA

QIONG WANG, University of Illinois Urbana-Champaign, Champaign, IL

YU XIANG, George Washington University, Washington, DC

ALAN D. YOUNG, P & Y Associates, LLC

LIANG ZHENG, City University of Hong Kong, Hong Kong, China

PART I
Smart Data Pricing in Today's Ecosystem

1 Will Smart Pricing Finally Take Off?

ANDREW ODLYZKO

1.1 INTRODUCTION

Will smart pricing dominate telecommunications? We certainly do see growth in sophisticated pricing in many areas of the economy. Congestion charges for cars entering central business districts and "smart" electric meter deployments are spreading. Airlines are even beginning to auction seat upgrades [1]. And there is no shortage of desire for smart pricing in telecommunications. For a survey of recent developments, see Reference 2. Many new technological developments, such as software-defined networking (SDN), are touted as facilitating differentiated services and differentiated pricing. The overwhelming consensus of the industry, as well as of the research community, and of regulators, is that flat rates are irrational. Thus, for example, in 2011, Jon Leibowitz, the then-Chairman of the US Federal Trade Commission could not "quite understand why something like metering hasn't taken off yet." (See Reference 3 for references to this and similar recent quotes, as well as for a summary of the arguments in favor of flat rates.)

Yet there are reasons for caution in the rush to smart pricing. After all, the modern consensus about its desirability is not new. It goes back centuries, to the days of snail mail. Furthermore, industry has often either stumbled onto flat or almost flat rates, or been forced into them, all against its will, and ended up benefiting. Thus, for example, US wireless service providers have been boasting of the low per-minute voice call revenues that reign in United States, much lower than in most of the world. What they universally neglect to mention is that these low prices are the result of the success of the block-pricing plan introduced by AT&T Wireless in 1998, which also eliminated roaming and long-distance charges. This plan, the result not of a careful study of historical precedents or the economics of communications but rather the fruit of a desperate carrier looking for a way to gain customers, was widely derided but proved unexpectedly popular. It forced the rest of the industry to follow suit with

Smart Data Pricing, First Edition. Edited by Soumya Sen, Carlee Joe-Wong, Sangtae Ha, and Mung Chiang.

similar plans and led to large increases in voice usage (see, e.g., the chart in Reference 4). The end result is that the United States has the world's highest per-subscriber voice usage, yielding those low average per-minute prices that the industry boasts of. Probably not coincidentally, US wireless service providers are among the world's most profitable. This story, and others similar to it, should make one cautious about rushing to follow the industry consensus. This is true even when such a consensus is fortified by scholarly studies, because those tend to be even more biased toward fine-grained pricing. The telecom industry and telecom researchers have historically been notorious for not understanding what is in the industry's own interests.

The traditional preoccupation with smart pricing is likely to be reinforced by the economics of telecom. Contrary to common opinion, it is not all that capital intensive. As is demonstrated in Section 1.8, telecom is simply not in the same category as such large and important industries as electricity or roads when it comes to the ratio of capital investment to revenues. Telecom is primarily about service, customer inertia, and territorial strategic plays (where the territories may be physical or virtual).

Although the telecom industry is not very capital intensive, communications is extremely valuable and any society is willing to pay astonishing amounts for it. As an example, by some measures, the United States spends almost 50% more on telecom services than it does for electricity. (See Section 1.5 for more data and references.) Furthermore, in spite of all the complaints from the industry about its supposedly impoverished state, there appears to be very large profits in many parts of it. As this passage is being written in the summer of 2013, Verizon is in the process of buying out Vodafone's 45% stake in the Verizon Wireless unit for $130 billion. This means that the whole of Verizon Wireless is being valued at almost $300 billion. As will be shown in Section 1.9, that is about four times the cost of replacing all the tangible assets of that enterprise. It is also almost enough to replace the entire US telecom infrastructure, both wireless and wired, with the latter redone in fiber. This is anomalous by traditional standards, but then, as will be discussed in Section 1.9, the entire economy is behaving anomalously, with very high corporate profits, low interest rates, and low capital investment. Whether this is a temporary aberration, or whether we are in a new economic era, remains to be seen. However, telecom is very much in the mainstream of this historically unusual behavior, and so many traditional yardsticks of financial performance may not apply.

While the telecom industry has often been blind to profitable opportunities, it has always been aware that high profits are possible. However, it has usually faced difficulties in using their favorite methods for profit extraction because of various combinations of legal and regulatory constraints and the peculiar nature of demand for communication services. Table 1.1 shows an approximation of current prices paid by users for varying amounts of data from various services.

This table demonstrates the main problem faced by telecom. The most valuable information can often be conveyed in just a few bits. Thus, for example, in the early days of postal services, when receivers paid on delivery, information would often be transmitted in the form of small modifications in the address. The addressee would then scan the envelope, figure out what the message was, and refuse to accept (and pay for) the letter.

TABLE 1.1 Price per Megabyte

SMS	$1000.00
Cellular voice	1.00
Wireline voice	0.10
Residential Internet	0.01
Backbone Internet	0.0001

Practices from two centuries ago may seem irrelevant, but in fact they are very instructive, as the basic economic issues have always been the same, even as technology has changed drastically, cf. [5]. Thus, for example, today, we have the telecom industry investing heavily in deep packet inspection. In the past, post offices had employees hold letters up against burning candles to make sure that there were no enclosures that were subject to extra fees. The basic incentive is to extract as much value as possible, and that usually requires fine-grained pricing to achieve successful price discrimination. But usually, in communication as well as in transportation, limits are placed on what service providers are allowed to do. The net neutrality debate is just another instance of the ancient conflict between economic efficiency and fairness in markets [6]. Giving unfettered control of any critical service to any provider, or an oligopoly of providers, either *de jure* or *de facto* (by allowing natural monopoly mechanisms to operate), is equivalent to abolishing property rights with the usual negative impacts on innovation and efficiency. Hence, we have almost always had constraints, such as those of common carriage. The real question is about the appropriate level of constraints.

Public talk of capacity limits is often just a public relations measure, designed to overcome opposition to service provider strategies. Thus, for example, in early 2013, Michael Powell, the President of the US cable industry association [and former Chairman of the Federal Communications Commission (FCC)] admitted, contradicting many earlier declarations by a variety of executives and experts, that "cable's interest in usage-based pricing was not principally about network congestion, but instead about pricing fairness" [7]. Whenever business leaders talk of "fairness," it is generally safe to assume that they are really after extracting more revenues through differential pricing. This is neither a novel nor is it nefarious. In fact, differential pricing was and is at the core of regulatory economics, as it can be used to promote social welfare, and has been frequently mandated by governments. However, historically, the degree of price discrimination that was allowed varied depending on economics, with more discrimination being allowed when the costs of providing those services have been large [8]. The question for the near future is whether modern telecom should be allowed more power to discriminate. Further, even if it is given that power, one should consider whether it would be wise to use it. The right answer depends on the balance between growth in demand and improvements in technology.

The main problem, past, present, and future, that is faced by telecom is that the most valuable information usually requires just a few bits to convey. The second main problem is that because of technological progress, transmission capacity is growing. Thus the industry is faced with the challenge of persuading users to pay for big pipes when the additional value that enlarging those pipes provides is not all that high. (There are arguments that the value of transmission capacity, as well as that of computing power and storage, should be measured on a logarithmic scale, so that going from what is now a slow 1 Mbps link to a 1 Gbps one corresponds only to an increase in value from 6 to 9, cf. [9].) At the moment, that additional capacity is consumed largely by video. But the value is still dominated by the low bandwidth voice and texting.

The general conclusion of this work, based on the study of trends in demand and supply, is that in wireline communication, the critical issue faced by the telecom industry is not handling overpowering exafloods of traffic, as has often been claimed, cf. [10–12], but stimulating demand to fill the growing capacity of transmission systems [13]. The most effective way to do that is to offer flat rates and open access to encourage innovation. To the extent that any market segmentation is needed, it is best handled by offering flat rate services with different peak speeds. Pricing by volume of traffic (whether using caps or other schemes) may be attractive at the moment to service providers preoccupied with trying to protect their traditional subscription video service revenues. However, it is an ineffective instrument that does not address any of the issues well and, in the long run, is likely to damage not only the economy as a whole but also the profits of service providers. Any truly "smart pricing" measures, such as congestion charges, are likely to be detrimental to the industry.

These general conclusions for wired communications apply directly mainly to the richer and more industrialized countries. Even in those, there is likely to be exceptional situations where the cost structure forces some "smart pricing" approaches. For poor countries, the best choices along the frontier of feasible technological and business models is likely to lean further toward "smart pricing." This would be consistent with the general observation, cf. [5], that at the consumer level, sophisticated pricing is most appropriate for large and relatively infrequent transactions, and simple pricing for small and frequent ones. This is also what we observe in the market today, with the greatest proliferation of "smart pricing" in less-developed countries, where the relative burden of telecommunications charges is higher.

In wireless communication, the optimal choice even in rich countries appears to be different than that for wireline, because of a different balance between feasible supply and potential demand. There have been widespread projections that wireless data traffic would continue to double each year, as it had done for several years. Those are now being disproved, as growth rates are declining (see Section 1.13). Still, those rates are high, and there is far more traffic that are likely to use the radio path if that were feasible, as wireless data traffic is under 5% of wireline. Coupled with the low value of most of this data, and the resulting low likelihood of service providers being able to extract large new revenues, it appears probable that the incentives for the industry will be to constrain usage and to implement differentiated quality of service to protect the most valuable low bandwidth applications. So somewhat finer-grained pricing is likely to prevail in this domain than in wireline. Still, the need to limit what Nick Szabo [14] has aptly called the *mental transaction costs* involved in fine-grained pricing, and related concerns, is likely to restrict the complexity of schemes that succeed. The sophisticated pricing plans so beloved of researchers are likely to be confined to areas such as business-to-business dealings and may be of limited applicability even there.

However, the strong prejudice in favor of "smart pricing" among both industry leaders and academic researchers guarantees that many schemes will be developed, and quite a few will be deployed. Chances are that, as was true of many sophisticated prioritization schemes developed for voice private branch exchanges (PBXs) or early data switches, they will not see much use. But for those cases where they might be used, it appears that most of current research, as well as academic instruction, is missing some important ingredients. As is discussed in Section 1.12, it will likely be important to explore the most effective ways to introduce noise and other impairments into communication systems to provide differential quality of service. (On the other hand, there will likely also be demand for methods to detect such actions.)

The next section reviews briefly some of the main fallacies that invigorate the push for "smart pricing." This is followed by a section on some missed opportunities in telecommunications, demonstrating how this industry tends to "stumble to success," pursuing mistaken goals, and prospering by accident. Section 1.4 has a very brief discussion of the reasons telecom has been so poor at innovating in services and is likely to remain poor in the future. Section 1.5 discusses this industry's place in the entire economy. Section 1.6 points out that high profits have not infrequently been obtained in this sector. Section 1.7 sketches the main changes that have taken place in the money flows in telecommunications in recent decades. Section 1.8 demonstrates that, contrary to general opinion, this industry is not all that capital intensive. Section 1.9 discusses some of the puzzles of the modern economy and the degree to which the telecom industry exhibits similar behavior. Section 1.12, cited earlier, discusses some missing ingredients in modern research and education, should "smart pricing" become widespread. Sections 1.10 and 1.11 take a historical look at some earlier work on telecom pricing and the degree to which it reflected the prejudices we observe today. Sections 1.13 and 1.14 then discuss the growth in the demand for data traffic and improvements in transmission technologies and what the contrasts are with those that for optimal pricing strategies. Finally, Section 1.15 provides the conclusions.

1.2 TELECOM MISTAKES

Many of the basic but general issues that have a bearing on the possible adoption of smart pricing have already been explored in the literature (see, e.g., [5, 15–17]) and so will be touched on very lightly here. However, they do need to be mentioned, because there are many misapprehensions about the nature of telecom and these issues often have an important bearing on the optimal choices of pricing policies. For example, we are constantly told that content is king. (Content is taken here to mean material prepared by professionals for wide distribution and not, as some use it, to denote anything in digital form.) But

Content is not king.

Yes, content, in the sense of material prepared by professionals for wide distribution, is important. But it is simply nowhere near as important as basic connectivity, and the revenues of various services reflect that. This is discussed in detail in References 5, 18. Evidence of this fundamental fact is all around, and some of this will show up later in this paper (e.g., in the observation that US wireless carriers have revenues about three times as large as those that the cable industry derives from subscription video). However, content has historically attracted disproportionate attention and continues to do so today. For example, an article in the *Economist* [19] stated

> A common saying in the industry is that Mexico's phone sector may be about four times more valuable than the television market, but the latter is four times as powerful.

What is especially perplexing about the centuries-old preoccupation with content is that content is not cheap. For telecom service providers to sell content, they generally have to buy it at high prices. (And so, net of what they pay to content producers, US cable networks appear to be getting more revenue out of Internet access and voice services than out of carrying subscription video and all on a far smaller slice of their transport capacity.) Back in 2005, Ed Whitacre, then the CEO of AT&T, caused a flare-up in the net neutrality debate with his threat that he would not let Google use his wires without payment. Strangely enough, it is not clear if anybody raised the question as to whether his basic premise was correct, that is, in the absence of any legal or regulatory constraint, it would be Google paying AT&T. Why should not AT&T have to pay Google? Perhaps Whitacre was right, and Bing might have been an acceptable substitute for Google search for AT&T customers. But perhaps not. Imagine that Whitacre had said he was not going to let ESPN or HBO use AT&T's U-Verse wires without payment. Instead of being called *evil* by small groups of advocates of an open Internet, he surely would have been called *insane* by almost everybody.

Because content is not king, the vast majority of papers and discussions about net neutrality, industry structure, and related issues are of doubtful relevance. For example, many academic papers start with the assumption that the Internet is a two-sided market. It simply is not. Most of the values that users get from it is not content but simple connectivity, such as being able to tell their friends and business partners they are stuck in traffic. Compared to old communication technologies, the Internet does provide many unique features and, in particular, allows for bridging content and connectivity. (The main search service of Google, which provides the bulk of that company's revenues and profits but very little traffic, is in this intermediate zone, as are most of the facilities of social networks that users care about.) However, the features that matter the most are not the ones that allow content providers to target individual consumers but the ones that allow for group formation and for individuals or groups to become creators and distributors.

Closely allied to the myth that content is king is another extremely widespread and extremely damaging notion, that of streaming video, [20]. However, all the evidence suggests that

True streaming video is, and will remain, a very small fraction of traffic.

Video does dominate current Internet traffic by volume, but it is almost exclusively transmitted as faster-than-real-time progressive downloads. That is the only method that makes sense technologically. (Video conferencing is completely different, but we now have enough experience to be able to predict safely that it will not be contributing giant amounts of traffic.) Furthermore, this was easily predictable and was predicted a long time ago. For example, George Gilder wrote about it two decades ago, and he attributes the idea to Nicholas Negroponte even earlier. Although their prediction has come true, almost everyone thinks that the floods of video they consume are true streaming video. This skews business decisions and public policy discussions, because networks dominated by real-time long-lived data flows of predictable size and with tight latency constraints do indeed lend themselves to many of the pricing and network management techniques that are so beloved by both top managers and telecom researchers, cf. [21].

The myth of real-time streaming video is so pervasive and strong that it also affects networking researchers. For the past decade, this author has been taking polls asking those in the audience to raise their hands if they saw any advantage at all, for anyone, in transmitting video faster than real time. Usually, even among networking researchers, at most, 10% have responded. The highest positive response rates were around 40%, in a couple of groups of audiences packed with researchers working on wireless ad-hoc networks, and who understand that one cannot count on connectivity being maintained, but can use buffers to compensate. (While one can envisage ultra-reliable wired networks, in the wireless arena, this is simply not achievable; there are far too many unpredictable sources of impairments.) This demonstrates that even networking researchers do not know what is happening in today's networks or why it is happening.

The preoccupation with real-time streaming video leads to the constant questioning about the potential demand for high speed access. Who needs gigabit in the home, is the question that is being asked, because the most that most observers can imagine is a few streams that might possibly come to 20 Mbps each in some future high definition (HD) television (TV). This perfectly illustrates the lack of vision not just for the future but on the present that afflicts this industry. After all, why are people buying 300 Mbps home WiFi access points if all they are after is streaming a few movies? Yet such routers are selling, and high speed home access is also selling (when offered at reasonable cost), because they allow for low transaction latency.

The main function of data networks is to cater to human impatience.

This is something that the computer industry, as well as many other competitive industries, whether online search or Internet commerce, understand well. If users do not get their web search results in a second, they go away. On the other hand, the telecom industry has a hard time assimilating this notion. Yet, if you want to download a 8 GB video to your portable device in less than a minute, you absolutely have to have a gigabit link. Hence,

Overprovisioning is not a bug but a feature, as it is indispensable
to provide low transaction latency, which is the main function of data networks.

Once you have overengineered your network, it becomes clearer that pricing by volume is not particularly appropriate, as it is the size and availability of the connection that creates most of the value. That is also what the users perceive directly. Generally speaking (and there are obviously exceptions, buffer bloat can lead to contrary experience), increased bandwidth means that things happen faster, the network is more responsive, etc. This is something immediately perceptible to users. It does not require them to engage in any mental transaction costs to figure out where they are with respect to violating some volume caps, for example.

In wireline, the vision of a largely empty network dominated (initially in value, and eventually likely also in volume) by cascades of mostly machine-to-machine transactions driven by human impatience that was easy to predict a long time ago, cf. [21], does appear to be realistic and likely inevitable. As George Gilder has said, "*You waste that which is plentiful*" and in most wired networks, bandwidth is plentiful. Wireless, though, appears to be different, as will be discussed later.

1.3 VOICE AND OTHER MISSED OPPORTUNITIES IN TELECOM

Correct technological predictions are hard in general, but telecom predictions seem to be worse when compared to other areas. Some of the many mistakes can be excused easily. For example, the popularity of wireless had been consistently underestimated by the industry for several decades. But this was understandable, because the service was novel, and the high value that people had placed on mobility was not easy to predict. (There is a saying that you cannot tell how many people will use a bridge by counting how many swim across a river.) But others are far more surprising and illustrate well how telecom has often "stumbled to success." As just one example, on an e-mail discussion list as recently as the summer of 2006, one of the top technical officers of a major US cable company insisted that the idea of taking some of the bandwidth away from video services and employing it for Internet access was impractical. He insisted that "[t]he vast majority of folk in this country watch analog tv and don't have electronics to consume them digitally, don't want them or can't afford them." Yet today, Internet access is already, or is about to become, the main business of the cable networks.

The most perplexing of the many mistakes that telecom has made is in neglect of voice. Even today, voice services provide the bulk of worldwide telecom revenues, but the industry has not been paying attention. When 3G was being prepared for deployment around the turn of the millennium, industry was touting it as an enabler of all sorts of fancy digital "content" services. But it was obvious that voice offered the greatest profit opportunities [22], and voice has indeed been the main revenue generator for 3G. However, while the industry did benefit from this easy-to-anticipate but unanticipated windfall, it has neglected other opportunities in voice [22]. Those opportunities include voice messaging, and, perhaps most important, high quality voice. Current wireless voice quality is poor, far poorer than the "toll quality" voice standard of wired services. (And that "toll quality" is also poor, given what is possible with modern codecs.) From this, and from the rapid expansion of wireless revenues, the industry appears to have concluded that the public does not care about voice quality. It is far more probable that the public accepted low quality wireless voice in order to gain mobility. But this does not mean that quality could not be sold as an added value feature. It might have provided large additional revenues and profits in the 3G world. There capacity was constrained, and therefore, it would have been possible to charge extra for higher quality. As it is, HD voice, which is part of the plan for long-term evolution (LTE), is likely to just become a standard service, as its resource requirements are low compared to the capacity of the new system.

It is impossible to prove that high quality voice, if deployed and marketed properly, would have been a great success. Soon we may obtain some indication from the public's reaction to HD voice in LTE. But even before that, there were a variety of reasons for believing that voice was promising, including the success of Integrated Digital Enhanced Network (iDEN) with its simple push-to-talk feature. Human culture is primarily an oral one, and we have the astonishing success of the telephone to look back to, which surprised many observers by attracting far more usage and spending than postal services and the telegraph.

Those who denigrate voice can point to data such as that of Table 1.2. It shows steady level of voice traffic on US wireless networks (based on the data from Reference 23), which represents a decline in voice usage on a per-user basis, because the number of subscriptions has been growing during the period covered by this table. It has been surmised that this decline was due to usage migrating from voice to texting. That may very well be true, but it does not necessarily mean voice is unimportant. Texting has major advantages (in particular, being asynchronous, and thus less intrusive than voice), and the phenomenon shown in this table may be an indicator of a substantial opportunity in voice messaging, one that possibly could have generated good revenues in the restricted 3G environment.

Moving forward, the opportunity to gain additional revenues with HD voice appears to be gone, but voice should not be neglected, as it is right now, in a variety of services. Furthermore, it appears that in the development of video services, the industry is neglecting social communication in the traditional preoccupation with content.

TABLE 1.2 Voice to Text Substitution (US)

Year	Voice Minutes billions	Texts billions
2005	1495	81
2006	1798	159
2007	2119	363
2008	2203	1005
2009	2275	1563
2010	2241	2052
2011	2296	2304
2012	2300	2190

1.4 THE TELECOM INDUSTRY AND INNOVATION

The telecom industry has repeatedly shown that it can perform well in increasing transmission capacity. It has also shown itself to be miserably poor at inventing new services. This may very well be the result of a basic cultural mismatch. The basic mission of telecom carriers is to provide ubiquitous connectivity. This is not an easy task, especially because it involves being able to respond to massive disasters, natural or man-made. Most likely, the skills, mindset, and the organization that can accomplish this are simply not tuned to anticipating what the capricious public will want. Even when very smart people with innovative ideas join such organizations, their initiatives tend to be blocked. From this perspective, it would be best, both for the society and for their shareholders, if telcos stuck to their expertise, which is that of providing "dumb pipes." Unfortunately, that is not likely to happen, as their managers (and shareholders) dream of "content" and other glamorous futures.

1.5 THE LARGE TELECOMMUNICATIONS REVENUES

Measuring revenues of the telecommunications sector is not simple. (For example, should one count the home WiFi access points people buy or the cost of the WiFi equipment in a personnel computer (PC) or tablet?) Even concentrating just on the revenues of service providers presents serious problems, as various bundles mix communications with content. However, any reasonable methodology shows that telecom attracts very large revenues. Here we cite some figures from Reference 24, which has extensive statistics (and discussion) based on data up to the year 2011. A very attractive feature is that those statistics cover all the advanced industrialized nations over about two decades and thus provide interesting international comparisons. (It should be mentioned that other sources sometimes show different estimates. For example, for 2011, Table 3.4 of Reference 24 shows US wireless telecom revenues of $210 billion, while CTIA, the industry association, computes it at $170 billion for that year.) In that year, telecom revenues inside the Organisation for Economic Co-operation and Development (OECD) countries came to $1.35 trillion, with United States accounting for $526 billion (others sometimes cite figures as low as $350 billion for the United States). Hence, it seems safe to estimate worldwide telecom revenues in 2011 as being close to $2 trillion. About half of the revenues (for OECD and, therefore, likely for the whole world) comes from wireless.

For comparison, worldwide advertising spending for 2013 was projected to come to $518 billion, which was only around a quarter of telecom revenues [25]. (In the United States alone, advertising is more significant, as at $172 billion it comes close to a third of telecom revenues.) As only about $100 billion of advertising goes into online forms, there is still plenty of room for Facebook, Google, and other companies to grow their advertising businesses. But there is no way that the telecom business can be supported by anything in its present size by ads alone.

Yet another interesting comparison (relevant to later discussions of capital intensity) is with the electric power industry. In the United States, total revenues

TABLE 1.3 US Wireless Industry Statistics

Year	Revenues ($ billions)	Capex ($ billions)	Capex/ Revenues
2004	102.1	27.9	27.3%
2005	113.5	25.2	22.2
2006	125.5	24.4	19.4
2007	138.9	21.1	15.2
2008	148.1	20.2	13.6
2009	152.6	20.4	13.3
2010	159.9	24.9	15.6
2011	169.8	25.3	14.9
2012	185.0	30.1	16.3

from electricity sales from end users, residential as well as commercial, came to $364 billion in 2012 (based on the statistics from the US Energy Information Administration). Of this amount, something like a third went to pay for fuel; so, the total amount this industry had to cover for maintenance and nonfuel operations, and provide for profits and interest was only about half of what the telecom industry received.

Yet another interesting comparison is with Google. In 2012, its worldwide revenues came to just about $50 billion. Its growth and profit rates were far higher than that for most telecom service providers, but still, it commanded just 2.5% of the telecom revenue stream. So, telcos will not get rich by squeezing Google alone. (Even squeezing Microsoft, with worldwide revenues of about $80 billion per year, would be of limited help.)

A few other figures are interesting. Some key statistics of the US wireless industry, drawn from Reference 23, are presented in Table 1.3.

Thus from 2004 to 2011, the cellular industry increased its revenues by 66%. The US cable industry increased its revenues during that period from $60.0 billion to $97.6 billion, or 63% [26]. However, residential video grew just from $41.8 billion to $56.9 billion, or 36%, and the bulk of the growth came from the "other" category (dominated by voice and Internet access), which went from $18.2 billion to $40.7 billion, a growth rate of 124%. Content may have all the glamor but this is not where the main action is.

1.6 THE HIGH POTENTIAL FOR PROFITS IN TELECOMMUNICATIONS

The telecom industry has often earned very high profits. For example, the British Post Office had an extraordinarily high net profit margin of 68% in 1839, on the eve of the Penny Post reform [5]. (This was a conscious move to tax first-class letters. It served primarily as just another tax to help pay for the general government expenses, and secondarily as a subsidy for the "content" inside newspapers, which were carried for free.) More recently, over the past few years, Carlos Slim Helú has been ranked as the richest person in the world. This resulted largely from the splendid profits of Telmex and Telcel, which still enjoy dominant positions in Mexican communications and, by most evaluations, manage to keep prices high and penetration of advanced services low in a poor country.

Monopolies have at times been very innovative and have worked to lower costs as well as promote usage. The examples of the pre-1840 British Post Office and of Mexican telecom industry today (as well as many others, including many governments in recent times which milked the telecom sector to support other activities) suggest that in telecom, the incentives may not always point in that direction. Instead, short-term profit maximization can often be achieved by raising prices and limiting usage. Advocates of the Penny Post reform in Britain, not infrequently, promised that the increase in business from the new, lower and simpler, postage rates would compensate for decreased revenue per letter. This did not happen, and the profits from this service declined drastically. Yet, no serious attempts to go back were made, as the reform was wildly popular, both for lowering the costs of communication and for the simplicity it brought, with the complex system of distance-dependent tariffs and limitations to a single sheet dispensed with.

On the other hand, the Penny Post reform did lead to a switch from a regime of static revenues to one of rapid growth. This is a phenomenon that has occurred a number of times when prices were simplified and lowered, a phenomenon that typically does not fit the economic models used to support "smart pricing," which tend to be static. It took a quarter century, but eventually British Post Office profits exceeded those attained before the Penny Post reform [5].

1.7 TELCO (R)EVOLUTIONS

The historical pattern, going back centuries, has been for telecommunications to grow faster than the economy as a whole [5]. That applied also at the end of the twentieth century. Among the OECD countries, telecommunications revenue as a fraction of gross domestic product (GDP) increased from 2.13% in 1985 and 2.36% in 1995 to 3.58% in 2001 (Table 3.2 on p. 77 of Reference 24). That was the high point, though, and over the past few years, it has been close to 3%. One of the contemporary justifications offered for the Internet bubble was that the creation of the Internet, allowing interconnection of the growing number of computers, would yield dramatic productivity improvements, and this would stimulate increased spending on telecom. Some analysts predicted that the fraction of GDP going to this sector would double. It did for some countries (Korea went from 2.05% in 1990 to 4.70% in 2002 and was at 4.36% in 2011), but overall the growth has been far more modest. The United States went from 2.71% in 1995 to 4.10% in 2001 and then down to 3.51% in 2011. Thus it appears that modern economies are only willing to expand around 3% of their output on telecommunications.

What is especially intriguing is that some countries that are not just rich, but have excellent telecom infrastructures, manage to spend only modest amounts on that industry. There are some outliers (Luxembourg and Norway, in particular, with 1.2%

of GDP going into telecom) that can be disregarded, because they have very high incomes per capita, so that looking at fractions of GDP conceals substantial total spending. However, Finland at 2.58% and Sweden at 1.51% (both for 2011) provide intriguing examples that deserve deeper investigations.

In addition to overall growth, there have been large additional changes within the industry. The most obvious one is the rise of wireless. In terms of the number of people served, and the revenues and profits, it dwarfs the Internet. (It was also built primarily on the value of the low bandwidth voice and messaging services, and until recently, the contribution of content to this growth has been negligible.) According to statistics in Reference 24 (Table 3.4), mobile revenues accounted for 47.8% of total telecom revenues in the OECD countries in 2011, 39.9% in the United States, and a record high of 84.4% in Japan. (This figure for Japan is suspiciously high, as it is hard to imagine how that country could maintain and expand its wired infrastructure on just 15.6% of telco revenues that came to 2.85% of GDP in 2011. According to Table 4.1.12 in Reference 24, in June 2012 Japan had almost half of the OECD's fiber connections, with 65% of its broadband subscribers on fiber.) Thus the share of GDP that goes to wireline has been decreasing. It appears that wired services survived largely because of a collapse in most of their costs.

In the US setting, a rough rule of thumb a couple of decades ago, before the rise of the Internet, was that access, switching, and long distance each accounted for about one-third of the total cost of the phone system. Today, only access is significant. This can be seen by looking at financials of two prominent companies. Level 3, especially after its absorption of Global Crossing, is universally regarded as the largest Tier-1 backbone carrier. Its share of world Internet traffic has been estimated at 10–20% [partially depending on how one counts its relatively new content delivery network (CDN) business]. Yet its revenues for 2012 were only $6.4 billion. In the worldwide telecom industry with revenues of $2 trillion (or even in the wireline sector of that industry with revenues of $1 trillion), this is extremely small. This demonstrates that long distance transport has become very inexpensive.

The other prominent company is Akamai, the largest CDN company. It has at various times claimed to deliver up to 20% of the world Internet traffic. But its revenues in 2012 were just $1.4 billion. Thus switching (of which Akamai has to do a lot) has also become inexpensive.

The same conclusions about the relatively low significance of long-distance transmission and switching in modern telecom can also be reached by looking at prices for Internet transit (in which large customers, whether Internet Service Providers (ISPs) or businesses or universities, pay for access to the Internet) or for CDN services. At current CDN prices for about $0.01–$0.02 per gigabyte (in large volumes, several petabytes per month), the whole volume of world Internet traffic, still under 50,000 PB per month in mid-2013, would cost only $6–12 billion per year to deliver.

The collapse in costs of switching and transport is what has led to the transformation of the effective architecture of the Internet documented in References 27, 28. (The excess fiber buildout of the Internet bubble was also an enabler of this transformation.) Tier-1 carriers such as Level 3 have become much less significant, as lower ranked ISPs have been interconnecting, and large content providers have been building out their own long-distance networks that allow them to reach the ISPs at the edges.

Various other changes have taken place, often ones that appear not to have been documented. For example, at least in the United States, businesses used to provide a disproportionate fraction of telecom revenues through a conscious and government-sanctioned price discrimination policy. That price discrimination has disappeared, or even reversed, as enterprises are able to obtain advantageous deals in many cases.

Several conclusions appear inescapable when one considers the figures cited earlier. One is that with practically all costs coming from the access piece, that is, (for wired services) installing and maintaining the wire to the end user, the marginal costs of carrying extra traffic are close to negligible. Hence, charging according to volume of traffic cannot easily be justified on the basis of costs.

An even more fundamental implication of the new cost structure is for network engineering and management. An important goal of much of telecom research has been to devise ways to increase the engineering efficiency of the system. We now have practical applications where this was achieved [29, 30]. However, there the high utilization occurred in controlled environments, with high volumes of predictable traffic, and with highly trained professional managing the network. (Something similar has happened to the backbone of the public Internet. The low utilizations that prevailed in the late 1990s, cf. [21], have been increased in many, perhaps most, networks, although there are no publicly available statistics on the subject. This was a result of more attention paid to traffic engineering, as well as slower rates of traffic growth, and slower progress in available transmission technologies.) However, on a global scale, and from the perspective of the welfare of the entire system, any efficiency gains at the core have to be balanced against the costs at the edges. Given the imbalance we have, with edge costs dwarfing those at the core, it makes sense to overprovision the core to an absurd degree in order to keep things simple (and thus inexpensive) for the users at the edges. But of course optimizations are done locally, not globally, so the temptation is always to introduce something more clever that ends up hurting the system.

The final point is that the collapse of costs means that even with a diminished flow of funding for the wireline sector, it is possible to build high capacity networks. The big question is whether one can induce incumbent service providers to do so.

1.8 CAPITAL INTENSITY

The telecom industry frequently boasts of its high capital investments. It is also widely accepted that this industry is characterized by very high fixed capital commitments. But neither of these notions is true. For examples of truly capital intensive businesses, one needs to look at industries such as electric power, railroads, or highways.

In the OECD countries, telecom investment as fraction of revenues was 13.9% in 2011 (Table 3.8 in Reference 24). US wireless service providers have also been investing about 15% of their revenues in recent years, as is shown by Table 1.3. This has been typical also for the wireline telephone industry for the past half a century at least, with the exception of the Internet bubble years. Then investment spiked up to 27.2% of revenues for OECD as a whole. Similarly, the US cellular industry invested close to 30% of its revenues in the early 2000s, during its rapid growth phase. But now it is down to the traditional telco level of about 15%. That is not very high by the standards of industries that are really capital intensive.

A 2006 estimate of what is needed to maintain US highways (not just the interstate system, but all major highways, but excluding local streets) at about their then level of service called for spending $223.4 billion in 2013 (Table 2.1 of Reference 31). Of this amount, $125.1 billion was to be for capital expenditures (with 40% of that for additional capacity) and $98.4 billion for operations and maintenance (ordinary repairs, snow removal, etc.). Actual expenditures have fallen short of these desiderata, but the main point is that transportation experts estimated that a healthy state of their industry required devoting 56% of the total annual expenditures to capital investments.

A good way to compare capital intensity of different industries is to look at replacement costs as compared to annual revenues. In 1913, US railroads had book value of $16.6 billion and annual revenues of $3.2 billion (in current dollars) [32]. Thus it would have taken 5 years of revenue to pay for the investment that went into creating that network. This comparison has to be treated with caution, in that the book value had a lot of what was then called *water* in it, and so was overstated. On the other hand, book value was also understated, in that it was based on historical costs, and a brand new replacement of various lines, with the need to tear down buildings in cities that had become densely populated in the meantime, would have cost far more.

US electric power industry had revenues of $364 billion in 2012, as was mentioned before, and of this something close to a third was accounted for by fossil fuel purchases. Hence, a fairer comparison to telecom, where most of the "fuel" that provides value is generated by users, is closer to $250 billion, or half of telecom revenues. Yet the book value of the US electric power industry is around $1 trillion. Replacement cost is likely to be far higher, as the NIMBY ("not in my backyard") opposition, environmental concerns and the like have driven costs of construction very high. A typical example is presented by nuclear power plants. Until recently, the United States had 104 operational nuclear reactors that provided about 20% of the nation's electricity. While there have been promises of novel designs for nuclear reactors that would be both safe and inexpensive to build, so far none have shown to satisfy both criteria, and current estimates of building new ones are on the order of $5–10 billion each. Hence, just replacing existing US reactors would cost $500–1000 billion.

Compared to roads, railroads, or the electric power industry, telecom is not very capital intensive. Just 1 year of the OECD estimate of $526 billion telecom revenue for the United States would suffice to provide a brand new infrastructure, with fiber to almost every house or business. For example, financial analysts estimate that wiring up from scratch every home in Kansas City with fiber would cost about $1200 per household [33]. At that rate, every one of the roughly 140 million households and business establishments could be covered for just $170 billion. Of course, not every place is as easy to operate as Kansas City, but if we exclude some small proportion of the nation (and the debates are whether this is 3%, 5%, or 10%), we could accomplish a complete rewiring for under $200 billion.

In wireless, industry statistics show that cumulative capital investment, from the start of service three decades ago, came to $365 billion by the end of 2012 [23]. Much of that investment has of course been written off, as old equipment gets replaced. So to replace everything (and it is far easier to replace telecom installations, even cell towers, than it is to replace electric power plants) would probably not cost more than half of the cumulative total, or about $180 billion. But just to be safe, let us assume it would take $240 billion.

When we add this up, we find that a modern telecom infrastructure for the United State, with fiber to almost every premise, would not cost more than $450 billion, which is well under 1 year's annual revenue. But there is no sign of willingness from the industry to spend that kind of money, even though Verizon is willing to pay $130 billion to buy out Vodafone's share of Verizon Wireless.

Hence, we can indeed conclude that modern telecom is less about high capital investments and far more a game of territorial control, strategic alliances, services, and marketing.

1.9 MYSTERIES OF INVESTMENT, COSTS, PROFITS, AND PRICES

The relatively low capital intensity of telecommunications has to be placed in a proper context. Industrialized countries in general have been reducing their capital investments. For example, in 2012, US investment came to only 16.2% of GDP, as compared to about 20% for most of the 1980s and 1990s [34]. Some sources put US fixed capital formation rate even lower. For example, the World Bank reports US investments at 20% of GDP as recently as 2006 but down to 14% in 2010 and 15% in 2011 [35]. (In contrast, China's capital investments came to 46.1% of GDP in 2012 [34].)

The low capital investments in the rich economies is one of the major modern puzzles. Another one is the very high level of profits at the same time when interest rates are very low. (This phenomenon, as well as the historically abnormally high fraction of profits going to the finance sector, predates the crash of 2008 and the resulting action by monetary authorities to drive down interest rates.) Why do not entrepreneurs take advantage of those record low bond rates and compete away those abnormally high profits? (An interesting perspective is that in the nineteenth century, it was taken for granted that, just as basic economic logic would predict, low interest rates go together with low corporate profits. It is only in modern times that the mantra of low interest rates boosting profits has become dominant.) That would reduce the high unemployment and increase growth rates. Yet, that is not happening, showing that the normal dynamics of capitalism are not operating the way theory and historical norms predict.

One worry is that a substantial part of the apparently high profits is an accounting mirage. After all, if profits can be moved around the world to escape all taxmen, so that even Starbucks in the United Kingdom shows up as not making any profits, while the parent company is thriving, how far can one trust any of these figures? But even if this is a factor, other statistics, such as the record lows of GDP going for wages, still present a conundrum.

An illustration of the these modern financial puzzles is provided by the valuation of Verizon Wireless. At the price of $130 billion for the 45% share owned by Vodafone, the entire business is valued at almost $300 billion. Yet, based on the capital investment figures for the US wireless industry cited in Section 1.8, the assets of this service provider could be replaced for something like $80 billion, a quarter of the market value. Traditionally, the Tobin Q (the ratio of market value to replacement cost) has been around 1 for most industries, and recently has been around 2. For wireless to be at twice, even the elevated level found elsewhere is remarkable. It likely reflects the inertia of the system. The limited radio spectrum, multi-year contracts, and the like keep this sector locked up.

Yet another puzzle is the claim that building out fiber networks to the home is impossibly expensive. Yet, at the cost of $1500 per household (in excess of the $1200 estimate of Reference 33 for the Google project in Kansas City, were it to reach every household), and at a cost of a capital of 8% (which we are told is an impossible target for state and municipal pension funds to reach), this would cost only $10 per house per month. The problem is that managers and their shareholders expect much higher rates of return than 8% per year. One of the paradoxes is that the same observers who claim that pension funds cannot hope to earn 8% annually are also predicting continuation of much higher corporate profit rates.

Associated with the puzzle of high profits and low interest rates is the decreased relation of visible costs relevant to the ostensible provision of goods and services to the prices and profits we see in the marketplace. Thus, for example, a recent investigation by a US Senate committee found that among 30 for-profit colleges, "an average of 22.4 percent of revenue went to marketing and recruiting, 19.4 percent to profits and 17.7 percent to instruction" [36]. The American for-profit higher education industry might be thought of as an outlier, in that its real business could be regarded as more to find naive and educationally unqualified people who will sign up for student loans and less to provide real education and training. However, it is not all that unusual (and it is not certain just how high the instructional spending is in the nonprofit educational sector). Consider the pharmaceutical sector. The high cost of research and development (R&D) there is supposed to be driving the economics of this industry. Yet, the statistics for the 19 largest pharmaceutical and biotech companies in the world showed that of their health care revenues of $498 billion, $71 billion went for R&D, $110 billion for profits (a mixture of figures for 2006 and 2008 from Reference 37), and the bulk for various other expenses.

Similarly, surprising statistics occur in telecom. A recent financial analyst report estimated that for Sprint and T-Mobile, the costs of running the network in 2012 came to 31.1% and 26.7%, respectively, while the costs of acquiring new customers (just selling and marketing expenses, and handset subsidies, but excluding overhead expenses, and resources devoted to retaining current customers) were 26.6% and 22.1%, respectively [38].

Statistics such as those mentioned above demonstrate that the focus in our modern world is on seizing strategic bottlenecks in the economy and squeezing high profits out of them with as little investment as possible. Apple is an example of having achieved outstanding success at this (and so is Google). Apple's own R&D expenditures are extraordinarily low for what is a shining light for technical innovation (much less than half of Nokia's R&D spending in 2010, e.g., cf. [39]). Most of the R&D that makes Apple products attractive is performed by suppliers who have to accept very modest profits. More than that, the key advantage that Apple now has is the plethora of apps for its devices, which have been built largely by independent suppliers. These app developers received $6.5 billion over 4 years [40], but there are about 300,000 of these developers, so the average payout has been very modest. On the other hand, Apple has been earning more than twice in profits each quarter as much as all the app developers earned in revenue in 4 years ($13.1 billion in the fourth quarter of 2012).

It is noteworthy that initially Apple had a closed app system, and it achieved success by opening it up to outside developers, but keeping tight control.

The conclusion of this discussion is that modern economy is full of paradoxes and does not fit the traditional model of how capitalism is supposed to function. But in any event, it appears that service providers are reluctant to invest much but are anxious to obtain high profits. Thus the drive for "smart pricing" is natural. On the other hand, the low marginal costs of handling actual traffic means that there is a lot of flexibility in pricing. Hence, user preferences can play a substantial role, no matter what management's favorite models tell them to do.

1.10 A HISTORICAL VIGNETTE: BRIDGER MITCHELL AND FLAT RATES

In evaluating prospects for various pricing plans, it is useful to take a historical view, especially because this is almost never done. For instance, modern arguments in favor of "smart pricing" for data networks are nowhere near as strong as those present a century ago for the voice telephone network. In those days, the industry faced high marginal costs (primarily those of the human operators involved in setting up calls), so the case for charging for calls was overwhelming in the standard model the industry used. Arguments can be made that moving to usage-based pricing was instrumental in promoting the spread of telephony in large cities, cf. [41]. However, what is most remarkable is that flat rates persisted in most of the United States. It does not appear that the industry ever tried to understand how it happened, or how they were prospering with the supposedly irrational and ruinous flat pricing model.

An excellent example of the strength of the "smart pricing" dogma is offered by an article by Bridger Mitchell from 1978 on charging for local voice calls [42]. It was based on an earlier and more extensive Rand report issued 2 years earlier, which acknowledged extensive comments and suggestions from numerous other researchers in the field, and was published in one of the most prestigious journals in economics. Hence, it can be taken as representing the consensus of the established telecom economics community of that time.

Unlike most papers and books in this area, Mitchell did not use derogatory terms for flat rates. However, a deep-seated bias against them pervades the paper. There is also a very obvious ignorance of their effects. Aside from the savings on the costs of the measurement and billing system, there is no hint that flat rates are advantageous for anyone, and it is hard not to draw impression from the paper that they are just a cancerous sore on the body politic and ought to be eliminated as soon as possible.

While Mitchell did mention that "most telephone subscribers prefer flat rates," that is as far as he went. Nowhere did he mention the experiments that showed that vast majorities of people who participated in the AT&T experiments (and later in experiments at GTE) and were offered a choice continued with flat rates, even though it cost them more than switching to what Mitchell referred to as *usage-sensitive pricing* (USP) (also in the literature of the period often called *local measured service* (LMS)). Nor did he mention the analyses of the reasons for the preference. To what extent he had access to such studies is not clear. Some of the people he thanks for "comments and suggestions" were involved in the AT&T studies, but actual publication did not come until after the Mitchell paper appeared (such as References 43–45). However, there was plenty of older historical evidence on telecom pricing and that is also not mentioned.

The Mitchell paper started off with

> Although payment for nearly all other goods and services, including toll (long distance) telephone calls, increases with greater consumption, nearly 90 percent of the residential telephone subscribers and more than half the business subscribers in the United States now pay a flat monthly rate for local calls ...
>
> Recently, however, the telephone companies and regulatory commissions have been moving cautiously towards imposing usage charges for local telephone calls ...

This was then followed by a long recitation, taking up the rest of the first page, of various reasons USP was supposedly advancing or about to advance in the United States. (With the benefit of hindsight, we can tell those reasons were not strong enough. The trend went the other way, with even long-distance voice telephony moving toward flat rates.)

What was the basic justification for the move to USP? "[T]he costs of local service [had] moved upward since the late 1960's at a rate not far below the general price index." (All quotes in this section not attributed to any other source are from the Mitchell paper.) So there was not a crisis caused by surging usage, just the general pressure to avoid raising rates in a high inflation environment that led to the Nixon wage-price freeze, which was followed by even more dramatic price escalation. On the other hand, "technological advances [had] benefited long distance far more than local telephone calling." And what were the advances that Mitchell cited? "Development in microwave communications, coaxial cable, satellites, and waveguides." Certainly, all these contributed to lowering costs of providing phone service. What is interesting is Mitchell listed only advances in transmission and said nothing at all about switching. Electronic switching systems were advancing rapidly by the mid-1970s. The Bell System introduced the 1ESS switch a decade earlier, for example. With the ongoing and widely recognized advances in computing technologies, it should have been obvious that switching costs would decline dramatically (even if they had not at that point) and that this would apply to local as well as long-distance telephony. Mitchell was not unaware of electronic switching; as in discussing the costs of metering local calls, he did mention that such costs were much lower in local offices with electronic exchanges than in those with electromechanical ones, which made USP more attractive. But he completely ignored the effect of the new technologies on the basic costs of switching and thus of providing voice services.

The one cost element that was clearly not going to decline to any appreciable extent was local transmission, which meant primarily maintenance of the copper wire from the local central office to the customer. This was the 1970s, long before the era of the Internet or the various online computer services, and before PCs started penetrating households in any numbers. Hence, only voice services were of interest to regulators, legislators, and the public. So the link to the home was expected to continue as a copper wire. And the costs of installing and maintaining that copper were almost completely independent of usage. Hence, all the discussion about advantages of USP in Mitchell's paper ignored the vital point that the one component of costs of service that could not be appreciably reduced would not be affected by switching to USP.

Mitchell noted that "[a] welfare analysis of two-part tariffs requires knowledge of demand at different prices." However, "no data are available as yet on calling rates under usage-sensitive pricing," so he proceeded to make a variety of assumptions (which he admitted were simplified). Some of the questionable features of those assumptions are noted in the following text. However, this is not a careful review of his paper, so not everything is examined. The point is just to point out some of the more strikingly peculiar features of his work.

Mitchell decided to "[ignore] ... the dynamic effects of the number of subscribers in the telephone network on the value of service to any one subscriber." Today, it appears to be widely understood that the number of users is extremely important in determining value, especially for new products and services. This is shown, for example, by the frequent invocations of Metcalfe's Law, which does convey the message about the value of connectivity, even though there are serious arguments that its quantitative form i s incorrect [46]. Interestingly enough, the first modern studies of network effects were carried out inside the phone industry by Jeff Rohlfs [47] some years before Metcalfe formulated his "Law" and before Mitchell wrote his paper. Still, as Mitchell explained, ignoring this effect was not unreasonable for voice telephony in the United States in the 1970s, where this service had achieved an almost complete penetration. He did cite some published papers that had considered the influence of this effect on the early development of the telephone system.

What Mitchell did not say, but was implicit in his analysis, is that he was also ignoring the intensity of usage as a determinant of value. This is now widely recognized as important. Financial analysts, investors, and the press cite figures from social networks, say, or from more neutral outside monitoring organizations such as comScore about the average length of time that users spend on those networks. The big change in this perception, that is, heavy usage increases value, took place in the early days of the

Internet bubble, when AOL, then the dominant online service provider, switched from metered rates to flat ones [5]. Management resisted the move and was only induced to make it by competition, as it feared (correctly) that the time spent online would jump. But once they moved to the new environment (assisted by improvements in technology, as well as presence of competitive local service providers who were able to lease modems for the AOL dial service at more affordable rates than the ILECs had charged), they came to appreciate the value of intensity of usage and proudly hailed measures showing it was increasing. But that was two decades later, so one cannot fault Mitchell too much for ignoring it, as there were not many examples of this effect playing a major role in telephony in his time. On the other hand, there was extensive historical evidence of such effects, going back to the early days of postal services. An interested researcher could easily have discovered this.

Overall, Mitchell's models are simplistic, but that is of necessity true of all economic models. Still, even given the low level of complexity that Mitchell limited himself to, there are many questionable aspects of his work. For example, there is little analysis of the sensitivity of the results to basic assumptions. (This can be excused to some extent, of course, because this was before the era of widespread of desktop PCs and spreadsheets, so such calculations were not as easy to perform then as they are today.) In particular, Mitchell did not exhibit any combination of basic parameters for his model under which flat rates are optimal. Yet such must exist. The logic of bundling (clearly understood to merchants since time immemorial and explicated in the economics literature in the 1960s and visible in Figure 1 of Mitchell's paper) argues that for some parameters, flat rates provide higher profits for service providers.

The strangest omission on Mitchell's part was of the basic comparison of US and foreign volumes of usage. He allowed that one should do some real studies of the effects of USP. He also wrote that "new insights for US practice could be gained from a comparative analysis of the telephone pricing and demand experience of foreign utilities that have long operated with various forms of measured service." (Note that what he regarded as potentially valuable were just the lessons about applying USP, an indication of the deep bias against flat rates.) And earlier he cited briefly some Norwegian experiences with peak-load pricing. But nowhere did he even hint at the most salient difference between the United States and the rest of the world in the 1970s, namely, that daily usage of a phone line was about three times higher in the United States than in the rest of the world (see Reference 5 for data and references). This dramatic difference (especially when combined with a look at revenues of telcos in different countries) casts doubts on various Mitchell's estimates of marginal costs, as well as his assumptions about marginal utilities and the effects that USP might have on usage.

Another major issue, although one probably inevitable given the kinds of economic models that Mitchell relied on, was the static nature of his analysis. Consumers were supposed to have certain willingness to spend, and once prices were changed, they would adjust quickly and settle into a new equilibrium usage pattern. Yet, what we observe in practice is that changes in pricing lead not only to an instantaneous response but also to changes in long-term growth rates. This occurred when AOL introduced flat rates in 1996. There was a quick jump, followed by vigorous growth, as opposed to the static level of usage under metered rates. Earlier, the same phenomenon took place with the introduction of the Penny Post in Britain in 1840, as well as with the drastic lowering of taxes on newspapers in 1836 [9].

To be fair to the many researchers in the telecommunications economics community, it should be said that there were some papers that did consider some of the long-range effects. Particularly noteworthy was the work of Jeff Rohlfs [47] on network effects. Also, about the time that the Mitchell's paper [42] appeared, results from the AT&T and GTE experiments with metered local rates were becoming available, and so the awareness of the limitations of the consensus view was becoming wider. But this consensus view continued, and continues, to dominate.

1.11 ANOTHER HISTORICAL VIGNETTE: FLAT RATES FOR DATA

Another interesting historical vignette is the advocacy of flat rates for data services by Loretta Anania and Richard Jay Solomon. They were among the early pioneers researching the pricing of broadband services, in the days of Integrated Services Digital Network (ISDN) and asynchronous transfer mode (ATM) [48–50]. They also appear to have been unique in that community in arguing for flat rates. Furthermore, they had very good arguments, based on excellent insights into data networks, in particular about the lack of direct relation between volume and value of data transfers.

There are at least two interesting aspects to the work of Anania and Solomon. One is the very limited impact it has had, as measured in citations. This appears to reflect the strong bias in favor of "smart pricing." The other interesting aspect of Anania's and Solomon's work is the limited range of arguments they presented, which reflects the lack of studies on flat rates. Their early arguments for flat rates were basically limited to pointing out that users would have large opportunities to bypass service provider controls [49]:

> ... dynamic allocation of network resources will become increasingly difficult for the carrier (or regulator) to track. So, with integrated digital networks, the flat-rate, or pay-in-advance subscription solution, may be the best method of pricing.

Later, they began to mention user preferences, but without much emphasis or detail [50]. This can again be taken as an indication of how little attention has been paid historically to flat rates. This tradition continues. One can cite a variety of recent papers, such as References 51, and 52, that fail to address many of the key advantages of flat rates.

Given all these precedents, it appears inevitable that the industry, and the networking research community, will continue to press for "smart pricing." As will be explained later, there are reasons to doubt whether they will be very successful. However, to the extent they are, it appears they are neglecting some important aspect of implementing "smart pricing."

1.12 DIRECTIONS FOR SMART PRICING RESEARCH AND DEPLOYMENT

Current research directions on smart pricing are probably not optimal for influencing applications. Most of the work appears to be inspired by the desire to control congestion and to maximize the engineering efficiency of networks. However, engineering efficiency has seldom been the main driving force behind telecom pricing in the past, and it is not now either. Far more important has been the incentive to maximize revenues through price discrimination. While the basic incentives and practices of price discrimination are ancient, they were first explicated by the French "econoengineers" of the middle of the nineteenth century [53]. Their work was motivated by the desire to understand business policies of the most revolutionary industry of that era, namely, railroads. There is a memorable and oft-quoted 1849 passage by Jules Dupuit [54] on this subject

> It is not because of the few thousand francs which would have to be spent to put a roof over the third-class carriages or to upholster the third-class seats that some company or other has open carriages with wooden benches. ... What the company is trying to do is to prevent the passengers who can pay the second-class fare from traveling third class; it hits the poor, not because it wants to hurt them, but to frighten the rich. ... And it is again for the same reason that the companies, having proved almost cruel to the third-class passengers and mean to the second-class ones, become lavish in dealing with first-class passengers. Having refused the poor what is necessary, they give the rich what is superfluous.

This description is not an exaggeration. In fact, many railroads did not have any seats in third-class carriages. Some refused to run any third-class carriages, and among those that did, such carriages went by special trains that were slow and ran at inconvenient times. Thus not only was "versioning" common, so was a form of what is now called *damaged goods* practice, where extra costs were incurred in order to offer an inferior service at a lower price. A contemporary account claimed that "the humbler order of passengers will not easily forget that a [railway] director once proposed to hire a number of chimney-sweeps to render—what, with the best company, are nothing better than locomotive hutches—perfectly untenable" [55]. (It should be said that no evidence has been found that chimney sweeps were actually hired to make third-class travel less attractive. This was just a suggestion, but a very revealing one that explains the motivations driving many policy decisions.)

The incentives that drove railroad managers to versioning and damaged goods have also been present in telecommunications. One of the fundamental obstructions to the introduction of fundamentally new networking technologies, or sophisticated pricing for the current ones, is that, in a slight paraphrase of Scott Bradner's memorably pithy quote [56],

> *The Internet is not reliably bad enough to drive the creation of a new network.*

But that could be changed through "damaged goods" practices. "Bufferbloat," which is now an accidental impairment on many networks, could be introduced systematically. Various types of noise or artificial packet drops (coupled with Deep Packet Inspection and statistical profiling, as is already done in various traffic-shaping practices) could be introduced much more widely. For example, because the most valuable types of communication tend to use little bandwidth (as will be discussed in more depth later), and voice is very latency sensitive, artificial increases in latency could decrease the quality of Voice over Internet Protocol (VoIP) services. If smart pricing is to spread, there should be far more research on such approaches and they should be part of the standard networking courses.

There is also too little research on bamboozling customers. The financial analyst Craig Moffett and his colleagues noted that "[f]or years, the telecom industry has thrived on obfuscation" [57]. Even earlier, back in 1998, Scott Adams in one of his Dilbert books talked of phone companies that "form confusopolies to make it impossible for the average individual to determine who has the lowest price," [58, p. 161]. Furthermore, there is now solid quantitative research that demonstrates the effectiveness of confusing users with complex pricing plans [59]. Hence, this feature of complicated pricing deserves more attention both in research programs and in the education of students for the workplace.

1.13 GROWTH IN DEMAND

To evaluate prospects of various pricing schemes, we have to consider the balance of demand and supply of data transport.

Wireline traffic growth has been decelerating over the past few years and is now taking more than 2 years to double. This is shown by the mostly widely cited project to estimate and predict traffic growth, Cisco's Visual Networking Index (VNI) [60]. The May 2013 VNI report projects that the traffic on the wired Internet will grow only 25% from 2012 to 2013 and will have a compound growth rate of 21% from 2012 to 2017. The sources for the VNI studies are confidential, though, and there are some estimates of growth rates that are somewhat higher than those of VNI.

Here we cite some public data that generally support the VNI observations. (References and URLs are available at Reference 61.) The European IX association reports their aggregate peak traffic growth has declined from 57% in 2008 to 45% in 2012. TeleGeography estimates that the "[g]rowth in worldwide international Internet capacity declined from 63% in 2009 to 33% in 2013," [62]. Australia is an outlier among industrialized countries in that it is still maintaining a high growth rate at 59% during the year ended June 2013, just about the same as the annual compound rate of 60% over the previous 4 years. However, Australia's Internet traffic per capita is still only about half that of the United States.

The most intriguing outlier in the available statistics is Japan. It has the most advanced infrastructure in the world, in terms of fraction of wireline broadband subscribers who have fiber. Along with Hong Kong and South Korea, it usually shows up at the top of rankings by effective speed of connections. Yet, Japanese Internet traffic is relatively low and is growing slowly. The latest measurement from the remarkable cooperative industry effort, covering about 40% of the Japanese market (see Reference 61), taken in May 2013 and kindly provided by Kenjiro Cho (see also Reference 63 for earlier but more detailed summary) shows a continuation of the 20–25% annual growth rate of wireline traffic that has prevailed over the past half a dozen years.

In wireless, the calls for urgent action to deal with the perceived spectrum shortage were and continue to be fueled by reports and predictions of data traffic more than doubling each year (cf. [64, p. 76] and [65, 66]).

In Australia, the data downloaded to mobile handsets grep 43% from the fourth quarter of 2012 to the second quarter of 2013, an annual rate of 105%. However, in the United States, the growth of 123% from 2010 to 2011 moderated to 69% in the following year. The Cisco VNI of May 2013 projects a 66% annual growth rate of mobile data between 2012 and 2017. While this is a substantial decline from earlier VNI projections, it is doubtful whether even that rate can be sustained for long, as is discussed in the next section.

While growth rates for wireless traffic far exceed those for wireline, it is important to remember that total volumes of data transmitted by mobile wireless technologies (so excluding WiFi) are still low. For both Australia in mid-2013 and the United States for year-end 2012, they were under 3% of total Internet traffic for those countries. Thus the potential demand just from people switching their usage from wireline to wireless is huge.

1.14 TECHNOLOGY TRENDS

The three main ingredients of the ICT (information and communication technologies) industry are computing, storage, and communication. They have all been riding the Moore's Law curve, with relatively steady and predictable rates of improvement. (See Reference 61 for some statistics and references.) Those rates have declined, but over the past few years, the amount of raw computing, or raw storage, or basic photonic transmission that can be performed for a unit cost has been doubling about every 2 years. While there have always been fears that progress might grind to a halt, that has not happened, even though there are various indications of slowdowns. But perhaps even more important than a general slowdown is that locality is becoming more pronounced. While magnetic storage densities are still increasing rapidly, the bandwidth to disks is growing much more slowly. Similarly, much of the improvement in raw computing power is now coming from putting more cores on a chip, but the bandwidth to the chip is an increasingly important barrier to the utilization of that power. Thus in both computing and storage, data moves less, or as the database community has been saying for a long time, is becoming "cooler."

The arrival of the Internet led to a dramatic jump in data traffic in the mid-1990s, when for a while there actually was the "Internet traffic doubling every 100 days" phenomenon. This was followed by several years of doubling every year. However, for most of the past decade, worldwide Internet traffic growth has been decelerating, as discussed in Section 1.13. It is not impossible that it may fall substantially below the level of growth in computing power and storage, as has been predicted by the Cisco measurement and forecasting effort [60], especially because that has happened in Japan.

On the other hand, the volumes of either broadcast video or magnetic disk storage dwarf the Internet's transmission capacity. Thus, should masses of people suddenly "cut the cord" and attempt to obtain their favorite TV programs over the Internet, networks would collapse under the load. However, there is no sign of that happening, and the switch to the more natural mode of on-demand viewing is proceeding fairly slow (facilitated by kludgy half-way solutions such as video recorders). Similarly, there is great promise in machine-to-machine communication, but it is also arriving at a measured pace. Therefore, wireline networks do have an incentive to encourage innovative uses of their facilities. This incentive is especially important because of the competition from wireless. As we have discovered over the past three decades, mobility is extremely attractive. Both usage and investment are moving toward wireless, and if application developers concentrate their energies on the low traffic, small screen size, and low power mobile devices, wireline could become a backwater.

As for mobile wireless, it is very difficult to estimate the growth of transmission capacity, because there are so many technological and economic dimensions to this problem. Currently, the shift to 4G and LTE offers a quantum jump in data capacity. However, beyond that, the possible improvements in modulation and related approaches appear to be rather limited. The industry is aiming for a 1000 × boost, and while it usually does say that the time frame for this jump in traffic is impossible to predict, it does cite prominently the recent 2 × annual growth rates, cf. [65, 66]. It appears very unlikely that even 50% per year growth rates could be maintained for long.

Note that one of the favorite routes to increasing capacity, namely, assigning more spectrum for mobile wireless, can offer only limited relief. The National Broadband Plan released by the Federal Communications Commission in March 2010 reports that the United States currently has 547 MHz that can be used for such purposes and calls for doubling that within 10 years [64]. However, such a doubling would only compensate for 1 year of 100% annual growth and for 2 years of 50% annual growth.

Substantial improvements in capacity could be achieved just by building more facilities, but that would require greatly increased capital spending. That could come from either higher revenues from users or from restructuring the industry so it spends less on marketing, lobbying, and other activities and more on construction. As neither is likely to happen, we are likely to see traffic limited by available capacity. Combined with the wide disparity of various types of bits, this suggests that pricing will play a significant role in balancing demand and supply.

For wired networks, the capacity limitations of mobile wireless, as well as the "mental transactions costs" that are likely to be imposed by pricing, leave a substantial opening. However, to exploit their advantage, they will need to encourage use, and that argues for the simplest possible pricing, which is flat rate.

1.15 CONCLUSIONS

As technology advances, implementing "smart pricing" is becoming easier. Further, the deep (and irrational) prejudice in favor of "smart pricing" is likely to drive the industry and academic researchers to continue pursuing a variety of schemes, some to the point of deployment.

At the same time, general technology advances are complicating life and intensifying users' desire for simplicity. In the wireline arena, we find rapid improvements in transmission capacity, and relatively slow growth in demand from consumers. This creates incentives for service providers to stimulate usage and thus argues for flat rates and simple networks, with large pipes, with any market segmentation determined by the size of the pipe.

In wireless, technology improvements are slower and therefore demand growth faster. Thus in this area, pricing is likely to play a larger role. But the human desire to avoid "mental transaction costs" will still argue for simplicity, at least at individual consumer level.

There may well be more sophistication at the business-to-business level, when companies contract with service providers to provide certain transmissions in ways invisible to consumers, just as today Amazon sells ebooks with the wireless delivery achieved seamlessly by an established carrier. The arguments for simplicity are not as important at the business-to-business level as it is with consumers, because companies come closer to the economic rationality so beloved by experts. However, it is not clear whether "smart pricing" will spread far even there, because managers appear to prefer strategic games to real market. Furthermore, the example of Apple's app store shows that even at the level of businesses, there are advantages to simplicity.

ACKNOWLEDGMENTS

This paper was prepared with the partial support of DARPA contract FA8750-13-2-0023 on "The evolution of control plane resilience."

REFERENCES

1. S. McCartney. Flier auctions: better seats, going once, going twice .... Wall Street Journal, April 25, 2013.
2. S. Sen, C. Joe-Wong, S. Ha, and M. Chiang. "A survey of broadband data pricing: past proposals, current plans, and future trends," *ACM Computing Surveys Volume* 46(2), November 2013, 10.1145/2543581.2543582. Preprint available at: ⟨http://arxiv.org/pdf/1201.4197v2.pdf⟩.
3. A. Odlyzko, B. St. Arnaud, E. Stallman, and M. Weinberg, Know Your Limits: Considering the Role of Data Caps and Usage Based Billing in Internet Access Service, Public Knowledge, 2012. Available at: ⟨http://publicknowledge.org/files/UBP%20paper%20FINAL.pdf⟩.
4. D. Levinson and A. Odlyzko, "Too expensive to meter: the influence of transaction costs in transportation and communication," *Philosophical Transactions of the Royal Society A*, 366(1872), 2008, 2033–2046. Preprint available at: ⟨http://ssrn.com/abstract=1735657⟩.
5. A. M. Odlyzko. "The history of communications and its implications for the Internet," 2000 unpublished manuscript. Available at: ⟨http://ssrn.com/abstract=235284⟩.

6. A. M. Odlyzko. "Network neutrality, search neutrality, and the never-ending conflict between efficiency and fairness in markets," *Review of Network Economics*, 8(1), 2009, 40–60. Preprint available at: ⟨http://ssrn.com/abstract=1095350⟩.
7. K. Bode. "Cable Industry Finally Admits Caps Not About Congestion After Insisting for Years Caps Were About Congestion," DSL Reports, Jan. 17, 2013. Available at: ⟨http://www.dslreports.com/shownews/Cable-Industry-Finally-Admits-Caps-Not-About-Congestion-122791⟩.
8. A. M. Odlyzko. "The evolution of price discrimination in transportation and its implications for the Internet," *Review of Network Economics*, 3(3), 2004, 323–346. Preprint available at: ⟨http://ssrn.com/abstract=596301⟩.
9. A. M. Odlyzko. "The volume and value of information," *International Journal of Communication*, 6, 2012. Available at: ⟨http://ijoc.org/ojs/index.php/ijoc/article/view/1570/740⟩.
10. R. N. Clarke. "Costs of neutral/unmanaged IP networks," *Review of Network Economics*, 8(1), 2009. Preprint available at: ⟨http://ssrn.com/abstract=903433⟩.
11. B. Swanson. The coming exaflood. *Wall Street Journal*, Jan. 20, 2007.
12. B. Swanson and G. Gilder, *Estimating the Exaflood*. Discovery Institute white paper, Jan. 29, 2008. Available at: ⟨http://www.discovery.org/a/4428⟩.
13. A. M. Odlyzko. "Threats to the Internet: too much or too little growth?" Internet Evolution, Feb. 25, 2008. Available at: ⟨http://www.internetevolution.com/author.asp?section_id=592&doc_id=146747&⟩.
14. N. Szabo. "The mental accounting barrier to micropayments," 1996 white paper. Available at: ⟨http://szabo.best.vwh.net/micropayments.html⟩.
15. A. M. Odlyzko. "Internet pricing and the history of communications," *Computer Networks*, 36, 2001, 493–517. Preprint available at: ⟨http://ssrn.com/abstract=235283⟩.
16. A. M. Odlyzko. "The many paradoxes of broadband," *First Monday*, 8(9), 2003, 1–15. Available at: ⟨http://firstmonday.org/htbin/cgiwrap/bin/ojs/index.php/fm/article/view/1072/992⟩.
17. A. M. Odlyzko. "Telecom dogmas and spectrum allocations," 2004 paper written for the *Wireless Unleashed* blog. Available at: ⟨http://www.dtc.umn.edu/~odlyzko/doc/telecom.dogmas.spectrum.pdf⟩.
18. A. M. Odlyzko. "Content is not king," *First Monday*, 6(2), 2001. Available at: http://firstmonday.org/htbin/cgiwrap/bin/ojs/index.php/fm/article/view/833/742.
19. Anonymous. Let Mexico's moguls battle. Economist, Feb. 4, 2012.
20. A. M. Odlyzko. The delusions of net neutrality. In *Telecommunications Policy Research Conference*, 2008. Available at: ⟨http://www.dtc.umn.edu/~odlyzko/doc/net.neutrality.delusions.pdf⟩.
21. A. Odlyzko. "The current state and likely evolution of the Internet. In *Proceedings Globecom'99, IEEE, 1999*, pp. 1869–1875. Preprint available at: ⟨http://www.dtc.umn.edu/~odlyzko/doc/globecom99.pdf ⟩.
22. A. M. Odlyzko. "Talk, Talk, Talk: so who needs streaming video on a phone? The killer app for 3G may turn out to be–surprise–voice calls," *Forbes*, Aug. 20, 2001, p. 28. Preprint available at: ⟨http://www.dtc.umn.edu//~odlyzko/doc/3g.accidental.success.txt⟩.
23. CTIA - The Wireless Association. Semi-Annual Year-End 2012 Top-Line Survey Results. Available at: ⟨http://files.ctia.org/pdf/CTIA_Survey_YE_2012_Graphics-_FINAL.pdf ⟩.
24. OECD. OECD Communications Outlook 2013. Available at: ⟨http://www.oecd-ilibrary.org/science-and-technology/oecd-communications-outlook-2013_comms_outlook-2013-en⟩.

25. Plunkett Research. Advertising & Banding Industry Overview. Available at: ⟨http://www.plunkettresearch.com/advertising-branding-market-research/industry-statistics⟩. Downloaded Sept. 26, 2013.
26. National Cable & Telecommunications Association. Cable industry revenue 1996–2011. Available at: ⟨http://www.ncta.com/Stats/CustomerRevenue.aspx⟩. Downloaded Sept. 28, 2012.
27. C. Labovitz, S. Iekel-Johnson, D. McPherson, J. Oberheide, F. Jahanian, and M. Karir. ATLAS Internet Observatory 2009 Annual Report. Available at: ⟨http://www.nanog.org/meetings/nanog47/presentations/Monday/Labovitz_ObserveReport_N47_Mon.pdf⟩.
28. C. Labovitz, S. Iekel-Johnson, D. McPherson, J. Oberheide, and F. Jahanian. Internet inter-domain traffic. In *ACM Computer Communication Review*, Oct. 2010, pp. 75–86. Available at: ⟨http://www.sigcomm.org/sites/default/files/ccr/papers/2010/October/1851275-1851194.pdf⟩.
29. C.-H. Hong, S. Kandula, R. Mahajan, M. Zhang, V. Gill, M. Nanduri, and R. Wattenholer. Achieving high utilization with software-driven WAN. In *ACM SIGCOMM 2013*, 2013. Available at: ⟨http://conferences.sigcomm.org/sigcomm/2013/papers/sigcomm/p15.pdf⟩.
30. S. Jain, A. Kumar, S. Mandal, J. Ong, L. Poutievski, A. Singh, S. Venkata, J. Wanderer, J. Zhou, M. Zhu, J. Zolla, U. Hölzle, S. Stuart, and A. Vahdat. B4: experience with a globally-deployed software defined WAN. *ACM SIGCOMM 2013*, 2013. Available at: ⟨http://conferences.sigcomm.org/sigcomm/2013/papers/sigcomm/p3.pdf⟩.
31. National Cooperative Highway Research Program (NCHRP). Future Dinancing Options to Meet Highway and Transit Needs, Dec. 2006. Available at: ⟨http://onlinepubs.trb.org/onlinepubs/nchrp/nchrp_w102.pdf ⟩.
32. U.S. Department of Commerce, Bureau of the Census. Historical Statistics of the United States: Colonial Times to 1970, 1975.
33. C. Kirjner and R. Parameswaran. Google Fiber–how much will Google spend to overbuild Kansas City? How much would it cost to expand? BernsteinResearch report, Financial Analyst report, April 8, 2013.
34. T. Orlik. Beijing falls short on rebalancing. *Wall Street Journal*, May 23, 2013.
35. World Bank. Gross fixed capital formation (% of GDP), interactive table at ⟨http://data.worldbank.org/indicator/NE.GDI.FTOT.ZS⟩, downloaded May 27, 2013.
36. T. Lewin. Senate committee report on for-profit colleges condemns costs and practices. New York Times, July 30, 2012.
37. Pharmaceutical industry. Wikipedia entry for the pharmaceutical industry. Available at: ⟨http://en.wikipedia.org/wiki/Pharmaceutical_industry⟩. Downloaded Oct. 7, 2013.
38. C. Moffett and N. Del Deo. Sprint: how long is long-term? MoffettNathanson Research report, Financial Analyst report, Oct. 11, 2013.
39. A. Troianovski and S. Grundberg. Nokia's bad call on smartphones. *Wall Streen Journal*, July 19, 2012.
40. D. Streitfeld. As boom lures app creators, tough part is making a living. *New York Times*, Nov. 18, 2012.
41. R. R. John. *Network Nation: Inventing American Telecommunications*. Harvard University Press, Cambridge, MA, 2010.
42. B. M. Mitchell. "Optimal pricing of local telephone service," *American Economic Review*, 68(4), 1978, 517–537.
43. J. G. Cosgrove and P. B. Linhart. Customer choices under local measured telephone service. *Public Utilities Fortnightly*, Aug. 30, 1979, pp. 27–31.

44. L. Garfinkel and P. B. Linhart. The transition to local measured telephone service. Public Utilities Fortnightly, Aug. 16, 1979, pp. 17–21.
45. T. F. Wong. "Identifying tariff induced shifts in the subscriber distribution of local telephone usage," in L. Courville, A. de Fontenay, and R. Dobell, eds., *Economic Analysis of Telecommunications: Theory and Applications*, North-Holland, Amsterdam, 1983, pp. 263–278.
46. B. Briscoe, A. Odlyzko, and B. Tilly. Metcalfe's Law is wrong. IEEE Spectrum, July 2006, pp. 26–31. Available at: ⟨http://www.spectrum.ieee.org/jul06/4109⟩.
47. J. Rohlfs. "A theory of interdependent demand for a communications service," *Bell Journal of Economics and Management Science*, 5, 1974, 16–37.
48. L. Anania and R. J. Solomon. "User arbitrage and ISDN," InterMedia, Jan. 1988, pp. 25–31.
49. L. Anania and R. J. Solomon. "Models of network infrastructure: pricing ISDN for access," in J. H. Alleman and R. D. Emmerson, eds., *Perspectives on the Telephone Industry: The Challenge for the Future*, Harper & Row, New York, 1989, pp. 287–303.
50. L. Anania and R. J. Solomon. "Flat—the minimalist price," in L. W. McKnight and J. P. Bailey, eds., *Internet Economics*, MIT Press, Cambridge, Mass. 1997, pp. 91–118.
51. J. M. Bauer and S. S. Wildman. The economics of usage-based pricing in local broadband markets. NCTA white paper, Dec. 2012. Available at: ⟨http://i.ncta.com/ncta_com/PDFs/Wildmanreport_web.pdf ⟩.
52. A. Nevo, J. L. Turner, and J. W. Williams. "Usage-based pricing and demand for residential broadband," preprint, Sept. 2013. Available at: ⟨http://ssrn.com/abstract=2330426 ⟩.
53. R. B. Ekelund Jr., and R. F. Hébert. *Secret Origins of Microeconomics: Dupuit and the Engineers*. University Chicago Press, Chicago, 1999.
54. R. B. Ekelund. "Price discrimination and product differentiation in economic theory: an early analysis," *Quarterly Journal of Economics*, 84(2), 1970, 268–278.
55. Anonymous. Railway comfort. *Household Words: A Weekly Journal Conducted by Charles Dickens*, Aug. 3, 1850, pp. 449–450.
56. S. Bradner. Will there be a next-generation network? Network World, July 21, 2003. Available at: ⟨http://www.networkworld.com/columnists/2003/0721bradner.html⟩.
57. C. Moffett, N. Del Deo, and A. Chan. Quick take – AT&T (T): market share(ing) Part II ... A quieter voice. BernsteinResearch report, Financial Analyst report, July 18, 2012.
58. S. Adams. *The Dilbert Future: Thriving on Business Stupidity in the 21st Century*. HarperCollins, New York, 1998.
59. M. Miao and K. Jayakar. "Bounded rationality and consumer choice: an evaluation of consumer choice of mobile bundles in China," Preprint available at: ⟨http://ssrn.com/abstract=2241581 ⟩.
60. Cisco Visual Networking Index. Available at: ⟨http://www.cisco.com/en/US/netsol/ns827/networking_solutions_sub_solution.html⟩.
61. MINTS. Minnesota Internet Traffic Studies project. Available at: ⟨http://www.dtc.umn.edu/mints/⟩.
62. TeleGeography. "Europe emerges as global Internet hub," Sept. 18, 2013 press release, ⟨http://www.telegeography.com/products/commsupdate/articles/2013/09/18/europe-emerges-as-global-internet-hub/⟩.
63. K. Cho. "Broadband traffic report: traffic trends over the past year," *Internet Infrastructure Review*, 16, 2012, 33–37. Available at: ⟨http://www.iij.ad.jp/en/company/development/iir/pdf/iir_vol16_report_EN.pdf ⟩.

64. U.S. Federal Communications Commission. *Connecting America: The National Broadband Plan,* March 2010. Available at: ⟨http://download.broadband.gov/plan/national-broadband-plan.pdf⟩.

65. 4G Americas. *Meeting the 1000x Challenge: The Need for Spectrum, Technology and Policy Innovation,* white paper, Oct. 2013. Available at: ⟨http://www.4gamericas.org/documents/2013_4G_20Americas_20Meeting_20the_201000x_20Challenge_2010_204_2013_FINAL.pdf⟩.

66. Qualcomm, Inc. *The 1000x Mobile Data Challenge: More Small Cells, More Spectrum, Higher Efficiency,* July 25, 2013 presentation deck. Available at: ⟨http://www.qualcomm.com/media/documents/tags/1000x-challenge⟩.

2 Customer Price Sensitivity to Broadband Service Speed: What are the Implications for Public Policy?

VICTOR GLASS, STELA STEFANOVA, and RON DIBELKA

2.1 INTRODUCTION

In the prebroadband era,[1] recovering a local telephone company's network costs for voice service was a relatively straightforward process. The users of the local network fell into two categories: end users who bought phone service and long-distance carriers who sold long-distance service to end users. The pricing arrangements were also relatively clear. The end user bought local service from the local telephone company and long-distance carriers such as AT&T and MCI paid the local phone companies for the use of their networks to complete long-distance calls. The relationships between the local and long-distance companies were clearly defined by Federal Communications Commission (FCC) rules and a standard tariff defined the services rendered by the local telephone company and the charges to long-distance carriers.

The broadband world has complicated network cost recovery considerably. In the broadband world, voice service is only one application traversing broadband networks, so it is no longer straightforward to separate it from other services. Accessing many applications over a broadband pipe produces diverse traffic patterns among users in comparison to traditional voice service, and to a large extent, usage is not closely tied to the prices they pay for broadband services. End users can buy a variety of broadband plans; the more expensive ones include higher speed access to the Internet, traditional voice, and cable TV. End users buying the same broadband plan often differ significantly in their use of the broadband network. The relationship between bandwidth use and price paid is far more tenuous among application

[1]Following the introduction of access charges with 47 C.F.R. Part 69 of FCC rules.

Smart Data Pricing, First Edition. Edited by Soumya Sen, Carlee Joe-Wong, Sangtae Ha, and Mung Chiang.

providers. Google and Yahoo are heavy bandwidth users because they essentially take pictures of the Internet to improve their web site performance.[2] YouTube, Netflix, and other video providers take up perhaps one-third of the Internet's bandwidth in the United States.[3] These applications providers pay a fraction of the bandwidth cost of an end user to the Broadband Service Provider (BSP). By contrast, small application providers such as a local restaurant face a usage/cost relationship close to those for end users.

Billing application providers is beyond the capabilities of many BSPs. If an application provider is not in a BSP's service area, it typically does not have or want a contractual relationship with the BSP. Without a contract, these application providers can terminate traffic free of charge. In case of default, BSPs recover their network costs from their customers. This is the so-called bill-and-keep arrangement.

A major policy issue is whether bill-and-keep is a welfare maximizing strategy from a government policy perspective. The answer to this question often depends on perspective. Application providers support bill-and-keep because they fear that BSPs could exercise monopoly power to raise terminating rates for their traffic. BSPs oppose bill-and-keep but face the almost impossible challenge of how to bill millions of application providers. The FCC supports a bill-and-keep system but recognizes that it needs a funding base to subsidize broadband service where private businesses cannot make a business case for offering broadband services at speeds widely available elsewhere. To raise support funds, the FCC recognizes that it may want to assess application providers and use the funds raised to make broadband service universally available. If, however, nothing is done to recover network costs from application providers, wireline BSPs may not build out to remote customers and may introduce usage limits such as the 10 GB per month consumption threshold already instituted by Verizon Wireless and AT&T Wireless more than which a customer pays an extra charge.

The focus of our paper is whether it is in the financial interest of application providers as a group to buy down end user charges. A reasonable case could be made that application providers, as a group, may be willing to subsidize end users if two conditions hold: First, end users are price sensitive to the cost of broadband speeds. If so, lowering the price of broadband service will encourage them to buy higher speed service and use the Internet more intensively. As a result, they would not throttle their use of Internet and will not be deterred from accessing applications with higher bandwidth requirements, which would foster new applications development. Second, subsidization of end users must be competitively neutral across providers and across networks. Research in the two-sided market literature shows that if subscribers to higher speed tiers are price elastic, it may be beneficial for content providers who do not enter in direct contractual relationships with end users to subsidize prices for

[2] Scott Cleland, December, 2008, "A First-Ever Research Study: Estimating Google's U.S. Consumer Internet Usage & Cost—2007–2010." Available at www.netcompetition.org/study_of_google_internet_usage_costs2.pdf.

[3] Scott Cleland, December, 2008, "A First-Ever Research Study: Estimating Google's U.S. Consumer Internet Usage & Cost—2007–2010." Available at www.netcompetition.org/study_of_google_internet_usage_costs2.pdf.

high speed offerings through the public support system (e.g., [1]). Buying down of end user charges could be done through a universal broadband fund such as the newly established Connect America Fund (CAF). The intended goal of CAF is "to enable all US households to access a network that is capable of providing both high-quality voice—grade service and broadband that satisfies the National Broadband Availability target," currently set at 4 Mbps downstream and 1 Mbps upstream.[4] To the extent that a private sector business case cannot be made to offer this minimum service, CAF funds should be available to fund the shortfall.[5] In fact, the FCC is currently evaluating whether end users, content providers, or both should contribute to CAF, which is designed to foster universal broadband connectivity.[6]

Estimating price elasticities of demand has a second and important benefit to BSPs whether the buy down occurs or not. If demand is price elastic to speed, it will affect their pricing strategies. Lower prices may boost demand for broadband services at higher speeds. Unfortunately, the empirical economic literature offers mixed results for price elasticity of demand for broadband. Some studies find own-price elasticity of broadband demand to be larger [2–4], while others estimate smaller price sensitivities [5, 6]. One simple explanation for this variation in estimates is that the choice to access to the broadband differs from the choice of which bandwidth service level to subscribe to once the customer chooses to buy broadband service. Specifically, households may have different price sensitivities for gaining access to the Internet and the service level they choose. Varian [4] estimates willingness to pay for speed using experimental data collected in 1998 and 1999. The own-price elasticities of demand for bandwidth range from −1.3 to −3.1. He finds that lower service bandwidths are perceived as substitutes for the level of bandwidth chosen by participants in the experiment and reports low willingness to pay for higher bandwidths. In a more recent experimental survey design study, Rosston et al. [7] estimate households' willingness to pay for speed and reliability. They find that households value broadband highly, but values for very fast speeds are only marginally higher than values for fast speeds. They explain that a typical household does not use applications requiring blazing speeds. Glass and Stefanova [8] find highly inelastic own-price elasticities of demand by looking at the total subscribers of a BSP and the lowest price for its introductory offer. The estimates are −0.66 using data from 2005 and −0.21 using data from 2009. These estimates are based on the introductory offer prices and can be interpreted as estimates of the price sensitivity of gaining access to the Internet. The lower price elasticity estimated from the 2009 dataset suggests that access to the Internet has become more of a necessity in the later period.

These studies suggest that once a household has subscribed to broadband, it may have different price sensitivity for access to the Internet as it relates to expected usage. A household may perceive bandwidth as a differentiated service, which is captured

[4]"Availability," Chapter 8, *Connecting America: The Broadband National Plan*, 2010, p. 163

[5]"Availability," Chapter 8, *Connecting America: The Broadband National Plan*, 2010, p. 163

[6]FCC has issued a Further Notice of Proposed Rulemaking on USF contribution reform on April 30, 2012, seeking comment on what services and service providers should contribute to the Fund and how contributions should be assessed.

to some extent by the choice of speed tiers. Even a low speed tier connection can be used for browsing e-mail, You Tube video, and near-real time video,[7] but customers who are heavy users, especially of near- and real-time video, are far more likely to purchase a high speed tier. In the current study, we use a detailed demand and price data collected in 2010 to estimate a demand model for Internet speed tiers offered by rural BSPs and use the parameter estimates to calculate price elasticities of demand for three differentiated speed tiers—with downstream speeds up to 3 Mbps, between 3 and 6 Mbps and with downstream speeds above 6 Mbps.[8]

2.2 MODEL

Our goal is to estimate price elasticities of demand for different speed service packages purchased by broadband users. The generalized Leontief [9], the Rotterdam model [10], the Translog model [11], and the Almost Ideal Demand System (AIDS) [12] are among the most popular attempts to provide flexible, but theoretically consistent, functional forms to study consumer purchases of different products. We use an AIDS model for our estimation.

The AIDS model is specified in the usual way:

$$w_i = \alpha_i + \sum_j \gamma_{ij} \ln P_j + \beta_i \ln \left(\frac{X}{P}\right),$$

where X is the total expenditure on broadband in a given market, P_j is the price of the jth service, w_i is the share of the total expenditures spent on the ith good, and P is the price index defined as

$$\ln P = \alpha_0 + \sum_j \alpha_i \ln P_i + \frac{1}{2} \sum_i \sum_j \gamma_{ij} \ln P_i \ln P_j.$$

We estimate the standard AIDS model specified above and an AIDS model with demographic variables. Following Heien and Wessells [13], demographic variables can be incorporated into the model by including them in the intercepts of the share equations.

$$\alpha_i = \rho_{i0} + \sum_k \rho_{ik} d_k,$$

where d_k denotes the kth demographic characteristic. α_i, γ_{ij}, β_i, ρ_{i0}, and ρ_{ik} are parameters to be estimated. Parameter restrictions are added to ensure that the demand equations are consistent with the theory:

$$\sum_i \alpha_i = 1 \text{ for adding up,} \quad \sum_i \gamma_{ij} = 0 \text{ and}$$

$$\sum_i \beta_i = 0 \text{ for linear homogeneity,} \; \gamma_{ij} = \gamma_{ji} \text{ for symmetry.}$$

[7]"Goals for high-performance America," Chapter 2, *Connecting America: The Broadband National Plan*, 2012, p. 17.

[8]FCC has used 3 and 6 Mbps as threshold speeds for reporting broadband services in its form 477.

These conditions guarantee that the sum of the budget shares is unity. Homogeneity restrictions imply that the budget shares will not change if all prices and expenditures are multiplied by the same positive constant. The symmetry restrictions require that compensated demand effects be symmetric to be consistent with consumer theory.

In order to have the budget shares for each market add up to 1 in the model with demographic variables, the adding up restriction was replaced by

$$\sum_i \rho_{i0} = 1 \text{ and } \sum_i \rho_{ik} = 0, \quad \text{for } k = 1, \ldots, K.$$

For a complete system of demand equations, the variance covariance matrix is singular because of the adding up property, so we drop one of the tiers and estimate the model using maximum likelihood. It is shown that maximum likelihood estimates are invariant to which product is dropped in the estimation [14].

Expenditure elasticities are computed at mean shares using the following expression,

$$e_i = 1 + \frac{\beta_i}{\overline{w}_i}.$$

Price elasticities are given by

$$e_{ij} = \delta_{ij} + \frac{\gamma_{ij} - \beta_i \left(\overline{w}_j - \beta_i \ln \left(\frac{X}{P} \right) \right)}{\overline{w}_i},$$

where $\delta_{ij} = 1$ if $i = j$ and 0 otherwise.

2.3 DATA

The data used in our analysis were obtained in 2010 from a nationwide survey of rate of return rural local exchange carriers (RLECs) participating in the National Exchange Carrier Association (NECA) pools and their affiliated Internet service providers. We collected data on Internet services purchased by rural households from these providers. Companies reported price, upload and download speed, and total number of subscribers for all Internet packages they offer. We used data for 194 broadband providers which reported at least 3 broadband services—one with download speed below 3 Mbps, one with download speed below 6 Mbps, and one with download speed greater than or equal to 6 Mbps. If a company offered more than one service in any of the three categories, we aggregated the services into the three representative service tiers.

2.4 VARIABLE DESCRIPTIONS

We aggregated the service tiers into three categories—services with downstream speed below 3 Mbps, services with downstream speed between 3 and 6 Mbps, and services with downstream speed above 6 Mbps. The resulting average speeds within these categories are 1.29 Mbps for the service tiers below 3 Mbps, 3.34 Mbps for the middle category, and 7.93 Mbps for the top category of services. Average prices increase with the speed offered and the average price of $75.37 paid per month for the top tiers is almost two times the average price of $37.93 paid per month for the lowest service tiers. The average number of subscribers for the tiers below 3 Mbps is 2016, while the average number of subscribers for the highest speed tiers is less than a quarter of that: 479. Average upload speeds also increase with the higher download speeds but are in general much lower than the download speeds. The average upload speed for the highest tier in our data of 6 Mbps and above is 1.29 Mbps. The average expenditure for all three service tiers per company is $135,335/month. The expenditure share for the lowest tier is 60% of total, for the middle tier is 28%, and for the top tier is 12%. Table 2.1 shows the summary statistics for the aggregated sample.

To derive the demographic variables, we matched the serving territories of the RLECs with census data as well as data published on the National Broadband Map web site.[9] We include variables to control for age, education, income, race, and density of the population of the markets in our study. Percentage of population of age less than 19 years and age greater than 60 years will be interpreted relative to the percentage of population of age between 19 and 60 years old. Educational achievement

TABLE 2.1 Statistics for Variables Used in the AIDS Model

Variable	Mean	Standard Deviation	Minimum	Maximum
Monthly price—tier 1	$37.93	$13.93	$15.00	$169.95
Monthly price—tier 2	$53.36	$19.07	$20.85	$199.95
Monthly price—tier 3	$75.37	$35.81	$20.85	$229.95
Subscribers—tier 1	2,016	2,840	9	16,608
Subscribers—tier 2	940	1,833	1	12,949
Subscribers—tier 3	479	1,493	1	12,697
Upspeed—tier 1 (in Mbps)	0.43	0.24	0.06	2.00
Upspeed—tier 2 (in Mbps)	0.76	0.49	0.26	4.00
Upspeed—tier 3 (in Mbps)	1.29	1.43	0.26	10.00
Downspeed—tier 1 (in Mbps)	1.04	0.44	0.13	2.38
Downspeed—tier 2 (in Mbps)	3.34	0.59	3.00	5.50
Downspeed—tier 3 (in Mbps)	7.93	2.63	6.00	20.10
Total expenditure	$135,335	$168,844	$3,898	$1,209,118
Share—tier 1	0.60	0.32	0.01	1.00
Share—tier 2	0.28	0.27	0.00	0.97
Share—tier 3	0.12	0.20	0.00	0.98

[9]The National Broadband Map is a joint project of the National Telecommunications and Information Administration (NTIA), the FCC, and all states, http://www.broadbandmap.gov/.

is often found to be important in studies of broadband adoption. We include a variable for the percentage of population with bachelor's degree or higher. People living in markets with lower household density may find it more beneficial to subscribe to higher speed broadband, if available, because of a larger need to telecommute, use telemedicine applications, or do online shopping, compared to households living in more densely populated areas. Percentage of population with incomes below federal poverty levels is also included in the model. White and Asian households are often found to be positively correlated with broadband adoption when compared to non-White, non-Asian populations, thus we include the percentage of population of White and Asian race to our model.

2.5 RESULTS

The explanatory variables include the logarithms of prices and expenditures for the three service tiers in both models and demographic characteristics of the markets in the second model. The parameter estimates are reported in Table 2.2. Most of the price coefficients achieve statistical significance at the 0.01 level, while the demographic variables in the AIDS model with demographics, are mostly insignificant. The exceptions are the negative coefficient on income below poverty levels for tier 3 and the negative coefficient for tier 3 on percentage of white population in the market.

Tables 2.3 and 2.4 reports the uncompensated own- and cross-price elasticities and expenditure elasticities for each service tier and corresponding asymptotic standard errors. All but two of the elasticities are statistically significantly different from zero at the 0.05 level.

All elasticities have the expected signs. Own-price elasticities of broadband services are negative and indicate elastic demand.[10] The elasticity estimates increase with speed with services below 3 Mbps having the lowest estimate (−1.721 with the standard AIDS model and −1.746 when demographics are included) and services above 6 Mbps having the highest estimate (−2.707 and −2.788, respectively). Subscribers to higher tiers, who we expect are more heavy users of the Internet (watching video, playing online precision games, etc.) are more sensitive to price, while lower tier subscribers, who we expect use the internet primarily for web browsing and checking e-mail, are less price sensitive.

Cross-price elasticities are positive, indicating that service tiers are substitutes. Cross-price elasticities of the higher speed tier with respect to the lower speed tier is always larger than the cross-price elasticity of the lower tier with respect to the higher speed tier. This observation is compatible with the idea that lower speed users are much less concerned by the monthly charges for higher speed services than higher

[10]We performed sensitivity analysis using different thresholds for definitions of the three tiers. Even though the magnitude of the elasticity estimates varied with the definition, we found the same quantitative relationships between the lowest, medium, and highest service tiers. One interesting finding was the inelastic demand for the lowest tier when it was defined to include only services below 1 Mbps. This is consistent with results and interpretation of our earlier study [8], as the introductory offer competes with the choice of not accessing the Internet.

TABLE 2.2 Econometric Estimates of the AIDS Models

Variable	Standard AIDS Model Expenditure Shares			AIDS Model with Demographics Expenditure Shares		
	Tier 1	Tier 2	Tier 3	Tier 1	Tier 2	Tier 3
Constant	1.318*	−0.129	−0.189	0.950	−0.257	0.306
	(0.197)	(0.188)	(0.138)	(0.683)	(0.57)	(0.4)
Log of Price 1	−0.506*	0.309*	0.196*	−0.490*	0.299*	0.192*
	(0.071)	(0.055)	(0.038)	(0.07)	(0.057)	(0.038)
Log of Price 2	0.309*	−0.321*	0.012	0.299*	−0.325*	0.026
	(0.055)	(0.061)	(0.037)	(0.057)	(0.064)	(0.038)
Log of Price 3	0.196*	0.012	−0.208*	0.192*	0.026	−0.218*
	(0.038)	(0.037)	(0.028)	(0.038)	(0.038)	(0.031)
Expenditures	−0.062*	0.036*	0.026*	−0.040	0.014	0.026

	(0.017)	(0.016)	(0.011)	(0.02)	(0.022)	(0.0141)
Age <19 years				0.432	−0.150	−0.282
				(1.215)	(1.034)	(0.789)
Age >60 years				−0.091	0.247	−0.156
				(0.78)	(0.659)	(0.528)
Bachelor degree or higher				−0.491	0.382	0.110
				(0.413)	(0.402)	(0.229)
Household density				−0.001	0.000	0.000
				(0.001)	(0.001)	(0.001)
Income below poverty				0.465	0.296	−0.760*
				(0.509)	(0.436)	(0.389)
White				0.090	0.205	−0.295*
				(0.257)	(0.241)	(0.148)
Asian				−5.462	9.451	−3.997
				(4.406)	(5.343)	(4.678)

Note: Values in parentheses are standard errors. Asterisk denotes significance at 0.05 level. Log likelihood ratio test fails to rejects the null hypothesis of no demographic effects in the model at 0.05 level but rejects it at 0.1 level.

TABLE 2.3 Own-Price and Cross-Price Elasticities of Broadband Service Tiers

	Standard AIDS Model Price			AIDS Model with Demographics Price		
Quantity	Tier 1	Tier 2	Tier 3	Tier 1	Tier 2	Tier 3
Tier 1	−1.721*	0.509*	0.315*	−1.746*	0.502*	0.310*
	(0.111)	(0.088)	(0.059)	(0.124)	(0.095)	(0.061)
Tier 2	0.978*	−2.164*	0.057	1.045*	−2.198*	0.101
	(0.194)	(0.216)	(0.13)	(0.23)	(0.235)	(0.139)
Tier 3	1.383*	0.104	−2.706*	1.390*	0.183	−2.788*
	(0.289)	(0.304)	(0.231)	(0.309)	(0.322)	(0.253)

Note: Values in parentheses are standard errors. Asterisk denotes significance at 0.05 level.

TABLE 2.4 Income Elasticities of Broadband Service Tiers

	Standard AIDS Model		AIDS Model with Demographics	
Tier	Income Elasticity	Standard Error	Income Elasticity	Standard Error
Tier 1	0.897	0.029	0.934	0.037
Tier 2	1.130	0.060	1.053	0.079
Tier 3	1.220	0.096	1.215	0.117

speed users are concerned with the cost of a lower speed service tier. Estimated expenditure elasticities show that all broadband services are normal goods and the demand is positively correlated with income levels.

2.6 CONCLUSIONS

This study utilizes cross-sectional data from rural rate of RLECs to estimate market demand functions for broadband service tiers. In addition to prices and number of subscribers by service tier, the data include demographic characteristics of the markets. We used two specifications of the AIDS to estimate the demand parameters and price and income elasticities. Our empirical findings indicate that DSL customers are price sensitive and price sensitivity increases for customers purchasing higher bandwidth packages. Together with our earlier study [8], the overall results suggest that DSL access has become a communications necessity, but customers are very price sensitive to the DSL service level they buy. Our results are consistent with other studies that find high household valuation for broadband but low incremental values for higher speeds. Moreover, the empirical results strongly suggest that application providers, as a group, would benefit from buying down the price of DSL for end users. Network providers would receive additional funds to expand network capacity and they would be less likely to introduce monthly capacity limits. Application providers, as a group, would benefit from greater use of the Internet and higher likelihood that end users would access applications with higher bandwidth needs.

REFERENCES

1. P. Hande, M. Chiang, A. R. Calderbank, and S. Rangan, "Network pricing and rate allocation with content provider participation," *Proceedings of IEEE INFOCOM*, Princeton, NJ, 2009.
2. D. Kridel, P. Rappoport, and L. Taylor "The demand for high-speed access to the Internet: the case of cable modems," in D. G. Loomis and L. D. Taylor, eds., *Forecasting the Internet: understanding the explosive growth of data communications*, Kluwer Academic Publishers, Norwell, MA, 2001.
3. A. Goolsbee, "Subsidies, the value of broadband, and the importance of fixed costs," in R. Crandall and J. H. Alleman, eds., *Broadband: should we regulate high-speed internet access?*, Brooking Institution Press, Washington, D.C., 2002, pp. 278–294.
4. H. R. Varian, *The demand for bandwidth: evidence from the INDEX project*, 2002. Available at http://www.sims.berkeley.edu/~hal/Papers/brookings.pdf. Accessed March 30, 2014.
5. R. K. Goel, E. T. Hsieh, M. A. Nelson, and R. Ram, "Demand elasticities for Internet services," *Applied Economics*, 38, 2006, 975–980.
6. K. T. Duffy-Deno, "Business demand for broadband access capacity," *Journal of Regulatory Economics*, 24, 2003, 359–372.
7. G. Rosston, S. Savage, and D. Waldman, "Household demand for broadband Internet service in 2010," *B. E. Journal of Economic Analysis and Policy Advances*, 10(1), 2010.
8. V. Glass and S. Stefanova, "An empirical study of broadband diffusion in rural America," *Journal of Regulatory Economics*, 38(1), 2010, 70–85.
9. W. E. Diewert, "An application of the Shephard duality theorem: a generalized Leontief production function," *Journal of Political Economy*, 79, 1971, 481–507.
10. H. Theil, "The information approach to demand analysis," *Econometrica*, 6, 1965, 375–380.
11. L. R. Christensen, D. W. Jorgenson, and L. J. Lau, "Transcendental logarithmic utility functions," *American Economic Review*, 65 1975, 367–383.
12. A. Deaton and J. Muellbauer, "An almost ideal demand system," *American Economic Review*, 70, 1980, 312–326.
13. D. Heien and C. R. Wessells, "Demand systems estimation with microdata: a censored regression approach," *Journal of Business and Economic Statistics*, 8, 1990, 365–371.
14. R. Pollak and T. Wales, "Estimation of the linear expenditure system," *Econometrica*, 37, 1969, 611–628.

3 Network Neutrality with Content Caching and Its Effect on Access Pricing*

FATIH KOCAK, GEORGE KESIDIS, and SERGE FDIDA

3.1 INTRODUCTION

The continuing network (net) neutrality debate (e.g., [1–3], 26) involves several different entities, such as Internet service providers (ISPs),[1] content providers (CPs), users, and governments (including partnerships). Although there are many different perceptions for the definition and the coverage of net neutrality, one succinct definition is provided in [4]: "[net neutrality] usually means broadband service providers charge consumers only once for Internet access, do not favor one CP over another, and do not charge CPs for sending information over broadband lines to end users."

CPs, such as Amazon, Google, Yahoo!, and eBay, typically support net neutrality because under nonneutral conditions, they expect additional access-networking expenses and additional limitations or exclusions on their access to their customers [5]. In contrast to CPs, ISPs (particularly residential ISPs) such as AT&T, Verizon, Comcast, and Deutsche Telekom typically believe that neutrality regulations threaten the profitability of their enormous infrastructure investments and maintenance costs [1, 5] and that CPs do not pay a fair share of these costs while profiting from advertising that is arguably not requested by consumers.[2] Also, flat-rate pricing frameworks leading to "all-you-can-eat" consumer behavior result in high transport costs and congestion in the ISPs' access networks, e.g., [6], which makes ISPs complain about this and leads them to take blocking (e.g., Comcast

*This research was supported by NSF CNS grant 1116626.

[1] Equivalently, network service providers or access providers.

[2] Note that neutrality regulations for wireless access in the United States have not been instituted at the time of writing of this chapter; see [6] for discussions of the mobile wireless access scenario.

Smart Data Pricing, First Edition. Edited by Soumya Sen, Carlee Joe-Wong, Sangtae Ha, and Mung Chiang.

blocking peer-to-peer (P2P) applications [7]) or pricing (e.g., [8]) measures. It has been argued that some of these problems can be compensated by side payments between CPs and ISPs [9–13]. Alternatively, the introduction of premium service classes for applications has been suggested for critical applications such as health monitoring and home security (which are being increasingly used [4]); streamed spectacle events such as sports activities or newly released movies [5]; and interactive real-time video-conferencing/video-phone sessions. Applications engaged in premium services will obviously receive a higher quality of service (QoS) than applications under best-effort network-access service and will need to pay usage-based costs (perhaps after a quota). Such payments are (content/application) neutral in nature [14] because of the willingness of the users to pay for the premium content [9]. Under net neutrality with flat-rate priced access,[3] ISPs may not have the incentive to improve their existing infrastructure by increasing capacity [15] (particularly the router/switch infrastructure to drive fiber to the home (FTTH)) or by improved security measures such as virus and spam filtering [5]. (Note that such usage-based costs may need to be authenticated to the human subscriber/end user.)

Regarding QoS management, the physical location of requested content is obviously important to the goal of decreasing delay experienced by the users [16]. This in turn underscores the importance of caching data proximal to the users, including their ISP. Some large CPs, such as eBay and Google, cache their content around the world on their own servers, while smaller CPs often use intermediary content distributors, such as Akamai, who have caching agreements with local ISPs at different locations [5]. Such agreements or more dedicated partnerships between ISPs and CPs (i.e., "eyeball" ISPs) lead to scenarios wherein ISPs may cache each other's content, which raises issues of transit pricing between them; see Figure 3.1. To achieve end-to-end QoS, Bornstaedt et al. [17] argued that a sending ISP should pay for the transport traffic over an interconnection between ISPs.

Notwithstanding arguments for and against side payments, the necessity for providing a single interface (single contract including mutual services) to the end user is emphasized by several presenters in [6]. Product offers to the end users are assumed to be made in mainly two different ways: pull (on-demand) or push. Product offers can be prepared in distributed (among ISPs), partially centralized (by any of the ISPs), or fully centralized (by an external single facilitator entity) ways [18]. We herein primarily consider the "pull" demand model for content product where content requested in the recent past is cached in anticipation of similar demand locally.

In the following, we first give a background on our problem setting in Section 3.2. A model involving two different eyeball ISPs connected at peering point(s), where revenue is generated corresponding to net traffic transmitted, is initially considered in Sections 3.3 and 3.4. We consider a caching model captured by a single parameter, Φ, affecting the revenue generated by transit traffic. We assume that there is no limit on the throughput downstream to the users of each ISP. In Section 3.5, we modify the model so that there is an upper bound on the throughput that the users can receive via their ISP. So, two possible mechanisms to distribute the allowed throughput among

[3]Limited only by uplink and downlink bandwidth of the service agreement.

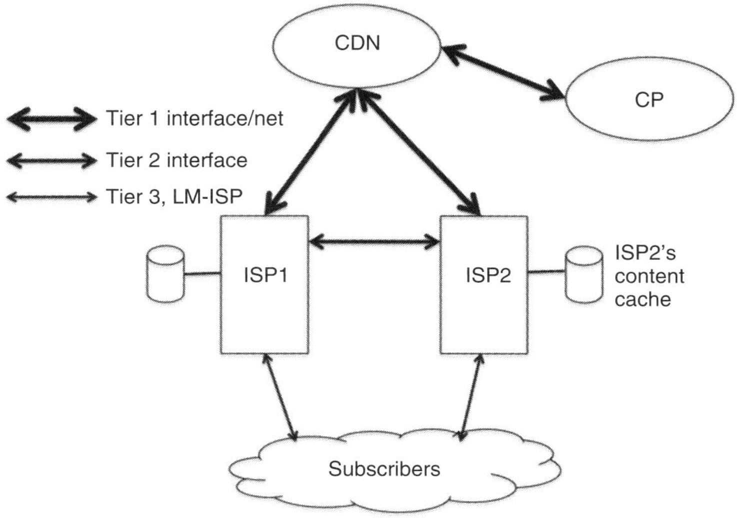

Figure 3.1 A CP may use a content distribution network (CDN) as depicted or may have a local caching agreement with a last-mile (LM) ISP, or neither.

the types of demands (local or remote content) are introduced. We next consider the scenario where there are multiple providers competing for the same group of users (without the throughput limit condition, as in the initial model). User/customer migration among competing ISPs because of the price difference between them is modeled by their "loyalties" to the ISPs. In Section 3.6, consideration of two ISPs competing for the same set of users is added to the model described in Section 3.4. We provide the results of numerical experiments on performance at Nash equilibrium in Section 3.7. We conclude with a discussion of future work in Section 3.8.

3.2 BACKGROUND

Suppose there is an eyeball-ISP provider whose revenue from its subscribers due to its local content is

$$pD, \tag{3.1}$$

where p is a usage-based price and D is the total demand at that price. Note that ISPs are continuing to depart from pure flat-rate pricing (based on access bandwidth) for unlimited monthly volume, for example, [19, 20].

Following [21], suppose that there are two broad classes of applications, one of which is significantly sensitive to congestion of access bandwidth, for example, delay-sensitive interactive real-time applications. Assume that applications of the other, best-effort type are unlikely to engage in usage-based pricing for access bandwidth. As pricing reduces, the demand for access-bandwidth reservation increases, causing additional congestion so that best-effort service will be increasingly inadequate for congestion-sensitive applications. Therefore, the demand for usage-priced access-bandwidth reservation may *accelerate* with reduced price. More specifically, say there is positive threshold

$$D_\theta < D_{\max}$$

such that overall demand sensitivity to price is greater when $D \geq D_\theta$ than when $D < D_\theta$. That is, for

$$d_{\max} > d_\theta,$$

a *convex*, piecewise-linear model for access bandwidth would be

$$D(p) = \max\{D_{\max} - d_{\max}p,\ \hat{D}_\theta - d_\theta p\}, \tag{3.2}$$

where

$$\hat{D}_\theta = D_\theta + \frac{(D_{\max} - D_\theta)d_\theta}{d_{\max}},$$

$$p_\theta = \frac{(D_{\max} - D_\theta)}{d_{\max}},$$

$$p_{\max} = \frac{\hat{D}_\theta}{d_\theta} = p_\theta + \frac{D_\theta}{d_\theta},$$

so that $D(p_\theta) = D_\theta$, see Figure 3.2.

So, in this model, the price range $[p_\theta, p_{\max}]$ (equivalently, demand range $[0, D_\theta]$) corresponds to low demand sensitivity to price, d_θ. The pricing range $[0, p_\theta]$ (demand range $[D_\theta, D_{\max}]$), when delay-sensitive applications typically need to

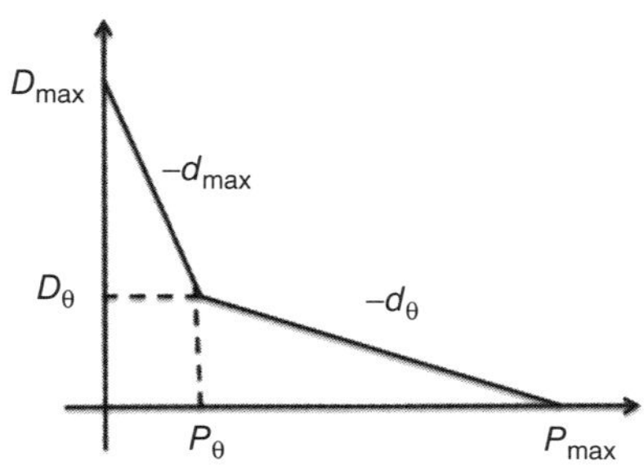

Figure 3.2 Convex piecewise-linear demand response.

adopt usage-priced (reserved or priority) access-bandwidth service, corresponds to higher demand sensitivity to price, $d_{\max}$.

Alternatively, suppose a convex, differentiable demand model that can approximate (3.2), specifically

$$D(p) = D_{\max}\left(1 - \frac{p}{p_{\max}}\right)^{\alpha}. \tag{3.3}$$

Here, $\alpha \geq 1$ and given $d_{\max} > d_\theta > 0$ and $0 < D_\theta < D_{\max}$, $p_{\max}$ may be found using $D'(0) = -d_{\max}$ and $D'((D)^{-1}(D_\theta)) = D'(p_\theta) = -d_\theta$. The specific forms of demand in Eqs. (3.2) and (3.3) are studied herein because they are tractable.

In [21], we explored the interior Nash equilibria resulting from such convex demand responses. Note how the above models reduce to linear demand response (e.g., by taking $\alpha = 1$), that is, revenue quadratic in prices, as assumed in many prior papers, for example, [22].

3.3 TWO DIFFERENT EYEBALL ISPs

Again, we consider a game focusing on two different eyeball ISPs, indexed a and b, on a platform of users and CPs, that is, the ISPs also serve as CPs so no separate pricing by CPs is modeled. For $k, j \in \{a, b\}$, the demand for ISP k's content is $D_k(p_j)$ when it is based on ISP j's access-bandwidth price p_j. In the following, the same price p_j will be used by ISP j irrespective of content source; that is, content is neutrally priced in this sense.

Suppose there are peering points between these two ISPs where the net transit traffic flow in one direction will correspond to the net revenue for the (net) receiving

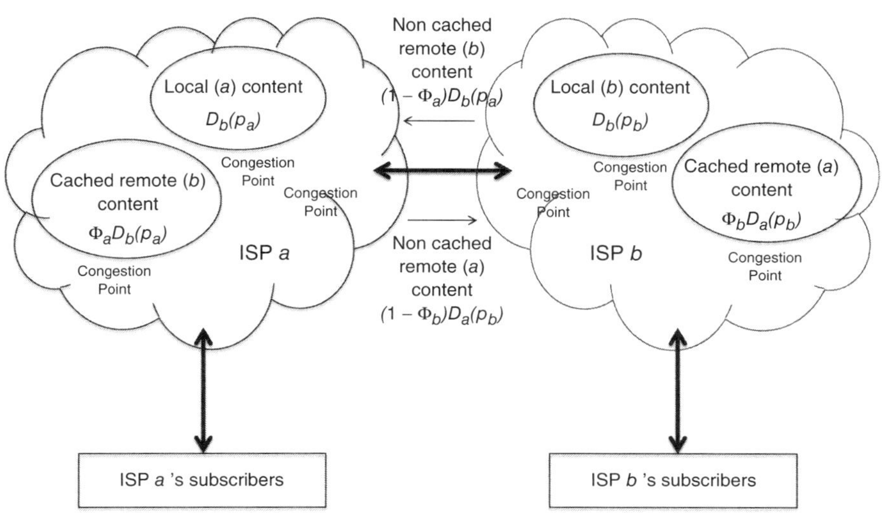

Figure 3.3 Caching remote content.

ISP at rate p_t from the (net) transmitting ISP. For example, France telecom charges p_t =\$3 per megabit, whereas pricing from the digital subscriber line access multiplexer (DSLAM) to core, that is, access bandwidth, for their *CPs* is \$40 per megabit [23]. This said, many existing peering agreements among nontransit ISPs have no transit pricing, that is, $p_t = 0$. See [24,25] for recent studies of models of transit pricing for a network involving a transit ISP between the CPs and end user ISPs.

Without caching, transit traffic volume is obviously maximal and remote content may be subject to additional delay possibly increasing demand (reducing demand sensitivity) for usage-priced bandwidth reservations. However, poorer delay performance may instead reduce demand for remote content or cause subscribers to change to ISPs that cache remote content. So, caching will result in reduced demand for premium services by transit traffic; in the following, we model this with a caching factor Φ_k. We assume fixed caching factors for each of the ISPs, which means the selected caching factors by the ISPs do not change no matter how their demand changes. The case where the caching factor is adapted because of the demand changes is among the future work.

3.4 THREE DIFFERENT CONGESTION POINTS PER ISP, FIXED CACHING FACTORS

By simply separately accounting for the demand for premium-access service by two different user populations with similar content preferences, we take the utilities as

$$\begin{aligned} U_a(p_a, p_b) =& D_a(p_a)p_a + \Phi_a D_b(p_a)p_a \\ &+ [(1-\Phi_a)D_b(p_a) - (1-\Phi_b)D_a(p_b)]^+ p_t, \\ U_b(p_a, p_b) =& D_b(p_b)p_b + \Phi_b D_a(p_b)p_b \\ &+ [(1-\Phi_b)D_a(p_b) - (1-\Phi_a)D_b(p_a)]^+ p_t, \end{aligned}$$

where $[x]^+ := \max\{x, 0\}$ in the second (transit revenue) terms. Note that $\Phi_k \leq 1$ will be chosen by ISP k at its *minimal* value, which we here assume to be strictly positive again because an ISP that does not cache any remote content may lose subscribers or demand for remote content may be reduced owing to poor delay performance, cf. Section 3.5. We will also assume that p_t is fixed and, by volume discount, $p_t < \min\{p_a, p_b\}$. Also, we have assumed different "upstream" congestion points for local and remote traffic and no revenue from cached (best-effort) traffic. Moreover, for $\alpha > 1$ (i.e., not linear demand response) note how this model assumes three different congestion points, one at the peering point, one at the local content source, and one at the cached content source, but *not* a single one further downstream toward the users, cf. Section 3.5. That is, in this section, we consider three separate congestion points per ISP for an example of convex demand (assumptions that include the linear demand-response scenario as a special case).

Again suppose, for $k \in \{a, b\}$, that

$$D_k(p) = D_{\max,k}\left(1 - \frac{p}{p_{\max}}\right)^{\alpha}, \tag{3.4}$$

where the maximal price $p_{\max} > 0$ and $\alpha \geq 1$ are also assumed to be common parameters for both ISPs to simplify the following expressions for Nash equilibria. Note that $U := pD(p)$ with $a > 1$ and $0 \leq p \leq p_{\max}$ is maximized at $p^* = p_{\max}/(1 + a)$, *i.e.*, $U'(p^*) = 0$ and $U''(p^*) < 0$. Without loss of generality, assume the demand ratio

$$\delta := \frac{D_{\max,b}}{D_{\max,a}} \leq 1, \tag{3.5}$$

that is, demand for ISP a's content is generally higher than that for ISP b.

The Nash equilibrium is a "stalemate" pricing point (p_a^*, p_b^*) at which *neither* ISP's utility will improve by a price change, that is,

$$\arg\max_{p_a} U_a(p_a, p_b^*) = p_a^* \quad \text{and} \tag{3.6}$$

$$\arg\max_{p_b} U_b(p_a^*, p_b) = p_b^*. \tag{3.7}$$

The first-order Nash equilibrium conditions and the solutions of these for three cases are provided in the following text.

Case 1: $(1 - \Phi_a)D_b(p_a^*) > (1 - \Phi_b)D_a(p_b^*)$.

$$\begin{aligned}\frac{\partial U_a(p_a, p_b)}{\partial p_a} &= D_a'(p_a)p_a + D_a(p_a) + \Phi_a[D_b'(p_a)p_a \\ &\quad + D_b(p_a)] + (1 - \Phi_a)D_b'(p_a)p_t = 0 \\ \frac{\partial U_b(p_a, p_b)}{\partial p_b} &= D_b'(p_b)p_b + D_b(p_b) + \Phi_b[D_a'(p_b)p_b \\ &\quad + D_a(p_b)] = 0.\end{aligned}$$

The solution is as follows:

$$p_a^* = \frac{p_{\max}}{1 + \alpha} - \frac{p_t(1 - \Phi_a)\delta\alpha}{(1 + \alpha)(1 + \Phi_a\delta)}, \tag{3.8}$$

$$p_b^* = \frac{p_{\max}}{1 + \alpha}. \tag{3.9}$$

The requirement $p_t < p_a^* < p_b^* < p_{\max}$ gives the following condition on p_t for an interior Nash equilibrium:

$$\frac{p_{\max}}{p_t} > 1 + \frac{\alpha(\delta + 1)}{1 + \delta\Phi_b}. \tag{3.10}$$

Another way to put the case condition $(1 - \Phi_a)D_b(p_a^*) > (1 - \Phi_b)D_a(p_b^*)$ is

$$1 < \frac{(1-\Phi_a)\delta}{1-\Phi_b}\left(\frac{p_{\max}-p_a^*}{p_{\max}-p_b^*}\right)^{\alpha} \text{ and} \tag{3.11}$$

$$1 < \frac{(1-\Phi_a)\delta}{1-\Phi_b}\left(1+\frac{(1-\Phi_a)\delta p_t}{(1+\Phi_a\delta)p_{\max}}\right)^{\alpha}. \tag{3.12}$$

Case 2: $(1-\Phi_a)D_b(p_a^*) < (1-\Phi_b)D_a(p_b^*)$.

$$\begin{aligned}\frac{\partial U_a(p_a,p_b)}{\partial p_a} &= D_a'(p_a)p_a + D_a(p_a) + \Phi_a[D_b'(p_a)p_a \\ &\quad + D_b(p_a)] = 0 \\ \frac{\partial U_b(p_a,p_b)}{\partial p_b} &= D_b'(p_b)p_b + D_b(p_b) + \Phi_b[D_a'(p_b)p_b \\ &\quad + D_a(p_b)] + (1-\Phi_b)D_a'(p_b)p_t = 0.\end{aligned}$$

The solution is as follows:

$$p_a^* = \frac{p_{\max}}{1+\alpha}, \tag{3.13}$$

$$p_b^* = \frac{p_{\max}}{1+\alpha} - \frac{p_t(1-\Phi_b)\alpha}{(1+\alpha)(\delta+\Phi_b)}. \tag{3.14}$$

The requirement $p_t < p_b^* < p_a^* < p_{\max}$ imposes the following condition on p_t:

$$\frac{p_{\max}}{p_t} > 1 + \frac{\alpha(\delta+1)}{\delta+\Phi_b} \tag{3.15}$$

The case condition $(1-\Phi_a)D_b(p_a^*) < (1-\Phi_b)D_a(p_b^*)$ can be rewritten as

$$1 > \frac{(1-\Phi_a)\delta}{1-\Phi_b}\left(\frac{p_{\max}-p_a^*}{p_{\max}-p_b^*}\right)^{\alpha} \text{ and} \tag{3.16}$$

$$1 > \frac{(1-\Phi_a)\delta}{1-\Phi_b}\left(1+\frac{(1-\Phi_a)\delta p_t}{(1+\Phi_a\delta)p_{\max}}\right)^{\alpha}. \tag{3.17}$$

Case 3: $(1-\Phi_a)D_b(p_a^*) = (1-\Phi_b)D_a(p_b^*)$.

$$\begin{aligned}\frac{\partial U_a(p_a,p_b)}{\partial p_a} &= D_a'(p_a)p_a + D_a(p_a) \\ &\quad + \Phi_a[D_b'(p_a)p_a + D_b(p_a)] = 0 \\ \frac{\partial U_b(p_a,p_b)}{\partial p_b} &= D_b'(p_b)p_b + D_b(p_b) \\ &\quad + \Phi_b[D_a'(p_b)p_b + D_a(p_b)] = 0.\end{aligned}$$

The solution of the above equations is as follows:

$$p_a^* = p_b^* = \frac{p_{\max}}{1+\alpha}. \tag{3.18}$$

The case condition reduces to

$$\frac{1-\Phi_b}{1-\Phi_a} = \frac{D_{\max,b}}{D_{\max,a}} = \delta. \tag{3.19}$$

3.5 ONE CONGESTION POINT PER ISP, FIXED CACHING FACTORS

In this scenario, at ISP a, the demands $D_a(p_a)$ (demand for local content) and $D_b(p_a)$ (demand for remote content) share a common significant congestion point proximal to the users, for example, in a wireless-access setting. Again, we consider a system where the players (eyeball ISPs) select access prices (plays) $p_a, p_b > p_t$.

Given the prices p_a for local content, we want an expression for demand $\hat{D}_{aa}$ (local content at ISP a) and $\hat{D}_{ba}$ (remote content at ISP a) that has the following intuitive property:

$$\lim_{D_{\max,b}\to 0} \hat{D}_{aa} = D_a(p_a) \text{ and } \lim_{D_{\max,a}\to 0} \hat{D}_{ba} = D_b(p_a) \tag{3.20}$$

and similarly for ISP b regarding $\hat{D}_{bb}$ and $\hat{D}_{ab}$ as a function of p_b.

The following assumed property is also intuitive because the presence of remotely originated traffic will congest locally originated traffic and vice versa:

$$\hat{D}_{aa} \le D_a(p_a) \quad \text{and} \quad \hat{D}_{ba} \le D_b(p_a) \tag{3.21}$$

and similarly for the other ISP b.

Proportion Rule. Suppose that the throughput limit downstream to the users is L_k for ISP $k \in \{a, b\}$. Then, at ISP a, the demands are as follows:

$$\hat{D}_{aa} = \begin{cases} \dfrac{D_a(p_a)}{D_a(p_a)+D_b(p_a)} L_a, & \text{if } D_a(p_a)+D_b(p_a) > L_a \\ D_a(p_a), & \text{else.} \end{cases}$$

and

$$\hat{D}_{ba} = \begin{cases} \dfrac{D_b(p_a)}{D_a(p_a)+D_b(p_a)} L_a, & \text{if } D_a(p_a)+D_b(p_a) > L_a \\ D_b(p_a), & \text{else.} \end{cases}$$

And similarly for ISP b.

Critical Price Rule. Another way to split the throughput among the demands is as follows. For ISP a, when $D_a(p_a) + D_b(p_a) > L_a$, a new price p_a^* is chosen so that

$$D_a(p_a^*) + D_b(p_a^*) = L_a. \tag{3.22}$$

If $p_a < p_a^*$, then congestion will occur.

So, the expressions for the ISP revenues here can be taken as

$$\begin{aligned} U_a(p_a) &= \hat{D}_{aa}p_a + \Phi_a \hat{D}_{ba}p_a \\ &\quad + [(1-\Phi_a)\hat{D}_{ba} - (1-\Phi_b)\hat{D}_{ab}]^+ p_t \\ U_b(p_b) &= \hat{D}_{bb}p_b + \Phi_b \hat{D}_{ab}p_b \\ &\quad + [(1-\Phi_b)\hat{D}_{ab} - (1-\Phi_a)\hat{D}_{ba}]^+ p_t. \end{aligned}$$

3.6 THREE DIFFERENT CONGESTION POINTS PER ISP, FIXED CACHING FACTORS, MULTIPLE PROVIDERS OF ONE OF THE TYPES

In this scenario, ISP a in Figure 3.3 is replaced by two ISPs, namely, ISP $a1$ and $a2$, which compete for the same group of subscribers. So, we need to consider three utility functions; U_{a1}, U_{a2}, and U_b; three demand functions, D_{a1}, D_{a2}, and D_b; and three access prices for each of the ISPs' own subscribers, p_{a1}, p_{a2}, and p_b. But the number of caching factors increases to four: $\Phi_{a1,b}$, $\Phi_{a2,b}$, $\Phi_{b,a1}$, and $\Phi_{b,a2}$ ($\Phi_{m,n}$ meaning willingness of ISP m to cache the content of ISP n). And, there are two transit prices, p_{t1} (for the traffic between ISP $a1$ and ISP b) and p_{t2} (for ISPs $a2$ and b).

$$\begin{aligned} U_{a1}(p_{a1}, p_b) &= \sigma_{a1} D_{a1}(p_{a1})p_{a1} + \sigma_{a1}\Phi_{a1,b} D_b(p_{a1})p_{a1} \\ &\quad + [\sigma_{a1}(1-\Phi_{a1,b})D_b(p_{a1})) \\ &\quad - (1-\Phi_{b,a1})D_{a1}(p_b)]^+ p_{t1} \\ U_{a2}(p_{a2}, p_b) &= \sigma_{a2} D_{a2}(p_{a2})p_{a2} + \sigma_{a2}\Phi_{a2,b} D_b(p_{a2})p_{a2} \\ &\quad + [\sigma_{a2}(1-\Phi_{a2,b})D_b(p_{a2})) \\ &\quad - (1-\Phi_{b,a2})D_{a2}(p_b)]^+ p_{t2} \\ U_b(p_{a1}, p_{a2}, p_b) &= D_b(p_b)p_b + \Phi_{b,a1} D_{a1}(p_b)p_b \\ &\quad + \Phi_{b,a2} D_{a2}(p_b)p_b + [(1-\Phi_{b,a1})D_{a1}(p_b) \\ &\quad - \sigma_{a1}(1-\Phi_{a1,b})D_b(p_{a1})]^+ p_{t1} \\ &\quad + [(1-\Phi_{b,a2})D_{a2}(p_b)) \\ &\quad - \sigma_{a2}(1-\Phi_{a2,b})D_b(p_{a2})]^+ p_{t2}, \end{aligned}$$

where

$$\sigma_{ai} = \frac{1/p_{ai}}{1/p_{a1} + 1/p_{a2}}, \quad \forall i \in \{1, 2\},$$

represents customer stickiness (loyalty, inertia) to the ith ISP (*e.g.*, [9]); that is, because $\sigma_{ai} \propto 1/p_{ai}$, the subscribers will not completely switch to the ISP with the lowest price.

The demand-response model provided in Eq. (3.3) is used here, now with $k \in \{a1, a2, b\}$.

3.7 NUMERICAL EXPERIMENTS

First, numerical results were obtained for the scenario where there are three congestion points per ISP (with fixed caching factors, as explained in Section 3.4) with $\alpha \in \{1, 2\}$, $D_{\max,a} = 20$, $D_{\max,b} = 10$, $p_{\max} = 5$, $p_t = 1$, $\Phi_a = 0.5$, and $\Phi_b = 0.3$ as the selected parameter values.

By using $U_a(p_a, p_b)$ (Figure 3.4) and $U_b(p_a, p_b)$ (Figure 3.5), the Nash equilibrium point (p_a^*, p_b^*) were found in the following way:

1. Uniformly at random over $(p_t, p_{\max})$ select an initial point $\gamma^{(0)} = (p_a^{(0)}, p_b^{(0)})$.
2. $\forall k \geq 1$, find the updated point $\gamma^{(k)} = (p_a^{(k)}, p_b^{(k)})$ by synchronous best-response updates, which are

$$\begin{aligned} p_a^{(k)} &= \arg\max_{p_a} U_a(p_a, p_b^{(k-1)}) \\ p_b^{(k)} &= \arg\max_{p_b} U_b(p_a^{(k-1)}, p_b). \end{aligned}$$

3. (a) If $\gamma^{(k-1)} \approx \gamma^{(k)}$, stop.
   (b) Else, return to step 2.

It was observed that the Nash equilibrium point found by using the above procedure is the same as the equilibrium point corresponding to the proper case solution provided in Section 3.4 (regardless of the randomly selected starting point) and it was found in just a few iterations (Figs. 3.4–3.7).

It can be observed in Figures 3.8 and 3.9 that $p_a^* > p_b^*$ and $U_a(p_a^*, p_b^*) > U_b(p_a^*, p_b^*)$ for both values of α. This is intuitive because $D_{\max,a} > D_{\max,b}$, which implies that the demand for ISP a's content will be larger than ISP b's at the same price. This immediately implies larger gain for ISP a, which also means that ISP a might have some margin for increasing p_a in order to gain even more utility. Therefore $p_a^* > p_b^*$ in this setting.

Next, numerical results were obtained for the model defined in Section 3.5, where one congestion point per ISP and fixed caching factors assumptions are used.

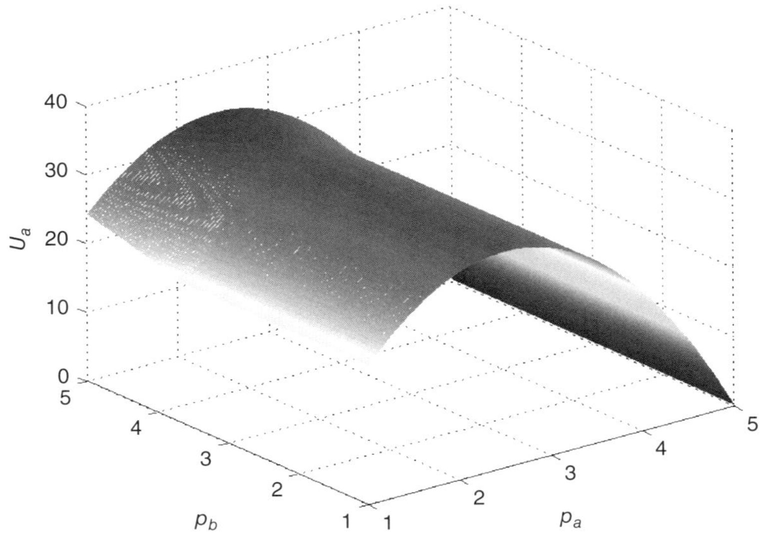

Figure 3.4 $U_a(p_a, p_b)$ (three congestion points for each ISP, fixed caching factors) ($\alpha = 1$).

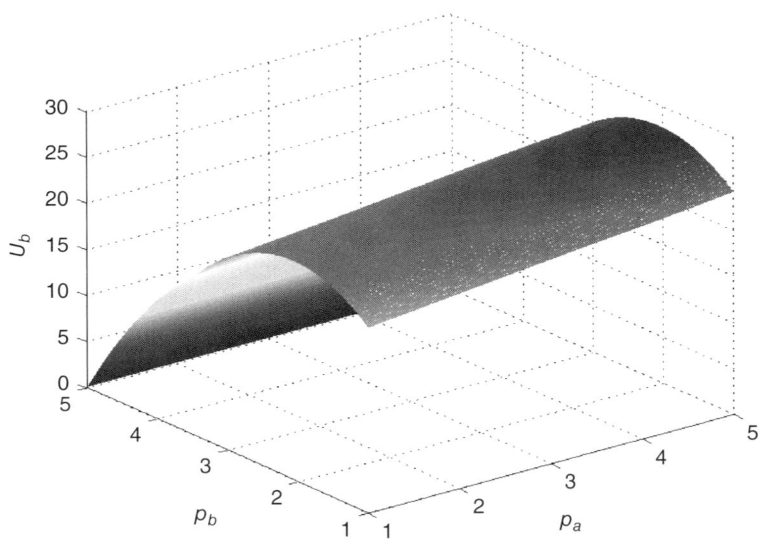

Figure 3.5 $U_b(p_a, p_b)$ (three congestion points for each ISP, fixed caching factors) ($\alpha = 1$).

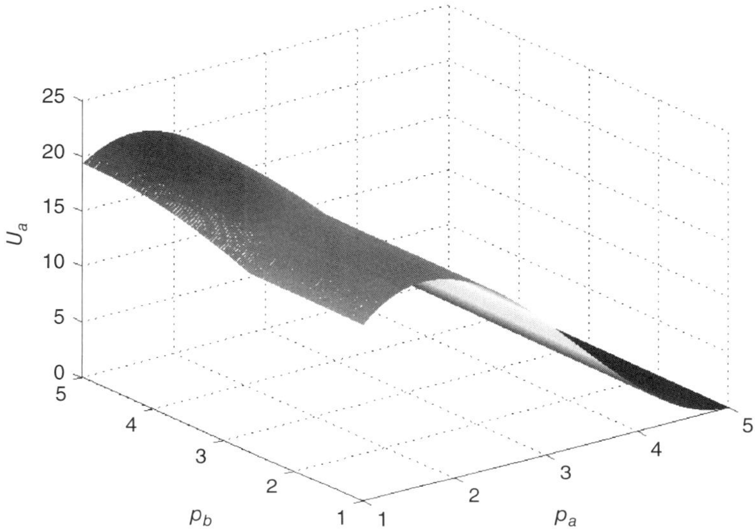

Figure 3.6 $U_a(p_a, p_b)$ (three congestion points for each ISP, fixed caching factors) ($\alpha = 2$).

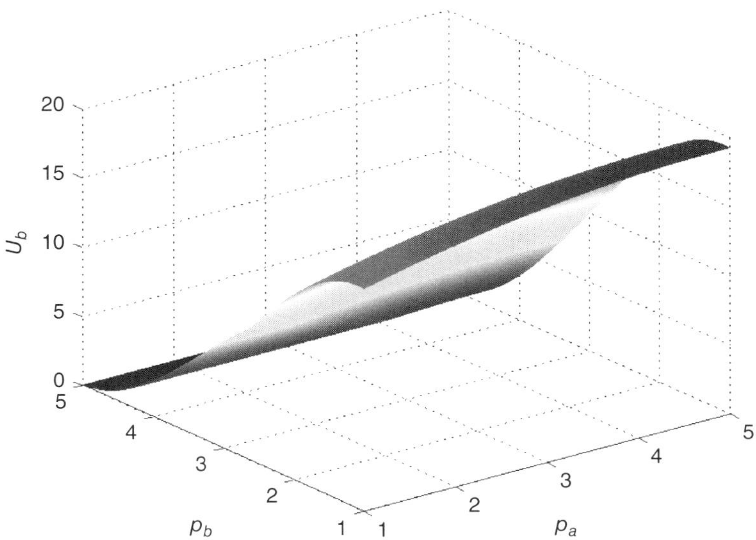

Figure 3.7 $U_b(p_a, p_b)$ (three congestion points for each ISP, fixed caching factors) ($\alpha = 2$).

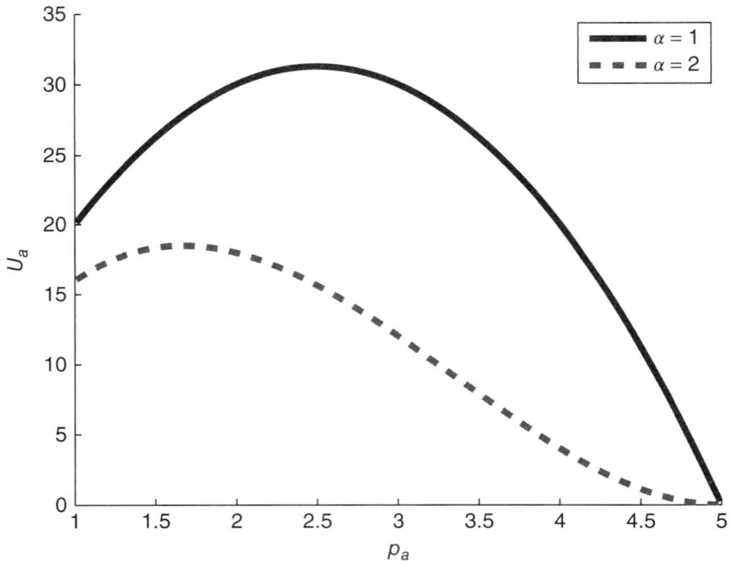

Figure 3.8 $U_a(p_a, p_b^*)$ (three congestion points for each ISP, fixed caching factors).

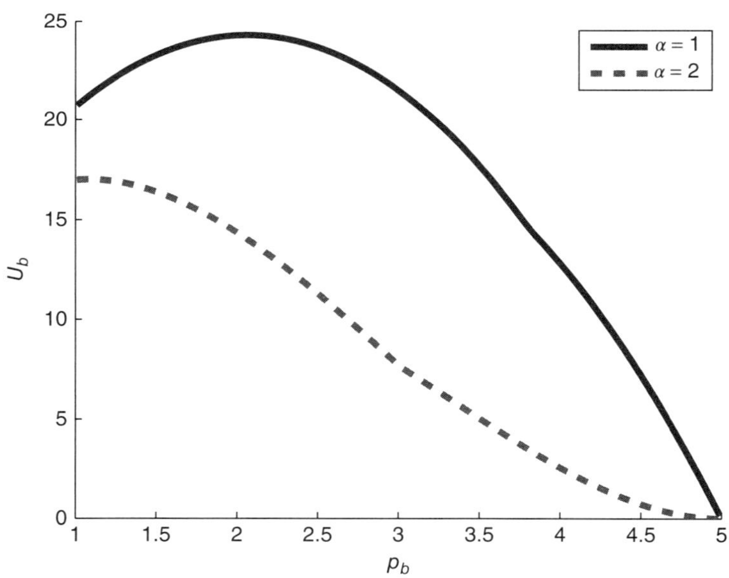

Figure 3.9 $U_b(p_a^*, p_b)$ (three congestion points for each ISP, fixed caching factors).

Here, the throughput limit is split among the ISPs according to the proportion rule, cf. Section 3.5. $\alpha \in \{1, 2\}$, $D_{\max,a} = 20$, $D_{\max,b} = 10$, $p_{max} = 5$, $p_t = 1$, $\Phi_a = 0.5$, $\Phi_b = 0.3$, $L_a = 50$, and $L_b = 5$ are the selected parameters values. Notice that one of the throughput limits (L_a) is selected, which is significantly larger than the other one (L_b) to analyze the scenario where congestion does not occur downstream to the users of ISP a, whereas it does occur for ISP b. If both of the throughput limits selected are very large, then the problem reduces to the three congestion points scenario (Section 3.4), because there will be no distribution of the throughput limit between the two different kinds of demand at the congestion point (of each ISP).

The Nash equilibrium point was again quickly found by using synchronous best-response updates.

In Figures 3.10 and 3.11, behaviors similar to those with Figures 3.8 and 3.9 are observed. But, it is worth noting that in Figure 3.11, for values of p_b where U_b is increasing (for both $\alpha \in \{1, 2\}$), the capacity L_b is fully utilized. In this region, increasing p_b does not lead to a decrease in the demand, which means that there is a linear increase in the utility of ISP b. But, after the peak, the total demand at ISP b is smaller than L_b; therefore, the increase in price p_b leads to decreases in both demand and utility.

Finally, numerical results were obtained for the case where there are multiple providers competing for the same group of subscribers (Section 3.6). Again, synchronous best-response updates are used, but for three utility functions ($U_{a1}(p_{a1}, p_{a2}, p_b)$, $U_{a2}(p_{a1}, p_{a2}, p_b)$, and $U_b(p_{a1}, p_{a2}, p_b)$) depending on the corresponding three access pricing parameters (p_{a1}, p_{a2}, and p_b). So, generally, for n

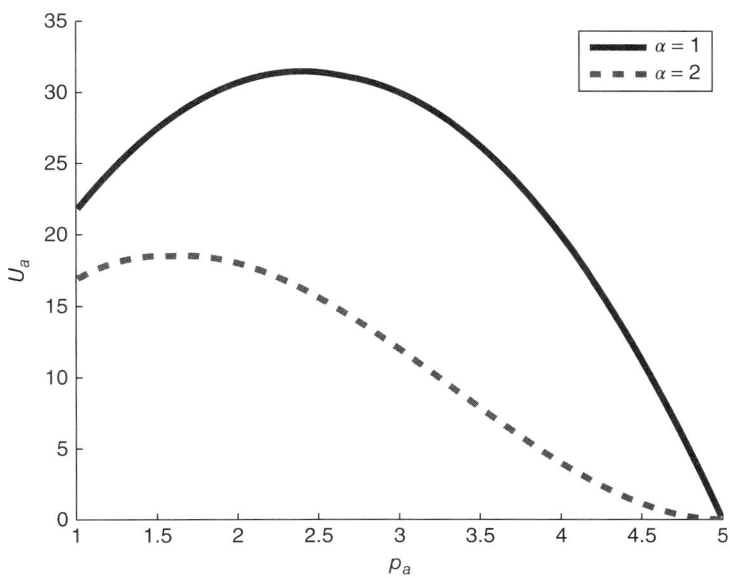

Figure 3.10 $U_a(p_a, p_b^*)$ (one congestion point for each ISP, fixed caching factors).

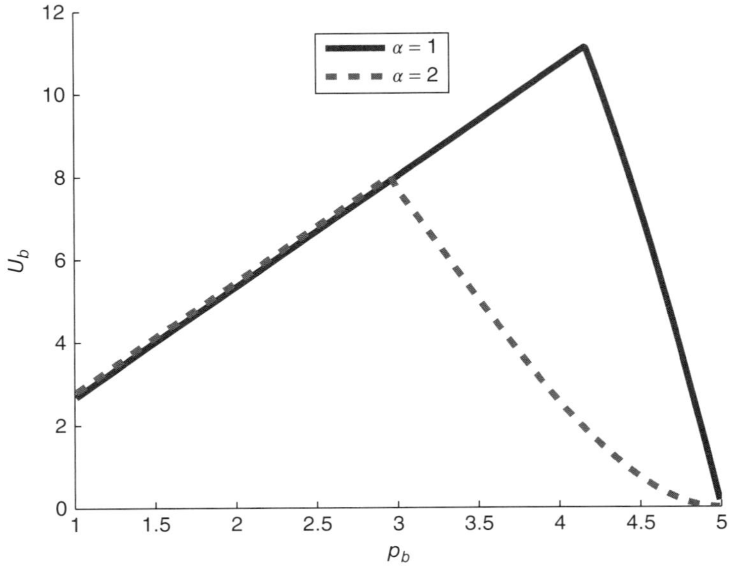

Figure 3.11 $U_b(p_a^*, p_b)$ (one congestion point for each ISP, fixed caching factors).

competing ISPs ($n = 2$ in our case of ISPs $a1$ and $a2$), the synchronous best-response update step ($n + 1$ player synchronous updates) is as follows:

$$p_i^{(k)} = \arg\max_{p_i} U_i(p_i, \underline{p}_{-i}^{(k-1)}), \quad \forall\, i,$$

where i is the index of the ISP (including the noncompeting ISP (in our case, ISP b)), p_i is the price used by ISP i, and $\underline{p}_{-i}$ is the set of prices used by the other ISPs.

The parameter values can be selected in various combinations. We used the parameters $D_{\max,a1} = 20$, $D_{\max,a2} = 20$, $D_{\max,b} = 10$, $p_{\max} = 5$, $p_{t1} = 1$, $p_{t2} = 1$, $\Phi_{a1,b} = 0.2$, $\Phi_{a2,b} = 0.8$, $\Phi_{b,a1} = 0.5$, and $\Phi_{b,a2} = 0.5$. These were selected so as to analyze the effect of (static but different) caching factors of competing ISPs (ISPs $a1$ and $a2$) on the utilities. It can observed from Figures 3.12 and 3.13 that the ISP with smaller Φ ($a1$) also has (again following intuition) a smaller utility compared to its competitor ISP ($a2$). The effect of α on the utilities and the equilibrium prices are in the same as the previous cases.

3.8 FUTURE WORK

In the future, we will extend our models of demand in the presence of a more complex mixture of applications with different service requirements and will do so by using more diverse, though naturally coupled, demand response models for each type of provider. Moreover, we will consider these problems in the context of competition

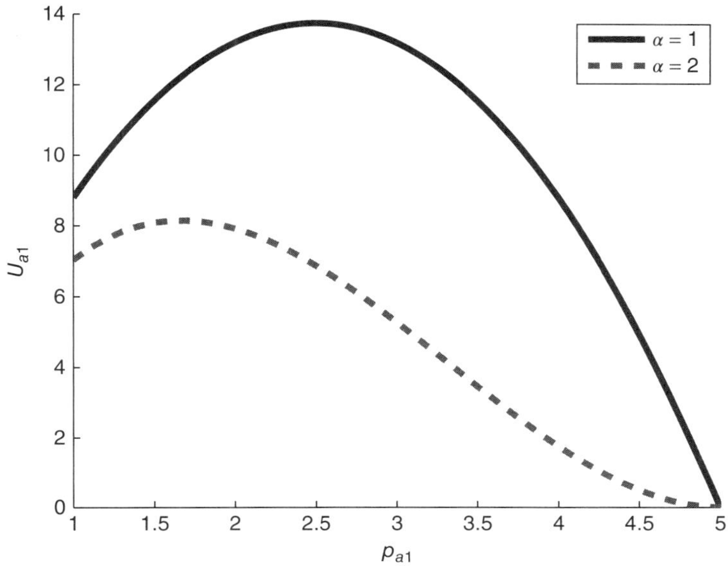

Figure 3.12 $U_{a1}(p_{a1}, p^*_{a2}, p^*_b)$ (three congestion points for each ISP, fixed caching factors, competing ISPs).

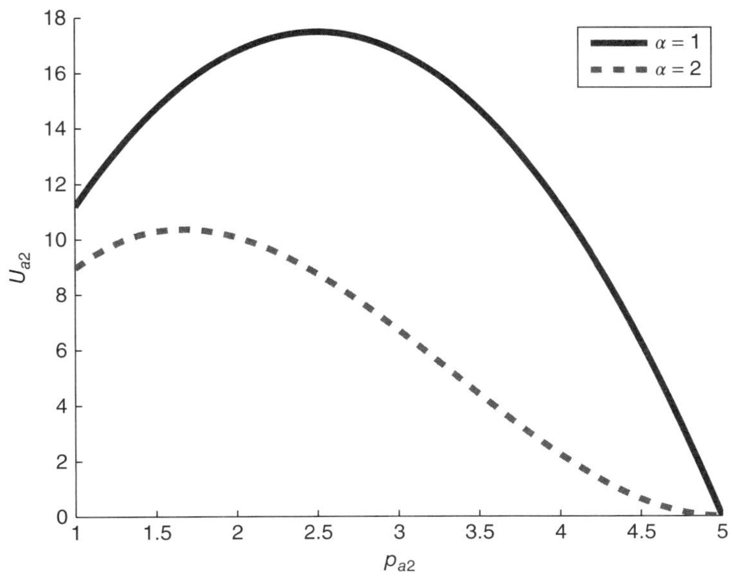

Figure 3.13 $U_{a2}(p^*_{a1}, p_{a2}, p^*_b)$ (three congestion points for each ISP, fixed caching factors, competing ISPs).

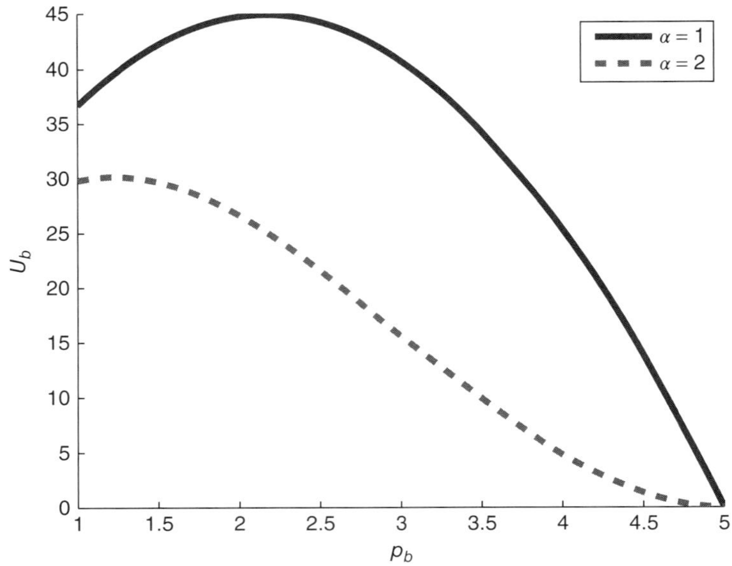

Figure 3.14 $U_b(p^*_{a1}, p^*_{a2}, p_b)$ (three congestion points for each ISP, fixed caching factors, competing ISPs).

(multiple providers of each type) and collaboration between providers of different types, where we could use, for example, Shapley values to decide how to share revenue in the latter case.

We will also consider the dynamic selection of a caching factor. For example, we can model the sensitivity of customer loyalty σ to the caching factor Φ in a way so as to reflect the loss in demand by user migration because of poor delay performance in the best-effort service class because of inadequate caching. So, the utilities of the ISPs will be affected by this potential loss because of user migration, and hence, they will depend on the caching factors as well as the access prices. As ISPs adjust their caching factors, in addition to their access prices, capturing the trade-off between the user migration and the transit traffic revenue.

REFERENCES

1. P. Waldmeir. The net neutrality dogfight shaking up cyberspace. *Financial Times, Mar. 23*, 2006.
2. P. Hande, M. Chiang, R. Calderbank, and S. Rangan. Network pricing and rate allocation with content provider participation. In *Proceedings IEEE INFOCOM*, 2009.
3. J. Musacchio, G. Schwartz, and J. Walrand. "A two-sided market analysis of provider investment incentives with an application to the net-neutrality issue," *Review of Network Economics*, 8(1), 2009, 22–39.

4. R. Hahn and S. Wallsten. “The economics of net neutrality,” *Economists’ Voice*, 3(6), 2006, 1–7.
5. P. Ganley and B. Allgrove. “Net neutrality: a user’s guide,” *Computer Law and Security Report*, 22(6), 2006, 454–463.
6. ETICS. First Economics and Technologies for Inter-Carrier Services (ETICS) Workshop Proceedings, 2010.
7. Comcast v. FCC, 600 F. 3d 642 (D.C. Cir. 2010).
8. S. Sen, C. Joe-Wong, S. Ha, and M. Chiang. “Pricing data: a look at past proposals, current plans, and future trends,” Arxiv preprint, arXiv:1201.4197, 2012.
9. S. Caron, G. Kesidis, and E. Altman. Application neutrality and a paradox of side payments. In *Proceedings ACM ReArch*, Philadelphia, Nov. 30,2010. See also http://arxiv.org/abs/1006.3894
10. E. Altman, P. Bernhard, S. Caron, G. Kesidis, J. Rojas-Mora, and S. Wong. A study of non-neutral networks. Telecommunication Systems Journal Special Issue on Socio-economic Issues of Next Generation Networks, June 2011.
11. E. Altman, A. Legout, and Y. Xu. Network non-neutrality debate: an economic analysis. In *Proceedings IFIP Networking*, 2011.
12. Y. Wu, H. Kim, P. H. Hande, M. Chiang, and D. H. K. Tsang. Revenue sharing among ISPs in two-sided markets. In *Proceedings of IEEE INFOCOM Mini Conference*, Shanghai, 2011.
13. P. Njoroge, A. E. Ozdaglar, N. E. Stier-Moses, and G. Y. Weintraub. Investment in Two Sided Markets and the Net Neutrality Debate. *Columbia Business School DRO (Decision, Risk and Operations)* Working Paper No. 2010-05, Oct. 2012.
14. R. B. Chong. “The 31 Flavors of Net Neutrality,” 12 Intellectual Property Law Bulletin, vol. 12, 2008.
15. H.K. Cheng, S. Bandyopadhyay, and H. Guo. The debate on net neutrality: a policy perspective. *Information Systems Research*, 22(1), 2011, 60–82.
16. J. Goldsmith and T. Wu. *Who Controls the Internet: Illusions of a Borderless World*. Oxford University Press, New York, 2006.
17. F. Bornstaedt, M. G. Roettgermann, F. T. Johansen, and H. Lønsethagen. “The Sending Party Network Pays”: a first step towards end-to-end quality of service. In *Proceedings IEEE ICIN*, 2011.
18. R. Douville. ETICS Architecture(s). In *Second Economics and Technologies for Inter-Carrier Services (ETICS) Workshop*, June 2011.
19. A. Shin. Who’s the bandwidth bandit? The *Washington Post*, Oct. 4, 2006. Available at: http://blog.washingtonpost.com/thecheckout /2006/10/bandwidth_bandit.html.
20. K. Bode. AT&T To impose caps. *Overages* Mar. 13,2011. Available at: http://www.dslreports.com/shownews/Exclusive-ATT-To-Impose-Caps-Overages-113149.
21. G. Kesidis. Side-payment profitability under convex demand-response modeling congestion-sensitive applications. In *Proceedings IEEE ICC, Ottawa*, June 2012.
22. N. Economides. “Net neutrality: non-discrimination and digital distribution of content through the internet,” *I/S: A Journal of Law and Policy*, 4(2), 2008, 209–233.
23. E.Pouyllau. Personal Communication *ARC MANEUR Meeting. INRIA*, Paris, May. 2011.
24. A. Dhamdhere and C. Dovrolis. Can ISPs be profitable without violating “Network Neutrality”? In *Proceedings ACM NetEcon*, Seattle, WA, 2008.

25. V. Valancius, C. Lumezanu, N. Feamster, R. Johari, and V. V. Vazirani. How many tiers? Pricing in the internet transit market. In *Proceedings ACM SIGCOMM*, 2011.
26. A.A. Gilroy. Net neutrality: background and issues. CRS Report for Congress, RS22444, 2008.

PART II
Technologies for Smart Data Pricing

4 Pricing under Demand Flexibility and Predictability

OZGUR DALKILIC, JOHN TADROUS, ATILLA ERYILMAZ,
and HESHAM EL-GAMAL

4.1 INTRODUCTION

The essence of *smart pricing* is exploiting the price responsiveness of the demand side to achieve objectives such as obtaining higher profit, improving customer experience, and sustaining reliable operation. The enormous pace of advances in technology and engineering and new economic practices cause rapid changes in consumer behavior that create new dynamics as well as problems affecting the physical infrastructures and the corresponding markets. Hence, more sophisticated and novel smart pricing methods are required to control and take advantage of the demand-side dynamics.

Smart pricing refers to various techniques such as charging consumers depending on the service usage time, setting location-based tariffs, and imposing prices based on consumer activity levels. For instance, hourly and daily fluctuations in demand are striking patterns that are common to both the Internet and the electricity grid as seen in Figure 4.1. Considering this pattern, Internet service providers and mobile operators charge their subscribers based on the amount of data communicated or the time of communication in order to alleviate congestion in the network [1, 2]. Similarly, electricity usage also can be priced based on the amount or the time of consumption so that daily fluctuations of the grid load are reduced [3].

The driving factors for smart pricing are the consumer characteristics and consumers' economic incentives. Consumption amount, service usage time, demand patterns over certain periods of time, and randomness of the load are several aspects of the overall consumer behavior and they consequently affect the price setters' strategies. On the other hand, consumer behavior is naturally responsive to the pricing decisions because humans oversee their own economic welfare. Therefore, the resulting closed loop interactions are critical to the operation of the physical infrastructure and the corresponding market in terms of reliability, stability, and efficiency.

Smart Data Pricing, First Edition. Edited by Soumya Sen, Carlee Joe-Wong, Sangtae Ha, and Mung Chiang.

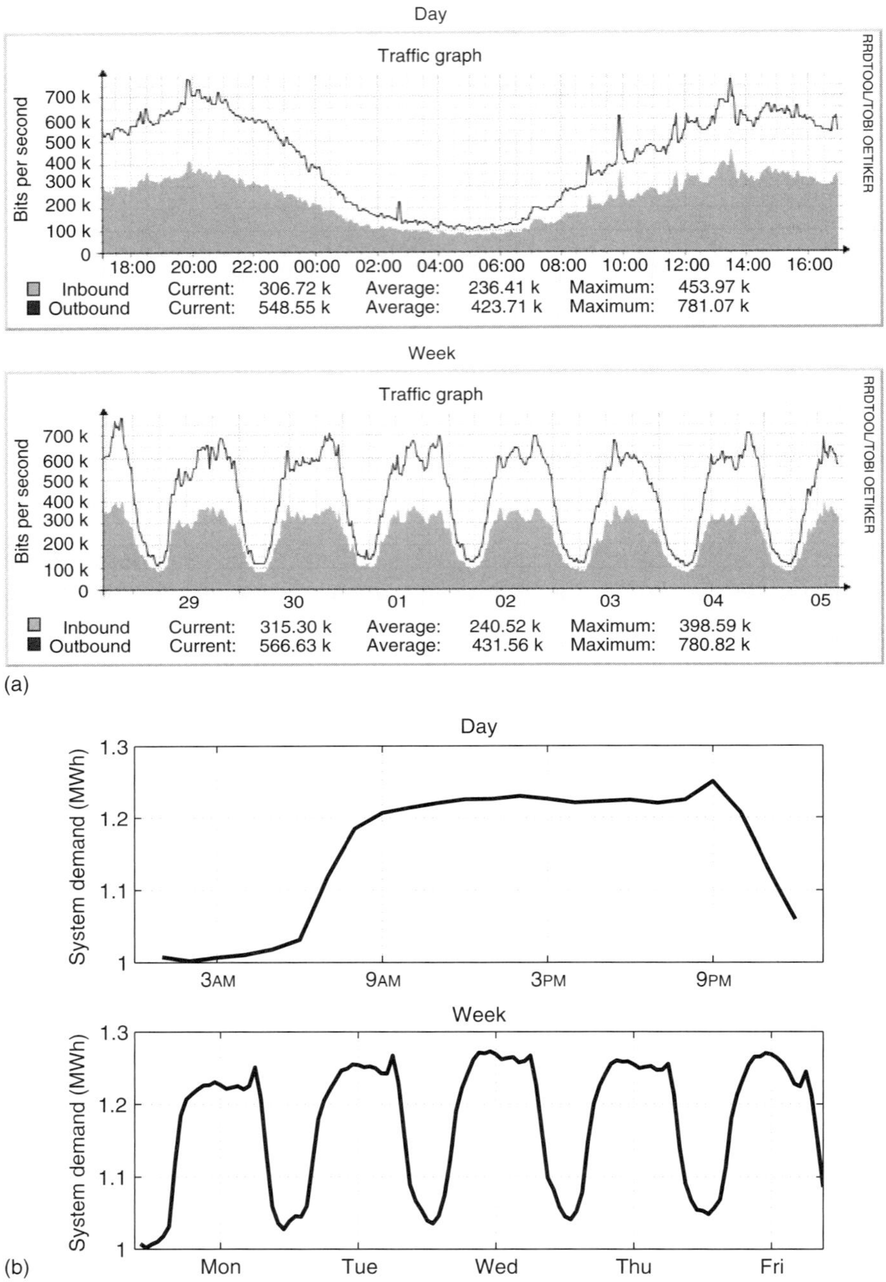

Figure 4.1 Hourly and daily fluctuations in demand in different systems. Consumer activity drops to minimum levels during late night, then increases significantly during the day to attain its peak, and again drops during night. (a) Demand pattern at a basestation of a wireless network operator, as measured by RRDTool [4, 5]. (b) Total hourly system demand as measured by ISO New England [6].

Demand-side characteristics are increasingly exposed to the market because of sophisticated control methods, ubiquitous communication capabilities, and innovative market practices. For instance, the trend for deregulation in energy markets, smart power meters for households, or advanced data gathering, tracking, and management methods over the Internet causes the consumers to be more active, more controllable, and more predictable. Therefore, the producers and intermediaries have more incentives and possibilities to influence or directly control the consumer behavior via smart pricing strategies in order to reduce costs and sustain stable operation. In this chapter, we focus on *demand-side flexibilities* and the *predictability of consumer activities* and discuss the means of exploiting these aspects of consumer behavior by using smart pricing techniques.

Demand-side flexibilities usually comprise temporal elasticities involving deferring loads until a specified deadline, shifting consumption within a certain period of time, and using a service intermittently. Consumers, who are motivated by their economic needs, are capable of exercising these flexibilities owing to technologies such as intelligent devices with two-way communication capabilities and sophisticated software applications that are able to manage user demand. Hence, consumers can be incentivized to defer or shift their demand by time- and consumption-dependent smart pricing methods as investigated in Section 4.2.

Predictability of consumer activities is a powerful capability reaped through the significant advancement of machine learning algorithms and sophisticated statistical modeling tools [7–11]. It captures the correlated and repeated user behavioral patterns over time, as well as the statistically anticipated future demand. The ability of suppliers/service providers to harness such an opportunity has a significant potential in smoothing-out demand fluctuations over time, reducing commodity costs, and enhancing the consumer satisfaction with high quality services. Essentially, suppliers/service providers can *proactively* serve portions of the peak-hour load, ahead of time, particularly during the off-peak hour, based on the anticipated demand of each consumer. Hence, at the actual peak-demand instant, a considerable amount of load will be already served and excessive costs will be avoided. Such a capability is referred to as *proactive service*.

In the rest of this chapter, demand-side flexibility features are introduced in Section 4.2, followed by an example design of pricing policies for the day-ahead electricity markets with flexible demand in Section 4.2.1. In Section 4.3, the notion of predictable user demand and proactive data service is discussed. Section 4.3.1 provides an example optimization of smart pricing policies for demand shaping and proactive data services.

4.2 PRICING UNDER DEMAND FLEXIBILITIES

Flexibilities and responsiveness to price are intimately related aspects of consumer behavior that have been present in almost all types of markets. In its simplest form, consumers, who have the elasticity to be served earlier or later than their intended time, can observe the market prices and decide on their time of consumption to obtain economic benefits. Recently, consumers have more ground to exhibit their temporal flexibilities in various markets, thanks to the rapid advances in technology and engineering. For example, energy consumption controllers can manage the electricity usage of households by deferring tasks like using washing machines to late night hours when the electricity price is lower [12]. Furthermore, as an increasingly common measure in mobile software industry, smart applications have the ability to regulate a mobile data user's downloads considering the type of network connection (e.g., Wi-Fi and 4G) so that user's payments are reduced or data plan limits are not exceeded [13, 14]. It should also be noted that consumers, who are economically driven entities, will have more motivation to employ their existing as well as emerging flexibilities in order to obtain economic benefits.

Although the increased elasticity of the demand side brings in advantages to the market, it also raises concerns that can be critical to the physical infrastructures and the corresponding markets. The closed-loop system resulting from supply demand interactions and the random fluctuating demand behavior have already created issues such as existence of equilibrium prices and system stability under random variations. These issues are further amplified by the increasing flexibility of the demand side. As an example, consider the electricity grid that exhibits daily load fluctuations as shown in Figure 4.1. It is beneficial for the grid to have shiftable demand that can be activated during low demand hours to shave the peak load. As discussed earlier, smart pricing policies can achieve this goal by incentivizing consumers to use their devices during late night hours. However, higher penetration of demand flexibilities into the market and increased price responsiveness of consumers may cause undesirable peaks in the grid load. Moreover, market prices may not reach equilibrium or equilibria because of the high responsiveness of flexible consumers. Consequently, the closed-loop dynamic system can exhibit increased volatility in price, supply, and demand [15, 16].

Smart pricing strategies are anticipated to exploit the increased flexibility of demand as well as alleviate the issues created by these flexibilities. Hence, smart pricing is essential in influencing consumers and incentivizing them to practice their flexibilities in order to cut down supplier costs and maintain reliable market operation by lessening abrupt changes in prices and system load. For instance, time-dependent pricing can be implemented to decrease the variability of the total load in mobile data networks by establishing peak and off-peak prices and communicating them to the subscribers to incentivize them to change their time of use and consumption amounts [17]. In this section, we focus on the smart electricity grid to explore demand-side flexibilities deeper with more specific examples. In the rest of the section, we study two smart pricing methods that harness the temporal flexibilities of consumers to obtain smoother grid load.

4.2.1 The Day-Ahead Electricity Market with Flexible Consumers

The smart electrical grid is coming into prominence as one of the systems that will be extensively reaping the benefits of flexibilities in the demand side. The penetration of information and communication technologies into the operation of the electrical network transforms it into the *smart grid* [18]. This evolution turns the demand side into a more controllable entity so that the resulting *demand-side flexibilities* can be intensively exploited for more efficient grid operation [19–21].

The flexibilities of electricity consumers such as end users and distributers arise in different forms. Changing the amount of consumption, *shaping the demand profile* over a day or a week, *delaying* the activation of electric loads such as household appliances or factory machines, and *shifting* the time of use of electricity throughout the day are the type of flexibilities that can be harnessed for various objectives [12, 22–24]. For example, smart meters in households can communicate to the load aggregators or electricity retailers the usage preferences and load requirements of consumers in response to electricity prices. Then, via energy consumption controllers the electricity usage of household appliances can be administered by changing the time of the activation of the individual devices as seen in Figure 4.2. In such a scenario, the objective of the operator would be to decrease its costs while preserving consumer satisfaction, and the consumers would benefit from lower prices or rewards for their contribution [12, 25].

The methods that take advantage of demand-side flexibilities for efficient and reliable grid operation can be collected under *demand-side management* or *demand response*, which have recently drawn considerable attention in the engineering community [20, 21, 26, 27]. The common objective of these techniques is to alleviate the fluctuations of the grid load in order to decrease the capital, operational, and maintenance costs and increase robustness and reliability of the electrical grid. For example, as a demand response technique, load aggregators and companies can participate in the grid operation; they help the grid operator in smoothing out the grid load profile by shedding their load when the operator signals them to do so [28]. It should be noted that the parties involved in demand response are economically driven entities. Therefore, a viable way to implement sophisticated *smart pricing* schemes is to incentivize the participants to expose their flexibilities such as the ability to shift or delay load or to change consumption levels.

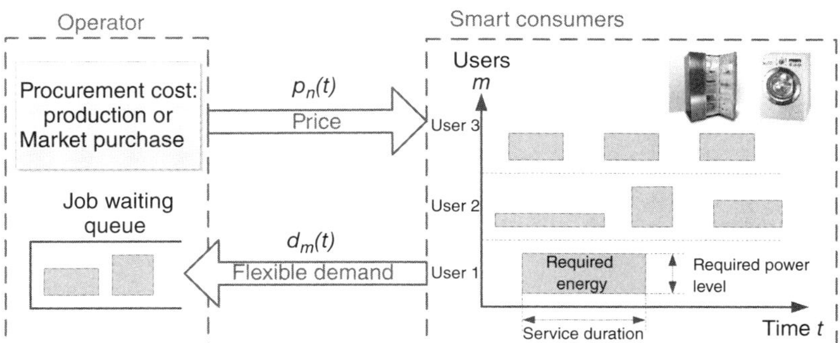

Figure 4.2 Intermediaries such as load aggregators and retailers can control their customers' demand by delaying the activation time of the individual tasks. Depending on the electricity prices, the customers can adjust their preferences such as service deadlines.

In the literature, there are numerous works that utilize *optimization* and *game theoretic* techniques to exploit the demand-side flexibilities in the electrical grid by the help of smart pricing, For example, [12, 23, 24, 29–33]. In various works, it is the key concept to build practically useful demand response models and to devise corresponding distributed smart pricing algorithms [23, 30]. In general, the objective is to achieve optimum social welfare, which is measured as the aggregate system utility minus the total costs incurred. Toward this objective, pricing algorithms are designed to incentivize the participating units to shift their demand while maximizing their own utilities. In these problems, the presence of supply uncertainty as well as the dynamics of the day-ahead and the real-time planning can also be taken into account [24, 32]. Furthermore, game theoretic approaches are also employed by formulating games between subscribers of utility companies in order to develop distributed algorithms to reach the Nash equilibrium [29, 31].

Although envisioned to significantly improve the grid efficiency, demand-side flexibilities and economic incentives of consumers can have adverse effects on the electrical grid operation [15]. For instance, the consumers with shiftable loads can act opportunistically and greedily by purchasing the maximum amount of electricity that they can at lower prices [16, 25, 34]. If a large number of such consumers schedule their loads in the hour of the day with the lowest price, the total load in the grid significantly deviates from its predicted daily pattern. In this case, the supply deficit should be settled in a balancing market that incurs additional costs to both suppliers and consumers [35]. Furthermore, if the market price quickly responds to changes in load, intelligent consumer behavior creates high supply and price volatility giving rise to a closed-loop system with stability problems [36].

Including the consumers with flexible demand together with the suppliers in the electricity market is a feasible strategy to mitigate the adverse effects of the demand-side flexibilities [35]. In the following, two smart pricing strategies are presented for the day-ahead electricity market comprising multiple number of suppliers and consumers where consumers can flexibly shift some of their total daily demand within a day.

4.2.1.1 Day-Ahead Electricity Market Day-ahead electricity market is implemented 1 day before the procurement of electric power. It involves transactions between generator companies as suppliers and intermediaries such as load aggregators, distributors, and retailers as consumers. At the termination of the market, the day-ahead prices are established.

The market is operated by a nonprofit third party called *Independent System Operator* (ISO). The ISO is responsible for administering the transactions between the market participants, setting the day-ahead prices, determining generation and consumption schedules, and ensuring the critical constraint of matching supply and load at all times. The day-ahead schedules for the load, supply, and the price are computed for the duration of a day that is divided into T time slots. The slotted time structure models the appropriate durations, possibly ranging from minutes to hours, over which the load-supply matching will be performed by establishing the day-ahead schedules.

The intermediaries are economically driven parties, so they naturally seek ways of attaining lower prices. In order to increase their profit, the intermediary parties exploit the demand-side flexibilities of their customers. Consequently, intermediaries appear as flexible consumers to the day-ahead market by reflecting the flexibilities of the end users. For instance, a load aggregator can shape its load by moving its customers' shiftable loads to the times of the day with lower market prices. Hence, the load aggregator becomes more price responsive as seen by the market operator.

We consider M generator companies (suppliers) and N intermediaries (consumers) with flexible loads in the day-ahead market that is operated by an ISO.

1. *Suppliers*. At each time slot, generator m is required to provide the amount of power that is determined by its supply schedule $\mathbf{s}^m \triangleq (s_t^m)_{t=1}^T$. By generating $\mathbf{s}^m$ amount of power over the time horizon, company m obtains $p^m(\mathbf{s}^m)$ of revenue from the market, but it also pays $C^m(\mathbf{s}^m)$ for generation costs.

Given the market prices, the goal of each generator company is to offer the supply schedule to the market that maximizes the company's profit. Accordingly, the optimization problem for generator company m is formulated as

$$\max_{\mathbf{s}^m \geq \mathbf{0}} \quad p^m(\mathbf{s}^m) - C^m(\mathbf{s}^m). \tag{4.1}$$

2. *Consumers*. Consumer n has a total D^n amount of load that has to be served completely by the end of the time horizon. The load schedule of consumer n is defined over the time slots as $\mathbf{x}^n \triangleq (x_t^n)_{t=1}^T$, and it has to satisfy $\sum_{t=1}^T x_t^n = D^n$. By consuming $\mathbf{x}^n$ amount of power, consumer n pays $p^n(\mathbf{x}^n)$ to the market and obtains $U^n(\mathbf{x}^n)$ amount of utility.

Consumer n aims to allocate its total daily load D^n at the most favorable time slots based on its utility functions and market price. This objective is formulated as the following optimization problem

$$\max_{\mathbf{x}^n \geq \mathbf{0}} \quad U^n(\mathbf{x}^n) - p^n(\mathbf{x}^n) \tag{4.2}$$

$$\text{s.t.} \quad \sum_{t=1}^{T} x_t^n = D^n. \tag{4.3}$$

Demand-Side Flexibilities. The flexibility of consumer demand is modeled by the utility function $U^n(\mathbf{x}^n)$ together with the constraint to satisfy the predetermined demand D^n over the scheduling horizon T. In particular, for the first algorithm, the utility function is separable and additive over time, that is, $U^n(\mathbf{x}^n) = \sum_{t=1}^T u_t^n(x_t^n)$, whereas for the second algorithm the utility function is arbitrary, that is, no structure such as convexity is assumed. For example, if a demand needs to be served over a limited interval, say between slots t_1 and t_2, we can set $u_t^n(x) = 0$ for $t \notin \{t_1, \ldots, t_2\}$. As another example, if a consumer has a higher preference for time slot t_0 than for time slot t, we set $u_{t_0}^n(x) \geq u_t^n(x)$. Such ordering over time between utility functions can be utilized to model the scenario where a user wants its demand to be served at

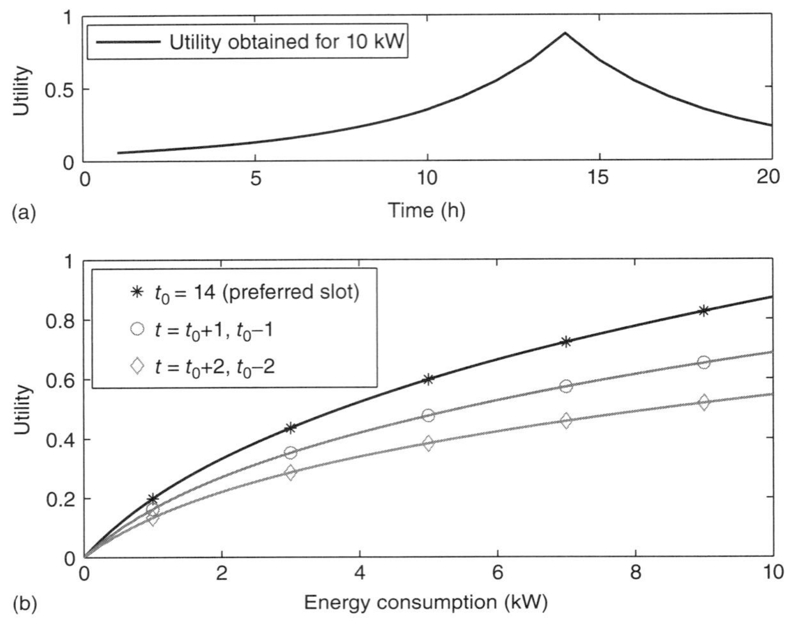

Figure 4.3 Utility obtained per 10 kW energy consumed at each time slot (a) and utility functions for different time slots (b) for consumer n with preferred time slot $t_0^n = 14$.

t_0 and a disutility is imposed for every time slot in which the serving time is delayed, as depicted in Figure 4.3.

3. *ISO*. The ISO's objective is to maximize the social welfare, which is defined as the total utility obtained by the participating entities in the market minus the total costs incurred. While maximizing social welfare, the ISO is also responsible for matching supply to load. This objective is formulated as

$$\max_{(\mathbf{x}^n)_{n=1}^N, (\mathbf{s}^m)_{m=1}^M} \quad \sum_{n=1}^N U^n(\mathbf{x}^n) - \sum_{m=1}^M C^m(\mathbf{s}^m) \tag{4.4}$$

$$\text{s.t.} \quad \sum_{n=1}^N x_t^n \leq \sum_{m=1}^M s_t^m, \quad \forall t \in \{1, \dots, T\}, \tag{4.5}$$

$$\sum_{t=1}^T x_t^n = D^n, \quad \forall n, \tag{4.6}$$

$$x_t^n \geq 0, s_t^m \geq 0, \quad \forall n, m, t, \tag{4.7}$$

where the constraint (4.5) is for matching supply to load and the constraint (4.6) is for satisfying each consumer's demand over the time horizon.

Remark 4.1 (Distributed solutions). *The knowledge of the cost and utility functions are required by the ISO to solve the welfare problem (4.4). However, this information is private to the market participants. Noting that the suppliers and the consumers are economically driven parties who aim to maximize their own welfare, it is more desirable to obtain distributed solutions.*

Next, two distributed algorithms that achieve the optimum solution of the social welfare problem are presented under different sets of assumptions. In the first algorithm, we assume that the cost and the utility functions possess various convexity and smoothness properties. In the second algorithm, on the contrary, we do not restrict ourselves to any particular class of cost and utility functions. Instead, we simply assume that the feasible set of the social welfare problem is finite and discrete.

4.2.2 Optimal Time-Dependent Pricing under Convexity Assumptions

In this setup, cost and utility functions assume convexity properties:

1. *Suppliers*. Electric power plants exhibit a two-level cost structure: generation costs and ramp up/down costs. Ramp up/down costs are associated with the changes in the power output of a generator. Furthermore, both types of costs can be time-varying. Hence, the following multi-time scale cost function can be used for the generator company m:

$$C^m(\mathbf{s}^m) = \sum_{t=1}^{T} f_t^m(s_t^m) + \sum_{t=1}^{T-1} g_t^m(|s_{t+1}^m - s_t^m|). \tag{4.8}$$

Assumption 4.1 (Convex costs). *Generation cost function $f_t^m(s)$ and ramp up/down cost function $g_t^m(s)$ are increasing and strictly convex in $s \geq 0$ for all t and m.*

Remark 4.2 (Ramp Costs). *The cost structure with convex ramp up/down costs also captures the hard limits on ramping amounts while giving a tractable mathematical formulation for analysis. For instance, quadratic functions penalize changes in generation between consecutive time slots with increasingly higher costs. Intuitively, the given cost structure is tantamount to the Lagrangian relaxation of equality constraints in duality theory.*

2. *Consumers:* Similarly to the suppliers, time-dependent utility functions can be used to capture the consumer satisfaction. Specifically, the total utility that consumer n gets from its load schedule $\mathbf{x}^n$ can be expressed by

$$U^n(\mathbf{x}^n) = \sum_{t=1}^{T} u_t^n(x_t^n). \tag{4.9}$$

We make the following assumption on utility functions:

Assumption 4.2 (Concave utilities). *Consumer utility function $u_t^n(x)$ is increasing and strictly concave in $x \geq 0$ for all t and n.*

The objective function in Eq. (4.4) is concave because of Assumptions 1 and 2, and the constraints (4.5)–(4.7) define a convex set. Hence, the optimization problem is a convex program and it has a unique global optimum. Furthermore, the objective is separable over the consumer set. These observations motivate the use of well-known dual optimization methods, and one such algorithm, which results in time-dependent market prices and achieves the optimal solution of problem (4.4), is given in References 37 and 38.

4.2.3 Optimal Bundle Pricing under Discreteness Assumptions

Although the cost and utility structures of Section 4.2.2 cover a wide range of supplier and consumer characteristics, convexity assumptions can be quite restrictive as well. For instance, power plants usually consist of multiple generating units; hence, activation and deactivation costs introduce discontinuities to the total cost [39]. On the other hand, obtaining zero utility when a task is not processed and obtaining a constant level of utility when a task is processed are a common consumer characteristics that result in discontinuous and nonconcave utility functions. Therefore, in this setup, the utility and cost functions are not assumed to have any particular structure. Instead, the following assumption facilitates the development.

Assumption 4.3 (Discrete and finite feasible set). *The feasible set for the social welfare problem is finite and discrete.*

This assumption is practically meaningful because the amount of supplied and demanded power can be well approximated by quantization with sufficiently small bins. For instance, the production bids from the generating units can be easily enforced to be multiples of kilowatt or megawatt.

The idea leading to the distributed bundle pricing algorithm is based on linearizing the problem (4.4) and then solving the linear program with distributed primal-dual methods. Note that the linearized problem would be still difficult to solve because of the arbitrary structures of the cost and utility functions. However, optimality of the primal-dual solutions for a linear problem can be established via complementary slackness (CS) conditions, which introduce further simplifications. In particular, the market participants' common objective is maximizing their own welfare; hence, the local solutions possess certain properties on the maximal values of the primal and dual variables. The reader is referred to References 37 and 38 for the detailed derivations of the algorithm presented here and to References 40 and 41 for a thorough study of primal-dual methods for linear programs.

The distributed bundle pricing algorithm, which achieves the optimal value of the problem (4.4), is depicted in Figure 4.4. In the algorithm, π_x^n and π_s^m are the surplus terms for consumer n and supplier m, respectively; π is the surplus term for the overall market transactions; and $p^n(\mathbf{x}^n)$ and $p^m(\mathbf{s}^m)$ are the bundle prices for the production and consumption schedules $\mathbf{x}^n$ and $\mathbf{s}^m$, respectively. Bundle prices are for

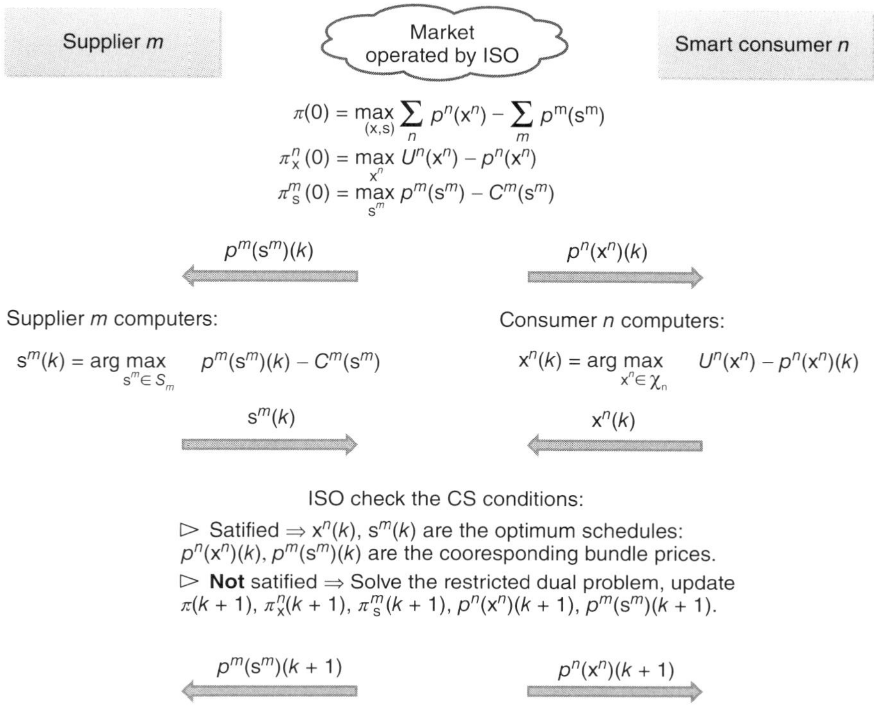

Figure 4.4 Distributed bundle pricing algorithm obtained by the primal-dual method under discreteness assumptions.

the schedules over the whole optimization horizon T, and they are specific to market participants. The initialization step of the algorithm and the fact that the market participants solve their own local problems simplifies the process of verifying CS conditions for the ISO [38].

Remark 4.3 (Complexity of solving the problem). *The market participants are required to solve their own welfare problems which can be hard to solve due to the arbitrary structure of the utility and cost functions. Nevertheless, the highly complex problem of the ISO in Eq. (4.4) is decomposed into smaller problems that can be more manageable for the individuals.*

4.2.4 Numerical Examples and Insights

For the numerical experiments, consumer utility functions of the form $u_t^n(x) = (x+1)^{1/\beta_t^n} - 1$ are used, where $u_t^n(x)$ is strictly concave in x for $\beta_t^n > 1$. For the generator company, functions $f_t^m(s) = \left(1/\alpha_t^m\right) s^{\alpha_t^m}$ for generation costs and $g_t^m(\delta_s) = \rho_t^m |\delta_s|$ for ramp costs are used, where $f_t^m(s)$ is strictly convex in s for $\alpha_t^m > 1$.

In the numerical setup, which represents a typical urban area and its surroundings, there are $N = 20$ consumers (load aggregators) with flexible loads, $M = 10$ generator companies, and the schedules are determined for the next 20 time slots. Parameters β_t^n, α_t^m, and ρ_t^m are selected randomly for each participant such that the convexity assumptions are satisfied. For simplicity, the generator cost parameters are set identical for all time slots, that is, $\alpha_t^m = \alpha^m$ and $\rho_t^m = \rho^m$ for all t. On the other hand, to model the flexibilities a preferred time slot t_0^n is randomly chosen for each consumer n and $\beta_{t_0^n}^n$ value is assigned randomly. Then, for other time slots, $t \neq t_0^n$ β_t^n is increased as the difference $|t - t_0^n|$ gets larger, that is, $\beta_t^n = \beta_{t_0^n}^n \nu^{|t-t_0^n|}$ with $\nu > 1$. With this construction, $u_{t_0}^n(x) > u_{t\pm1}^n(x) > u_{t\pm2}^n(x) > \ldots$ for all $x > 0$ is obtained. The described characteristics of the utility functions used in the numerical examples are demonstrated in Figure 4.3. It is worth noting that the specific flexibilities of different loads or devices that belong to a single consumer are not considered in this utility model.

Furthermore, in order to model a realistic market environment and to investigate the effects of the flexible load on electricity market, inflexible load, which does not shift between time slots, is introduced. Inflexible load has a sinusoidal pattern that effectively models fluctuations in electric load during the course of a day. In the numerical example, the total flexible load is set to approximately 5% of the inflexible load.

Figure 4.5a displays the load allocation achieved by the time-dependent distributed pricing algorithm. As expected, the flexible consumers are incentivized to shift their demand to the time slots where inflexible load has the lowest concentration and, hence, the lowest prices occur. Moreover, prices at these time slots show a fairly smooth behavior. This is similar to the desired output of *demand response* mechanisms (e.g., [30]) that stack more load or shed load when a surplus or deficit in supply, respectively, is predicted.

For comparison, a baseline day-ahead pricing scheme is also examined in which flexible consumers do not participate; price schedule is determined based on the inflexible load. In this scheme, each flexible consumer decides on its load schedule that maximizes its utility based on the already-settled day-ahead prices. Resulting price schedule and load allocation are presented in Figure 4.5b, where it is observed that the flexible load concentrates at the time slots where the lowest prices are witnessed. Nevertheless, this is an expected outcome because all the flexible consumers can opt to serve their demand at time slots with low price as long as their flexibilities are loose enough. Furthermore, although the consumers with flexible demand maximize their welfare, the generator companies suffer high production costs that are not compensated by the market price.

4.3 PRICING UNDER PREDICTABLE DEMAND

Recent studies on human behavioral patterns have asserted that the activities of humans are highly predictable [7–11, 42]. These findings are strengthened by the success of existing machine learning and statistical modeling tools that harness such a predictability in characterizing customer interests and preferences. Yet, several prediction capabilities are considerably underutilized in the direction of resource allocation and demand management.

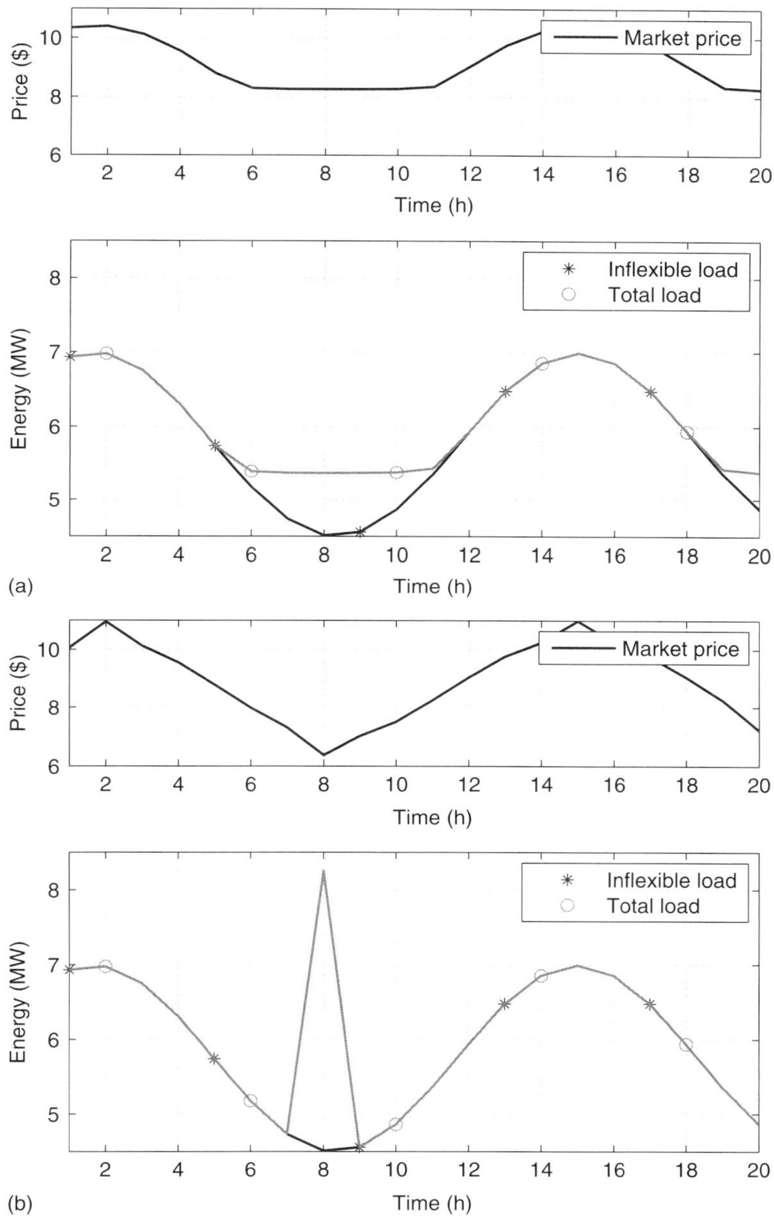

Figure 4.5 Smoother load allocation is achieved by the distributed time-dependent pricing algorithm. (a) Price and load schedules obtained by the distributed time-dependent pricing algorithm. (b) Price and load schedules obtained by the baseline scheme.

In data networks, for example, service providers incur a proportional cost because of the resources harnessed in data delivery. These costs typically vary on a timely basis depending on the consumer activities. For instance, Figure 4.1 a shows that the demand on data services consistently drops to minimum levels during late night and early morning hours. Peak demand, however, occurs during the noon and early night hours. Coupling this observation with the fact that energy costs grow superlinearly with the total load (see, e.g., [43–48]), service providers are urged to substantially shape the users demand over time in order to render it evenly distributed and minimize excessive peak costs. This can be realized through *proactive data services*, a resource allocation paradigm that utilizes the predictability of the user demand in load balancing as follows.

Essentially, data items with high popularity can be partially delivered and stored on the devices of respective consumers during the off-peak hour, that is, when the demand levels are reasonably low. These portions can then be pulled directly from the devices at the actual instant of demand, which is projected to occur during the peak hour. Thus, only the unserved parts will need to be supplied as actual demand during the peak hour, and service providers effectively cut down extreme peak-hour costs, while attaining a higher level of consumer satisfaction [49–55]. Besides cost minimization, consumers will experience a better quality of service (QoS) performance in terms of service delays, and buffering. Figure 4.6 provides an illustration of the load patterns under proactive data service.

The quality of proactive service, however, relies heavily on the accuracy of predictions made. While the randomness associated with the demand patterns may result in an undesired confusion at the service provider, pricing incentives can be offered so as to enhance the certainty about future demands. While the direct approach for demand shaping offers discounts on the original price at the off-peak time to regulate the demand fluctuations (cf. [1, 17]), the pricing approach considered in this section aims to render the users demand more deterministic, rather than changing their activity times.

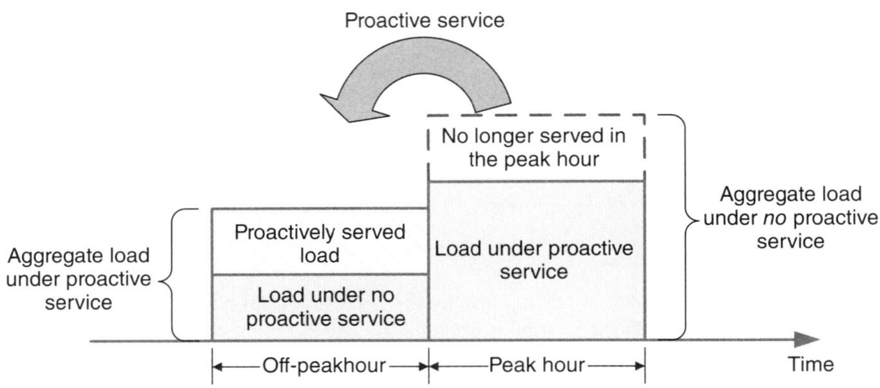

Figure 4.6 Through proactive service capabilities, service providers can even the aggregate load over time by partially serving the peak-hour demand during the preceding off-peak hour.

Such indirect approach yields price allocation strategies over data items and varies from a user to another, that is, user and data-item-dependent pricing. With proactively served items receiving lower prices, consumer demand can be efficiently shaped to increase the likelihood of these items [50, 51], as discussed in the following sections.

4.3.1 Pricing for Demand Shaping and Proactive Download in Data Networks

This section is concerned with the joint design of proactive data service strategies and pricing policies to minimize the expected service provider costs and to enhance of the certainty about the *random* user demand.

To utilize the predicable demand capabilities, service providers need to have sufficient knowledge regarding several aspects of their respective customers, a requirement that can be fulfilled through intelligent learning and tracking of the user interactions. Such interactions include the following.

1. *Activity Patterns.* Each user has an intrinsic activity profile determined by the *time instants* at which he/she requests data. Such activities depend on a variety of elements, including age, type of work, education level, economical status, and so on. The aggregation of user activity patterns by the service providers allows for a strong acquaintance with the load levels over time (like in Fig. 4.1) and gives meaningful estimates about the cost dynamics. This can particularly be captured by the parameter $q_{n,t}$, which is the probability that user n will remain silent at time t. Thus, the user requests a data item with probability $1 - q_{n,t}$.

2. *Interests and Preferences.* These represent a key element that characterizes the demand of each user when active. For instance, between two lectures, a student may decide to watch a YouTube video, thus becoming *active* with respect to the YouTube service provider. Now, among the available collection of YouTube videos, he/she selects a specific comedy movie. Such a choice is governed by the user's perception of this video, which ultimately is tied to his/her own *interests and preferences*. Several service providers, such as Netflix, Hulu, and YouTube, already employ sophisticated collaborative filtering techniques (cf. [56–59]) in order to infer these quantities. Nevertheless, they can further be utilized in proactive data services, as discussed in the following section. In the sequel, the vector $\mathbf{v}_n := (v_n(m))_{m=1}^{M}$, with $v_n(m) \in [0, 1]$, represents the valuation (or rating) of data item m ($m = 1, \cdots, M$) as recognized by user n.

3. *Economical Responsiveness.* As users belong to a variety of economic classes, their responses to a service price differs from one to another, even though they might share the same interest in such a service. Rich users, for example, may pay for an expensive data item as long as it is well preferred. On the other hand, the rest of the users, who equally prefer this data item, may decide not to buy this item in favor of buying another cheaper one with a similar or slightly less quality. Service providers can leverage the economical responsiveness to enhance the proactive service performance.

The responsiveness of user n at time t depends on the price vector $\mathbf{y}_{n,t} := (y_{n,t}(m))_{m=1}^{M}$, where $y_{n,t}(m) \in [0, y_{\max}]$ is the price of data item m as set by the service provider to user n, and time t. The maximum allowable price for a data item is $y_{\max}$. Now, the economical responsiveness is measured by a function $\phi_{m,t} : [0, y_{\max}]^M \rightarrow [0, 1]$ that maps pricing vectors into demand profiles. An example mapping considered here is

$$\phi_{m,t}(\mathbf{y}_{n,t}) = (1 - q_{n,t})\frac{v_n(m)/y_{n,t}(m)}{\sum_{j=1}^{M} v_n(j)/y_{n,t}(j)}, \quad \forall m, n, t. \tag{4.10}$$

The ratio of $v_n(m)/y_{n,t}(m)$ quantifies the utility of choosing item m, while dividing by the sum of utilities for all data items captures the *relative* utility of choosing item m among the M data items.

Figure 4.7 displays the above three aspects on a diagram illustrating the high level operation of cost minimization and demand shaping through: proactive services and pricing. While each user has a wide set of characteristics such as culture, age, education, work, and economic state, the service provider can only observe his/her demand and payments.

Examples for such service provider networks are Netflix and YouTube, which are supposed to have a collection of M data items and N end users.

4.3.1.1 *Demand Profiles* The service provider can divide the operational time axis into numerable time slots, $t = 0, 1, \cdots$ to facilitate the cost optimization process and efficiently enable proactive data services. To each time slot t, a *demand profile* for user n ($n = 1, \cdots, N$) is constructed and denoted by $\mathbf{p}_{n,t}$, where $\mathbf{p}_{n,t} := (P_{n,t}(m))_{m=1}^{M}$ is a collection of probabilities quantifying the demand of this user over the set of M

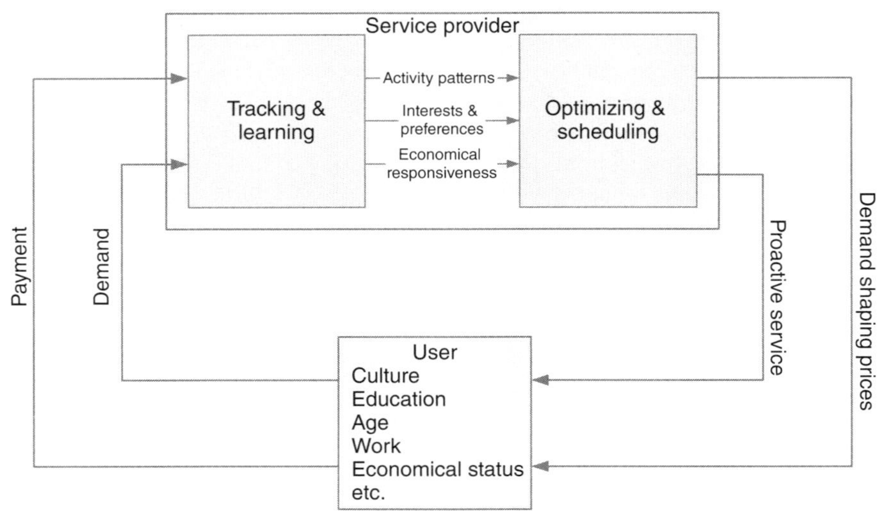

Figure 4.7 The cyclic interactions between the user and service provider taking place during cost minimization under proactive data services.

data items, at that specific time slot. Further,

$$P_{n,t}(m) := \phi_{m,t}(\mathbf{y}_{n,t}).$$

Thus, $P_{n,t}(m)$ is the probability that user n requests item m at time t when offered a pricing vector $\mathbf{y}_{n,t}$. Naturally, the mapping $\phi_{m,t}$ decreases in $y_{n,t}(m)$ and depends implicitly on $q_{n,t}$ and $\mathbf{v}_n$ as manifested in Eq. (4.10).

The random variable $\mathbf{I}_{n,t}(m)$ describes the demand of user n to data item m at time t, which is equal to 1 if such an event happens and is 0 otherwise. That given, the distribution of $\mathbf{I}_{n,t}(m)$ is

$$\mathbf{I}_{n,t}(m) = \begin{cases} 1, & \text{with probability } P_{n,t}(m), \\ 0, & \text{with probability } 1 - P_{n,t}(m). \end{cases}$$

As a number of works suggest (cf. [9, 11, 17, 51]), the user activity patterns exhibit a periodic behavior that typically repeats itself over a course of a single day, a result that has been further manifested by the measurements depicted in Figure 4.1a and extended in References 4 and 5. The service provider can divide the day into a cycle of T time slots over which $q_{n,t}$ varies from one slot to another, but with $q_{n,t} = q_{n,kT+t}$, for any positive integer k and time slot $t = 0, \cdots, T-1$. In other words, the activity profile for every user n is *cyclostationary* with a period of T slots.

Remark 4.4 (Cyclostationary demand profiles). *With the service provider targeting new pricing schemes in response to the users activity patterns, interests and preferences, and economical responsiveness, the optimized prices are periodic and the demand profile of every user is* ***cyclostationary*** *with period T.*

Following Remark 4.4, the demand profiles satisfy $\mathbf{p}_{n,t} = \mathbf{p}_{n,t+kT}$ for any user n, time t, and nonnegative integer k. Figure 4.8 illustrates the periodic nature of user demand statistics over time. With the introduced notation, the total load encountered at slot t for a *nonproactive* network is given by

$$L_t := \sum_{m=1}^{M} \sum_{n=1}^{N} S(m)\mathbf{I}_{n,t}(m), \quad t \geq 0, \tag{4.11}$$

with $S(m)$ being the size of data item m.

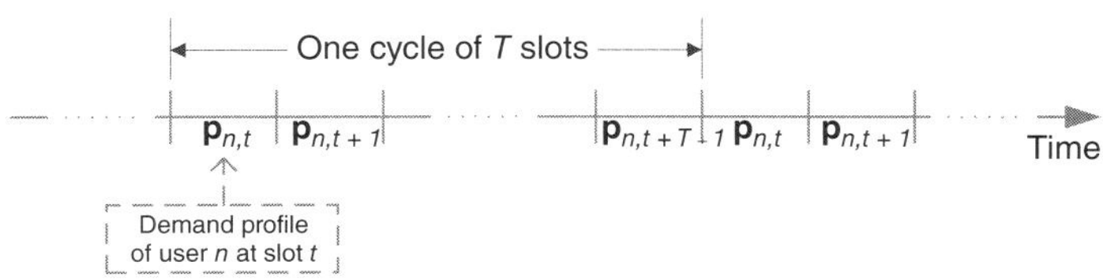

Figure 4.8 Cyclostationary demand profiles.

In the following, the constructed demand profiles are harnessed in cost minimization. Section 4.3.2 addresses the cost profile and proactive data service allocation, while Section 4.3.3 discusses the pricing strategies leveraged to attain the new profiles.

4.3.2 Cost Minimization via Proactive Data Service and Demand Shaping

The service provider seeks an optimal allocation of *proactive data services, demand profiles,* and *pricing policies* attaining the optimized profiles to minimize the time average cost of data delivery. The cost at each slot is typically a superlinear function of the aggregate load denoted by C. Proactive downloads constitute a key enabling parameter to the optimization framework and are modeled as follows. The proactive download value of data item m at time $t+1$ is denoted by $x_{n,t+1}(m) \leq S(m)$ and served to user n at time slot t. Such proactive downloads are assumed to take place only for the one-slot-ahead data; hence, take into account the potential items receiving fast updates (as some YouTube channels, and social networks experience fast update rates). The vector $\mathbf{x}_{n,t+1} := (x_{n,t+1}(m))_{m=1}^{M}$ captures all the proactive downloads made to user n at time t.

The resulting aggregate load at the service provider in time slot t can be written in terms of L_t as

$$L_t + \sum_{m,n} x_{n,t+1}(m) - x_{n,t}(m)\mathbf{I}_{n,t}(m). \tag{4.12}$$

Figure 4.9 illustrates the dynamics of proactive downloads for some user n at time t.

The service provider now can optimize over the proactive downloads $\mathbf{x}$, demand profiles $\mathbf{p}$, and pricing policies $\mathbf{y}$ to minimize its time average expected cost as

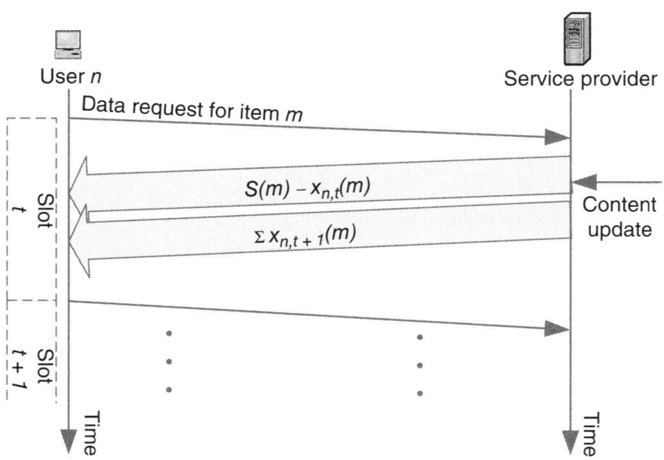

Figure 4.9 Dynamics of proactive downloads for a single user.

$$\begin{aligned}
\underset{\mathbf{x},\mathbf{p},\mathbf{y}}{\text{minimize}} \quad & \frac{1}{T}\sum_{t=0}^{T-1} \mathbf{E}\left[C\left(L_t + \sum_{m,n} x_{n,t+1}(m) - x_{n,t}(m)\mathbf{I}_{n,t}(m)\right)\right] \\
\text{subject to} \quad & 0 \le x_{n,t}(m) \le S(m), \quad \forall m, n, t = 0, \cdots, T-1, \\
& \text{User flexibility measures} && (4.13) \\
& \text{Economical responsiveness} && (4.14) \\
& \mathbf{x}_{n,0} = \mathbf{x}_{n,T}, \quad \forall n. && (4.15)
\end{aligned}$$

The following remarks about the above formulation can be made.

Remark 4.5 (User flexibility). *The constraints* (4.13) *are posed to provide a guarantee on the user satisfaction under demand shaping. Essentially, the output modified profiles are not allowed to considerably deviate from the initial profiles (those obtained under no demand shaping). These constraints protect the users from drastic pricing policies, for otherwise service providers may abruptly influence the economical responsiveness to render the demand fully deterministic.*

To elaborate further on the user flexibility, one can think of a service provider that targets all users demand to be directed to the data item with the smallest size. It can substantially raise the prices of the rest of data items. This will lead to a major change in the demand profiles of those users who do not essentially prefer such an item. Thus, users will feel unsatisfied about the quality of the cheap data items, with the rest being overpriced.

One of the potential metrics that capture the user flexibility is the *entropy* of his/her initial profile [60]. Given that user n is active at time t, then the initial probability of requesting item m is denoted by $\tilde{\pi}_{n,t}(m)$, with the normalized profile $\tilde{\boldsymbol{\pi}}_{n,t} := (\tilde{\pi}_{n,t}(m))_{m=1}^{M}$. The entropy of such a profile $H(\tilde{\boldsymbol{\pi}}_{n,t})$ quantifies the certainty about the users demand. High values of entropy reflect an indeterministic user who equally prefers any of the offered data items. On the other hand, small values of entropy reveal a deterministic user, who is more specific about the content to choose. Hence, high entropy users exhibit more flexibility to change their demand in response to pricing policies, whereas low entropy ones are typically less flexible.

Coupling this observation with the notion that deterministic users facilitate proactive data downloads, the user flexibility measure can be realized by an M-dimensional ball centered at the initial normalized profile with a radius proportional to the *entropy* of that profile [51]. This is called the *entropy-ball constraint* (EBC). Figure 4.10 depicts an illustrating shape for the EBC with $M = 3$.

Remark 4.6 (Economical responsiveness). *By considering the economical responsiveness constraints* (4.14)*, price allocation is required to attain the optimized demand profiles through the mapping $\phi_{m,t}$. As the objective function does not directly involve pricing, the optimization can be conducted first over proactive downloads and demand profiles and then price allocation step can be carried out separately to achieve the optimal profiles.*

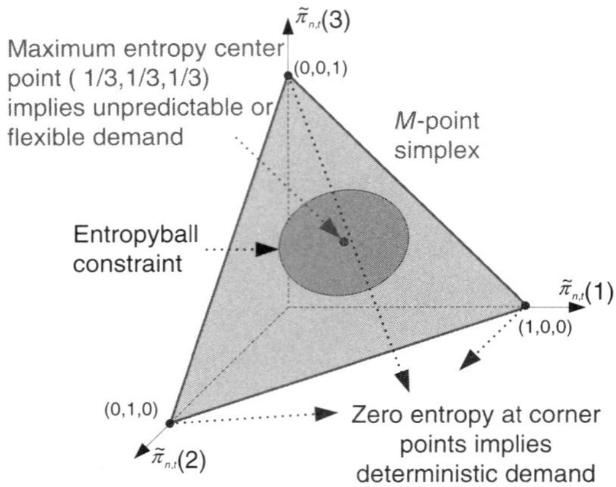

Figure 4.10 An example of $M = 3$ data items that user n can choose from in slot t. Entropy ball shrinks to a zero radius at the corner points and attains a maximum radius at the center point.

Price allocation strategies that utilize the economical responsiveness knowledge are addressed in more detail in Section 4.3.3.

Remark 4.7 (Infinite-horizon cost minimization). *The set of constraints* (4.15) *posed above ensures that the optimization over one cycle (or a day) is sufficient to minimize the infinite-horizon time average expected cost.*

Detailed proof of this result is detailed in Reference 50.

Remark 4.8 (Inflexible users). *Under no demand shaping, proactive downloads only can be employed to smooth-out the network traffic over time and leverage significant cost reduction gains.*

In Reference 50 the cost reduction via proactive downloads only is proven to scale with the number of users at least as the first derivative of the cost function does. Figure 4.11 depicts the effect of proactive downloads on the network load. The expected load is considerably balanced through optimal proactive services as compared to that of reactive allocation strategies.

Remark 4.9 (Nonconvexity of the optimization). *In general, the optimization problem is nonconvex, even under a convex cost function C [61]. The reason is the product form of* $x_{n,t}(m)\mathbf{I}_{n,t}(m)$ *with both* $x_{n,t}(m)$ *and* $\mathbf{I}_{n,t}(m)$ *being optimization variables. This gives rise to potentially suboptimal, but efficient, solutions through successive approximations of convex problem.*

In Reference 51, an iterative algorithm has been developed, under a convex cost function C. The algorithm starts with a point $(\tilde{\mathbf{x}}, \tilde{\mathbf{p}})$, where $\tilde{\mathbf{p}}$ is the initial demand

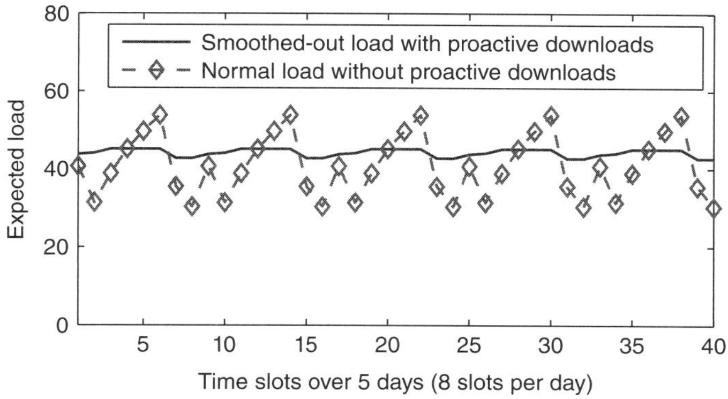

Figure 4.11 Expected total load under proactive downloads only with quadratic cost function.

profiles and $\tilde{\mathbf{x}}$ is the associated globally optimal[1] proactive download. In other words, the initial point is the result of optimal proactive downloads only without demand shaping.

The algorithm applies a series of successive solutions to approximate convex optimization problem, until it converges to a locally optimal solution of the original problem $(\hat{\mathbf{x}}, \hat{\mathbf{p}})$, with a strictly lower cost than that of $(\tilde{\mathbf{x}}, \tilde{\mathbf{p}})$. That is, with demand shaping, the system can leverage a strictly reduced cost beyond that of proactive downloads only. A numerical example on this algorithm is provided in the following section.

4.3.3 Pricing Policies Attaining Modified Profiles

Upon the calculation of the optimized profiles $\hat{\mathbf{p}}$, the service provider assigns new pricing policies, $\mathbf{y}$, so that the users adjust their demand accordingly. Figure 4.12 illustrates this step for some user n at time slot t.

Generally, the mapping functions $\phi_{m,t}$ capturing the economical responsiveness can fit the optimized profiles through multiple pricing policies. The service provider

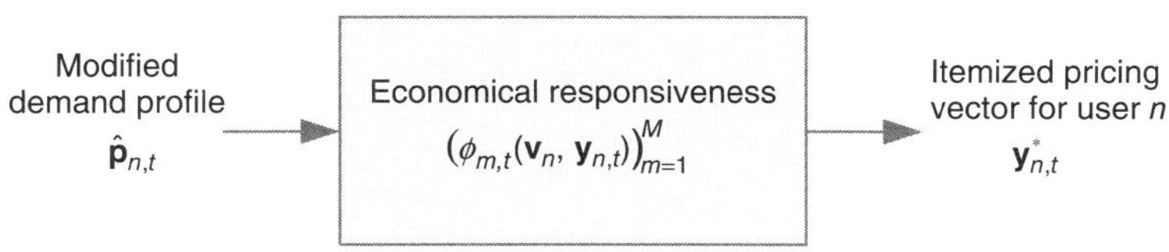

Figure 4.12 Price allocation step utilizes the economical responsiveness of the users to attain the optimized profiles.

[1]When the demand profiles are given, a convex cost function C yields a convex optimization problem in $\mathbf{x}$. Hence, globally optimal $\tilde{\mathbf{x}}$ is attainable. Full proof is given in Reference 50.

then needs to carefully select the best of such policies. One possible objective is the maximization of the proportionally fair price allocation over all data items, users, and time slots. That is,

$$\begin{aligned} \underset{\mathbf{y}}{\text{maximize}} \quad & \sum_{t=0}^{T-1} \sum_{m=1}^{M} \sum_{n=1}^{N} \log y_{n,t}(m) \\ \text{subject to} \quad & \hat{P}_{n,t}(m) = \phi_{m,t}(\mathbf{y}_{n,t}(m)), \quad \forall m, n, t, \\ & 0 \le y_{n,t}(m) \le y_{\max}, \quad \forall m, n, t. \end{aligned} \tag{4.16}$$

The optimization (4.16) can further be decoupled across time and users. With the instance realization in Eq. (4.10), the service provider may solve a geometric programming problem per user n and time slot t.

Remark 4.10 (General form of economical responsiveness). *The choice of $\phi_{m,t}$ in Eq.* (4.10) *and the objective function of Eq.* (4.16) *are not the only possible choices for the pricing problem. Instead, any general mapping and objective functions can be incorporated depending on several factors including the system size, the type of service, and so on. In the case of choices that yield a nonconvex pricing problem, efficient local optimal solutions can still be achieved through approximate solutions.*

The following example illustrates smart pricing for proactive download and demand shaping under a quadratic cost function.

Example 4.1 Quadratic Cost Function

In this example, the cost incurred at the service provider takes a quadratic form of the aggregate load, that is, $C : \mathbb{R}_+ \to \mathbb{R}_+$, with $C(L) = L^2$. The following assumptions are used throughout the example.

(i) For any time slot t, $\mathbf{I}_{n,t}(m)$ is independent of $\mathbf{I}_{n,t+1}(j)$ for all m, j. (ii) For any two users n, k such that $n \neq k$, $\mathbf{I}_{n,t}(m)$ is independent of $\mathbf{I}_{k,t}(j)$ for all m, j.

Setup. Suppose that the number of daily time slots is $T = 2$, one slot captures the peak-demand dynamics, whereas the other captures the off-peak. Also, consider a small system of $N = 2$ users and $M = 3$ data items with sizes $\mathbf{S} = (3, 2, 4)$. The activity of the users during the off-peak hour are $q_{n,0} = 1 - p_0, n = 1, 2$ with $p_0 = 0.1$. At the peak hour, the activity is $q_{n,1} = 1 - p_1$, $n = 1, 2$. The initial profiles of both users during the off-peak hour are $\tilde{\mathbf{p}}_{1,0} = p_0 \cdot (0.8, 0.1, 0.1)$, $\tilde{\mathbf{p}}_{2,0} = p_0 \cdot (0.3, 0.1, 0.6)$. Likewise, the same profile structure applies during the peak hour with p_0 replaced by p_1. The parameters p_0 and p_1 represent the user activity during the off-peak and peak hours.

Proactive Data Download without Demand Shaping. With the service provider applying proactive data download to the initial profiles but not shaping them, the allocation of such downloads is a convex problem. Under $p_1 = 0.9 > p_0$, only the peak-hour load is partially served during the off-peak hour. That is, $\tilde{\mathbf{x}}_{n,1} \succcurlyeq \mathbf{0}$ but $\tilde{\mathbf{x}}_{n,0} = \mathbf{0}$. Solving this problem for the above setup yields the initial feed $(\tilde{\mathbf{x}}, \tilde{\mathbf{p}})$ for an iterative algorithm that jointly assigns proactive downloads and demand profiles.

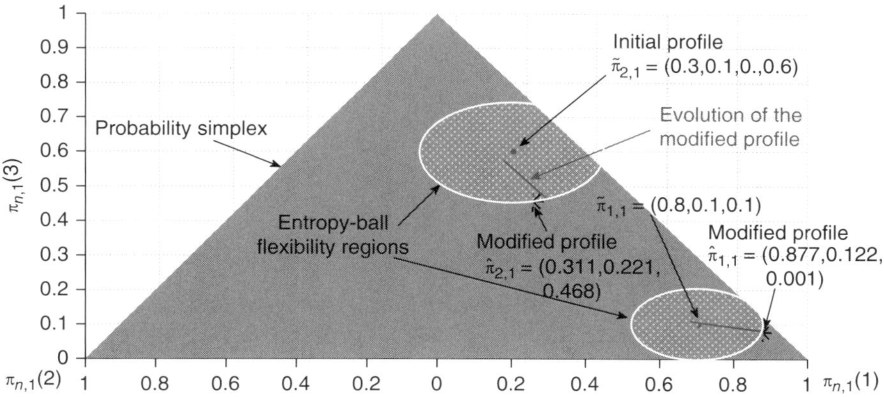

Figure 4.13 Evolution of demand profiles throughout the iterative algorithm. Optimized profiles always lie on the boundary of the flexibility regions.

Joint Proactive Download and Demand Shaping. Under the user flexibility metric featuring the entropy-ball constraints developed, the iterative algorithm utilizes successive solutions to approximate convex problems[2] to guarantee that the resulting cost is strictly less than that obtained under $(\tilde{\mathbf{x}}, \tilde{\mathbf{p}})$. Figure 4.13 depicts the evolution of the demand profiles under such an algorithm. The resulting profiles are always pushed to the boundary of the entropy-ball regions. Further, convergence takes place in a relatively small number of steps.

Smart Pricing. Upon obtaining the new peak-hour profiles $\hat{\mathbf{p}}_{1,1}$ and $\hat{\mathbf{p}}_{2,1}$, the smart pricing algorithm of Eq. (4.16) is invoked to assign new prices for the available items. Assuming economical responsiveness $\phi_{m,t}$ of Eq. (4.10), user preference vectors $\mathbf{v}_1 = (0.8, 0.1, 0.1)$ and $\mathbf{v}_2 = (0.3, 0.1, 0.6)$, and maximum price $y_{\max}$ =\$5, the results of the simulations are summarized in Table 4.1. It should be noted that the modified profiles always lie on the boundary of the entropy ball, as the service provider is interested in pushing the profiles in the direction of the most deterministic behavior that minimizes the cost. Interestingly, the second user receives the smallest price for the second data items, which originally was the least likely to be requested. The reason is that the size

TABLE 4.1 Modified Profiles and Corresponding Prices

Item	Modified Profile	Modified Profile	Prices	Prices
Index	$\hat{\mathbf{p}}_{1,1}/p_1$	$\hat{\mathbf{p}}_{2,1}/p_1$	$\mathbf{y}^*_{1,1}$, \$	$\mathbf{y}^*_{2,1}$, \$
1	0.8772	0.3111	0.0274	3.7592
2	0.1222	0.2211	0.0246	1.7632
3	0.0006	0.4678	5.0000	5.0000

[2]Detailed algorithm is provided in References 50 and 51.

of this data item is smallest, so the service provider finds it more efficient to increase the likelihood of this item for the second user than improving the certainty about the third item, with a positive proactive download, because it has the largest size and, hence, most contribution to the cost.

REFERENCES

1. C. Joe-Wong, S. Ha, and M. Chiang. "Time-dependent broadband pricing: feasibility and benefits. In *2011 31st International Conference on Distributed Computing Systems (ICDCS)*. IEEE, 2011, pp. 288–298.
2. S. Sen, C. Joe-Wong, S. Ha, and M. Chiang, "Incentivizing time-shifting of data: a survey of time-dependent pricing for internet access," *IEEE Communications Magazine*, 50(11), 2012, 91–99.
3. A. J. Wood and B. F. Wollenberg, *Power generation, operation, and control*. Wiley-Interscience, Hoboken NJ, 2012.
4. IIT Wireless Services, Illinois Institute of Technology, 2013 [online]. Available: http://www.iit.edu/ots/wireless.shtml
5. RRDtool Gallery, RRDtool, 2013 [online]. Available: http://oss.oetiker.ch/rrdtool/gallery/index.en.html
6. Iso New England, historical data. Available at: http://www.iso-ne.com/markets/hstdata/index.html
7. N. Eagle and A. Pentland. "Reality mining: sensing complex social systems," *Personal and Ubiquitous Computing*, 10, 2006, 255–268.
8. K. Farrahi and D. Gatica-Perez. Discovering human routines from cell phone data with topic models. In *The 12th IEEE International Symposium on Wearable Computers*, pp. 29–32, 2008.
9. B. S. Jensen, J. E. Larsen, K. Jensen, J. Larsen, and L. K. Hansen. "Estimating human predictability from mobile sensor data. *2010 IEEE International Workshop on Machine Learning for Signal Processing (MLSP)*, pp. 196–201, Sept. 2010.
10. R. Kwok. "Personal technology: phoning in data," *Nature*, 458, 2009, 959–961.
11. C. Song, Z. Qu, N. Blumm, and A. Barabas. "Limits of predictability in human mobility," *Science*, 327, 2010, 1018–1021.
12. C. Joe-Wong, S. Sen, S. Ha, and M. Chiang, "Optimized day-ahead pricing for smart grids with device-specific scheduling flexibility," *IEEE Journal on Selected Areas in Communications*, 30(6), 2012, 1075–1085.
13. S. Higginbotham, "Mobile operators want to charge based on time and apps," *gigaOm*, December 14 2010.
14. "The Mother of Invention: Network Operators in the Poor World Are Cutting Costs and Increasing Access in Innovative Ways," Special Report, *The Economist*, September 24 2009.
15. D. Materassi, M. Roozbehani, and M. A. Dahleh. Equilibrium price distributions in energy markets with shiftable demand. In *2012 IEEE 51st Annual Conference on Decision and Control (CDC)*. IEEE, 2012, pp. 3183–3188.

16. M. Roozbehani, A. Faghih, M. I. Ohannessian, and M. A. Dahleh, The intertemporal utility of demand and price elasticity of consumption in power grids with shiftable loads. In *2011 50th IEEE Conference on Decision and Control and European Control Conference (CDC-ECC)*. IEEE, 2011, pp. 1539–1544.
17. S. Ha, S. Sen, C. Joe-Wong, Y. Im, and M. Chiang. Tube: time-dependent pricing for mobile data. In *Proceedings of the ACM SIGCOMM 2012 Conference on Applications, Technologies, Architectures, and Protocols for Computer Communication*. ACM, 2012, pp. 247–258.
18. H. Farhangi. "The path of the smart grid," *IEEE Power and Energy Magazine*, 8(1), 2010, 18–28.
19. D. Kirschen. "Demand-side view of electricity markets," *IEEE Transactions on Power Systems*, 18(2), 2003, 520–527.
20. P. Palensky and D. Dietrich. "Demand side management: demand response, intelligent energy systems, and smart loads," *IEEE Transactions on Industrial Informatics*, 7(3), 2011, 381–388.
21. G. Strbac. "Demand side management: benefits and challenges," *Energy Policy*, 36(12), 2008, 4419–4426.
22. A. Papavasiliou and S. S. Oren. "Supplying renewable energy to deferrable loads: algorithms and economic analysis," In *2010 IEEE Power and Energy Society General Meeting*. IEEE, 2010, pp. 1–8.
23. N. Li, L. Chen, and S. Low. Optimal demand response based on utility maximization in power networks. In *2011 IEEE Power and Energy Society General Meeting*. IEEE, 2011, pp. 1–8.
24. L. Jiang and S. Low. Multi-period optimal energy procurement and demand response in smart grid with uncertain supply. In *2011 50th IEEE Conference on Decision and Control and European Control Conference (CDC-ECC)*. IEEE, 2011, pp. 4348–4353.
25. S. Chen, N. Shroff and P. Sinha. Heterogeneous Delay Tolerant Task Scheduling and Energy Management in the Smart Grid with Renewable Energy. *Selected Areas in Communications, IEEE Journal on*, 31(7), 2013, 1258–1267.
26. M. H. Albadi and E. El-Saadany. Demand response in electricity markets: an overview. In *IEEE Power Engineering Society General Meeting, 2007*. IEEE, 2007, pp. 1–5.
27. F. Rahimi and A. Ipakchi, "Demand response as a market resource under the smart grid paradigm," *IEEE Transactions on Smart Grid*, 1(1), 2010, 82–88.
28. R. Faranda, A. Pievatolo, and E. Tironi, "Load shedding: a new proposal," *IEEE Transactions on Power Systems*, 22(4), 2007, 2086–2093.
29. A. Mohsenian-Rad, V. Wong, J. Jatskevich, and R. Schober. Optimal and autonomous incentive-based energy consumption scheduling algorithm for smart grid. In *Innovative Smart Grid Technologies (ISGT), 2010*. IEEE, 2010, pp. 1–6.
30. L. Chen, N. Li, S. Low, and J. Doyle. Two market models for demand response in power networks. In *IEEE SmartGrid Comm*, 2010.
31. A. Mohsenian-Rad, V. Wong, J. Jatskevich, R. Schober, and A. Leon-Garcia, "Autonomous demand-side management based on game-theoretic energy consumption scheduling for the future smart grid," *IEEE Transactions on Smart Grid*, 1(3), 2010, 320–331.
32. P. Samadi, A. Mohsenian-Rad, R. Schober, V. Wong, and J. Jatskevich. Optimal real-time pricing algorithm based on utility maximization for smart grid. In *2010 First IEEE International Conference on Smart Grid Communications (SmartGridComm)*. IEEE, 2010, pp. 415–420.

33. M. He, S. Murugesan, and J. Zhang. Multiple timescale dispatch and scheduling for stochastic reliability in smart grids with wind generation integration. In *2011 Proceedings IEEE INFOCOM*. IEEE, 2011, pp. 461–465.

34. A. Faghih, M. Roozbehani, and M. Dahleh. Optimal utilization of storage and the induced price elasticity of demand in the presence of ramp constraints. In *2011 50th IEEE Conference on Decision and Control and European Control Conference (CDC-ECC)*. IEEE, 2011, pp. 842–847.

35. C. Su and D. Kirschen, "Quantifying the effect of demand response on electricity markets," *IEEE Transactions on Power Systems*, 24(3), 2009, 1199–1207.

36. M. Roozbehani, M. A. Dahleh, and S. K. Mitter. "Volatility of Power Grids Under Real-Time Pricing," *IEEE Transactions on Power Systems*, 27(4), 2012, 1926–1940.

37. O. Dalkilic, O. Candogan, and A. Eryilmaz. Pricing algorithms for the day-ahead electricity market with flexible consumer participation. In *The 2nd IEEE International Workshop on Smart Data Pricing (SDP)*. IEEE, 2013.

38. O. Dalkilic, O. Candogan, and A. Eryilmaz. Pricing Algorithms for the Day-Ahead Electricity Market with Flexible Consumer Participation. Available at: http://www2.ece.ohio-state.edu/ dalkilio/pricealg2012.pdf, Tech. Rep., 2012 [online]. The Department of Electrical and Computer Engineering, The Ohio State University.

39. N. P. Padhy. "Unit commitment-a bibliographical survey," *IEEE Transactions on Power Systems*, 19(2), 2004, 1196–1205.

40. C. Papadimitriou and K. Steiglitz. *Combinatorial Optimization: Algorithms and Complexity*. Dover Publications, New York, 1998.

41. D. Acemoglu, O. Candogan, A. Ozdaglar, and P. Parrilo. "Iterative auction design for graphical valuations," 2012.

42. H. Yu, D. Zheng, B. Y. Zhao, and W. Zheng. Understanding user behavior in large-scale video-on-demand systems. In *EuroSys'06*, 2006.

43. A. Nag, M. Tornatore, and B. Mukherjee. "Energy-efficient and cost-efficient capacity upgrade in mixed-line-rate optical networks," *IEEE/OSA Journal of Optical Communications and Networking*, 4(12), 2012, 1018–1025.

44. Y. Wu, P. A. Chou, and S.-Y. Kung. "Minimum-energy multicast in mobile ad hoc networks using network coding," *IEEE Transactions on Communications*, 53(11), 2005, 1906–1918.

45. K. Bicakci, H. Gultekin, and B. Tavli. "The impact of one-time energy costs on network lifetime in wireless sensor networks," *IEEE Communications Letters*, 13(12), 2009, 905–907.

46. M. K. Karray. "Analytical evaluation of QoS in the downlink of OFDMA wireless cellular networks serving streaming and elastic traffic," *IEEE Transactions on Wireless Communications*, 9(5), 2010, 1799–1807.

47. L. Z. Ribeiro and L. A. DaSilva. "A framework for the dimensioning of broadband mobile networks supporting wireless Internet services," *IEEE Wireless Communications*, 9(3), 2002, 6–13.

48. X. Zheng and Y. Cai. Reducing electricity and network cost for online service providers in geographically located internet data centers. In *IEEE/ACM International Conference on Green Computing and Communications (GreenCom), 2011*, pp. 166–169, 4–5 Aug. 2011.

49. J. Tadrous, A. Eryilmaz, and H. El Gamal. "Proactive resource allocation: harnessing the diversity and multicast gains," *IEEE Transactions on Information Theory*, 2013, DOI: 10.1109/TIT.2013.2257911.
50. J. Tadrous, A. Eryilmaz, and H. El Gamal. "Proactive data download and user demand shaping for data networks," *IEEE Transactions on Information*, 59(8), 2013, 4833–4854. submitted. Also available on: http://arxiv.org/pdf/1304.5745.pdf.
51. J. Tadrous, A. Eryilmaz, and H. El Gamal. Pricing for demand shaping in proactive data networks. In *The 2nd IEEE International Workshop on Smart Data Pricing (SDP)*, April 2013.
52. J. Tadrous, A. Eryilmaz, and H. El Gamal. Proactive content distribution for dynamically changing content. In *Proceedings of 2013 IEEE International Symposium on Information Theory (ISIT)*, July 2013.
53. J. Tadrous, A. Eryilmaz, and H. El Gamal. Proactive resource allocation in cognitive radio networks. *2011 Conference Record of the Forty Fifth Asilomar Conference on Signals, Systems and Computers (ASILOMAR)*, pp. 1425–1429, 6–9 Nov. 2011.
54. J. Tadrous, A. Eryilmaz, and H. El Gamal. Proactive multicasting with predictable demands. *2011 IEEE International Symposium on Information Theory Proceedings (ISIT)*, pp. 239–243, July 31 2011-Aug. 5 2011.
55. H. El Gamal, J. Tadrous, and A. Eryilmaz. Proactive resource allocation: Turning predictable behavior into spectral gain. In *2010 48th Annual Allerton Conference on Communication, Control, and Computing (Allerton)*, pp. 427–434, Sept. 29 2010-Oct. 1 2010.
56. Z. Huang, D. Zeng, and H. Chen. "A comparison of collaborative-filtering recommendation algorithms for e-commerce," *IEEE Intelligent Systems*, 22(5), 2007, 68–78.
57. T. L. Wickramarathne, K. Premaratne, M. Kubat, and D. T. Jayaweera. "CoFiDS: a belief-theoretic approach for automated collaborative filtering," *IEEE Transactions on Knowledge and Data Engineering*, 23(2), 2011, 175–189.
58. J. Salter and N. Antonopoulos. "CinemaScreen recommender agent: combining collaborative and content-based filtering," *IEEE Intelligent Systems*, 21(1), 2006, 35–41.
59. L. Kozma, A. Ilin, and T. Raiko. Binary principal component analysis in the Netflix collaborative filtering task. In *IEEE International Workshop on Machine Learning for Signal Processing, 2009. MLSP 2009*, pp. 1–6, 1–4 Sept. 2009.
60. T. Cover and J. Thomas. *Elements of Information Theory*. Wiley – Interscience, Hoboken NJ, 2006.
61. S. Boyd and L. Vandenberge. *Convex Optimization*. Cambridge University Press, United Kingdom, Cambrdige, 2004.

5 Dual Pricing Algorithms by Wireless Network Duality for Utility Maximization

CHEE WEI TAN and LIANG ZHENG

5.1 INTRODUCTION

As wireless networks are becoming more heterogeneous and ubiquitous in our life, it is also becoming more difficult to allocate wireless resources and price them. A wireless service provider, through the dynamic measurement of network conditions, for example, interference, channel variation, number of users, and users' mobility, can put in place a data pricing policy that determines a price for wireless resource allocation according to the type of demand for these resources. Recently, smart wireless data pricing has become an important topic of research [1–11]. An effective pricing algorithm can lead to more efficient use of the limited resources in the system and increase the overall utility or revenue [12–14]. A pareto optimal utility can be achieved using suitable pricing that reflects the amount of resources being consumed [11]. The goal of pricing is thus to encourage users to adopt a social behavior that maximizes social welfare rather than a purely noncooperative, that is, myopic and greedy, approach to resource allocation, which can lead to unfairness or unstable network conditions. How should wireless resources in large-scale heterogeneous wireless networks be analyzed and designed with clearly defined fairness and optimality in mind?

Wireless network optimization optimally matches the demand and supply of wireless resources subject to network constraints [15, 16]. Roughly speaking, it requires solving a constrained optimization problem to maximize a system objective subject to the demand and supply constraints of data services in addition to the diverse QoS constraints, interference, congestion, fairness, heterogeneity in wireless technologies, market power factors, and so on. It is common to specify the objective as the total social welfare of all the users, and a widely used social welfare is the sum of utilities.

Smart Data Pricing, First Edition. Edited by Soumya Sen, Carlee Joe-Wong, Sangtae Ha, and Mung Chiang.

Another possible objective is the egalitarian fairness of the system. If only a portion of the users experiences better network conditions than other users, the pricing algorithm can drive the prices very high for users in the system at a time of congestion and high interference, exposing some users to unfairness. It is thus important that an optimal data pricing algorithm be designed through a judicious optimization of the wireless network utility.

Wireless network utility maximization is intimately related to the signal-to-interference-and-noise-ratio (SINR) assignment for all users. A *shadow price* associated with SINR assignment constraints determines the performance received by the user. Given a set of power budget and interference temperature constraints, the SINR assignment of all users must be jointly coordinated, but there are two major hurdles that need to be overcome. First, algorithms that adapt the transmit power and interference temperature based on allocated SINR targets assume that the SINR targets are within the feasibility region, which, however, requires a centralized admission controller. Second, the algorithms have to be decentralized, practical to deploy, and be fast enough with minimal or, preferably, no parameter tuning. This is especially important because secondary users can arrive and depart in a dynamic setting, and so resources have to be adapted fast enough to converge to a new optimal operating point whenever the network conditions change. This, however, is challenging because of the tight coupling between primary and secondary users in the SINR assignment. It is desirable that the resource allocation for primary and secondary users be distributed with minimal overhead.

There are several works that partially address these two challenges in the literature. Hande et al. [17] study the SINR assignment in the context of wireless cellular networks by a reparameterization of the feasible SINR region and propose a load and spillage algorithm that jointly updates the SINR assignment and power. This algorithm, however, confines the optimality by considering a reduced feasible SINR region. Huang et al. [18] propose algorithms to coordinate the secondary users by sensing a feedback signal from the primary users' interference temperature condition to reduce outage but do not address the joint optimal SINR assignment and power allocation. Lotfinezhad et al. [19] propose a radio admission control and scheduling policy, which, however, does not handle the interference temperature constraint. To coordinate the interference temperature by controlling the SINR assignment, Ragan and Madan [20] propose a belief propagation framework by wireless scheduling to solve a nonconvex utility maximization problem. Utility maximization algorithms using fractional frequency reuse have been proposed in References 21 and 22 for wireless cellular networks. Krishnan and Luss [23] consider a joint SINR assignment and power control to maximize the worst-case SINR with an interference temperature constraint. In Reference 24, this optimization of the egalitarian fairness of SINR is solved for individual power constraints using a nonlinear Perron–Frobenius theory.

The main contributions of this chapter are summarized as follows.

- We solve the wireless network utility maximization problem by parameterizing the global optimality in terms of SINR assignment, power and interference temperature allocation as optimization problems having spectral radius constraints.

These spectral radius constraints capture fundamental characteristics of the underlying competition for resources. A special case that maximizes the egalitarian fairness of all the SINRs as the utility is solved optimally using a tuning-free geometrically fast convergent algorithm.

- We develop a wireless network duality and characterize analytically the power and interference temperature of the primal and dual networks. The relationship between them and the gradients of the spectral radius constraints in the utility maximization problem is established. The wireless network duality is thus useful for the computation of *shadow price* for SINR assignment.
- The utility maximization problem is solved using an optimization technique that can be interpreted as iteratively minimizing the *interference load* in the network. In particular, the egalitarian fairness SINR problem is adapted iteratively to solve the general utility maximization problem, which can further be solved distributively by leveraging the wireless network duality.
- Our algorithms can be practically implemented in today's wireless networks (3GPP systems) as they reuse a power control submodule already widely implemented. Numerical evaluations show that our algorithms have good performance, often yielding the optimal solution in tens of iterations even for a large number of users.

The rest of this chapter is organized as follows. In Section 5.2, we present the system model and reformulate our utility maximization problem in the SINR domain with spectral radius constraints. Then, using the nonlinear Perron–Frobenius theory, we first solve a special case of the utility maximization problem, the weighted max–min SINR, in Section 5.3.1. Next, using nonnegative matrix theory, we present the wireless network duality that we use to design a distributed algorithm to solve the utility maximization problem. We evaluate the performance of our algorithms numerically in Section 5.4. Finally, we conclude the chapter in Section 5.5. All the proofs can be found in our paper [25].

The following notation is used. We let $\mathbf{e}_l$ denote the lth unit coordinate vector and $\mathbf{I}$ denote the identity matrix. The superscripts $(\cdot)^\top$ denotes transpose. We denote $\mathbf{x}\circ\mathbf{y}$ as a Schur product of $\mathbf{x}$ and $\mathbf{y}$, that is, $\mathbf{x}\circ\mathbf{y} = [x_1y_1, \dots, x_Ly_L]^\top$, and denote $\mathbf{x}/\mathbf{y}$ as the component-wise division between $\mathbf{x}$ and $\mathbf{y}$, that is, $\mathbf{x}/\mathbf{y} = [x_1/y_1, \dots, x_L/y_L]^\top$. Let $\mathbf{1} = [1, \dots, 1]^\top \in \mathbb{R}^L$. For a vector $\mathbf{x} = [x_1, \dots, x_L]^\top$, $\mathrm{diag}(\mathbf{x})$ is its diagonal matrix $\mathrm{diag}(x_1, \dots, x_L)$. Let $e^{\mathbf{x}}$ denote $e^{\mathbf{x}} = (e^{x_1}, \dots, e^{x_L})^\top$ and $\log \mathbf{x}$ denote $\log \mathbf{x} = (\log x_1, \dots, \log x_L)$. The Perron–Frobenius eigenvalue of an irreducible nonnegative matrix $\mathbf{F}$ is denoted as $\rho(\mathbf{F})$, and the Perron right and left eigenvector of $\mathbf{F}$ associated with $\rho(\mathbf{F})$ are denoted by $\mathbf{x}(\mathbf{F})$ and $\mathbf{y}(\mathbf{F})$, respectively. An equality involving eigenvectors is true up to a scaling constant.

5.2 UTILITY MAXIMIZATION

The network utility is a network-wide quality-of-service (QoS) measurement that quantifies the efficiency of the wireless resource allocation. Before talking about the pricing strategies that serve to maximize the network utility, in this section, we first introduce the concept of utility function and its dependence on all the users' resource allocation.

We consider wireless networks with L users transmitting simultaneously on a shared spectrum. Owing to the shared nature of the wireless medium, the interference is unavoidable when users are transmitting simultaneously, that is, each user's transmission leads to interference for other users. The signal-to-interference-and-noise ratio (SINR) is an important performance metric used to measure the quality of wireless transmission. Let $\mathbf{G} = [G_{lk}]_{l,k=1}^{L} > 0_{L\times L}$ represent the channel gain, where G_{lk} is the channel gain from the kth transmitter to the lth receiver, and $\mathbf{n} = [n_1, \ldots, n_L]^\top > \mathbf{0}$, where n_l is the noise power at the lth user. The vector $\mathbf{p} = [p_1, \ldots, p_L]^\top$ is the transmit power vector. Now, the SINR of the lth user in the primary network can be given in terms of $\mathbf{p}$:

$$\mathsf{SINR}_l^P(\mathbf{p}) = \frac{p_l G_{ll}}{\sum\limits_{k\neq l} p_k G_{lk} + n_l}. \tag{5.1}$$

From Eq. (5.1), we can observe the intimate connection between the SINR and users' transmit power: a user would like to achieve a high SINR by increasing its transmit power, while this increase, in turn, gives rise to a higher interference level for other users. We also define a nonnegative matrix $\mathbf{F}$ with entries:

$$F_{lk} = \begin{cases} 0, & \text{if } l = k \\ G_{lk}, & \text{if } l \neq k \end{cases} \tag{5.2}$$

and the vector

$$\mathbf{v} = \left(\frac{1}{G_{11}}, \frac{1}{G_{22}}, \ldots, \frac{1}{G_{LL}}\right)^\top. \tag{5.3}$$

Assuming that $\mathbf{F}$ is irreducible, that is, each link has at least an interferer, we denote vector $\boldsymbol{\gamma}$ as the SINR for all users for brevity and its equivalent form is given by

$$\gamma_l = \frac{p_l}{\big(\operatorname{diag}(\mathbf{v})(\mathbf{F}\mathbf{p} + \mathbf{n})\big)_l}, \quad l = 1, \ldots, L, \tag{5.4}$$

where we use $\mathbf{F}$ and $\mathbf{v}$ in Eqs. (5.2) and (5.3), respectively. Thus, the utility function can be expressed as

$$U(\boldsymbol{\gamma}) = \sum_{l=1}^{L} \omega_l u_l(\mathsf{SINR}(\mathbf{p})), \tag{5.5}$$

which is a sum of individual utility of each user. The positive vector $\boldsymbol{\omega}$ is used to reflect a priority of the utility of each user. For example, the utility function can be the weighted sum rate [26–28] and the α-fairness utility [29].

As the SINR is a nontrivial nonlinear function of power, the joint SINR assignment and power control is a great challenge of resource allocation policy. It is also important to consider the power constraints for the system that reflect the available resource budget in the wireless networks. A general power constraint set $\mathscr{P}$ is given by

$$\mathscr{P} = \{\mathbf{p} \mid \mathbf{w}_l^\top \mathbf{p} \leq \overline{p}_l, \quad l = 1, \ldots, L\}, \tag{5.6}$$

where $\overline{\mathbf{p}} \in \mathbb{R}^L$ and $\mathbf{w}_l \in \mathbb{R}^L,\ l = 1, \ldots, L$ are, respectively, the upper bound and positive weight vector for the weighted power constraints. In particular, when $\mathbf{w}_l = \mathbf{e}_l$, we have the individual power constraint $p_l \leq \overline{p}_l$ for all l, and when $\mathbf{w}_l = \mathbf{1},\ l = 1, \ldots, L$, we have a single total power constraint $\mathbf{1}^\top \mathbf{p} \leq \min_{l=1,\ldots,L} \overline{p}_l$. Next, let q denote the interference temperature vector, consisting of the total interference and the noise for each user, i.e., q = diag(v)(Fp + n). To mitigate the interference coming from the other users in the wireless networks, we also impose an individual interference temperature constraint to each user, and the interference temperature constraint set $\mathscr{Q}$ is given by

$$\mathscr{Q} = \{\mathbf{q} \mid \mathbf{e}_l^\top \mathbf{q} \leq \overline{q}_l, \quad l = 1, \ldots, L\}. \tag{5.7}$$

Using the system model introduced above, the utility maximization problem is then given by

$$\begin{array}{ll} \text{maximize} & U(\boldsymbol{\gamma}) \\ \text{subject to} & \mathbf{p} \in \mathscr{P}, \\ & \mathbf{q} \in \mathscr{Q}, \\ & \mathbf{q} = \operatorname{diag}(\mathbf{v})(\mathbf{F}\mathbf{p} + \mathbf{n}), \\ & \gamma_l = p_l / q_l, \quad l = 1, \ldots, L, \\ \text{variables:} & \boldsymbol{\gamma}, \mathbf{p}, \mathbf{q}. \end{array} \tag{5.8}$$

Denote the optimal $\mathbf{p}$ in Eq. (5.8) as $\mathbf{p}^\star$. To proceed further, we reformulate Eq. (5.8) as an equivalent optimization problem in the SINR domain with a set of spectral radius constraints, whose problem structure is useful for developing the wireless network duality and proposing a distributed pricing algorithm.

Then, we have the following result.

Theorem 5.1 *The optimal value in Eq.* (5.8) *is equal to the optimal value of the following problem:*

$$\begin{array}{ll} \textit{maximize} & U(\boldsymbol{\gamma}) \\ \textit{subject to} & \rho\big(\operatorname{diag}(\boldsymbol{\gamma} \circ \mathbf{v})(\mathbf{F} + (1/\overline{p}_l)\mathbf{n}\mathbf{w}_l^\top)\big) \leq 1, \quad l = 1, \ldots, L, \\ & \rho\big(\operatorname{diag}(\mathbf{v})(\mathbf{F}\operatorname{diag}(\boldsymbol{\gamma}) + (1/\overline{q}_l)\mathbf{n}\mathbf{e}_l^\top)\big) \leq 1, \quad l = 1, \ldots, L, \\ \textit{variables:} & \boldsymbol{\gamma}. \end{array} \tag{5.9}$$

Let us denote the optimal solution of Eq. (5.9) *by* $\boldsymbol{\gamma}^\star$. *In addition, we have*

$$\rho\big(\operatorname{diag}(\boldsymbol{\gamma}^\star \circ \mathbf{v})(\mathbf{F} + (1/\overline{p}_i)\mathbf{n}\mathbf{w}_i^\top)\big) = 1, \tag{5.10}$$

where

$$i = \arg\max_{l=1,\ldots,L} \rho\big(\operatorname{diag}(\boldsymbol{\gamma}^\star \circ \mathbf{v})(\mathbf{F} + (1/\overline{p}_l)\mathbf{n}\mathbf{w}_l^\top)\big), \tag{5.11}$$

or

$$\rho\big(\operatorname{diag}(\mathbf{v})(\mathbf{F}\operatorname{diag}(\boldsymbol{\gamma}^\star) + (1/\overline{q}_j)\mathbf{n}\mathbf{e}_j^\top)\big) = 1, \tag{5.12}$$

where

$$j = \arg\max_{l=1,\ldots,L} \rho\big(\operatorname{diag}(\mathbf{v})(\mathbf{F}\operatorname{diag}(\boldsymbol{\gamma}^\star) + (1/\overline{q}_l)\mathbf{n}\mathbf{e}_l^\top)\big). \tag{5.13}$$

Using a logarithmic mapping of a variable, Eq. (5.9) can be transformed to an optimization problem with a convex constraint set. For $\boldsymbol{\gamma} = (\gamma_1, \ldots, \gamma_L)^\top > \mathbf{0}$, let

$$\tilde{\gamma}_l = \log \gamma_l, \quad l = 1, \ldots, L,$$

That is, $\boldsymbol{\gamma} = e^{\tilde{\boldsymbol{\gamma}}}$. Then, Eq. (5.9) is equivalent to

$$\begin{array}{ll} \text{maximize} & U(e^{\tilde{\boldsymbol{\gamma}}}) \\ \text{subject to} & \log \rho(\operatorname{diag}(e^{\tilde{\boldsymbol{\gamma}}} \circ \mathbf{v})(\mathbf{F} + (1/\overline{p}_l)\mathbf{n}\mathbf{w}_l^\top)) \leq 0, \quad l = 1, \ldots, L, \\ & \log \rho\big(\operatorname{diag}(\mathbf{v})(\mathbf{F}\operatorname{diag}(e^{\tilde{\boldsymbol{\gamma}}}) + (1/\overline{q}_l)\mathbf{n}\mathbf{e}_l^\top)\big) \leq 0, \quad l = 1, \ldots, L, \\ \text{variables:} & \tilde{\boldsymbol{\gamma}}. \end{array} \tag{5.14}$$

Let us denote the optimal solution of Eq. (5.14) by $\tilde{\boldsymbol{\gamma}}^\star$. Note that $\tilde{\boldsymbol{\gamma}}^\star = \log \boldsymbol{\gamma}^\star$ for all l.

Remark 5.1 *For an irreducible nonnegative matrix* $\mathbf{F} \in \mathbb{R}_+^{L\times L}$, $\log \rho(\operatorname{diag}(\boldsymbol{e}^{\tilde{\boldsymbol{\gamma}}})\mathbf{F})$ *is a convex function [30], because of the log-convexity property of the Perron–Frobenius eigenvalue [31]. Therefore, the constraint set in Eq.* (5.14) *is convex.*

Now, we make the following assumption on the objective function that is useful in our proposed distributed pricing algorithm later in Section 5.3.8.

Assumption 5.1 *The utility function* $U(\boldsymbol{\gamma})$ *is concave in* $\log \gamma_l$ *for all* l.

We briefly introduce the concept of Pareto optimality and Pareto dominance on the utility feasible set and then propose a pricing algorithm that is Pareto optimal. An SINR vector $\hat{\boldsymbol{\gamma}}$ Pareto dominates the another SINR vector $\boldsymbol{\gamma}$ if $u(\hat{\gamma}_l)$ is greater than $u(\gamma_l)$ for all l, and at least one of these inequalities is strict. In this case, the network utility computed by $\hat{\boldsymbol{\gamma}}$ is sure to be greater than the network utility computed by $\boldsymbol{\gamma}$. An SINR vector $\hat{\boldsymbol{\gamma}}$ is Pareto optimal if there does not exist any $\boldsymbol{\gamma}$ that Pareto dominates it. In other words, no user can increase its utility without hurting any other users when the current resource allocation is Pareto optimal. Notably, the optimal resource allocation is Pareto optimal; however, it does not Pareto dominate all the resource allocation. Figure 5.1 illustrates the concept of Pareto optimality and Pareto dominance by a two-user example. The gray region in Figure 5.1 is the utility feasible region. Any points above the dashed line in the utility feasible region dominate the resource allocation vector $\mathbf{y}$, for example, the feasible point $\mathbf{x}$. The resource allocation vector $\mathbf{z}$ is Pareto optimal; however, although $u_1(\mathbf{z}) > u_1(\mathbf{x})$ and $u_1(\mathbf{z}) > u_1(\mathbf{y})$, it Pareto dominates neither $\mathbf{x}$ nor $\mathbf{y}$. An optimal pricing algorithm can coordinate the

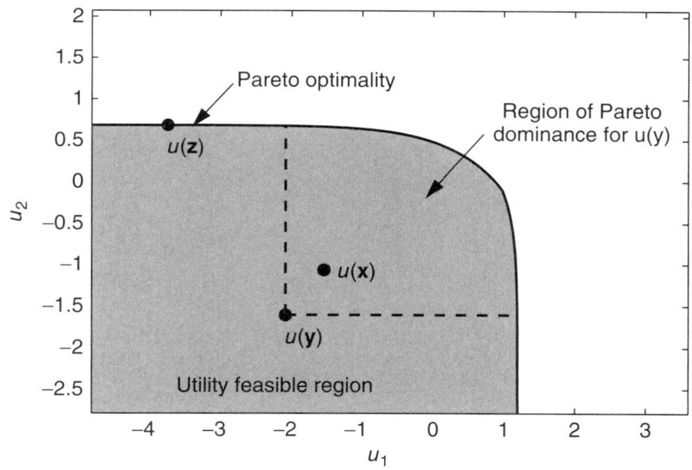

Figure 5.1 Illustration of the Pareto optimality and Pareto dominance. The utility feasible region is for a two-user example with the utility function $U(\boldsymbol{\gamma}) = 0.6 \log \gamma_1 + 0.4 \log \gamma_2$, that is, $u(\gamma_1) = 0.6 \log \gamma_1$ and $u(\gamma_2) = 0.4 \log \gamma_2$, respectively. The channel gains are given by $G_{11} = 0.76$, $G_{12} = 0.19$, $G_{21} = 0.11$, and $G_{22} = 0.78$. The weights and upper bound for the weighted power constraints are $\mathbf{w}_1 = (0.78, 0.12)^\top$, $\mathbf{w}_2 = (0.46, 0.85)^\top$, and $\overline{\mathbf{p}} = (7.5, 6.0)^\top$.

users to maximize the overall network utility based on a feedback signal to balance between the change (increase or decrease) of its own utility and the impact on other users' utility caused by this change. In the next section, we use the wireless network duality to yield such a pareto optimal pricing algorithm.

5.3 THE WIRELESS NETWORK DUALITY

The heterogeneity of a wireless network further complicates the network optimization as our wireless network technologies such as cellular networks, cognitive radio networks, and femtocell networks are changing rapidly. An important feature in wireless networks that can be exploited for utility maximization is the notion of wireless network duality. Network duality is a fundamental concept in multiuser communication [32, 33]. Network dualities for wireless cellular and ad-hoc networks were investigated in References 34 and 35 for a total power minimization problem (a convex problem) using linear programming duality. A network duality was later developed for a max–min weighted SINR problem (a nonconvex problem) in References 26, 36, and 37 using nonnegative matrix theory and geometric programming duality. These network dualities (including the one in this chapter) assert that two different networks, respectively, a primal network and a dual network, can be construed to attain an identical SINR performance. For example, the well-known uplink–downlink duality in cellular network states that the uplink network reverses the link directions in the downlink network, as illustrated in Figure 5.2. Hence, a feasible SINR for one

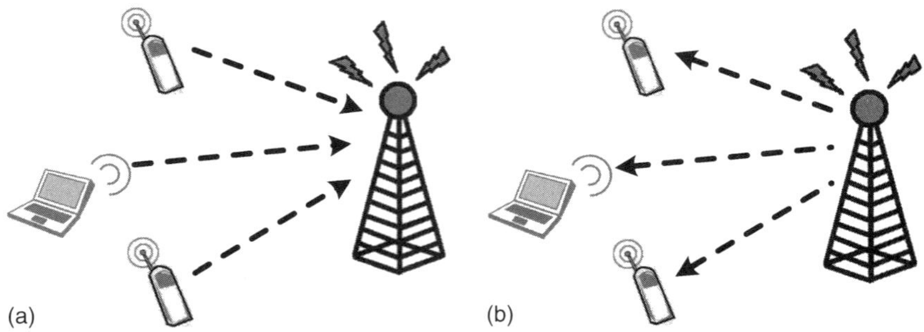

Figure 5.2 Uplink–downlink duality. (a) The uplink network. (b) The downlink network.

is also feasible for the other. There are many applications for the uplink–downlink duality, for example, beamforming and power control optimization in Reference 34 and optimizing the energy-robustness trade-off in cellular network in Reference 38.

In this section, we develop a wireless network duality involving the primary network and the dual network, as illustrated in Figure 5.3(a) and (b), respectively, which facilitates a distributed pricing algorithm design. The primary network and the dual network, comprising the wireless network duality, are constructed to attain identical SINR performance. In particular, the dual network reverses the link directions in the primary network (cf. Figure 5.3), that is, the channel gain from the kth transmitter to the lth receiver is G_{kl} in the dual network. This will be used to propose and analyze distributed algorithms for jointly optimal SINR assignment, power and interference temperature control. Our algorithms are based on a novel *wireless network duality* that decouples SINR assignment, power and interference temperature. Further, the

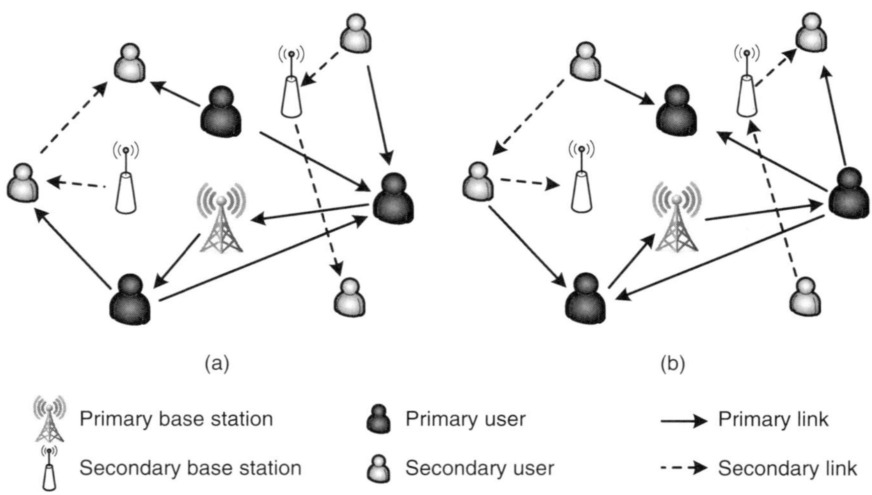

Figure 5.3 Wireless network duality. (a) The primal network. (b) The dual network.

power and interference temperature in the primal and the dual networks are jointly optimized to solve the utility maximization problem formulated for the primal network subject to the power budget and interference temperature constraints. Through this wireless network duality, the egalitarian SINR fairness problem can be interpreted as solving the general utility maximization problem in a distributed manner when the weights in the egalitarian SINR fairness problem are tuned appropriately.

5.3.1 Wireless Network Duality and Algorithms

In this section, we consider Eq. (5.14) for both smooth and nonsmooth utility functions. In particular, in Section 5.3.3, a max–min weighted SINR problem (for egalitarian SINR fairness) will be first solved using a nonlinear Perron–Frobenius theory.[1] This is then used together with the wireless network duality in Section 5.3.4 to solve Eq. (5.14). It will be shown that the transmit power and interference temperature can be analytically expressed as the Perron right eigenvectors of the specially constructed matrices associated with the spectral radius constraints in Eq. (5.14). This leads to the development of a wireless network duality involving the dual network as illustrated in Figure 5.3b, which facilitates a distributed algorithm design to maximize the primal network utility.

5.3.2 Smooth and Nonsmooth Utility

The reformulation introduced in Section 5.2 allows us to decompose the utility maximization problem in Eq. (5.8) into first optimizing $\boldsymbol{\gamma}$, that is, optimizing the SINR assignment and then optimizing the power $\mathbf{p}$ and interference temperature $\mathbf{q}$. In this section, we discuss the assumption for the objective function, which will also be useful in our proposed distributed algorithm in Section 5.3.6 to solve Eq. (5.8) for both smooth and nonsmooth utility functions.

For example, the α-fairness utility [29] satisfies Assumption 5.1, given by

$$U(\boldsymbol{\gamma}) = \begin{cases} \sum_{l=1}^{L} \log \gamma_l, & \text{if } \alpha = 1, \\ \sum_{l=1}^{L} (1-\alpha)^{-1} \gamma_l^{1-\alpha}, & \text{if } \alpha > 1. \end{cases}$$

Note that Eq. (5.14) includes the sum rate maximization problem studied in References 27 and 28, when $U(\boldsymbol{\gamma}) = \sum_{l=1}^{L} \log(1 + \gamma_l)$, but this objective function does not satisfy Assumption 5.1, and henceforth, it is a nonconvex problem that requires global optimization techniques, for example, those studied in References 27 and 28.

[1] There are several extensions to the classical (linear) Perron–Frobenius theorem in nonnegative matrix theory for classes of nonlinear maps. We use the finite dimensional nonlinear Perron–Frobenius theory developed in Reference 39.

If $U(\boldsymbol{\gamma})$ is differentiable and separable, that is, $U(\boldsymbol{\gamma}) = \sum_{l=1}^{L} U_l(\gamma_l)$, let $\nabla U_l(\gamma_l) = \partial U_l/\partial \gamma_l$ and $\nabla^2 U_l(\gamma_l) = \partial^2 U_l/\partial^2 \gamma_l$ denote the first-order and second-order derivatives of $U_l(\gamma_l)$ with respect to γ_l, respectively. Then, $U_l(\gamma_l)$ is concave in $\log \gamma_l$ if and only if the curvature is sufficiently large [17]:

$$\nabla^2 U_l(\gamma_l) \leq -\frac{\nabla U_l(\gamma_l)}{\gamma_l}. \tag{5.15}$$

In the general case (when $U(\boldsymbol{\gamma})$ can be nonsmooth), we consider the subgradient of $U(\boldsymbol{\gamma})$, whose definition is given as follows.

Definition 5.1 (cf. [40]). *The subgradient* $\mathbf{g} \in \mathbb{R}^L$ *of* $U(\boldsymbol{\gamma})$ *at* $\hat{\boldsymbol{\gamma}}$ *satisfies*

$$U(\boldsymbol{\gamma}) \leq U(\hat{\boldsymbol{\gamma}}) + \mathbf{g}^\top(\boldsymbol{\gamma} - \hat{\boldsymbol{\gamma}})$$

for any feasible $\boldsymbol{\gamma}$. *If* $U(\boldsymbol{\gamma})$ *is concave and differentiable at* $\boldsymbol{\gamma}$, *the subgradient is unique and given by its gradient* $\mathbf{g} = \nabla U(\boldsymbol{\gamma})$. *On the other hand, if* $U(\boldsymbol{\gamma})$ *is concave but not differentiable, its subgradient is in the set*

$$\bigcap \{\mathbf{g} \mid U(\boldsymbol{\gamma}) \leq U(\hat{\boldsymbol{\gamma}}) + \mathbf{g}^\top(\boldsymbol{\gamma} - \hat{\boldsymbol{\gamma}})\}$$

for any feasible $\boldsymbol{\gamma}$.

5.3.3 Nonsmooth Special Case: $U(\boldsymbol{\gamma}) = \min_{l=1,\ldots,L} \frac{\gamma_l}{\beta_l}$

In this section, let us consider the max–min weighted SINR problem (for egalitarian SINR fairness), which is a special case of Eq. (5.9) that has a nonsmooth concave objective function:

$$\begin{array}{ll} \text{maximize} & \min_{l=1,\ldots,L} \dfrac{\gamma_l}{\beta_l} \\ \text{subject to} & \rho\big(\operatorname{diag}(\boldsymbol{\gamma} \circ \mathbf{v})(\mathbf{F} + (1/\bar{p}_l)\mathbf{n}\mathbf{w}_l^\top)\big) \leq 1, \\ & \qquad\qquad l = 1, \ldots, L, \\ & \rho\big(\operatorname{diag}(\mathbf{v})(\mathbf{F}\operatorname{diag}(\boldsymbol{\gamma}) + (1/\bar{q}_l)\mathbf{n}\mathbf{e}_l^\top)\big) \leq 1, \\ & \qquad\qquad l = 1, \ldots, L, \\ \text{variables:} & \boldsymbol{\gamma}, \end{array} \tag{5.16}$$

where $\boldsymbol{\beta}$ is a positive vector with the entry β_l used to reflect a priority of the lth link. A larger β_l indicates a higher priority.

Let us define the following set of nonnegative matrices:

$$\mathbf{B}_l = \operatorname{diag}(\mathbf{v})\left(\mathbf{F} + \frac{1}{\bar{p}_l}\mathbf{n}\mathbf{w}_l^\top\right), \quad l = 1, \ldots, L, \tag{5.17}$$

$$\mathbf{D}_l = \left(\mathbf{I} + \frac{1}{\bar{q}_l - n_l v_l}\operatorname{diag}(\mathbf{v})\mathbf{n}\mathbf{e}_l^\top\right)\operatorname{diag}(\mathbf{v})\mathbf{F}, \quad l = 1, \ldots, L. \tag{5.18}$$

By applying the nonnegative matrix theory and the nonlinear Perron–Frobenius theory, we obtain a closed-form solution to Eq. (5.16) as well as the corresponding optimal solution in Eq. (5.8), which is unique.

Lemma 5.1 *The optimal value of Eq.* (5.16) *is given by*

$$\frac{1}{\max\limits_{l=1,\ldots,L}\{\rho(\mathrm{diag}(\boldsymbol{\beta})\mathbf{B}_l), \rho(\mathbf{D}_l\,\mathrm{diag}(\boldsymbol{\beta}))\}}. \tag{5.19}$$

The optimal solution to Eq. (5.16) $\boldsymbol{\gamma}^\star$ *is a vector with* $\gamma_l^\star/\beta_l$ *equal to a common value*

$$\gamma_m^\star/\beta_m = \frac{1}{\max\limits_{l=1,\ldots,L}\{\rho(\mathrm{diag}(\boldsymbol{\beta})\mathbf{B}_l), \rho(\mathbf{D}_l\,\mathrm{diag}(\boldsymbol{\beta}))\}}$$

for all l, where

$$m = \arg\max_{l=1,\ldots,L}\left\{\rho\left(\mathrm{diag}(\boldsymbol{\beta})\mathbf{B}_l\right), \rho\left(\mathbf{D}_l\,\mathrm{diag}(\boldsymbol{\beta})\right)\right\}. \tag{5.20}$$

If the optimal value is $1/\rho(\mathrm{diag}(\boldsymbol{\beta})\mathbf{B}_m)$ *for m in Eq.* (5.20), *the optimal power and interference temperature in Eq.* (5.8) *is, respectively, given by*

$$\mathbf{p}^\star = \frac{\overline{p}_m}{\mathbf{w}_m^\top\mathbf{x}\left(\mathrm{diag}(\boldsymbol{\beta})\mathbf{B}_m\right)}\mathbf{x}\left(\mathrm{diag}(\boldsymbol{\beta})\mathbf{B}_m\right) \tag{5.21}$$

and

$$\mathbf{q}^\star = \mathrm{diag}(\boldsymbol{\gamma}^\star)^{-1}\mathbf{p}^\star,$$

and if the optimal value is $1/\rho(\mathbf{D}_m\,\mathrm{diag}(\boldsymbol{\beta}))$ *for m in Eq.* (5.20), *the optimal interference temperature and power in Eq.* (5.8) *is, respectively, given by*

$$\mathbf{q}^\star = \frac{\overline{q}_m}{\mathbf{e}_m^\top\mathbf{x}\left(\mathbf{D}_m\,\mathrm{diag}(\boldsymbol{\beta})\right)}\mathbf{x}\left(\mathbf{D}_m\,\mathrm{diag}(\boldsymbol{\beta})\right) \tag{5.22}$$

and

$$\mathbf{p}^\star = \mathrm{diag}(\boldsymbol{\gamma}^\star)\mathbf{q}^\star.$$

We next give an intriguingly simple algorithm to compute the analytical solution in Lemma 5.1. In particular, by applying the nonlinear Perron–Frobenius theory in Reference 39, the following algorithm computes $\mathbf{p}^\star$ given in Lemma 5.1.

Theorem 5.2 *Starting from any initial point* $\mathbf{p}(0)$, $\mathbf{p}(k)$ *converges geometrically fast to the power* $\mathbf{p}^\star$ *given in Lemma 5.1.*

Algorithm 5.1

Max–min weighted SINR Algorithm

Initialize $\mathbf{p}(0)$.

1) Each lth user updates its power $p_l(k+1)$ as follows:

$$p_l(k+1) = \frac{\beta_l}{\mathsf{SINR}_l^P(\mathbf{p}(k))} p_l(k).$$

2) Normalize $\mathbf{p}(k+1)$:

$$\mathbf{p}(k+1) \leftarrow \frac{\mathbf{p}(k+1)}{\max\limits_{l=1,\ldots,L} \left\{ \dfrac{\mathbf{w}_l^\top \mathbf{p}(k+1)}{\bar{p}_l}, \dfrac{\mathbf{e}_l^\top \operatorname{diag}(\mathbf{v})\mathbf{F}\mathbf{p}(k+1)}{\bar{q}_l - n_l v_l} \right\}}.$$

Remark 5.2 *At Step 1, the Foschini–Miljanic power control algorithm update in Reference 41 (a power control submodule widely implemented in 3GPP systems) is used. At Step 2, the computation of* $\mathbf{w}_l^\top \mathbf{p}(k+1)$ *and the normalization of* $\mathbf{p}(k+1)$ *can be computed by a gossip algorithm in a distributed manner [42]. Notice that* $\mathbf{e}_l^\top \operatorname{diag}(\mathbf{v})(\mathbf{F}\mathbf{p}(k+1)+\mathbf{n})$ *and* n_l *are, respectively, the interference and the noise power at lth receiver, which can be measured from the interference temperature. Thus,* $\mathbf{e}_l^\top \operatorname{diag}(\mathbf{v})\mathbf{F}\mathbf{p}(k+1)$ *can be locally obtained.*

Interestingly, using the Friedland–Karlin inequalities in References 43 and 27, Eq. (5.16) is equivalent to Eq. (5.9) with a smooth objective function given by

$$U(\boldsymbol{\gamma}) = \sum_{l=1}^{L} \left(\mathbf{x}(\boldsymbol{\Omega})\circ\mathbf{y}(\boldsymbol{\Omega})\right)_l \log \gamma_l, \tag{5.23}$$

where $\boldsymbol{\Omega} \in \{\operatorname{diag}(\boldsymbol{\beta})\mathbf{B}_l, \mathbf{D}_l \operatorname{diag}(\boldsymbol{\beta})\}$ is the matrix defined for the mth user with m given in Eq. (5.20): if the optimal value for Eq. (5.16) is $1/\rho(\operatorname{diag}(\boldsymbol{\beta})\mathbf{B}_m)$ for m in Eq. (5.20), then $\boldsymbol{\Omega} = \operatorname{diag}(\boldsymbol{\beta})\mathbf{B}_m$; if the optimal value for Eq. (5.16) is $1/\rho(\mathbf{D}_m \operatorname{diag}(\boldsymbol{\beta}))$ for m in Eq. (5.20), then $\boldsymbol{\Omega} = \mathbf{D}_m \operatorname{diag}(\boldsymbol{\beta})$.

5.3.4 Wireless Network Duality

In this section, we develop a dual network having identical SINR performance as the primal network but with all the link directions reversed. The concept of the dual network is actually a virtual network using the same topology of the primal network. Thus, we use $\mathbf{F}^\top$ as the channel fading matrix in the dual network (because of the reversed link directions).

We now turn to solve Eq. (5.14) for general utility functions that satisfy Assumption 5.1 using a projected subgradient method [40]. Interestingly, this method can be made distributed by connecting the gradients of the spectral radius functions in Eq. (5.14) with the power and the interference temperature in both the primal and the dual networks (cf. Figure 5.3). This is achieved by applying the wireless network duality that exploits the structure of the spectral radius functions in Eq. (5.14).

Recall that we have already defined the nonnegative matrices $\mathbf{B}_l$ and $\mathbf{D}_l$ given in Eqs. (5.17) and (5.18), respectively. The gradients $\mathbf{g} \in \mathbb{R}^L$ of $\log \rho(\text{diag}(\boldsymbol{\gamma})\mathbf{B}_l)$ and $\log \rho(\mathbf{D}_l\, \text{diag}(\boldsymbol{\gamma}))$ at $\boldsymbol{\gamma}$ are given, respectively, by [27, 43]

$$\mathbf{g} = \mathbf{x}\big(\,\text{diag}(\boldsymbol{\gamma})\mathbf{B}_l\big)\circ\mathbf{y}\big(\,\text{diag}(\boldsymbol{\gamma})\mathbf{B}_l\big) \tag{5.24}$$

and

$$\mathbf{g} = \mathbf{x}\big(\mathbf{D}_l\,\text{diag}(\boldsymbol{\gamma})\big)\circ\mathbf{y}\big(\mathbf{D}_l\,\text{diag}(\boldsymbol{\gamma})\big), \tag{5.25}$$

normalized such that $\mathbf{1}^\top\mathbf{g} = 1$. We already know from the reformulation in Section 5.2 that the Perron right eigenvectors of the matrices in Eqs. (5.10) and (5.12) are the optimal transmit power $\mathbf{p}^\star$ and the optimal interference temperature $\mathbf{q}^\star$ of the primal network, respectively, which also appear in Eq. (5.24) for $l = i$ in Eqs. (5.11) and (5.25) for $l = j$ in Eq. (5.13), respectively. Observe that the gradient for the spectral radius function is the Schur product of the Perron right and left eigenvectors (recall that the Schur product is the component-wise product of two vectors). This interesting characterization leads naturally to the development of a dual network, where a physical interpretation can be given to the Perron left eigenvectors of the matrices in Eqs. (5.10) and (5.12). It enables a distributed method to compute $\mathbf{g}$ in Eqs.(5.24) and (5.25). As shown in Figure 5.3b, the dual network reverses the link direction of the primal network, that is, the channel gain from the kth transmitter to the lth receiver is G_{kl} in the dual network.

Definition 5.2 *Let the power* $\mathbf{s}$ *and the interference temperature* $\mathbf{t}$ *in the dual network be given, respectively, by*

$$\mathbf{s} = \text{diag}(\boldsymbol{\gamma})\mathbf{t} \tag{5.26}$$

and

$$\mathbf{t} = \text{diag}(\mathbf{v})(\mathbf{F}^\top\mathbf{s} + \mathbf{w}_\iota), \tag{5.27}$$

where the weight vector $\mathbf{w}_\iota$ *for the transmit power constraint in the primal network is assumed to be a virtual received noise in the dual network, and the index* ι *corresponds to either* i *in Eq.* (5.11) *or* j *in Eq.* (5.13), *which indicates the power or interference temperature constraint that is tight at optimality of Eq.* (5.14). *Now, the SINR of the* l*th user in the dual network can be given in terms of* $\mathbf{s}$*:*

$$\mathsf{SINR}_l^D(\mathbf{s}) = \frac{s_l}{\big(\text{diag}(\mathbf{v})(\mathbf{F}^\top\mathbf{s} + \mathbf{w}_\iota)\big)_l}. \tag{5.28}$$

Observe that the channel matrix in Eq. (5.1) is replaced by its transpose in Eq. (5.28). However, this dual network must achieve all possible SINR values that are feasible in the primal network. Combining Eqs. (5.26) and (5.27), we have

$$\mathbf{s} = \mathrm{diag}(\boldsymbol{\gamma}\circ\mathbf{v})(\mathbf{F}^{\top}\mathbf{s} + \mathbf{w}_{\iota}) \tag{5.29}$$

and

$$\mathbf{t} = \mathrm{diag}(\mathbf{v})\big(\mathbf{F}^{\top}\,\mathrm{diag}(\boldsymbol{\gamma})\mathbf{t} + \mathbf{w}_{\iota}\big). \tag{5.30}$$

Corresponding to the primal network power and interference temperature constraints, the dual network power and interference temperature constraints can be given, respectively, by (note the use of ι, where ι can be i in Eq. (5.11) or ι can be j in Eq. (5.13))

$$\mathbf{n}^{\top}\mathbf{s} \le \bar{p}_{\iota},$$

and

$$\mathbf{n}^{\top}\mathbf{t} \le \bar{q}_{\iota}.$$

Moreover, it is still true that either the dual network power constraint or the dual network interference temperature constraint is tight at optimality. We then have the following result that connects the transmit powers and interference temperatures of both the primal and dual networks, which is used in a distributed algorithm design in Section 5.3.6.

Lemma 5.2 *The Perron right and left eigenvector of the nonnegative matrix* $\mathrm{diag}(\boldsymbol{\gamma}^{\star})\mathbf{B}_i$, *where* $\mathbf{B}_i$ *is given in Eq.* (5.17) *with i in Eq.* (5.11), *satisfy*

$$\mathbf{x}\big(\,\mathrm{diag}(\boldsymbol{\gamma}^{\star})\mathbf{B}_i\big) = \mathbf{p}^{\star} \tag{5.31}$$

and

$$\mathbf{y}\big(\,\mathrm{diag}(\boldsymbol{\gamma}^{\star})\mathbf{B}_i\big) = \mathrm{diag}(\boldsymbol{\gamma}^{\star}\circ\mathbf{v})^{-1}\mathbf{s}^{\star}, \tag{5.32}$$

respectively. The Perron right and left eigenvector of the nonnegative matrix $\mathbf{D}_j\,\mathrm{diag}(\boldsymbol{\gamma}^{\star})$, *where* $\mathbf{D}_j$ *is given in Eq.* (5.18) *with j in Eq.* (5.13), *satisfy, respectively,*

$$\mathbf{x}\big(\mathbf{D}_j\,\mathrm{diag}(\boldsymbol{\gamma}^{\star})\big) = \mathbf{q}^{\star} \tag{5.33}$$

and

$$\mathbf{y}\big(\mathbf{D}_j\,\mathrm{diag}(\boldsymbol{\gamma}^{\star})\big) = \mathrm{diag}(\boldsymbol{\gamma}^{\star}/\mathbf{v})\mathbf{t}^{\star}. \tag{5.34}$$

Finally, Table 5.1 summarizes the wireless network duality that characterizes the transmit power and interference temperature in the primal and the dual networks as the Perron right and left eigenvectors of appropriately constructed nonnegative matrices.

TABLE 5.1 The Wireless Network Duality Illustrates the Connection between the Primal and the Dual Networks in Terms of Both the Perron Right and Left Eigenvectors of the Nonnegative Matrices Associated with the Spectral Radius Constraints in Eq. (5.14)

Wireless Network Duality			
	Primal Network		Dual Network
Power budget constraint	$\mathbf{p}^\star = \mathbf{x}\left(\mathrm{diag}\left(\boldsymbol{\gamma}^\star \circ \mathbf{v}\right)\left(\mathbf{F} + \left(1/\bar{p}_i\right)\mathbf{n}\mathbf{w}_i^\top\right)\right)$	$\leftrightarrow$	$\mathbf{s}^\star = \mathrm{diag}\left(\boldsymbol{\gamma}^\star \circ \mathbf{v}\right)\mathbf{y}\left(\mathrm{diag}\left(\boldsymbol{\gamma}^\star \circ \mathbf{v}\right)\left(\mathbf{F} + \left(1/\bar{p}_i\right)\mathbf{n}\mathbf{w}_i^\top\right)\right)$
	$= \left(\mathbf{I} - \mathrm{diag}\left(\boldsymbol{\gamma}^\star \circ \mathbf{v}\right)\mathbf{F}\right)^{-1}\mathrm{diag}\left(\boldsymbol{\gamma}^\star \circ \mathbf{v}\right)\mathbf{n}$		$= \left(\mathbf{I} - \mathrm{diag}\left(\boldsymbol{\gamma}^\star \circ \mathbf{v}\right)\mathbf{F}^\top\right)^{-1}\mathrm{diag}\left(\boldsymbol{\gamma}^\star \circ \mathbf{v}\right)\mathbf{w}_i$
	$\mathbf{w}_i^\top \mathbf{p}^\star \leq \bar{p}_i$	$\leftrightarrow$	$\mathbf{n}^\top \mathbf{s}^\star \leq \bar{p}_i$
Interference temperature constraint	$\mathbf{q}^\star = \mathbf{x}\left(\left(\mathbf{I} + \frac{1}{\bar{q}_j - n_j v_j}\mathrm{diag}\left(\mathbf{v}\right)\mathbf{n}\mathbf{e}_j^\top\right)\mathrm{diag}\left(\mathbf{v}\right)\mathbf{F}\,\mathrm{diag}\left(\boldsymbol{\gamma}^\star\right)\right)$	$\leftrightarrow$	$\mathbf{t}^\star = \mathrm{diag}\left(\boldsymbol{\gamma}^\star/\mathbf{v}\right)^{-1}\mathbf{y}\left(\left(\mathbf{I} + \frac{1}{\bar{q}_j - n_j v_j}\mathrm{diag}\left(\mathbf{v}\right)\mathbf{n}\mathbf{e}_j^\top\right)\mathrm{diag}\left(\mathbf{v}\right)\mathbf{F}\,\mathrm{diag}\left(\boldsymbol{\gamma}^\star\right)\right)$
	$= \left(\mathbf{I} - \mathrm{diag}\left(\mathbf{v}\right)\mathbf{F}\,\mathrm{diag}\left(\boldsymbol{\gamma}^\star\right)\right)^{-1}\mathrm{diag}\left(\mathbf{v}\right)\mathbf{n}$		$= \left(\mathbf{I} - \mathrm{diag}\left(\mathbf{v}\right)\mathbf{F}^\top\mathrm{diag}\left(\boldsymbol{\gamma}^\star\right)\right)^{-1}\mathrm{diag}\left(\mathbf{v}\right)\mathbf{w}_j$
	$\mathbf{e}_j^\top \mathbf{q}^\star \leq \bar{q}_j$	$\leftrightarrow$	$\mathbf{n}^\top \mathbf{t}^\star \leq \bar{q}_j$

5.3.5 Interference Load Minimization

In Section 5.2, the constraints of the utility maximization problem in Eq. (5.8) can be succinctly reformulated as spectral radius constraints, that is, $\rho\big(\operatorname{diag}(\boldsymbol{\gamma})\mathbf{B}_l\big) \leq 1\ \forall\, l$ and $\rho\big(\mathbf{D}_l \operatorname{diag}(\boldsymbol{\gamma})\big) \leq 1\ \forall\, l$, where $\mathbf{B}_l$ and $\mathbf{D}_l$ are given in Eqs. (5.17) and (5.18), respectively. These spectral radius functions capture the effect of interference on the feasibility of SINR assignment. The spectral radius thus plays the role of a useful measure for interference, which we call the *interference load*. This means that a smaller $\rho\big(\operatorname{diag}(\boldsymbol{\gamma})\mathbf{B}_l\big)$ or $\rho\big(\mathbf{D}_l \operatorname{diag}(\boldsymbol{\gamma})\big)$ indicates a smaller interference load on the network, which leads to a larger feasible SINR region that can be optimized. On the other hand, the interference load increases with interference and, therefore, reduces the feasible SINR region. This connection between the interference load and our utility maximization problem in Section 5.2 is made precise in the following. By leveraging both the wireless network duality in Section 5.3.4 and the *interference load minimization* problem (to be introduced in the following), a distributed algorithm will then be proposed to solve the utility maximization problem in Eq. (5.9).

First, let us consider the following convex optimization problem given by

$$\begin{array}{ll} \text{maximize} & \boldsymbol{\alpha}^\top \tilde{\boldsymbol{\gamma}} \\ \text{subject to} & \rho\big(\operatorname{diag}(e^{\tilde{\boldsymbol{\gamma}}})\mathbf{B}_l\big) \leq 1, \quad l = 1, \ldots, L, \\ & \rho\big(\mathbf{D}_l \operatorname{diag}(e^{\tilde{\boldsymbol{\gamma}}})\big) \leq 1, \quad l = 1, \ldots, L, \\ \text{variables:} & \tilde{\boldsymbol{\gamma}}, \end{array} \tag{5.35}$$

where $\mathbf{B}_l$ and $\mathbf{D}_l$ are given in Eqs. (5.17) and (5.18), respectively, $\tilde{\boldsymbol{\gamma}}$ is the logarithmic mapping by $\tilde{\boldsymbol{\gamma}} = \log \boldsymbol{\gamma}$, and $\boldsymbol{\alpha} \in \mathbb{R}_+^L$ is a given probability vector that is used to approximate $U(e^{\tilde{\boldsymbol{\gamma}}})$ using its Taylor series expansion up to the first-order terms (cf. proof of Theorem 5.3).

In the following, it is fruitful to consider the *interference load minimization* problem that is intimately related to Eq. (5.35) and instead minimizes a spectral radius function subject to a single linear constraint:

$$\begin{array}{ll} \text{minimize} & \max\limits_{l=1,\ldots,L} \{\rho(\operatorname{diag}(e^{\tilde{\boldsymbol{\eta}}})\mathbf{B}_l), \rho(\mathbf{D}_l \operatorname{diag}(e^{\tilde{\boldsymbol{\eta}}}))\} \\ \text{subject to} & \boldsymbol{\alpha}^\top \tilde{\boldsymbol{\eta}} \geq 0, \\ \text{variables:} & \tilde{\boldsymbol{\eta}}, \end{array} \tag{5.36}$$

where $\tilde{\boldsymbol{\eta}}$ is the logarithmic transformation: $\tilde{\boldsymbol{\eta}} = \log \boldsymbol{\eta}$. The following result connects Eqs. (5.35) and (5.36).

Lemma 5.3 *Let $\boldsymbol{\gamma}^\star$ and $\boldsymbol{\eta}^\star$ be the optimal solution of Eqs.* (5.35) *and* (5.36), *respectively, and let $\xi^\star$ and $\zeta^\star$ be the optimal value of Eqs.* (5.35) *and* (5.36), *respectively. Then, $\boldsymbol{\gamma}^\star$ and $\boldsymbol{\eta}^\star$ satisfy*

$$\boldsymbol{\gamma}^\star = \frac{1}{\zeta^\star}\boldsymbol{\eta}^\star. \tag{5.37}$$

Furthermore, because $\boldsymbol{\alpha}$ is a probability vector, $\xi^\star$ and $\zeta^\star$ satisfy

$$\xi^\star = -\log \zeta^\star.$$

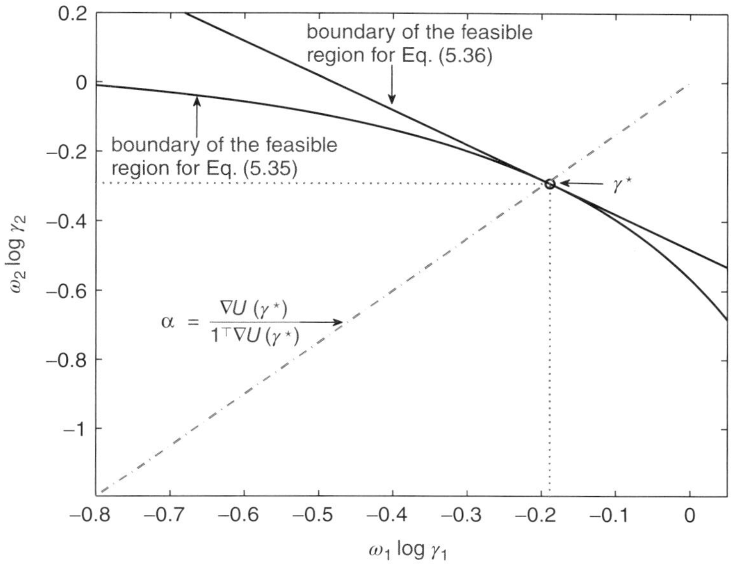

Figure 5.4 Illustration of the connection between Eqs. (5.35) and (5.36). Achievable region for a two-user example with objective function $\sum_l \omega_l \log \gamma_l$. The channel gains are given by $G_{11} = 0.69$, $G_{12} = 0.12$, $G_{21} = 0.13$, and $G_{22} = 0.70$ and the weight is $\omega = [0.40,\ 0.60]^\top$. The maximum power and interference temperature for users are $\overline{\mathbf{p}} = [1.50,\ 1.00]^\top$ W and $\overline{\mathbf{q}} = [2.50,\ 3.00]^\top$ W, respectively. The noise powers for both users are 1 W. The intersection point of the direction $\boldsymbol{\alpha}$ and achievable region is the optimal solution. Moreover, the minimization of Problem (5.36) also intersects with the optimal solution of Eq. (5.35) at the boundary of the feasible region.

The formulation of Eq. (5.36) that minimizes the interference load thus provides a connection (by choosing $\boldsymbol{\alpha}$ to be proportional to the subgradient of the utility function) between the general utility maximization problem in Eq. (5.8) and its special case of egalitarian SINR fairness optimization in Eq. (5.16). An interesting interpretation of Lemma 5.3 is that the optimal SINR in the general utility maximization can be scaled relative to the optimal SINR achieved under the egalitarian SINR fairness. Figure 5.4 illustrates the geometric interpretation of the connection between Problems (5.35) and (5.36) by an example using $U(\boldsymbol{\gamma}) = \sum_{l=1}^{L} \omega_l \log \gamma_l$ with a positive $\boldsymbol{\omega}$.

5.3.6 Utility Maximization Algorithm

In Eq. (5.36), we have shown that it is possible to find a scaling factor that connects the SINR assignment in problem (5.35) (which is related to Eq. (5.14) through a Taylor's series first-order approximation) with the egalitarian SINR fairness. This means that Eq. (5.35) can be first solved by considering Eq. (5.36) to find the scaling factor and then to scale the optimal solution of Eq. (5.36) to finally obtain the solution of Eq. (5.14).

We use the projected subgradient method to solve Eq. (5.36). The parameter $\boldsymbol{\alpha}$ in Eq. (5.36) is updated iteratively. In particular, at the $(k+1)$th iteration, we update $\boldsymbol{\alpha}(k)$ as the subgradient of $U(e^{\tilde{\boldsymbol{\gamma}}})$ at $\tilde{\boldsymbol{\gamma}}(k)$. Thus, instead of solving Eq. (5.14) directly, we replace the objective function of Eq. (5.14) in a neighborhood of a feasible point $\tilde{\boldsymbol{\gamma}}(k)$ by its Taylor series (cf. proof of Theorem 5.3), which is a successive convex approximation technique. Meanwhile, the algorithm iterates according to the subgradient of both the objective function and the constraint functions. However, computing the gradient of the spectral radius functions, that is, the Schur product of the Perron right and left eigenvectors, requires centralized computation in general. This Schur product is the *shadow price* of the SINR assignment for utility maximization.

However, by exploiting the wireless network duality, this task can be made distributed. Making use of the results in Section 5.3.4, we can then obtain a distributed algorithm to solve Eq. (5.14). Observe that the gradient $\mathbf{g} \in \mathbb{R}^L$ of $\rho\big(\operatorname{diag}(\boldsymbol{\eta})\mathbf{B}_l\big)$ and $\rho\big(\mathbf{D}_l \operatorname{diag}(\boldsymbol{\eta})\big)$ in terms of $\mathbf{p}$, $\mathbf{q}$, $\mathbf{s}$, and $\mathbf{t}$ are, respectively, given by (cf. Lemma 5.2)

$$\mathbf{g} = \mathbf{p} \circ \big(\operatorname{diag}(\boldsymbol{\eta} \circ \mathbf{v})^{-1}\mathbf{s}\big) \tag{5.38}$$

and

$$\mathbf{g} = \mathbf{q} \circ \big(\operatorname{diag}(\boldsymbol{\eta}/\mathbf{v})\mathbf{t}\big), \tag{5.39}$$

normalized such that $\mathbf{1}^\top\mathbf{g} = 1$. Furthermore, Eqs. (5.38) and (5.39) can be rewritten, respectively, as

$$g_l = p_l\left(\frac{1}{\eta_l v_l}\right)s_l \tag{5.40}$$

and

$$g_l = q_l\left(\frac{\eta_l}{v_l}\right)t_l. \tag{5.41}$$

Now, in Eqs. (5.40) and (5.41), the respective variables p_l, q_l, s_l, t_l, and η_l can be locally obtained, thus making the gradient computation distributed. We next use Eqs. (5.40) and (5.41) to obtain a distributed algorithm based on the projected subgradient method to solve Eq. (5.8).

Theorem 5.3 *Starting from any initial point $\boldsymbol{\eta}(0)$, if the step size $\nu(k)$ satisfies*

$$\sum_{k=0}^{\infty} \nu(k) = \infty, \quad \sum_{k=0}^{\infty} (\nu(k))^2 < \infty,$$

then $\mathbf{p}(k)$ and $\mathbf{q}(k)$ in Algorithm 5.2 converge to the optimal solution $\mathbf{p}^\star$ and $\mathbf{q}^\star$ of Eq. (5.8), respectively.

Furthermore, if a constant step size is used in Step 4, Algorithm 5.2 is guaranteed to converge to a neighborhood of the optimal solution.

Remark 5.3 *If $U(\boldsymbol{\gamma})$ is smooth, Algorithm 5.2 solves Eq. (5.36) with $\boldsymbol{\alpha}$ given by $\frac{\nabla U(\boldsymbol{\gamma}^\star)}{\mathbf{1}^\top \nabla U(\boldsymbol{\gamma}^\star)}$, where $\boldsymbol{\gamma}^\star$ is the optimal solution to (5.8) (cf. Figure 5.4).*

Algorithm 5.2

Utility Maximization Algorithm

Initialize $\boldsymbol{\eta}(0)$, set the step size $\nu(0) \in (0, 1)$.
1. Compute the weight $\boldsymbol{\alpha}(k)$:
if $U(\boldsymbol{\eta})$ is smooth,

$$\boldsymbol{\alpha}(k) = \frac{\nabla U(\boldsymbol{\eta}(k))}{\mathbf{1}^\top \nabla U(\boldsymbol{\eta}(k))},$$

else

$$\boldsymbol{\alpha}(k) = \frac{\hat{\mathbf{g}}}{\mathbf{1}^\top \hat{\mathbf{g}}},$$ where $\hat{\mathbf{g}}$ satisfies
$U(\boldsymbol{\eta}) \leq U(\boldsymbol{\eta}(k)) + \hat{\mathbf{g}}^\top(\boldsymbol{\eta} - \boldsymbol{\eta}(k))$ for any feasible $\boldsymbol{\eta}$.
end if
2. In the primal network, set the power and interference temperature output of Algorithm 5.1 with $\boldsymbol{\beta} = \boldsymbol{\eta}(k)$, which upon its convergence solves the primal network optimization problem:

$$\begin{array}{ll} \text{maximize} & \min_{l=1,\ldots,L} \dfrac{\mathsf{SINR}_l^P(\mathbf{p})}{\eta_l(k)} \\ \text{subject to} & \mathbf{w}_l^\top \mathbf{p} \leq \bar{p}_l, \quad l = 1, \ldots, L, \\ & \mathbf{e}_l^\top \mathbf{q} \leq \bar{q}_l, \quad l = 1, \ldots, L, \\ & \mathbf{q} = \mathrm{diag}(\mathbf{v})(\mathbf{F}\mathbf{p} + \mathbf{n}), \\ \text{variables:} & \mathbf{p}, \mathbf{q} \end{array}$$

as $\mathbf{p}(k)$ and $\mathbf{q}(k)$, respectively.
The computation of this step also provides in addition the value of $\imath_k$, that is, the i_kth power or j_kth interference temperature constraint that is tight in the primal network optimization problem, where $i_k = \arg\max_{l=1,\ldots,L} \rho\big(\mathrm{diag}(\boldsymbol{\eta}(k) \circ \mathbf{v})(\mathbf{F} + (1/\bar{p}_l)\mathbf{n}\mathbf{w}_l^\top)\big)$ and $j_k = \arg\max_{l=1,\ldots,L} \rho\big(\mathrm{diag}(\mathbf{v})(\mathbf{F}\,\mathrm{diag}(\boldsymbol{\eta}(k)) + (1/\bar{q}_l)\mathbf{n}\mathbf{e}_l^\top)\big)$.
In the dual network, set the power and interference temperature output of Algorithm 5.1 with $\boldsymbol{\beta} = \boldsymbol{\eta}(k)$ and with $\mathsf{SINR}^P(\mathbf{p})$ replaced by $\mathsf{SINR}^D(\mathbf{s})$ at Step 1 of Algorithm 5.1 and the normalization at Step 2 of Algorithm 5.1 $\max_{l=1,\ldots,L}\{\mathbf{w}_l^\top \mathbf{p}(k+1)/\bar{p}_l, \mathbf{e}_l^\top \mathrm{diag}(\mathbf{v})\mathbf{F}\mathbf{p}(k+1)/(\bar{q}_l - n_l v_l)\}$ replaced by $\max\{\frac{\mathbf{n}^\top \mathbf{s}(k+1)}{\bar{p}_{\imath_k}}, \frac{\mathbf{n}^\top \mathrm{diag}(\mathbf{v})\mathbf{F}^\top \mathbf{s}(k+1)}{(\bar{q}_{\imath_k} - \mathbf{n}^\top \mathrm{diag}(\mathbf{v})\mathbf{w}_{\imath_k})}\}$ which on its convergence solves the dual network optimization problem:

$$\begin{array}{ll} \text{maximize} & \min_{l=1,\ldots,L} \dfrac{\mathsf{SINR}_l^D(\mathbf{s})}{\eta_l(k)} \\ \text{subject to} & \mathbf{n}^\top \mathbf{s} \leq \bar{p}_{\imath_k}, \\ & \mathbf{n}^\top \mathbf{t} \leq \bar{q}_{\imath_k}, \\ & \mathbf{t} = \mathrm{diag}(\mathbf{v})(\mathbf{F}^\top \mathbf{s} + \mathbf{w}_{\imath_k}), \\ \text{variables:} & \mathbf{s}, \mathbf{t} \end{array}$$

as $\mathbf{s}(k)$ and $\mathbf{t}(k)$ respectively.

Algorithm 5.2

3. Each lth user updates its gradient $g_l(k)$ as follows:
   if $\boldsymbol{\alpha}(k)^\top \log \boldsymbol{\eta}(k) < 0$
   $g_l(k) = \alpha_l(k)$,
   else
   if $\imath_k = i_k$
   $g_l(k) = p_l(k)s_l(k)/(\eta_l(k)v_l), \quad l = 1, \ldots, L,$
   if $\imath_k = j_k$
   $g_l(k) = q_l(k)t_l(k)\eta_l(k)/v_l, \quad l = 1, \ldots, L.$
   end if
4. Update $\boldsymbol{\eta}(k+1)$:
   $\boldsymbol{\eta}(k+1) \leftarrow \boldsymbol{\eta}(k)e^{-\nu(k)(\mathbf{g}(k)/\mathbf{1}^\top\mathbf{g}(k))}$.
5. Update $\nu(k+1)$ according to Theorem 5.3 and go to Step 1.

Remark 5.4 *As we run Algorithm 5.1 at each iteration of Algorithm 5.2 as an inner loop, Algorithm 5.2 is a two time-scale algorithm. At Step 2, we obtain the primal network power* $\mathbf{p}(k)$ *and the primal network interference temperature* $\mathbf{q}(k)$ *from the output of Algorithm 5.1 by using the input weight parameter* $\boldsymbol{\eta}(k)$*. Similarly, we can also obtain the dual network power* $\mathbf{s}(k)$ *and the dual network interference temperature* $\mathbf{t}(k)$ *in the same way. This means that* $\mathbf{p}(k)$*,* $\mathbf{q}(k)$*,* $\mathbf{s}(k)$*, and* $\mathbf{t}(k)$ *are the optimal solutions of Algorithm 5.1 for a given* $\boldsymbol{\eta}(k)$*. The computation of* $\mathbf{1}^\top\nabla U(\boldsymbol{\eta}(k))$*,* $\boldsymbol{\alpha}(k)^\top \log \boldsymbol{\eta}(k)$*, and* $\mathbf{1}^\top\mathbf{g}(k)$ *can be made distributed by a gossip algorithm [42].*

5.3.7 A Software Implementation

The spectral radius function is an important feature that characterizes the optimal wireless network utility. We next discuss how to compute this function numerically when it is used in optimization problems. Our spectral radius function is implemented as a Matlab routine using the cvx software. The cvx software tool in Reference 44 is a Matlab toolbox software for rapid definition, manipulation, and solution of convex optimization problems and is freely available for download. This cvx software solves optimization problems that are appropriately modeled using a set of libraries known as *atom library* in cvx. The modularity of the *atom library* software facilitates new functionalities that are specific to some optimization problems with unique features by adding new (user-defined) cvx atoms in cvx. In addition, new cvx atoms need to follow the disciplined convex programming ruleset (DCP ruleset for short), which are drawn from basic principles of convex optimization. This means that a violation of a set of modeling rules can lead to parsing software error.

We now discuss how the convex spectral radius function used in the previous sections given by

$$f(\tilde{\boldsymbol{\gamma}}) = \rho\big(\operatorname{diag}(e^{\tilde{\boldsymbol{\gamma}}})\mathbf{B}\big), \tag{5.42}$$

can conform to the cvx DCP ruleset, thereby facilitating its cvx implementation to solve optimization problems. A trick to numerically evaluate Eq. (5.42) is an application of the Perron–Frobenius theorem to rewrite Eq. (5.42) as

$$f : \mathbb{R}^L \to \mathbb{R}_+, \quad f(\tilde{\gamma}) \triangleq \inf\{\lambda \mid \mathrm{diag}(e^{\tilde{\gamma}})\mathbf{B}\mathbf{z} \leq \lambda \mathbf{z}\},$$

where $\mathbf{z} \in \mathbb{R}^L$ is an (auxiliary) optimization variable. Thus, Eq. (5.42) can be evaluated numerically by solving a geometric program (a special class of convex optimization problems). Using the cvx software, the implementation of the spectral radius function in cvx is given by

```
function cvx_optval = spectral_radius( x, B )
s = size( B, 1 );
cvx_begin gp
   variables rho z( s )
   minimize( rho );
   subject to
      diag( exp(x) ) * B * z <= rho * z;
cvx_end
```

The above function cvx_optval is then put into a subdirectory named as @cvx in the cvx software package, and Matlab can automatically recognize the function call of spectral_radius whenever it is used in the constraints or objectives of a convex optimization problem that follows the cvx DCP ruleset. This cvx software implementation can be useful for verifying the solution computed by the distributed dual algorithms and the nonlinear Perron–Frobenius-theory-based fixed-point algorithm in the previous sections.

5.3.8 Connection between Dual Algorithm and Pricing Function in Game Theory

There are several approaches in the literature to solving the wireless utility maximization problem. A well-studied approach to maximize the utility using distributed power control is to design a noncooperative power control game where users maximize their utility. The outcome of the game results in a Nash equilibrium that is often inefficient.

In the seminal work [11], pricing of transmit powers was introduced to obtain Pareto improvement of the noncooperative power control game, that is, to obtain improvements in user utilities relative to the case with no pricing. In particular, the lth user solves a local utility maximization problem given by

$$\begin{array}{ll} \text{maximize} & u_l(\mathsf{SINR}(\mathbf{p})) + c_l(\mathbf{p}) \\ \text{subject to} & \mathbf{p} \in \mathscr{P}, \\ & \mathbf{q} \in \mathcal{Q}, \\ \text{variables:} & p_l, q_l, \end{array} \tag{5.43}$$

where $c_l(\cdot)$ is a general pricing function, which is not restricted to any particular form. In Eq. (5.43), the lth user treats the transmit powers of other users as fixed and only optimizes its own transmit power. In Reference 11, the pricing function is linearly proportional to the transmit powers in Eq. (5.43), that is, $c_l(\mathbf{p}) = c\alpha_l p_l$, where c and $\{\alpha_l\}$ are positive scalars. Clearly, this pricing function imposes a price that increases with the transmit power of the user. The pricing factor, that is, c, is a common parameter shared by all the users in the game. It is important to note that c has to be appropriately tuned (possibly by a centralized controller through multiple rounds of the noncooperative power control game) such that user's self-interest leads to best possible improvement in overall utility maximization.

Saraydar et al. [11] also explored the noncooperative game equilibrium of equalizing the SINR (called the equal-SIR equilibrium in Reference 11), which is equivalent to solve the max–min SINR fairness optimization in Eq. (5.16). It is interesting to note that the max–min SINR fairness optimization (when used appropriately as an iterative computational submodule) can achieve the social optimum in terms of utility maximization (as used in our dual algorithms). In the following, we state a result that establishes a connection between our dual algorithms for solving Eq. (5.8) in the previous sections and the noncooperative game-theoretic pricing function introduced in Reference 11.

Theorem 5.4 *The optimal solution of Eq.* (5.43) *is equal to the optimal solution of Eq.* (5.8), *if the pricing function* $c_l(\mathbf{p})$ *is given by*

$$c_l(\mathbf{p}) = -(1/\omega_l) \sum_{j \neq l}^{L} \omega_j u_j(\mathbf{p}). \tag{5.44}$$

Proof: Denote the optimal power of Eq. (5.8) by $\mathbf{p}^\star$, and $\mathbf{p}'$ is a vector with $p'_j = p^\star_j$, $j \neq l$ and p'_l being any feasible value satisfying $\mathbf{p}' \in \mathscr{P}$. Then, we have

$$\sum_{j=1}^{L} \omega_j u_j(\mathbf{p}^\star) \geq \sum_{j=1}^{L} \omega_j u_j(\mathbf{p}'). \tag{5.45}$$

Dividing both sides of Eq. (5.45) by ω_l and separating $u_l(\mathbf{p}^\star)$ and $u_l(\mathbf{p}')$, respectively, from the summation, we can obtain

$$u_l(\mathbf{p}^\star) + \frac{1}{\omega_l} \sum_{j \neq l}^{L} \omega_j u_j(\mathbf{p}^\star) \geq u_l(\mathbf{p}') + \frac{1}{\omega_l} \sum_{j=1}^{L} \omega_j u_j(\mathbf{p}'). \tag{5.46}$$

Substituting Eq. (5.44) into Eq. (5.46), we can rewrite Eq. (5.46) as $u_l(\mathbf{p}^\star) - c_l(\mathbf{p}^\star) \geq u_l(\mathbf{p}') - c_l(\mathbf{p}')$, which is the definition of a Nash equilibrium [11]. ■

5.4 NUMERICAL EXAMPLES

In this section, we provide numerical examples to illustrate the performance of Algorithm 5.1 in Section 5.3.3 and Algorithm 5.2 in Section 5.3.6, illustrating the convergence properties of Algorithms 5.1 and 5.2. Recall that an optimization problem having two different utility functions can be constructed to yield the same optimal solution for egalitarian SINR fairness (cf. (5.23) in Section 5.3.3). By setting the weight of Eq. (5.23) appropriately, and applying Algorithms 5.1 and 5.2 to solve the max–min SINR and the weighted sum of logarithmic SINR utility, respectively, we evaluate the numerical convergence of the power and interference temperature iterates.

We use the following channel gain matrix

$$\mathbf{G} = \begin{bmatrix} 0.69 & 0.12 & 0.14 \\ 0.13 & 0.70 & 0.13 \\ 0.14 & 0.15 & 0.75 \end{bmatrix},$$

and the following weights for the power constraints:

$$\mathbf{w}_1 = [\ 0.8491\ 0.9340\ 0.6781\ ]^\top$$
$$\mathbf{w}_2 = [\ 0.7577\ 0.7431\ 0.3922\ ]^\top$$
$$\mathbf{w}_3 = [\ 0.6555\ 0.1712\ 0.7060\ ]^\top.$$

We set $\overline{\mathbf{p}} = [1.50\ 1.00\ 1.20]^\top$ W and $\overline{\mathbf{q}} = [2.50\ 3.00\ 2.20]^\top$ W. The noise power of each user is 1 W.

Figure 5.5 plots the evolution of the power and interference temperature for three users that run Algorithm 5.1 with $\boldsymbol{\beta}$ equal to $\mathbf{1}$. We set the initial power vector to $\mathbf{p}(0) = [0.5\ 0.5\ 0.5]^\top$ W and run Algorithm 5.1 for 10 iterations before it terminates.

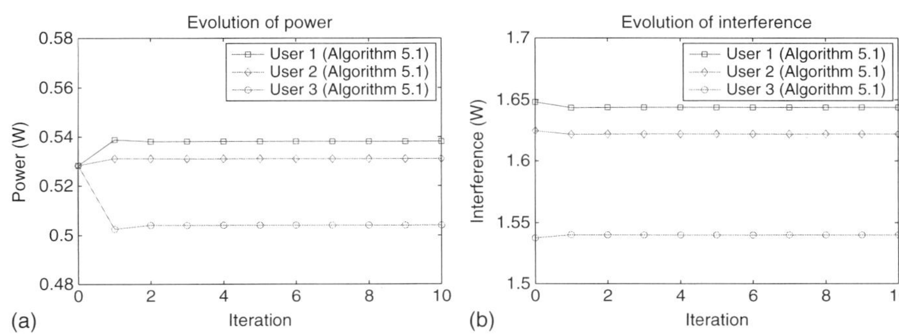

Figure 5.5 Illustration of the convergence of Algorithm 5.1 with randomly chosen initial $\mathbf{p}(0)$. We plot the trajectory of power for each of the three users in (a) and the trajectory of interference temperature for each of the three users in (b). We can observe from the figures that the convergence of Algorithm 5.1 is geometrically fast.

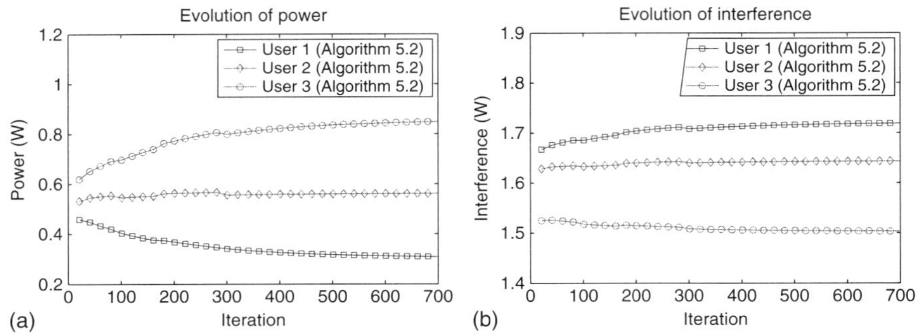

Figure 5.6 Illustration of the convergence of Algorithm 5.2 with randomly chosen initial $\boldsymbol{\eta}(0)$. Algorithm 5.2 is a distributed algorithm. $\mathbf{p}(k)$ and $\mathbf{q}(k)$ are obtained from Step 2, which terminates when $\epsilon = 10^{-13}$. The number of iterations in the figure corresponds to the outer loop (slower time scale). We plot the trajectory of power for each of the three users in (a) and the trajectory of interference temperature for each of the three users in (b).

Figure 5.5 shows that Algorithm 5.1 converges geometrically fast to the optimal solution (verifying Theorem 5.2). The optimal $\boldsymbol{\gamma}^\star$ is $[0.3273\ 0.3273\ 0.3273]^\top$ (verifying Lemma 5.1).

Figure 5.6 plots the evolution of the power and interference temperature for three users that run Algorithm 5.2. The objective function that we use in this numerical example is $U(\boldsymbol{\gamma}) = \sum_{l=1}^{L} \omega_l \log \gamma_l$, where $\boldsymbol{\omega}$ is $[0.23\ 0.41\ 0.36]^\top$. The sum of weighted logarithmic SINR satisfies Assumption 5.1. We set the initial $\boldsymbol{\eta}$ to $\boldsymbol{\eta}(0) = [0.82\ 0.90\ 1.11]^\top$ and run Algorithm 5.2 for 700 iterations and run Algorithm 5.1 as an inner loop. The stopping criterion for the inner loop is $\epsilon = 10^{-13}$. We use a constant step size, that is, $\nu(k) = 0.04$ for all k. Figure 5.6 shows that Algorithm 5.2 converges to the optimal solution (verifying Theorem 5.3). The optimal $\mathbf{p}^\star$ is $[0.31\ 0.57\ 0.86]^\top$ W and the optimal $\mathbf{q}^\star$ is $[1.72\ 1.65\ 1.51]^\top$ W so that $\mathbf{w}_2^\top \mathbf{p}^\star$ is equal to $\bar{p}_2$ (verifying Theorem 5.1). We also adjust the value of ϵ to verify the robustness of our algorithm. Figure 5.7 shows the evolution of the power and

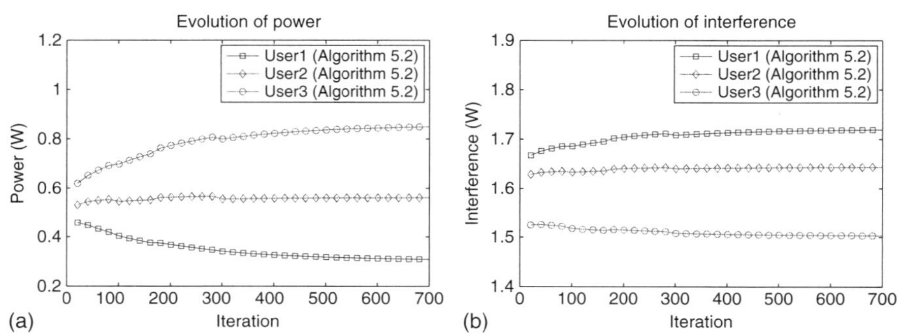

Figure 5.7 An illustration of Algorithm 5.2. All the other parameters are the same as Figure 5.6 except that $\epsilon = 10^{-3}$.

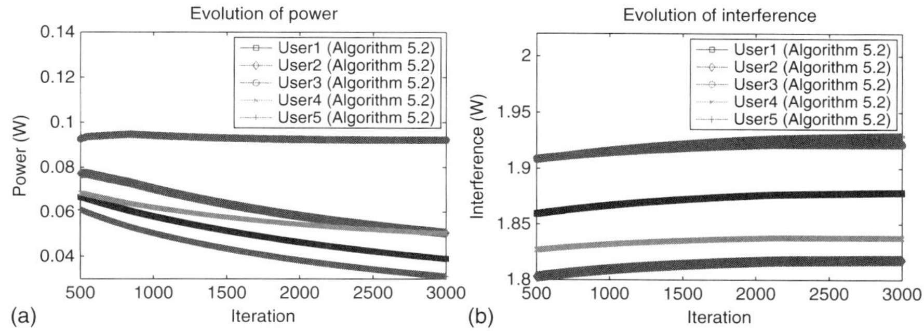

Figure 5.8 An illustration of Algorithm 5.1 for 30 users. In this figure, we show the power and interference temperature evolution for only five users.

interference temperature when the inner loop of Algorithm 5.2, that is, Algorithm 5.1, terminates with $\varepsilon = 10^{-3}$ keeping all the other parameters the same as those used in Figure 5.6. Furthermore, we also run Algorithm 5.2 with a suitably large number of primary and secondary users. Figure 5.8 shows the evolution of the power and interference temperature for 5 users out of 30 users.

Next, we compare the convergence of the power and interference temperature in two different problems that have the same optimal solution. From the second numerical example, we already know in advance that the second power constraint is tight at optimality. Thus, we can set $U(\boldsymbol{\gamma}) = \sum_{l=1}^{L} \left(\mathbf{x}(\mathbf{B}_2)\circ\mathbf{y}(\mathbf{B}_2)\right)_l \log \gamma_l$. Although the solution of these two different problems are the same, Figure 5.9 shows that Algorithm 5.1 converges faster than Algorithm 5.2.

To illustrate, we list down the `cvx` Matlab routine to solve the following utility maximization problem example using the `cvx` atom mentioned in Section 5.3.7:

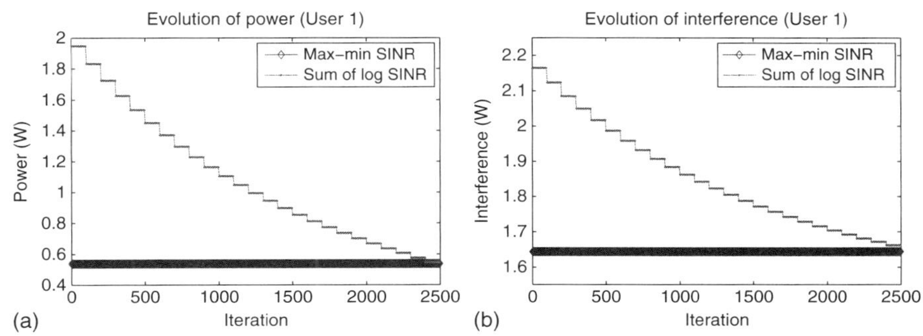

Figure 5.9 Performance comparison of Algorithms 5.1 and 5.2 by solving two equivalent problems, namely, Eqs. (5.16) and (5.9) with a utility given by Eq. (5.23). (a) and (b) show the trajectories of the power and the interference temperature iterates, respectively, for User 1. The number of iterations in the figure corresponds to the inner loop (faster time scale). With the weight $\boldsymbol{\omega}$ chosen as in Eq. (5.23), these two algorithms converge to the same solution. As shown, Algorithm 5.1 converges much faster than Algorithm 5.2.

$$\begin{array}{ll} \text{minimize} & \boldsymbol{\omega}^\top \tilde{\boldsymbol{\gamma}} \\ \text{subject to} & \log \rho\big(\operatorname{diag}(e^{\tilde{\boldsymbol{\gamma}}})\mathbf{B}_l\big) \geq 0, \quad l = 1, \ldots, 3 \\ \text{variables:} & \tilde{\boldsymbol{\gamma}}. \end{array} \tag{5.47}$$

Using the software atom developed for the spectral radius function in Section 5.3.7, the cvx code to solve Eq. (5.47) numerically is given by

```
cvx_begin
   variables gamma_tilde(3)
   maximize( w' * gamma_tilde );
   subject to
      log( spectral_radius( gamma_tilde, B1 ) ) <= 0;
      log( spectral_radius( gamma_tilde, B2 ) ) <= 0;
      log( spectral_radius( gamma_tilde, B3 ) ) <= 0;
cvx_end
```

In the above, the matrices B1, B2, and B3 are evaluated using Eq. (5.17), and they contain the problem parameters, that is, the channel gains, the weight and weighted power constraints, and so on.

5.5 CONCLUSION

We studied the network utility maximization in a wireless network with both power budget constraints and interference temperature constraints in this chapter. We first reformulated the problem as one in the SINR domain that has appropriately constructed spectral radius constraints. The advantages of our reformulation were that, firstly, it captured the entire feasible SINR region and, secondly, it decoupled the SINR assignment for primary and secondary users from power and interference temperature control. We also studied a special case of egalitarian fairness utility, which is the max–min weighted SINR problem using the nonlinear Perron–Frobenius theory, and a geometrically fast convergent (with no parameter tuning) algorithm was proposed to solve this max–min weighted SINR problem. Then, we developed a wireless network duality and established its connection to the gradient of the spectral radius function as shadow prices for SINR assignment in the general utility maximization problem. We leveraged the wireless network duality and an interference load minimization problem, which interestingly is related to the max–min weighted SINR problem, to develop a distributed algorithm to solve the utility maximization problem. Extensive numerical examples demonstrated the good performance of our algorithms, particularly the flexibility, the fast convergence time, and the robustness to different numbers of primary and secondary users. We also discussed a connection between this shadow price dual algorithm approach for utility maximization with a noncooperative power control game in obtaining improvements in user utilities relative to the case with no pricing.

REFERENCES

1. S. Ha, S. Sen, C. Joe-Wong, Y. Im, and M. Chiang. TUBE: time-dependent pricing for mobile data. In *Proceedings of ACM SIGCOMM*, 2012.
2. C. Joe-Wong, S. Ha, and M. Chiang. Time dependent broadband pricing: feasibility and benefits. In *Proceedings of IEEE ICDCS*, 2011.
3. S. Sorooshyari, C. W. Tan, and M. Chiang. "Power control for cognitive radio networks: axioms, algorithms, and analysis," *IEEE/ACM Transactions on Networking*, 20(3), 2012, 878–891.
4. S. Ren and M. van der Schaar. "Data demand dynamics in wireless communications markets," *IEEE Transactions on Signal Processing*, 60(4), 2012, 1986–2000.
5. Y. Cui, T. Ma, and X. Cheng. Multi-hop access pricing in public area WLANs. In *Proceedings of IEEE INFOCOM*, 2011.
6. P. Hande, M. Chiang, R. Calderbank, and S. Rangan. Network pricing and rate allocation with content provider participation. In *Proceedings of IEEE INFOCOM*, 2009.
7. P. Hande, M. Chiang, R. Calderbank, and J. Zhang. Pricing under constraints in access networks: revenue maximization and congestion management. In *Proceedings of IEEE INFOCOM*, 2010.
8. E. Altman, P. Bernhard, S. Caron, G. Kesidis, J. Rojas-Mora, and S. Wong. "A model of network neutrality with usage-based prices," *Telecommunication Systems*, 52(2), 2011, 1–9.
9. S. Caron, G. Kesidis, and E. Altman. "Application neutrality and a paradox of side payments," *In Proc. ACM ReARCH*, 2010.
10. S. Ren, J. Park, and M. van der Schaar. "Entry and spectrum sharing scheme selection in femtocell communications markets," *IEEE/ACM Transactions on Networking*, 21(1), 2012, 218–232.
11. C. U. Saraydar, N. B. Mandayam, and D. J. Goodman. "Efficient power control via pricing in wireless data networks," *IEEE Transactions on Communications*, 50(2), 2002, 291–303.
12. S. Shakkottai and R. Srikant. "Economics of network pricing with multiple ISPs," *IEEE/ACM Transactions on Networking*, 14(6), 2006, 1233–1245.
13. S. Shakkottai, R. Srikant, A. Ozdaglar, and D. Acemoglu. "The price of simplicity," *IEEE Journal on Selected Areas in Communications*, 26(7), 2008, 1269–1276.
14. T. Basar and R. Srikant. Revenue-maximizing pricing and capacity expansion in a many-users regime. In *Proceedings of IEEE INFOCOM*, vol. 1, 2002.
15. S. Shakkottai and R. Srikant. "Network optimization and control," *NOW Monograph in Foundations and Trends in Networking*, 2(3), 2007, 271–379.
16. M. Chiang, P. Hande, T. Lan, and C. W. Tan. "Power control in wireless cellular networks," *NOW Monograph in Foundations and Trends in Networking*, 2(4), 2008, 381–533.
17. P. Hande, S. Rangan, M. Chiang, and X. Wu. "Distributed uplink power control for optimal SIR assignment in cellular data networks," *IEEE/ACM Transactions on Networking*, 16(6), 2008, 1420–1433.
18. S. Huang, X. Liu, and Z. Ding. "Decentralized cognitive radio control based on inference from primary link control information," *IEEE Journal on Selected Areas in Communications*, 29(2), 2011, 394–406.

19. M. Lotfinezhad, L. Ben, and E. S. Sousa. Optimal control of constrained cognitive radio networks with dynamic population size. In *Proceedings of IEEE INFOCOM*, 2010.
20. S. Rangan and R. Madan. "Belief propagation methods for intercell interference coordination in femtocell networks," *IEEE Journal on Selected Areas in Communications*, 30(3), 2012, 631–640.
21. A. L. Stolyar and H. Viswanathan. Self-organizing dynamic fractional frequency reuse for best-effort traffic through distributed inter-cell coordination. In *Proceedings of IEEE INFOCOM*, 2009.
22. A. L. Stolyar and H. Viswanathan. Self-organizing dynamic fractional frequency reuse in OFDMA systems. In *Proceedings of IEEE INFOCOM*, 2008.
23. K. R. Krishnan and H. Luss. Power selection for maximizing SINR in femtocells for specified SINR in macrocell. In *Proceedings of IEEE WCNC*, 2011.
24. C. W. Tan, M. Chiang, and R. Srikant. "Fast algorithms and performance bounds for sum rate maximization in wireless networks," *IEEE/ACM Transactions on Networking*, 21(3), 2012, 706–719.
25. L. Zheng and C. W. Tan. "Cognitive radio network duality and algorithms for utility maximization," *IEEE Journal on Selected Areas in Communications*, 31(3), 2013, 500–513.
26. C. W. Tan, M. Chiang, and R. Srikant. "Maximizing sum rate and minimizing MSE on multiuser downlink: optimality, fast algorithms and equivalence via max–min SINR," *IEEE Transactions on Signal Processing*, 59(12), 2011, 6127–6143.
27. C. W. Tan, S. Friedland, and S. H. Low. "Nonnegative matrix inequalities and their application to nonconvex power control optimization," *SIAM Journal on Matrix Analysis and Applications*, 32(3), 2011, 1030–1055.
28. C. W. Tan, S. Friedland, and S. H. Low. "Spectrum management in multiuser cognitive wireless networks: optimality and algorithm," *IEEE Journal on Selected Areas in Communications*, 29(2), 2011, 421–430.
29. J. Mo and J. Walrand. "Fair end-to-end window-based congestion control," *IEEE/ACM Transactions on Networking*, 8(5), 2000, 556–567.
30. J. F. C. Kingman. "A convexity property of positive matrices," *Proceedings of the American Mathematical Society*, 12(2), 1961, 283–284.
31. S. Boyd and L. Vanderberghe. *Convex Optimization*. Cambridge University Press, Cambridge, U.K, 2004.
32. P. Viswanath and D. N. C. Tse. "Sum capacity of the vector Gaussian broadcast channel and uplink–downlink duality," *IEEE Transactions on Information Theory*, 49(8), 2003, 1912–1921.
33. W. Yu. "Uplink–downlink duality via minimax duality," *IEEE Transactions on Information Theory*, 52(2), 2006, 361–374.
34. F. Rashid-Farrokhi, K. J. Ray Liu, and L. Tassiulas. "Transmit beamforming and power control for cellular wireless systems," *IEEE Journal on Selected Areas in Communications*, 16(8), 1998, 1437–1449.
35. B. Song, R. L. Cruz, and B. D. Rao. "Network duality for multiuser MIMO beamforming networks and applications," *IEEE Transactions on Communication*, 55(3), 2007, 618–630.
36. D. W. H. Cai, T. Quek, and C. W. Tan. "A unified analysis of max–min weighted SINR for MIMO downlink system," *IEEE Transactions on Signal Processing*, 59(8), 2011, 3850–3862.

37. D. W. H. Cai, T. Quek, C. W. Tan, and S. H. Low. "Max–min SINR coordinated multipoint downlink transmission - duality and algorithms," *IEEE Transactions on Signal Processing*, 60(10), 2012, 5384–5395.
38. C. W. Tan, D. P. Palomar, and M. Chiang. "Energy-robustness tradeoff in cellular network power control," *IEEE/ACM Transactions on Networking*, 17(3), 2009, 912–925.
39. U. Krause. "Concave Perron–Frobenius theory and applications," *Nonlinear Analysis*, 47(2001), 2001, 1457–1466.
40. D. P. Bertsekas. *Nonlinear Programming*. Athena Scientific, Belmont, MA, USA, 2nd edition, 2003.
41. G. J. Foschini and Z. Miljanic. "A simple distributed autonomous power control algorithm and its convergence," *IEEE Transactions on Vehicular Technology*, 42(4), 1993, 641–646.
42. S. Boyd, A. Ghosh, and D. Shah. "Randomized gossip algorithms," *IEEE Transactions on Information Theory*, 52(6), 2006, 2508–2530.
43. S. Friedland and S. Karlin. "Some inequalities for the spectral radius of non-negative matrices and applications," *Duke Mathematical Journal*, 42(3), 1975, 459–490.
44. M. Grant and S. Boyd. CVX: Matlab software for disciplined convex programming, version 1.21, 2011. http://cvxr.com/cvx.

6 Human Factors in Smart Data Pricing

SOUMYA SEN, CARLEE JOE-WONG, SANGTAE HA,
and MUNG CHIANG

6.1 INTRODUCTION

Smart data pricing (SDP) explores mechanisms beyond traditional flat-rate or usage-based pricing models to alleviate network congestion, monetize bandwidth, and help consumers save on their monthly bills. But realizing such mechanisms requires the creation of the right economic incentives and tools to understand and modify consumer behavior in a desired manner. The need for such a holistic approach makes SDP a truly interdisciplinary research topic that draws on principles from network engineering, economics, and human–computer interaction (HCI). Therefore, understanding how to enable consumers to respond to SDP incentives remains the most important factor in the successful adoption of such data pricing plans. HCI researchers have explored related issues in the context of computer networks and electricity markets, and in this chapter, we introduce these earlier findings and build on them to address the needs of mobile networks. In mobile networks, the feasibility of SDP is particularly promising because of the variance in users' content consumption behaviors, varying degrees of elasticity of demand of different content types, and the richness of capabilities afforded by smart devices to make them a part of the network congestion management infrastructure.

Past research has shown that consumers prefer simplicity in billing, predictability in their expense, and ease in understanding feedback (e.g., pricing signals) from the network. Hence, for SDP to be successful, these user concerns must be addressed in both the realization of the data plan (i.e., creation of economic incentives) and the means of communicating with users (i.e., the design of the user interfaces (UIs) for the feedback-control loop between the users and the network operators). To achieve this, we need to understand user psychology through HCI techniques such as focus

Smart Data Pricing, First Edition. Edited by Soumya Sen, Carlee Joe-Wong, Sangtae Ha, and Mung Chiang.

group studies, testing of user reaction to designs of interfaces that convey the pricing signals, development of experimental testbeds and application prototypes for offering new pricing plans, and field trials of these pricing plans.

In this chapter, we first discuss some of the basic principles and methodologies of HCI research in Section 6.2, followed by a detailed overview of findings from previous HCI works that have studied user psychology in energy markets (Section 6.3), home network management (Section 6.4), and bandwidth pricing (Section 6.5). We then discuss in Section 6.6 how *day-ahead dynamic time-dependent pricing* (DDTDP) has the potential to account for the lessons learned from previous HCI research and meet the design goals of creating a "win—win" for both consumers and operators. We follow this introduction to DDTDP with a discussion of the perspectives of various stakeholders of the Internet ecosystem (i.e., network operators, users, application developers, content providers, and regulators) on the feasibility and promise of DDTDP in Section 6.7. In Section 6.8, we then discuss our trial of DDTDP, the system implementation and the graphical user interface (GUI) features, followed by the results from the trial using both HCI-based qualitative findings (e.g., from focus groups, trial debriefing) and quantitative validation (e.g., from usage logs). Finally, we conclude in Section 6.9 with a discussion of the results learned from this trial and past HCI works, as well as future directions that can be explored as SDP practices continue to evolve.

6.2 METHODOLOGY

In this section, we give an overview of HCI research techniques relevant to SDP. We focus on the process of designing and testing the effectiveness of SDP systems, for example, designing UIs for offering prices and then evaluating them through expert analysis and field or laboratory studies. For a more detailed overview of HCI in general, including its theoretical underpinnings, we refer the reader to Reference 1. Readers familiar with the basic design principles and methodologies may skip ahead to Section 6.3 for information on prior HCI works on related topics.

6.2.1 Designing Systems with Users in Mind

Arguably, the most important factor in the successful adoption of SDP mechanisms is the usability and convenience of the application interfaces designed to modify user behavior. As many SDP mechanisms involve direct user interaction with the prices [e.g., selecting a desired quality-of-service (QoS) or delaying traffic to later times], complete SDP systems usually require a UI to facilitate this interaction. This interface should effectively communicate relevant information and simplify actions desired by users, for example, making it easy for them to delay a particular app until a cheaper time.

Interface design begins by considering the goals of target users. In the context of SDP, these would be SDP customers, with the general goal of saving as much money as possible without compromising their experience. More specifically, we can construct *scenarios* of how a user might interact with the interface; for instance, suppose an Internet Service Provider (ISP) offers a QoS-based pricing plan, in which a user can choose to pay more for higher QoS on a given session. One scenario might be that "Alice is streaming YouTube on her Galaxy tablet during her lunch break and wants to view an HD video with higher QoS. She opens the pricing app on her tablet, scrolls through a list of apps to the YouTube app, and checks the box with 'higher QoS.' She then clicks on the YouTube icon, opening the YouTube app, and starts playing the video with higher QoS."

Scenarios allow designers to imagine tasks that a user wishes to perform, which express the dynamics of the interface and allow problems to be spotted in transitioning from one stage of a task to another. Designers can also imagine scenarios in which users make errors and wish to correct them—for instance, if a user accidentally selects the wrong app for higher QoS, is there a clear sequence of actions for unselecting it? By focusing on a user's perspective, scenarios allow a designer to simplify user interactions and transitions between app screens.

To facilitate navigation among screens, a designer needs to ensure that links to different screens are clearly labeled and visible. In particular, users should understand the effect of clicking any given button, for example, whether it will simply bring them to another screen or take some action that affects the amount spent on their data plan. If a user is required to provide input to the interface, such as specifying particular apps for higher QoS, then the interface must first indicate the types of information being requested. We illustrate some key principles of interface design by discussing DataWiz,[1] a free mobile application for data monitoring for iOS and Android platforms, which was developed by the authors and colleagues at DataMi.[2] DataWiz is a monitoring and alerting application with several desirable functionalities, but does not include pricing components. These GUIs, shown in Figure 6.1, were based on several focus group studies, feedback from multiple network operators, and the help of HCI researchers and professional graphic designers and app developers:

- *Consistency.* Navigation sequences, layout, and terminology should be as consistent as possible throughout the app, for example, similar displays for daily and monthly information. DataWiz's main screen (Fig. 6.1a) allows users to tap the weekly and monthly bars to display weekly or monthly usage on the center pie chart, in a manner analogous to the daily usage shown in the figure.
- *Shortcuts.* Users should be able to quickly execute repeated sequences of actions, for example, selecting some actions as "default." For instance, tapping the monthly bar of the main screen in Figure 6.1a displays the monthly usage on the pie chart, while the default was set to daily usage to give the user an immediate update on how much data he/she has used up from his/her "virtual" daily quota.
- *Informative Feedback.* Should a user input something into the interface, there should be some confirmation, for example, highlighting an app after it is

[1] DataWiz, http://scenic.princeton.edu/datawiz/.
[2] DataMi, http://www.datami.com.

Figure 6.1 DataWiz screenshots, illustrating basic design principles. (a) Home screen, (b) usage graphs, (c) usage prediction, (d) usage details,(e) maps screen, and (f) settings screen.

selected, to confirm that this input was received. In Figure 6.1b, users can switch between displaying 3G and WiFi, as well as daily, weekly, and monthly usage. Their selection is highlighted accordingly, for example, 3G monthly usage in the figure.

- *Closure.* Users should receive an indication when an action has been completed, for example, confirmation that they have finished selecting higher QoS. For instance, when toggling between DataWiz's graph displays in Figure 6.1b, a new graph is displayed when the selection has been completed.
- *Error Prevention and Handling.* Users should be able to easily reverse actions or go back to a previous screen, aiding them in correcting errors and reducing user anxiety at accidentally pressing a wrong button. For instance, users at DataWiz's

usage prediction screen (Fig. 6.1c) can press the back button to easily return to the home screen of Figure 6.1a.

- *Simplicity.* Few screens and simple displays are easier for users to understand and take less storage on the interface's physical device. In DataWiz, the number of screens is minimized, with the app's main screens easily accessible from the four links on the top of the screens in Figure 6.1c–e.

To achieve these goals, designers can manipulate the following basic visual tools of interface design to suggest appropriate actions to users.

- *Font and Color.* Font style and color can be used to further differentiate different groupings. For instance, one section might have a white and another a dark gray background. On DataWiz's usage prediction screen (Fig. 6.1c), colors separate the past from the future: the white bars indicate future (i.e., predicted) usage, whereas green bars indicate past usage. The larger font of the dates of the usage details screen in Figure 6.1d shows how font size can be used to draw attention to blocks of text, while the orange icons in the map screen of Figure 6.1e draw attention to the locations where the user has consumed data.
- *Grouping and Structure.* Information that is logically separate should be separated either by space or by placing it on different lines, for example, prices for different times of the day might be separated by extra space between them. On DataWiz's settings page in Figure 6.1f, we see this grouping with thicker lines dividing the billing data, data cap, and alert settings.
- *Alignment and Order.* Users generally read from top to bottom and left to right; thus information should be placed to reflect this order. For instance, in a list of prices and corresponding times, we might put the most recent time at the top to reflect the fact that it is most immediately important. In the DataWiz app, the monthly data cap setting is listed above the weekly and daily caps on the settings screen (Fig. 6.1f) to reflect its greater importance.
- *Icons.* Design of intuitive icons is an essential component of a well-designed interface as it helps users to navigate with ease. Figure 6.1c–e shows navigational icons on the top (from left to right) for (a) "back" to the home screen (Fig. 6.1a), (b) the "usage prediction" screen (Fig. 6.1c), (c) the "usage details" screen (Fig. 6.1d), and (d) the "usage location" or map screen (Fig. 6.1e). Similarly, on the home screen (Fig. 6.1a), the user has access to two additional icons at the bottom of the screen for modifying "settings" and viewing the "usage location." In between these icons is an icon at the bottom center of the screen that can be swiped upwards to pull out the usage graphs.
- *White Space.* White space allows the designer to separate blocks of related elements, enforcing the desired grouping and structure. Including extra space around a particular button or piece of information also highlights this information. On DataWiz's main screen (Fig. 6.1a), we see white space used around the center circle to draw attention to its information.

Throughout the design process, a designer should continually evaluate the interface prototype and incorporate feedback to improve the design. Even at the earliest iterations, evaluation is important, because it is much easier to correct major mistakes at an early stage. In the case of SDP, experts might review the design to anticipate any problems users would have or users might be asked to explore the interface in a laboratory or a natural setting. In the remainder of this chapter, we consider these methods for evaluating the success of SDP UIs.

6.2.2 Expert Evaluations

While trials with future users would provide the most accurate evaluation of a prototype interface design, it is infeasible to conduct such trials at every design iteration. Thus, we first focus on methods for expert analysis, which have lower overhead and can help identify obvious problems. These fall into two categories: ensuring that designs adhere to accepted cognitive principles and that they are consistent with known empirical results.

An expert or designer can conduct a *cognitive walkthrough* with the aid of scenario descriptions like the ones used in the interface design. First, the expert specifies a user's desired task (e.g., delaying an app store download from 8 PM until midnight to take advantage of lower prices at midnight). Given this task, the evaluator writes down the specific list of actions required to accomplish the task—for instance, clicking on the "Deferral" tab in the app, selecting the "App Store" application, and entering "midnight" as the desired time for the App Store to start downloading. For each task, the evaluator also writes down the system response (e.g., the App Store icon is highlighted after it is selected by the user). The evaluator then asks the following four questions in order to assess whether a user can easily complete the given task.

- *Does the Action Achieve the User's Goal at That Point?* Is it reasonable to assume that, as part of achieving the larger task, a user would set the action's effect as an intermediate goal? For instance, the user might make "selecting the App Store" a subgoal of delaying App Store downloads.
- *Will the User See That the Action Is Available?* With each action, the user will generally need to press a button or provide input. An undesirable design would require intermediate steps such as navigating to a separate screen on the app to select the appropriate button.
- *Will Users Be Able to Identify the Appropriate Action?* Given that the action's effect is a subgoal and that the user can see that the action is available, the designer must ensure that users can recognize how to execute this action, for example, pressing the correct button.
- *Will Users Understand the Feedback from the Action?* After a user takes an action and feedback is given, the user must understand the meaning of this feedback, for example, that highlighting a button means it was successfully selected.

More generally, an evaluator can examine the entire system to assess whether it conforms to general design guidelines, called *heuristics*. While many different heuristics have been suggested, most generally conform to the design principles suggested in Section 6.2.1. Additionally, the designer should create a help screen allowing users to search for documentation and instructions should they require them.

6.2.3 Conducting a Field Trial

A trial with real users is the most effective way to test an interface, as the interface's success ultimately rests on its reception by users. While user responses can be best tested in the field, that is, in a user's natural environment, in some situations a controlled laboratory setting may yield more insight. For instance, it is easier to test specific features of the design when users work with the interface in a controlled setting while the features are varied. Field studies are also usually more expensive, as a researcher may need to travel to participants' homes or workplaces and purchase portable recording equipment. Laboratory studies, on the other hand, require participants to take the time to come to a research laboratory. In the following discussion, we outline different aspects of the trial design for both options.

Participants. It is important to recruit participants representative of actual interface users, and ideally to choose people who will become actual users. In particular, participants' knowledge of computing devices and their experience with similar interfaces should be similar to that of real users. The number of participants recruited depends on the purpose of the trial: if a statistical analysis is used, at least 10 participants are necessary, while 3–5 are generally adequate to identify usability problems [2].

Variables and Hypotheses. In a laboratory setting, researchers should identify the dependent and independent variables. For instance, are researchers comparing the speed of user response with two different designs? These variables are especially important in a laboratory setting, where experiments can be set up to evaluate the dependent variables for all possible combinations of independent variables. The experiments are designed to test hypotheses or predictions of how the independent variables affect dependent ones. For instance, we might hypothesize that one design yields a faster user response than another. The experimental results will then either validate the hypothesis or validate the null hypothesis, that is, the baseline assumption that the dependent variable is not affected by the independent one. Hypothesis testing is especially powerful when combined with a careful experimental design that allows statistical methods to be used for calculating quantitative confidence metrics of the hypothesis's validity.

Experimental Design. Once a hypothesis has been chosen, the experimenter must choose whether to vary the dependent variables between subjects or within subjects. Between-subject variation involves randomly sorting subjects into different groups, each of which will experience a different combination of the independent variables. One of the groups, designated as the "control" group, experiences no manipulation of the experimental variables, which helps to ensure that observed differences in the dependent variables are in fact due to changes in the independent variables. This type of variation eliminates any learning effect, in which a user's experience with one set of variables affects his or her experience with another.

The learning effect can be a serious disadvantage of within-subject or repeated experiments, in which the same group of subjects is exposed to a variety of experimental conditions in sequence. By varying the sequence, the learning effect can be somewhat controlled, though not completely. Within-subject experiments have the advantage of controlling for *subject* variation; they also require fewer total subjects and thus fewer experimental resources. Another possibility is a mixed design, in which one variable is manipulated between groups and another within groups.

Quantitative Analysis. Quantitative methods can be combined with standard statistical tools for hypothesis testing on quantitative variables. SDP's main quantitative aspect is the user's willingness to consume less data during congested hours, which may be tested for different interface designs and pricing plans. However, depending on the type of SDP in question, other quantitative metrics may also be relevant, for example, measuring the average amount of time that a user delays apps given a set of time-dependent prices. Standard statistical tests and comparisons can be used to conduct such tests [3].

Qualitative Analysis. We identify two main methods of soliciting qualitative data on participants' experiences: protocol analysis and interviews or questionnaires. Protocol analysis can involve analyzing audio or video recordings of the user interacting with the interface; however, in the context of SDP, it is perhaps more applicable to think of recording keystrokes within the interface, as pricing interfaces generally do not require active engagement aside from pressing buttons and typing in input. These keystrokes can tell us how often users press the wrong buttons, how often they use different features of the interface, and how rapidly they perform different tasks. One potentially significant disadvantage is the volume of data collected, especially from very engaged users; this can be mitigated by automated analysis that summarizes the raw keystroke logs.

The quantitative data provided by protocol analysis can be supplemented by more subjective data provided directly by experiment participants. Structured postexperiment interviews can provide high level insights into participants' experience and identify any unexpected design flaws. These are generally conducted according to an interview template of topics to cover and specific questions to ask. The interviewer usually begins with a very general question and then focuses on specific points, either following the question script or jumping off of comments in the subject's responses.

A more fixed form of questioning uses prepared questionnaires or surveys. While less flexible, these methods also have less overhead and are thus especially effective with large numbers of subjects or ones in remote locations. Given these limitations, most experimenters conduct pilot studies with the questionnaire to identify possible points of confusion or misunderstanding before releasing it to more respondents. Questionnaires usually need to be distributed to large numbers of people in order to yield a reasonable number of responses.

Questionnaires should begin by explaining the type of information sought and then proceed to ask questions tending to this purpose. Given that survey respondents are often self-selecting, the survey should include general questions on a user's background, to put the other responses in context for the experimenter and ensure a representative sample population. These will then be followed by open-ended, scalar, multiple-choice, or ranking questions. Open-ended questions are the most flexible question format but also require more effort on the part of the participant and are the least likely to yield consistent answers. Thus, it is usually best to employ other types of questions. Scalar or rank questions ask a user either to assign numbers, for example, on a scale from 1 to 10, to describe the accuracy of a specific statement;

or to rank several choices, usually in order of preference. These types of questions give more consistent results;however, it is important to specify what the numbers on a scale mean. Moreover, while too many choices can paralyze users or result in arbitrary decisions, too few can skew responses and miss subtleties in user opinions. If scalar or ranking questions are not appropriate, multiple-choice questions are usually employed.

6.2.4 Choosing an Evaluation Method

In the above discussion, we identified five methods for evaluating an interface: cognitive walkthroughs, heuristic analysis, statistical methods, laboratory studies, and field trials. In the latter two cases, we can identify two types of user feedback: interview- or questionnaire-based and protocol analyses. In Table 6.1, we summarize key features of these five types of user feedback, as described in Reference 1, to aid researchers in choosing the evaluation method most appropriate for their situation.

In practice, experimenters would likely combine several of the techniques in Table 6.1 to gain a comprehensive evaluation. For instance, one could use statistical and protocol methods to yield objective, evidence-based quantifiable information and then use information from cognitive walkthroughs, heuristics, or interviews to ensure that subjective user impressions match this empirical evidence. Interviews and heuristic analyses would focus more on high level information insights, while other methods yield insights into the specific effects of different interface features. In terms of necessary resources, statistical data collection and keystroke logging for protocol analysis require the most resources, while cognitive walkthroughs require the most expertise. None of these methods is very intrusive or obvious to users during their interaction with the interface, introducing only minimal bias into the trial results.

In the next three sections, we explore HCI research works that have utilized the above methodologies to investigate SDP and related topics in various contexts. We first discuss lessons on interface design from HCI works in energy markets and then give an overview of HCI studies on Internet usage and data pricing.

6.3 HCI LESSONS FROM THE ENERGY MARKET

Like broadband networks, electricity grids suffer from periods of peak demand. To relieve this peak demand, schemes such as time-dependent pricing (TDP) and usage visualization have been practiced in the electricity market, and hence, lessons

TABLE 6.1 Features of Different Evaluation Methods

	Cognitive	Heuristic walkthrough (w.t.)	Statistical	Interview	Protocol
Objectivity	No	No	Yes	No	Yes
Information	Low level	High level	Low/high level	High level	Low level
Equipment	Low	Low	Medium	Low	Medium
Expertise	High	Medium	Medium	Low	High

from it can be useful in the context of data networks. HCI works that have studied users' energy consumption have ranged from power strips that change color to show the energy used by individual electrical sockets [4] to a large-scale media art installation visualizing energy consumption in an office building [5]. These studies on user-friendly interface design bring out a key trade-off in HCI design, which is whether to visualize the information on energy usage pictorially or numerically.

The energy consumption monitoring application *WattBot* used color coding as a qualitatively simple way to indicate the usage amounts to the users [6]. However, users have also been found to quantitatively track changes in their energy usage behavior by viewing their usage history [7] and to care about the convenience of monitoring their usage. For example, researchers testing a desktop widget that showed computer energy efficiency found that users appreciated the inconspicuous, easy-access nature of the widget [8]. These lessons on using color coding for qualitative feedback and numerical usage history for quantitative feedback can be transferred to the design of client-side GUIs in broadband pricing applications, as discussed in Section 6.8.3.

6.4 USER PSYCHOLOGY IN HOME NETWORKS

In Sections 6.4.1 and 6.4.2, we discuss related research efforts from HCI researchers on understanding the interaction between users and networks. We first explore works that have focused on studying user psychology and its role in interface designs for developing home network management and control tools and then investigate how users respond to nonmonetary penalty mechanisms (e.g., throttling and capping) imposed by ISPs on their home bandwidth usage. In Section 6.5, we cover lessons learned about user behavior and psychology from HCI research that involved incentives and pricing signals from the network to the end users to modify their usage behavior.

6.4.1 Network Management and QoS Control

Today, bandwidth is becoming a new type of resource that has to be shared and managed by multiple agents, often operating in a distributed framework. This effect can be observed in the case of shared data plans offered by wireless network operators in which a data cap is shared across different members and devices. Similarly, in the case of wired networks, multiple household members and Internet-enabled devices need to share the same broadband access link. This sharing burdens users with the task of "digital housekeeping" in which they have to set up, maintain, and troubleshoot their connectivity as well as understand, monitor, and share limited bandwidth speed or a data cap. HCI and networking researchers have been grappling with this issue [9, 10] and have created a few useful tools that can help household users understand, diagnose, and manage their home bandwidth usage. One such domestic tool for bandwidth management is the ***Home Watcher***.

The Home Watcher project [11] provides an appliance that shows who is using the bandwidth as well as how much each user is using. It also allows users to limit each other's bandwidth (and therefore, their Internet-related activities) from a publically situated open access display. This allowed the researchers to study the social consequences of revealing real-time resource usage and contention in the household.

The Home Watcher field study with 24 occupants in six households revealed several interesting insights. Making bandwidth usage activities visible reinforced existing beliefs about other household occupants' computing activities and increased people's awareness of their routines and activities, for example, alerting people about who else is online at given times of the day. Thus, Home Watcher raised some questions of user privacy: information about household members' bandwidth consumption patterns can affect daily routines and relationships. As Home Watcher also allows members to control each other's bandwidth, it revealed the politics of visibility and control. Participants desired more control to limit usage based on the time of day and importance of each user's activities on shared machines. They also wanted mechanisms to let others know when and why they were being throttled. However, there are concerns that existing power hierarchies in the home (e.g., between parents and children) may be threatened by the option to control others' Internet usage. These consequences and desires for finer granular control and communication between users are thus relevant to developing usage control applications in wired and wireless bandwidth and data cap sharing environments.

Other HCI researchers have focused on creating interactive visual tools for home network access as well as bandwidth controls. ***Eden*** [12] is an "interactive, direct manipulation home network management system" that aims to make it easier for household users to visualize the physical location of different devices in their home and perform membership management, access authentication, network monitoring, and QoS policy setting for bandwidth prioritization. It uses a customized Linux-based wireless router/access point and provides users with personalized visual representations of network devices and settings that can be managed with simple drag-and-drop moves. The Eden system also provides users with an interface element called a *badge*, which associates particular network control and prioritization properties with individual or groups of devices (e.g., a sites restriction badge, application restriction badge, faster badge, and slower badge). A trial of Eden with 20 participants showed that most users were able to understand the mapping of spatial home boundaries to the logical boundary of their home network and were able to use the badges effectively for parental control and other bandwidth prioritization activities.

While the previous systems worked within the limits of the underlying physical infrastructure by treating it as a preordained fixed set of facilities, the ***Homework*** project [13] took a new approach to provide users with access to parts of the home network infrastructure that have traditionally remained closed or hidden. One of the motivating reasons for this was that traditional Internet protocols and architecture were designed for configuration by experienced system administrators. In a home network setting, however, bandwidth access is locally negotiated between household members and requires immediate resolution of problems, so these users sometimes need direct control over their home network. Homework adopts a gateway model to change the home network infrastructure by developing a customized dedicated router that can capture traffic information and make it available to applications. It provides users with information and control over which machines can receive an Internet Protocol (IP) address to join the home network by implementing a Dynamic Host Configuration Protocol (DHCP) sever within a customized router. It also implements a dynamic name service (DNS) proxy into the router to give fine-granular control over which devices can access which specific sites. The DNS proxy intercepts domain name resolution requests for sites and drops requests if the requesting device is not allowed to access that domain, thus enabling parental controls. The Homework system was tested with 12 households in the United Kingdom and confirmed many of the observations in other works, including increases in social discord and privacy concerns. Many users were found to like having access to such control over their network infrastructure, but they viewed these features as mechanisms to issue threats rather than substitutes for negotiations within the family about network usage. Nevertheless, these results show the potential for reshaping the home network by allowing a range of interactive possibilities to manage and control the infrastructure.

Such systems for home network management and control provide an understanding of the features that users want, how such information can be presented to them using effective UIs and how user actions can be mapped onto the underlying physical infrastructure. These studies can inform our design choices for user-friendly applications for wired and wireless network pricing and user-mediated bandwidth prioritization and control.

6.4.2 Implications of Throttling

Many US operators have been throttling user bandwidth in an effort to control network congestion [14]. Only 30% of online Americans are reported to receive the "advertised" speed, even though ISPs charge them higher fees for higher speeds [15]. This issue of throttling, along with the net-neutrality issues involved in creating a tiered Internet service, has led HCI researchers to develop a visual network probe to educate and empower consumers about their data speeds. ***Kermit*** [16] is a probe that collects data from users' routers and calculates bandwidth counts and the online status of each machine. Kermit provides users with a visual interface to estimate their broadband connection speed, identify network bottlenecks, and control and prioritize their usage by displaying real-time and historical information about bandwidth usage for all household devices on the home network.

The field trial of Kermit with 10 households provided many design guidelines for future home broadband tools. It showed that users benefit from visualizations that allow for a less cluttered and personalizable display as more devices are added to the home network, as well as from more context about the network upload, download, bandwidth, and speed they are receiving. Users also need to be able to perceive visual differences to understand how their actions of limiting or prioritizing devices are affecting the network. They also like the options of controlling, scheduling, and limiting actions, as well as tools to remind them to reverse such actions. Thus, designers should incorporate these functionalities into the interface over a system that securely stores and handles the large amounts of traffic measurement data collected. These tools can not only educate and empower consumers but can also help them become broadband speed "watchdogs" to effectively participate in net-neutrality debates.

6.4.3 Response to Capping

Besides throttling and poor performance due to network congestion, users are also facing bandwidth caps imposed by ISPs. How users use and interact with the Internet in a metered bandwidth environment is of interest to web designers, application and service developers, and content providers in customizing their user experience. Many countries such as Africa, India, and Australia have had low bandwidth caps in digital subscriber line (DSL) for years, while most US operators introduced data caps after 2010. In Africa, Chetty et al. [17] conducted studies of users who have only known capped plans for Internet access to find out how data caps change usage patterns and what strategies users use to cope with the limitations. They found that caps affect how, when, and for how long users access the Internet. For example, in Kenya, Internet users deliberately plan their online activities before going online at Internet cafes [18]. In South Africa, some users avoid high bandwidth-consuming sites at the beginning of the month so as to stretch their caps through the monthly billing cycle. Caps, therefore, effectively force people to put a price on their Internet access, as in usage-based pricing. Additionally, to help navigate their bandwidth caps, users require new visual tools that reveal which applications consume their bandwidth the most and the cost incurred from visiting different sites.

Further studies on the effects of data caps by Chetty et al. [19] revealed that home users face difficulties with three types of uncertainties about their bandwidth usage, namely, invisible balances (inability to track usage and leftover bandwidth), mysterious processes (unidentifiable users and background processes), and multiple users. A qualitative study of 12 households living with data caps in South Africa showed that caps limit browsing activities and users preferred limitless consumption to faster but capped connections. Users generally have inadequate tools and warnings to prevent them from going overboard or to identify freeloaders on their wireless networks. Users also liked parental controls to limit children's access to gaming and multimedia-rich content and were even found to avoid downloading software updates, thus increasing security risks to the network. Moreover, because of the caps many households were found to create informal data sharing networks; families without caps or with workplace connections would download videos, movies, and music to share with their capped friends and relatives through CDs, DVDs, and so on. Family members would sometimes even visit friends and relatives to get online once when their Internet connection was capped. Another interesting behavior was that while users self-censor their usage early in the month, they go on "binge downloads" at the end of the billing cycle to use up most of the data cap because of the "use-it-or-lose-it" nature of today's capped plans, thus potentially increasing network congestion. These behaviors suggest that many usage and pricing assumptions relevant in unlimited bandwidth scenarios may not hold in capped environments, which in turn will affect the design of applications and devices operating in a usage-based, tiered, and capped plans.

Researchers have also developed another system, ***BISMark*** (Broadband Internet Service BenchMARK) [20], with similar features that helps users not only understand but also actively control how different devices in a home can share the available bandwidth cap. BISMark is a router firmware running on network gateways that gathers usage statistics and reports them to a controller to display usage information. BISMark provides users with visibility into which users and applications are consuming more bandwidth and policies to control how the cap is allocated across different users, devices, application, time of day, and so on. BISMark implements and enforces the user-specified policies via a secure OpenFlow control channel to each gateway device. It also allows users within a single household to trade caps with each other to increase overall resource utilization efficiency.

6.5 USER PSYCHOLOGY IN BANDWIDTH PRICING

Section 6.4 focused on HCI research studying user behavior under unlimited, capped, or bandwidth-throttled data plans without explicitly invoking pricing as a tool to modify user behavior. In this section, we discuss works that have studied user psychology and behavior when the network provides explicit pricing signals to users.

6.5.1 Effects of Variable Pricing

One of the early works on user reaction to pricing was conducted by the Berkeley Internet Demand Experiment (*INDEX*) [21, 22] in which users were exposed to different usage-based pricing models, including flat-rate and pay-per-byte pricings. INDEX is a prototype to offer differentiated-quality service on demand, with prices that reflect resource cost. The INDEX Control Center is an interface running on a desktop through which customers specify their choices and control access to the Internet. The interface also shows users a "spending meter" to reveal the cost incurred till the present time of an active session. This meter can be toggled to show the cost incurred till that time of the day or the entire month. The spending meter is updated every minute to keep users aware of their spending behavior in real time. INDEX also allows customers to instantaneously shift between different service qualities, even during an ongoing session, by clicking on the appropriate service quality. The INDEX project experiments exposed the participants to various forms of pricing plans, including volume pricing, volume plus capacity charge, self-selecting tariff, and so on. The experimental data showed that users are highly price sensitive. Moreover, in asymmetric bandwidth settings in which users can separately choose and pay for bandwidth in the upstream and downstream directions, it was found that most users are aware of this asymmetry, and they do reduce their bill. The authors also compared user behavior when their costs are determined by megabytes of data transferred versus minutes of connected time and found that the connected time goes up dramatically under volume pricing compared with connect time charges. The most important findings from the INDEX project are [22] that (i) the demand is very sensitive to price and to quality; (ii) differences in demand among users are persistent and large; and (iii) the commodity form of the service (e.g., transport of traffic by volume or time) has a big impact on demand.

6.5.2 Effects of Speed-Tier Pricing

Recent research has focused on creating experimental testbeds for mobile data pricing. *SpeedGate* [23] from AT&T is one such testbed allowing experimentation with different dynamic pricing schemes without accessing or modifying the core network elements (e.g., Policy charging and rules function (PCRF) and Policy and charging enforcement function (PCEF)). SpeedGate allows mobile users from any carrier with any type of smart phone to participate in pricing trials (i.e., carrier independence) and maintains persistent connections to smartphones over a virtual private network (VPN) even as users roam between different wireless networks (3G, 4G/LTE, WiFi). Most importantly, it allows dynamic speed-tier assignments: the testbed can redirect smartphone users to different ports on proxy servers in real time and adjust their maximum available bandwidth per user session (i.e., QoS level) dynamically based on the data pricing strategies used.

A trial of Speedgate to understand user's willingness to pay (WTP) for various speed tiers with a total of 29 users revealed many interesting results. It showed that users' WTP depends on the amount of bandwidth-intensive applications they used, their access to WiFi networks, and the speed of the 3G/4G connections, which depends on the coverage in the user's location. In particular, some users showed less willingness to pay for higher speed tiers, mainly because there was no real guarantee that they would indeed receive the maximum speed, and hence, their WTP depended on their experience because of varying network coverage. The results also suggested that most participants showed a limited dynamic range of WTP for wireless services regardless of the speed tiers offered. A second laboratory study with 12 participants for measuring WTP through rating of applications (stock, Pandora, YouTube) showed that WTP for higher speed tiers was higher for more bandwidth-intensive applications.

The researchers also conducted a separate trial [24] at the Purdue University campus by developing a smartphone application to study users' responses to incentives and disincentives and to observe the impact of their psychological characteristics. Their results showed that pricing schemes with probabilistic payments of higher incentive amounts have more positive results compared to schemes with deterministic payments of lower incentive amounts, even when the total payout is statistically similar in both schemes. This result is rather interesting given that customers in wired networks have been known to be risk averse, with a preference for flat-rate fees over the uncertainties involved with variable pricing. This new observation may, therefore, indicate that consumers in resource-constrained mobile networks are more risk seeking and want to save more on their monthly bills and, hence, potentially are more responsive to dynamic pricing plans. Furthermore, their study suggested a positive relationship between compliance with incentives and a user's psychological agreeableness and a negative relationship between compliance and a user's neuroticism (i.e., the extent of negative emotions experienced).

6.5.3 Effects of Dynamic Time-Dependent Pricing

The most basic form of TDP in practice is a two-period plan that charges different rates during the daytime and night time. For example, BSNL in India offers unlimited night time (2–8 AM) downloads on monthly data plans of Rs 500 ($10) and above. Another variation of TDP is one in which users can choose the time of day when they want higher network performance. For example, in 2010, the European operator Orange introduced a "Dolphin Plan" for £15 ($23.50 USD) per month that allows unlimited web access during a "happy hour" corresponding to users' morning commute (8–9 AM), lunch break (12 noon to 1 PM), late afternoon break (4–5 PM), or late night (10–11 PM) [25, 26]. This plan lets customers self-sort into different times of the day, temporally distributing the peak demand.

More dynamic forms of TDP have been offered for voice calls in India and Africa. In Africa, MTN Uganda's dynamic tariffing plans have shown that fiscally conscious users do respond to dynamic pricing plans that vary the incentives depending on the network conditions. Their customers can check the price discounts available on their handsets and decide on their usage accordingly. At 4 AM, when the network would otherwise be little used, usage was found to be as high as 99% if discounts were offered [27]. These dynamic discounts were even found to modify the temporal usage patterns and peak demand periods. For example, in addition to the normal peak hour at 8 AM, a new peak hour at 1 AM was observed as more people took advantage of cheaper prices for calls after midnight. Customers in developing countries were thus very price sensitive and were willing to stay up late in the night to save money, hinting at similar potential for TDP for mobile data. Another important effect observed with time-dependent price discounts was that incentives can in fact increase the overall demand because of a "sales-day effect" psychology. For example, when Vodacom introduced a similar TDP scheme in Tanzania, overall call volumes increased by 20–30% in areas where dynamic tariffing was practiced.

Similar levels of effectiveness of TDP for voice calls have been reported by networking researchers from TDP field trials conducted in the UC Berkeley student dormitories using a computer-telephony service [28]. The experimental subjects could use the service to make and receive phone calls from their computers or phones. While the trial did not involve real money, users' budget constraints were simulated through allocating a limited number of tokens per user per week. This experimental setup was then used to conduct a different pricing experiment each week to understand how prices can be used to change user behavior (e.g., entice them to talk less, talk at another time, or use a lower quality connection). The authors found that with their token scheme as a budget constraint, static pricing policies can be used to influence users' behaviors, but a simple congestion pricing scheme fails to encourage users to talk less. For example, using TDP, they encouraged users to shift 30% of their calls from the peak to the off-peak hours. Similarly, by using call-duration pricing (a variant of usage-based pricing) and charging a higher rate as a call lasts longer, the authors could encourage three times as many calls (18% instead of 6%) to terminate after a price increase. However, when using a simple congestion pricing scheme that charges a rate depending on the number of people calling (i.e., a real-time dynamic congestion pricing), they found that this pricing plan did not get users to terminate their calls earlier. They conjecture that users did not adjust their behavior because they did not know how long the price increases or decreases would last.

These observations hint that to make a congestion pricing scheme effective, the price changes need to be of relatively longer duration to incentivize users to change their behavior. The results from these experiments hint at two key lessons: (i) networking and HCI researchers do need to jointly identify effective ways to communicate the pricing signals back to the users and (ii) for dynamic TDP to change user behavior, users need to be made aware of the prices (and how long they will last) in advance, through plans such as DDTDP.

Our own experience from field trials of the *TUBE* system [29], which offered dynamic day-ahead TDP in the case of mobile data traffic, confirmed similar user behaviors, as discussed in Section 6.8. Our experiments also found a 30% reduction in the maximum observed peak-to-average usage ratio and a 130% increase in usage, particularly in the discounted, off-peak hours. These results are promising in that they indicate the feasibility of introducing such plans if they are designed carefully to facilitate users' understanding of and reactions to pricing signals from the network.

6.6 DAY-AHEAD DYNAMIC TDP

As alluded to above, DDTDP has certain features that make it an attractive pricing option to explore.

1. *Price Guarantees.* Users' general preference for flat-rate pricing is primarily driven by their preference for certainty in their monthly usage bills. Dynamic real-time pricing is, therefore, not very conducive to users as it is harder for them to react to changing prices at such a fine timescale [28]. Instead, DDTDP provides prices to users a day in advance, giving them an opportunity to plan their usage ahead of time.[3]
2. *Advance Notice.* In Reference 28, the authors noted that unless users are informed of the prices in advance, they usually do not change their usage behavior as they do not know for how long the offered prices are going to last. In DDTDP, the offered prices are for specified durations (e.g., hourly rates), and hence, users can determine how long they need to wait for discounted rates and decide accordingly whether to stop using data for noncritical applications.
3. *Granular Price Offerings.* In contrast to simple two-period pricing, DDTDP provides operators with the opportunity to vary prices at a finer time granularity (e.g.,1-h or 30-mi time periods). Many mobile applications that may not wait for half a day can wait for an hour or less (sometimes even without

[3]As the prices are offered in advance, an important component of offering DDTDP is determining the optimal prices to offer while accounting for the estimated time elasticity of demands for various price points from historical data. An analytical model and algorithms to enable such computation of optimized prices has been described in our previous work [29].

the user being involved, as in the case of downloads, backup, caching, M2M applications, etc.). The operator can take advantage of the inherent time elasticity of demand of such applications to set the time-varying incentives.

4. *Adaptive Pricing.*The operator can adapt the prices offered for each day based on the observed user behavior given the incentives offered previously [29]. This daily adaptation of prices introduces a "dynamic" aspect into the pricing plan. It also helps the operator spread out the demand temporally without shifting the peak demand from one time of the day to another, as has been observed to happen under simpler two-period pricing (e.g., all subscribers waiting to make longer calls at 1 AM [27]).
5. *User Feedback Mechanism.* DDTDP will create a control-feedback loop between the network operator and its clients, with the users having GUIs that will help them better understand and react to the prices on offer. It will thus enable end users to easily and conveniently respond to these time-varying pricing signals. DDTDP divides the congestion management functionalities between the network core and the end user devices, thus making these smart devices a part of the network management solution suite.

Despite these advantages of DDTDP, introducing any new pricing scheme must account for its possible impact on the entire Internet ecosystem. In Section 6.7, we, therefore, discuss how various stakeholders of this ecosystem that we interacted with through SDP forums and workshops (www.smartdatapricing.org)—namely ISPs, vendors, content providers, software developers, and regulators—view the potential of DDTDP.

6.7 PERSPECTIVES OF INTERNET ECOSYSTEM STAKEHOLDERS

Changes in pricing policy can affect multiple stakeholders in the Internet ecosystem, including network operators, consumers, content providers, software developers, and regulators. In this section, we provide the perspectives of these different stakeholders to understand the drivers of current changes in broadband pricing and the benefits that new plans such as DDTDP can bring.

6.7.1 Operator Perspectives

By 2016, ISPs are expected to carry 18.1 PB per month in managed IP traffic.[4] While this growth promises more revenue for ISPs, many ISPs are concerned about handling such traffic volumes on their network, as seen by Comcast's initiative to cap their wired network users at 300 GB per month [14]. Rural local exchange carriers (RLECs) suffer from an especially acute problem: although the cost of providing middle-mile bandwidth to their users has declined over the years, because

[4]Cisco's definition of "managed IP" includes traffic from both corporate IP wide area networks and IP transport of television and video-on-demand.

of an increase in the DSL demand needed to fill the middle mile, the bandwidth requirements of home users have increased quite sharply [30]. Indeed, rural customers' average home broadband speed is still less than the Federal Communications Commission's (FCC) 4 Mbps target, but the digital expansion required to meet this target is hampered by the high cost of middle-mile upgrades in rural areas [30]. New access pricing mechanisms that bring down middle-mile investment costs by incentivizing users to reduce RLEC peak demand and overprovisioning needs can thus be helpful in bridging the digital divide.

ISPs' current penalty mechanisms (e.g., overage fees, throttling, and termination) to reduce peak traffic, however, can be harmful to the entire Internet ecosystem and ineffective in bringing down peaks. For example, usage-based pricing does not incentivize users to avoid congested, peak-usage times, and thus does not reduce peak usage relative to average usage over the day [31]. As Clark pointed out, *"the fundamental problem with simple usage fees is that they impose usage costs on users regardless of whether the network is congested or not."* [32] Even two-period, for example, daytime and night time, pricing cannot always reduce peak usage, as many applications can be deferred for a few hours but are unlikely to wait for half a day to get discounted rates.

Dynamic day-ahead TDP does not suffer from many of the problems posed by these other pricing plans. Many US ISPs at the National Exchange Carrier Association's Expos in 2011 and 2012 thought that DDTDP could benefit both ISPs and customers, informing us that *"Clearly, customers do not want any form of usage measurement or control under traditional definitions. I'd be very interested in seeing a pitch for time-dependent pricing"* (Delhi ISP, email communication, September 2011). An Indian telecom executive was more specific, telling us that TDP can benefit ISPs by reducing peak traffic as well as customers by letting them save money during lower price times. Given these advantages to ISPs, it is therefore necessary to examine customer viewpoints on TDP, as user responses to the prices offered will ultimately determine TDP's efficacy.

6.7.2 Consumer Viewpoints

Consumers today face increasing costs of Internet subscriptions as well as harsh penalties from network operators in overage fees. For example, Verizon's move in 2012 to only offer shared data plans for all new consumers increased most customers' monthly bills [33]. To avoid overage fees from going over their monthly data caps, consumers are increasingly relying on usage-tracking and data compression apps (e.g., Onavo, 3G Watchdog Pro, and DataWiz). These trends are also observed outside the United States; in South Africa, for instance, consumers use ISP-provided usage-tracking tools [19] to stay within their data caps and often plan their Internet activities in advance to avoid wasting time online.

Empowering users to monitor and control their spending on Internet usage is emerging as a new area of research that needs to consider economic incentives and HCI aspects in a holistic manner [34]. For instance, under dynamic TDP users' interactions with the prices offered can be made more convenient by automatically scheduling applications such as large downloads and cloud synchronization to discounted periods even without user intervention.

To better understand user acceptability of TDP plans, we conducted pretrial surveys in India and the United States. The US surveys were conducted online and by the authors in person in Philadelphia and on college campuses, while the survey in India was conducted by a professional marketing firm in five major cities (Kolkata, Mumbai, Delhi, Madras, and Bangalore), as seen in Figure 6.2. We received 155 responses to the US survey and 546 responses to the survey in India.

In both surveys, we asked respondents for the maximum length of time for which they would delay different types of apps in exchange for a given discount on their data plan. More than 50% told us that they would wait 10 minutes to stream YouTube videos and that they would wait 3–5 hours for file downloads. In India, we also found that individuals without mobile phones hesitated to use mobile data because of the high cost of data plans but did realize the usefulness of having a mobile data plan. Residents in rural areas of the United States express similar sentiments [35], suggesting that lower cost time-dependent broadband access fees would boost user Internet adoption and help to bridge the United States' digital divide.

6.7.3 Content Provider Considerations

Content providers are becoming increasingly sensitive to users' desire to avoid excessive data usage and avoid paying overage fees for their Internet usage. These concerns have generally manifested as extra options for users to downgrade their quality of experience in exchange for using less data. For instance, Netflix offers such an option for users to stream lower quality videos [36], as well as another option for iPhone

Figure 6.2 Consumer survey in India.

users to use only WiFi. ISPs' measures to penalize demand are thus driving changes to not just their users' behavior but also the business plans of content providers.

6.7.4 Application Developer Concerns

Incentive-based pricing plans, such as TDP, require a feedback-control loop between the network backend and client-side devices, with the latter requiring new mobile applications that will support such functionalities. In developing these apps, however, developers must contend with the different mobile platforms prevalent today (i.e., iOS, Android, and Windows Mobile). The openness of these different platforms can make developing such applications difficult; for example, until the release of iOS 7 in 2013, such pricing apps could not run in the background and could not access the usage of specific apps [37, 38]. The Android and Windows platforms, however, have generally been more open.

New TDP pricing plans will create an opportunity for developers of nonpricing apps to optimize their apps according to changing pricing conditions. For instance, many apps can preload content during lower price times, for example, magazine apps preloading articles and pictures. Such preloading can not only save users money but can also reduce the delay in displaying content when users open an app. Moreover, the temporal flattening of demand eases congestion, allowing users to achieve higher throughput and better performance. Video applications, which tend to have the highest usage volumes, can especially benefit from such usage shifting. Implementing such measures, however, would require open APIs that allow apps to run in the background and access network prices in real time. Apps and systems that accommodate such features can thus help developers provide a better quality of experience and savings for their users.

6.7.5 Policy Evolution

Changes in pricing policies often lead to politically charged debates on network neutrality in the United States, but academics have cautioned that this ongoing debate overlooks the need for service providers to have flexibility in exploring different pricing regimes [39]: *"Restricting network providers' ability to experiment with different protocols may also reduce innovation by foreclosing applications and content that depend on a different network architecture and by dampening the price signals needed to stimulate investment in new applications and content."*

However, more recently, US regulators have accepted *"the importance of business innovation to promote network investment and efficient use of networks, including measures to match price to cost"* [40], particularly for ISPs operating in the wireless sector. In "New Rules for an Open Internet," the former Federal Communications Commission Chairman J. Genachowski explicitly specified that [41]: *"The rules also recognize that broadband providers need meaningful flexibility to manage their networks to deal with congestion... And we recognize the importance and value of business-model experimentation."*

6.8 LESSONS FROM DAY-AHEAD DYNAMIC TDP FIELD TRIALS

6.8.1 Trial Objectives

The *TUBE* (Time-dependent Usage-based Broadband price Engineering) system is an experimental testbed used to study user responses to day-ahead dynamic time-dependent usage-based pricing of bandwidth in mobile networks. For this trial, a prototype was developed that can compute optimized prices in the ISP backend (as described in References 29 and 42). These prices were displayed to the users through client-side UIs, which also let them monitor, control, and modify their behavior in response to the offered prices. The key steps in carrying out the trial were the following.

- Create a "virtual" ISP (i.e., bandwidth reseller) for the purpose of the trial by operating as a middleman between a real mobile operator and its participating clients, who are offered time-varying prices.
- Develop a communication infrastructure with server-side (ISP) implementation for computing and sending price information to a client-side app running on client (end user) mobile devices.
- Conduct pretrial surveys and focus group studies to incorporate feedback in the design of client-side GUIs.
- Validate results based on qualitative posttrial debriefings and quantitative results collected during the trial.

6.8.2 Trial Structure

6.8.2.1 Pretrial Focus Groups Focus group studies conducted in the design phase helped us to incorporate feedback from users to create intuitive and simple to use UIs. During these five-person sessions conducted at the Princeton University campus, we identified several functionality goals for the TDP app and solicited opinions on our app designs and the form factor of the devices used. We also received feedback on the designs from leading industry experts at AT&T, MTA, NECA, Bell Labs, and Reliance Communications. This iterative process of seeking user feedback and modifying the interfaces accordingly continued for a 4-month period to create a prototype of the mobile application for the client-side devices. Many of the functionalities that we incorporated had been found to be desirable by users in the context of energy markets and home network management solutions (e.g., usage monitoring and history, color coding of information, parental control) as described in Section 6.4.

These pretrial studies suggested that participants liked the idea of day-ahead dynamic time-varying prices because it empowers them to choose not only *how much* but also *when* to use their device so as to save on their monthly bills. The focus group studies provided insights about the following features of DDTDP.

- *Time Granularity.* Participants typically preferred TDP prices that varied at an hourly granularity rather than on the order of minutes (e.g., real time). Real-time price changes are harder for users to track; moreover, uncertainties in the duration of the offered prices make it harder for users to adapt to price changes [28].

- *Time Horizon.* The idea of "day-ahead" prices, as practiced in some electricity markets, was found to be appealing as it allows for advance planning of usage. The feasibility of this behavior is supported by reports that price-conscious Internet cafe users in Kenya plan their online activities before going online [18].
- *Information Visualization.* Our participants found it useful to view superimposed usage and price history so as to be able to determine when they used how much of their data cap.
- *Color Coding.* The participants found viewing the TDP price discounts in a color-coded form (e.g., red (<10%), orange (10–19%), yellow (20–29%), and green (>30%)) to be an intuitive way to communicate the prices. This idea for color-coding TDP prices or discounts was inspired by the traffic light signaling system and previous works in electricity sockets [4, 6]. The price discounts were also made available as numerical values to accommodate users who wished to see more details and as a secondary signal for color-blind users.
- *Application Control.* The idea to have a parental control feature to block certain apps at certain times of the day when prices are high was well received. Moreover, participants also wanted information on the cost of each application's data usage, or alternatively, a display showing the top bandwidth- (budget) consuming applications on their device.

This feedback from pretrial focus group studies was incorporated in the design of the mobile TDP application.

6.8.2.2 Participants, Billing, and Timeline To conduct this day-ahead TDP trial, we recruited 10 trial primary participants (the "primary participant" paid the monthly bills) in the Princeton, New Jersey, area through word of mouth and email lists. The primary trial participants were each given a jailbroken iPad2 with a 2-GB data plan from AT&T and our client-side app installed.[5] We set the devices' WiFi connectivity to "off" by default, because the goal of the trial was to determine if the 3G traffic can be made to wait for discounted periods when economic incentives are offered. However, to avoid skewing participants' usage patterns, we also asked participants to use the devices as they normally would, with the result that many of the primary participants shared the device with their family members.

The iPads were the only tablets owned by the participating families, and they were actively engaged with the iPads throughout the trial. Our trial setup did not impede the devices' mobility in any way (Section 6.8.2.3 discusses the setup implemented with AT&T to realize this). The demographic information of the primary participants and any secondary users who shared the device is given in Table 6.2.

During the trial period, we paid all participants' monthly AT&T bills for the iPad's 2-GB data plan and overage fees. Participants paid us according to a DDTDP plan

[5]The devices were returned after the end of the trial, with an option of purchasing it at a discounted rate.

TABLE 6.2 Demographics of the DDTDP Trial Participants

	Age & Gender	Occupation	Secondary Participants
P1	Female, 40	Personal assistant	None
P2	Male, 33	Graduate student	None
P3	Female, 34	Office staff	Female, 3
P4	Female, 50	Accounts manager	Female, 26 (waitress); male, 23 (waiter) male, 18 (student); female, 5
P5	Female, 21	Student	None
P6	Female, 48	Admin. assistant	Female, 22 (restaurant cook)
P7	Female, 35	Office manager	Male, 41 (software engineer)
P8	Male, 31	Graduate student	None
P9	Female, 30	HR manager	Male, 32 (theater technician)
P10	Female, 43	Office support	Male, 43 (systems technician); male, 14; female, 8

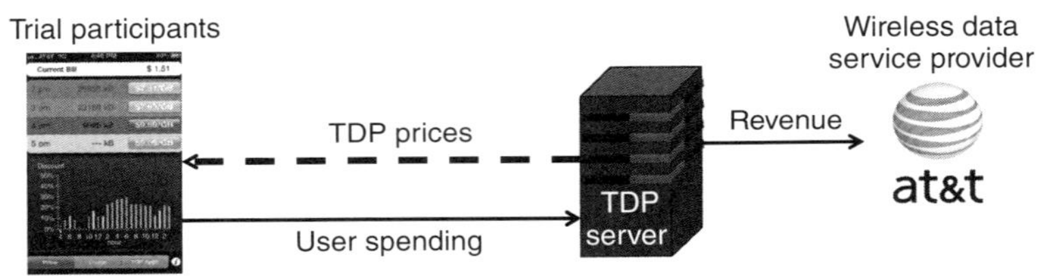

Figure 6.3 Money flow diagram for the DDTDP trial.

with time-varying price discounts on a baseline usage-based fee of $10/GB. Thus, we effectively became a resale ISP of AT&T's connectivity for these participants, as illustrated in Figure 6.3. To avoid creating multiple sources of incentives, we did not give any additional monetary incentive for participating in the trial.

We performed extensive in-laboratory testing of the TDP system between April and July 2011. The trial was conducted from July 2011 to April 2012, during which we visited participants three to four times. In the first phase of the trial, from July to September 2011, we handed out the iPads without the TDP app[6] installed to let our participants familiarize themselves with the device and to monitor their pre-DDTDP usage. In the second phase of the trial, from October to November 2011, we installed the DDTDP client app on participants' iPads and provided basic operational instructions. We also began offering DDTDP price discounts. From December 2011 to January 2012, we tested participants' changes in usage in response to different types of price displays. From February to March 2012, we continued to offer DDTDP, visiting the participants again in April for a posttrial debriefing to obtain feedback on their experience during the trial.

[6]We use the terms "TDP app" and "DDTDP app" interchangeably as DDTDP was the only form of TDP tested in our trial.

Three researchers coded the audio data and transcriptions of all participant interviews. The excerpts presented here are representative of the participants' interview responses and represent mutually agreed upon themes that emerged from their answers. Where applicable, we support our qualitative findings with observed data on participants' quantitative usage patterns and pricing history.

6.8.2.3 Implementation and Setup Becoming a resale ISP requires architecting and prototyping a system with two main interfaces: one with the real network operator (here, AT&T) and one with trial participants' devices. The first interface consisted of an APN (Access Point Name) between AT&T's mobile network core and our laboratory server facilities, which redirected trial participants' uplink and downlink data traffic to our DNS-enabled and network address translation (NAT)-enabled servers before rerouting it back to the Internet. Thus, participants' *data traffic*, but not voice traffic, flowed through our laboratory severs, as illustrated in Figure 6.4.

This APN setup allowed our server to monitor participants' aggregate usage at different times and adjust the future offered prices accordingly. We assigned a unique IP address to each device and created a Netfilter rule to measure its usage and store it in a MySQL database. In each hour, a price point for the next twenty-fourth hour was calculated using a Matlab-based prediction algorithm that estimated future demand as a function of the prices offered. All prices were announced 24 hours in advance (i.e., DDTDP). To interface with the apps on users' devices, the client-side apps were configured to pull the new prices each hour, so that at any time, a participant could launch our TDP app to see the prices for the next 24 hours and plan their usage accordingly. The server supports JavaScript Object Notation (JSON)/Hypertext Transfer Protocol (HTTP) and can exchange information with any device with browsing capability. All server modules were implemented on a Linux platform with an Intel Xeon 2.0 GHz CPU and 8 GB of RAM.

The client-side (iPad) system module is the mobile application (DDTDP app) with which trial participants interact. In order to bypass iOS platform restrictions that constrained many of the possible TDP functionalities, we jailbroke the iPads to gain root

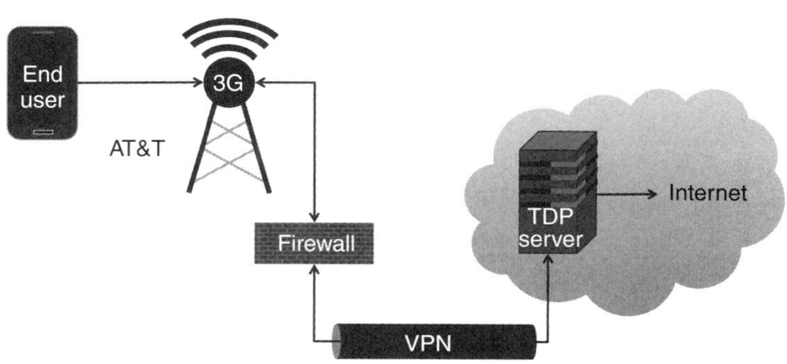

Figure 6.4 Schematic of the data flow in the DDTDP trial setup.

access to the operating system. The jailbreaking allowed us to implement such features as monitoring per-application usage volumes, blocking apps on users' request, and displaying a price indicator next to the battery indicator on the iPad home screen. We tracked each application's usage by hooking internal functions and running a daemon process to send requests and display the DDTDP prices, as well as to block apps if needed. The app increased the iPads' battery consumption by no more than 4% compared to typical usage without the DDTDP app.

6.8.3 Application User Interface

We next discuss the UI features of the DDTDP app, grouped by the following functionality goals.

1. *Data Monitoring.* Finding intuitive ways to communicate price and usage information to users for educating them about their data consumption behavior.
2. *Consumer Empowerment.* Identifying information that will empower consumers to have better control over their usage.
3. *Automation.* Understanding if and when consumers would like to automate their usage and scheduling decisions using an "autopilot" mode of operation.

We first focus on the monitoring features, whose purpose is to display price and usage information.

1. *Indicator Bar.* Participants could conveniently check the current price by glancing at an indicator bar next to the battery icon on the top icon tray of the iPad's home screen, as shown in Figure 6.5.[7] The indicator bar is color coded, with red denoting discounts <10% and green discounts >30%, and shows the percentage discount from the baseline price ($10/GB in this trial).
2. *Price Information.* The home screen of the DDTDP app displays price and usage information in a split screen, as shown in Figure 6.6a. The user's current DDTDP bill for the month is displayed on top, followed by a scrollable list of color-coded percentage price discounts and usage information. The list includes data on the past and future 24 hours, highlighting the current hour's information; clicking on the entry for an hour toggles the price display between $/GB prices and percentage discounts from the $10/GB baseline.

Figure 6.5 Color-coded price discount indicator bar on the iPad home screen for easy reference.

[7]Incorporating this feature in the iPad app required jailbreaking the device. Implementing it on more open platforms, such as Android, does not require jailbreaking.

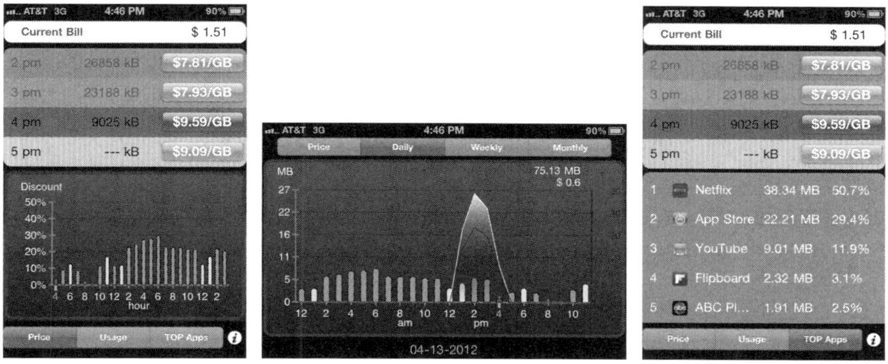

(a) (b) (c)

Figure 6.6 Screen views of information displayed by the app. The main screen (a) is split to show a scrollable history list on top and either the color-coded future prices and history or (c) the top five bandwidth-consuming apps below. The landscape view (b) shows the prices and usage superimposed on the same graph.

The bottom of the main screen displays the future prices as an easily understandable color-coded bar graph, helping users compare the discounts in different hours. Using the navigation buttons at the bottom of the screen, users can instead view the past price information, overlaid with their usage history by day, week, or month; and a list of the apps that consume the most data.

3. *Usage History.* As described earlier, users can monitor their usage from the list and graphs on the DDTDP app's main screen. Interested users can also turn the iPad horizontally to activate a larger landscape view of their usage, as seen in Figure 6.6b. Users can swipe left through their superimposed price and usage history, allowing them to quickly determine how much data they consumed at different price levels and at different times. This data can be viewed on a daily, weekly, or monthly granularity.
4. *Top-Five Apps.* In the pretrial focus groups, most people did not know which apps consumed the most data. We thus provided an option on the bottom half of the app's main screen, shown in Figure 6.6c, that displays the data usage volume of the five apps that use the most data.
5. *High Usage Notifications.* We tested users' responsiveness to pop-up notifications of high prices and accompanying usage-deferral recommendations (e.g., notices of low price future times).

We next discuss features designed to empower users in scheduling their usage according to the DDTDP prices.

1. *Parental Control.* As many of the primary participants shared their devices with younger secondary users, we provided a blocking feature to help the primary participant control others' usage. The user can select an app from a list of all

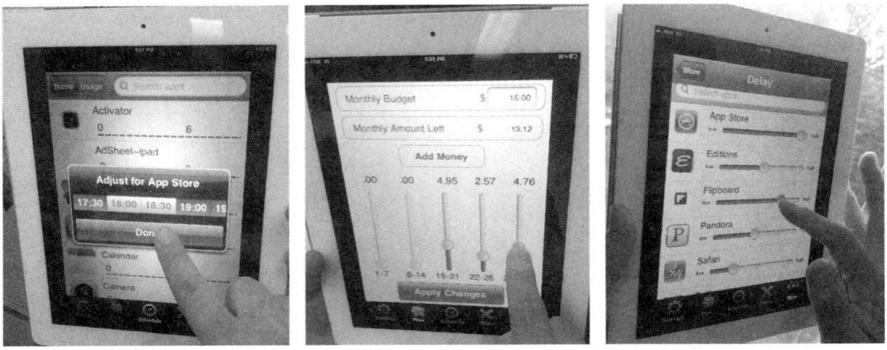

(a) (b) (c)

Figure 6.7 Screen views of different usage control features in the app. Users can (a) manually schedule apps for different times, (b) allocate their budget to different weeks, and (c) adjust delay tolerances for different apps for autopilot scheduling.

installed apps and then select times (on a half-hour granularity) in which that app should be blocked from using data, as seen in Figure 6.7a. For instance, participants could prevent their children from streaming media-rich content with high data volumes at high price times.

2. *Budget Adjustment.* Users could set a monthly budget for each billing cycle. As seen in Figure 6.7b, they could further control their spending by adjusting weekly budget sliders to divide their monthly budget among the weeks of their monthly billing cycle. By default, the budget was divided equally among the weeks of the billing cycle, with progressive adjustment of the remaining monthly budget across the remaining weeks as the month progresses. Users could also add money to their monthly budget.
3. *App-Delay Sensitivity Settings.* Users could set different delay sensitivities for each installed app by adjusting sliders between "high" and "low" tolerance, as shown in Figure 6.7c. When combined with automated application scheduling, this feature allowed the scheduling to account for individual delay preferences while scheduling enough data in low price periods to keep the user within the specified weekly budget.

To fully exploit the potential of user-specified delay sensitivities and the available budget, we also provided an optional autopilot mode of operation. In this mode, the TDP app computed a usage schedule for different apps to keep users within their specified budgets, based on the given prices for the next day and their predicted usage of different apps. The autopilot mode attempted as much as possible to defer only apps marked as delay insensitive. The autopilot mode provided warnings and blocked apps until their scheduled period and displayed the schedule to users on a schedule screen.

6.8.4 Trial Results

We now discuss users' reactions to the DDTDP prices offered in our trial setup. We evaluated the trial results in two ways: quantitatively, by measuring the change in usage in response to prices offered, and qualitatively, by conducting posttrial interviews with the trial participants. In these interviews, we sought to understand participants' general attitudes toward pricing before questioning them about their experience with the trial app and how the trial affected their opinions of TDP's potential impact on the Internet ecosystem.

6.8.4.1 Opinion on Smart Devices About 70% of our participants had used some form of smart devices before (e.g., Android and iPhones) and were positive about the convenience and usefulness of these devices. One participant (P7) described how the iPad is very useful for leisure-time activities: "*in the morning when I exercise I just have Netflix going*," while another (P3) told us the educational purpose that it serves for her toddler: "*She loves apps, and now she chooses her own YouTube videos, like nursery rhymes and stuff.*" But in spite of these benefits, not everyone views such smart devices as a necessary gadget: "*Do I like having them, are they convenient? Absolutely, but it is definitely a luxury item.*" (P6).

6.8.4.2 Understanding of Data Plans and Usage Before participating in our DDTDP trial, most participants were relatively unaware of their data usage and data plans. About 80% reported that they did not know how much data they used or which apps consumed the most data. A few, however, had some idea: "*I would think any kind of video and music sites*" (P6) and "*I assume that anything that's running constantly or streaming, you know, like a movie is going to use more than just browsing the Internet I would think*" (P9). Another participant knew that her tablet consumed more data than her phone: "*I also know I've used the iPad a lot more liberally for things like that [TV shows] than I would on my iPhone.*"

Despite their relative lack of knowledge about mobile data usage, most of our participants were concerned about recent data plan changes. When asked whether current data plans are reasonable, participants expressed concern over ISPs' shift from unlimited flat rate to tiered plans with $10/GB overages. One of them explained: "*To me, I think it's a fair price, but I know when my children uses it, if they use Netflix or something, then it gets to be too expensive.*"(P10). These pricing changes prompted some participants to pay more attention to their data usage: "*I didn't really have to pay attention to that until this iPad came up because the data plan I have for the family phones is unlimited data for one price.*"

6.8.4.3 Effectiveness of DDTDP DDTDP's goal is to use economic incentives to even out network usage over the day by exploiting the different degrees of time sensitivity of different applications. Such shifting can greatly benefit ISPs, as traffic data from our partner ISPs shows that the network traffic can vary by a factor of 10 over the day and by a factor of 2 within a few minutes. However, its success depends on users' willingness to pay attention and respond to the prices offered. Our trial results indicate that such economic incentives when offered through intuitive GUIs can indeed help in modifying usage behavior and shifting traffic to off-peak hours. Moreover, the incentives offered can further improve spectrum utilization in these off-peak hours by generating additional new demand.

All of our participants thought DDTDP would be viable *"as long as the interface is simple to use"* (P2). Some of our participants, however, were concerned about planning for future prices: participant P3, a mother of a toddler, thought that *"[I]t's definitely useful, but for families it really needs to be predictable."* Day-ahead pricing, in which prices for each hour are announced 1 day in advance, can provide some predictability, which was enough for some participants. Price-sensitive users *"made a conscious effort to look for the discounts"* (P10), adapting their usage in order to save money: *"I think it is a nice option to have where I can get a discount per month depending on when I use it, and I can schedule my day that way"* (P10). Moreover, they did not feel that paying attention to the prices was overly burdensome: *"I go to my bank account everyday, so I would think that this would just become a natural thing"* (P6). In fact, the DDTDP app helped users avoid unnecessary usage at high price periods: *"Yes, and [I] was less likely to goof off and waste more time and data"* (P5).

By measuring usage with DDTDP, we found that participants' efforts to pay attention to the prices do flatten out the traffic level over the day. To measure the reduction in peak traffic, we calculated the *peak-to-average ratio* (PAR), that is, the ratio of usage in the peak period to average per-period usage, for each day. We then compared the PARs from a control period of pre-DDTDP, that is, no variation in prices over the day, to those observed with DDTDP. We find that DDTDP reduces the PAR from pre-DDTDP usage, with a 30% reduction in the maximum observed PAR: Figure 6.8a shows the distribution of peak-to-average ratios on pre-DDTDP and DDTDP days of the trial.

In addition to reducing the PAR, we observed that participants' awareness of high discounts induced a "sales-day" effect among several participants; that is, they started using more than they otherwise would have. Figure 6.8b shows the average daily traffic for pre-DDTDP and DDTDP traffic; we see that traffic with DDTDP is much larger than pre-DDTDP, with a 130% overall increase. Combined with Figure 6.8a's result that the PAR decreases, we conclude that participants consumed much more traffic during off-peak periods when DDTDP discounts were offered. When asked about this, participant P9 told us *"[laughs] Kind of! But that also goes toward my personality of if it's on sale I must buy it!"* This result benefits ISPs: with DDTDP, they can offer discounts to shift traffic from peak to off-peak periods, as well as increase demand in off-peak hours. The result is mutually beneficial, as ISPs benefit from "valley filling" and users gain by consuming more at the discounted rates in off-peak times.

We also measured the types of applications shifted to lower price times. Participants generally delayed streaming and media services: Figure 6.8c shows the distribution of application usage among movies, web traffic, downloads, music, news and magazines, and other types of apps. We see a very significant increase in movie traffic with TDP, and smaller increases in the "music" and "other" categories. When asked about the apps they were willing to shift, participants' responses were consistent with

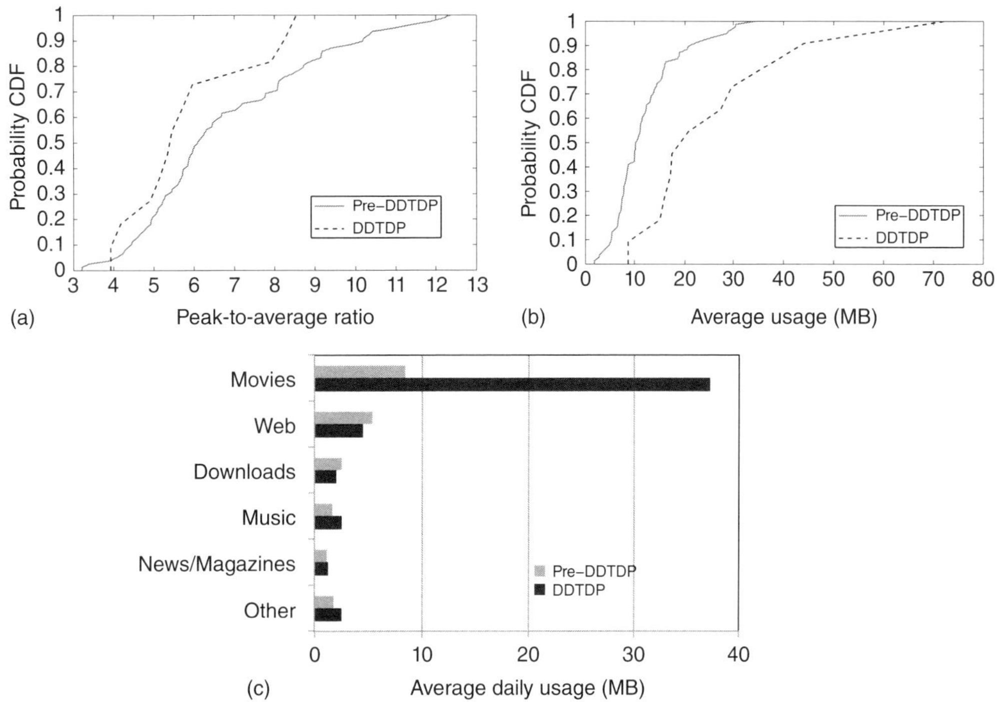

Figure 6.8 Usage statistics for TIP and TDP. (a) Peak-to-average ratios, (b) average daily usage, and (c) app distribution.

these measurements, with most participants willing to defer social networking and streaming. Participants P6 gave the examples of *"Social networking, emailing, or Skyping, I would definitely wait for that,"* while participant P9 said that *"If I'm surfing the web, I'm doing it now, I am not waiting, but for my movie usage it's kind of specific, so yeah."* Similarly, participant P1 was willing to shift nonurgent activities: *"If I'm trying to look up directions [GPS], it probably couldn't wait. But if I were trying to research on buying something and the discounts weren't good, I could wait a couple of hours to do it later."*

We found that TDP helps address consumers' concerns about the rising cost of mobile data plans; although many users are not fully aware of their monthly usage, they do want to save on their data plans. Given the right economic incentives, users are willing to modify their usage behavior in order to save money, which can also lower network congestion for ISPs. Our DDTDP trial yielded "flatter" traffic usage over the day, with lower peaks and higher off-peak demand.

6.8.4.4 Effectiveness of the User Interface We next examine (i) the effectiveness of different features of our DDTDP app's UI, (ii) a comparison of the effectiveness of different ways of displaying prices to users (e.g., color coding vs numerical values), and (iii) the effect of showing their usage history. Although the app's main function was to display prices, our participants viewed the app as a more general informational tool: *"To me, this application educates you"* (P4). Participants unanimously agreed on the usefulness of three key features: (i) a price indicator bar on the iPad's home screen, (ii) color coding the price discounts, and (iii) displaying usage history over time.

We first consider participants' responses to the "passive" price indicator displays (i.e., when no explicit warning or notification message is sent to the users). For this part of the trial, we used a color-coding scheme with orange for a price discount of less than 29% and green for discounts of 30% or more. To quantify the actual effects of the price indicator, we analyzed the usage in periods in which we intentionally varied the discounts by 1% to switch the color from orange to green. We defined two experimental stages and three period types: in the first experimental stage, all period types had a 10% discount and orange price indicator. In the second stage, type 1 periods switched to a higher (28%) discount with orange indicator, type 2 periods to a slightly higher (30%) discount with green indicator, and type 3 periods to a lower 9% discount with orange indicator. We then compared type 2 and type 3 periods to type 1 periods to see if users paid attention to the range of numerical discount values or just the color coding. Table 6.3 summarizes the three period types.

TABLE 6.3 Period Types in the Price Display Experiment

Period Type	Second Stage Color	Second Stage Discount, %
1	Orange	28
2	Green	30
3	Orange	9

Figure 6.9a plots each participant's average change in usage in Type 2 versus Type 1 periods. The price discounts in both periods increased by comparable amounts in the second experiment stage (28% and 30% respectively), but the indicator color changed from orange to green only for Type 2 periods. We found that participants increased their usage more when the color changed (Type 2 periods): each point on the graph in Figure 6.9a represents one participant, and the size of the point is proportional to their usage amount. We see that most participants' data points lie above the reference line, so that usage increased more (or decreased less) in Type 2 when compared to Type 1 periods. Wilcoxon's signed rank test yields only a 9.8% probability that the observed differences between the percent changes come from a symmetric distribution about zero. These results indicate that participants changed their usage because of the color of the indicator, despite the comparable numerical discounts.

Figure 6.9b shows the average percentage change in usage for each user in Type 1 periods versus Type 3 periods. For both period types, the color did not change, but the discount in Type 1 periods increased significantly (i.e., 28% in Type 1 versus 9% in Type 3 periods). Thus, if participants had reacted to the numerical price discounts, we would expect usage to increase in Type 1 and decrease in Type 3 periods: participants' data points should lie above the reference line. However, participants' changes in usage were comparable for both periods, with only half of users' data points above the reference line in Figure 6.9b. Some participants increased their usage dramatically in both types of periods, while most decreased their usage in both types of periods. Thus, participants were mostly agnostic to the numerical values of the discounts, likely because they only paid attention to the indicator color, which remained the same for both types of periods.

Our posttrial interviews confirmed that the observed usage changes represent conscious decision making on the part of the trial participants. Participants generally did pay attention to the price indicator color and chart of future prices: participant P4 *"paid attention to the little color icon that you guys have up on the top, and I tried to tell everyone not to use it unless it was in green [high-discount periods]... I do look at the chart and see at what point the discounts might come in,"* while participant P7 checked the price indicator before using data: *"I would see if it's a good color for me"* (P7). A few participants also checked the numerical discount: *"I do look at the percentage. I will go to the app and just see where it's gonna be over the next few hours"* (P10).

In a separate phase of the trial, we explicitly reminded participants of high price periods with pop-up notifications that were sent every 10 min if participants used a lot of data in high price periods. While many participants liked being notified, for example, *"As a bill-payer I would think it is very useful"* (P4), others thought that the notifications interfered with using their devices: *"the pop-ups were a bit annoying. I think it would be nice if the warning messages appear on the top of the screen"* (P2). The notifications, however, were generally successful: 90% of participants either decreased or did not increase their usage after one notification. About 60–80% of participants decreased their usage in response to multiple consecutive notifications. Figure 6.10 shows the distribution of observed changes in usage in response to the notifications sent; multiple notifications were only sent if participants consumed more

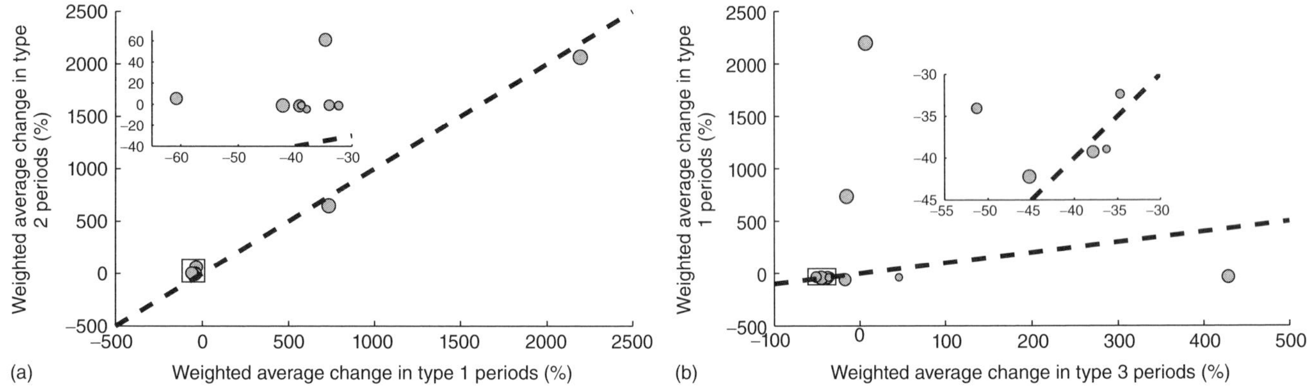

Figure 6.9 Changes in usage for period types 1, 2, and 3. (a) Period types 1 and 2 and (b) period types 1 and 3.

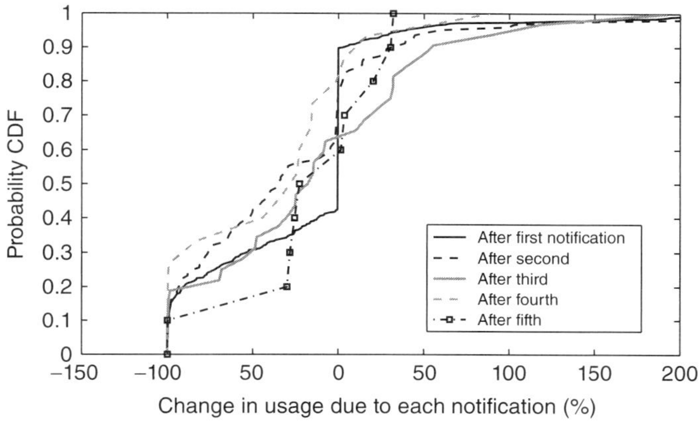

Figure 6.10 Changes in participant usage volumes in response to notifications sent.

than 2 MB of data in consecutive 10-min intervals. We see that only a small minority of participants increased their usage after receiving a notification.

Some participants suggested that prediction-based pop-ups with usage suggestions, for example, warnings such as *"[i]f you watched this movie, you would be over your usage quota"* (P9) would be more useful than reminders of high prices. Even alerting users to future low price times would be useful, as that would allow users to plan ahead: *"I think it's a great idea, when the iPad would say 'If you wait for half an hour, you can have 25% discount.' I thought it was incredibly useful for my decision"* (P6).

In fact, participants not only paid attention to the prices offered but also used the usage history displays to track their usage and amount spent over time. A mother of three explained how the app allowed her to monitor and control her children's usage when she was at work: *"I can go back and see when they were using it!"* (P4). She also gave an anecdote on having used this feature to check whether people at an iPad repair shop had used her data plan while fixing her broken screen. Another user (P3) suggested including a usage indicator on the home screen, similar to the price indicator, showing the remaining monthly budget. Our results thus imply that users find icons on this top task bar convenient for monitoring the current price and amount spent but prefer notifications to contain explicit suggestions for better ways to use mobile data.

6.8.4.5 Empowering User Choice We finally turn to app features that help users decide how to use mobile data so as to save money while still enjoying the use of their device. In addition to the price displays and usage history discussed earlier, we showed users which apps consumed the most data, allowing them to better choose which apps to stop using so as to use less data. We also allowed users to control the usage of secondary users such as children with an explicit scheduling functionality and to control their own usage with an autopilot feature that automatically delayed time-insensitive apps to lower price periods.

We displayed the data used by the top five data-consuming apps to educate our participants about which apps use the most data. Participants found this information useful for keeping track of their usage and controlling how much total data they consumed: *"Absolutely. I mean it would be nice if there were an indicator on the app icon themselves."* Even the displays within our TDP app increased their awareness of per-app data usage: *"usually it's always the same type of thing—the Internet, YouTube, Facebook, those type of things"* (P4). This awareness prompted participant P4 to pay more attention to her usage in general, including that not shown on our app: *"If I connect to my email and work and I have a thousand emails in my mailbox and it downloads all that stuff that's a big consumer as well."*

While looking at their usage history allowed participants to monitor the usage of secondary users, many participants wanted to control this usage as well. Most were concerned about their children's usage; as participant P6 stated: *"I don't think they realize how long they've been on there."* Participants liked the idea of explicitly blocking certain apps at certain times in order to curtail their children's usage: *"I think there's a lot of parents that would use it"* (P4). Indeed, as discussed in Section 6.4, some studies have found that parents liked the idea of controlling their children's data usage even on wired networks [11–13].

Finally, we allowed our participants to control their own usage with an "autopilot mode" in which the DDTDP app scheduled usage based on users' predicted behavior, their usage history, and the future prices. Although this autopilot mode was designed to help users save money by delaying the least time-sensitive apps, our participants preferred to manually control their usage: *"I like to control my device manually"* (P2). Some were hesitant to use the autopilot mode because of concerns about the loss of control: *"It's really annoying to be put in the autopilot mode. I would not want a computer to tell me what to use and when to use it"* (P3). As participant P8 pointed out, an autopilot mode would only be accepted by users *"if it is sufficiently intelligent to figure out the importance of each application's usage."* In fact, although our app allowed users to input and adjust different apps' delay sensitivities within the scheduling algorithm, participants were still concerned that automatic scheduling would interfere with their desired usage: *"[T]hat might not be the way you want it scheduled"* (P4). Participants were more accepting of automated scheduling of noncritical apps: *"If you could put it in a queue and let the system figure out when the cheapest time to do it is"* (P7). Effectively implementing TDP thus requires HCI researchers to balance users' desire to control their usage decisions with their willingness to automatically schedule noncritical apps.

6.9 DISCUSSIONS AND CONCLUSIONS

Our investigations reveal that consumers today are concerned about the increasing cost of data plans but are neither fully aware of their monthly usage nor equipped to effectively manage and control it to stay within their budget. DDTDP offers a potential "win—win" for both network operators and consumers, but its efficacy rests on if and how it can modify user behavior. We develop a system that offers DDTDP and conduct a trial that provides an initial validation of DDTDP's promise as a future direction of Internet pricing, following similar trends in other markets such as electricity and transport networks. We find that when economic incentives are offered to users in a simple and comprehensible format, many will be willing to wait for such a feedback signal from the network and change their behavior for noncritical mobile applications.

In our trial, the maximum PAR of data demand decreased by 30% while the total traffic increased by up to 130% with DDTDP, mostly because of the filling up of discounted valley periods from the sales-day effect. This observation bolsters the intuition that DDTDP can lead to a "flatter" demand profile with lower peaks and better spectrum utilization in off-peak hours. Additionally, the availability of such a client-side DDTDP application not only enables users to make deferral decisions but also helps them to self-educate, monitor, and control their usage and expenses. Our trial participants liked UI information displays, such as the color-coded price discounts, the indicator bar on the home screen, usage history, and so on. The trial also revealed that while users in general are not willing to wait for critical applications, they are quite willing to delay noncritical applications when TDP incentives are provided.

The trial also revealed interesting insights on the user-control aspects of DDTDP. First, our participants viewed parental control at the granularity of apps as being useful in managing their usage in a TDP regime. Second, they showed a desire to control their usage manually instead of delegating control to an autopilot mode. When coupled with the desire for parental control, we see that users want to take charge of their consumption behaviors, for themselves and their families, in ways that require transparency and flexibility. Therefore, realizing automated DDTDP would require careful consideration of this trade-off. This also leads to a direction for future research on the trade-off between users' willingness to implicitly reveal their delay tolerances to network operators and the privacy and trust aspects related to such disclosures.

As HCI tackles increasingly complex sociotechnical ecosystems, for example, mobile Internet, cloud computing, and smart grids, incorporating economic analysis as a part of user behavior studies is becoming important. Out work on investigating user behavior in a new TDP regime is an initial building block in this area of HCI research. Our work also introduces a framework for realistic experimentation with the HCI aspects of network economics. By interposing between ISPs and their customers, we act as a resale ISP with user-facing apps and ISP-side economic analysis. Future research on Internet pricing will need to address related issues on how similar HCI design principles can be effective in enabling other pricing variants, such as app-based or QoS-based pricing. Additionally, HCI within SDP will need to address the latest challenges brought about by the introduction of shared data plans in the United States, which requires both real-time rating and charging for all subscribers and new mechanisms for users to allocate and control the shared data caps. Many of the findings and design principles laid out in this chapter will be relevant in addressing these new challenges as SDP continues to evolve.

ACKNOWLEDGMENTS

The authors would like to acknowledge Mr. R. Rill and Ms. D. Butnariu for their assistance with application development and Ms. J. Bawa for her help during the trial participant debriefing process. We are also thankful to all our industry partners, including AT&T, MTA, Reliance, and NECA. The work has also immensely benefited from extensive discussions with participants of the Workshop on Smart Data Pricing (SDP 2012, Princeton, US & SDP 2013, Turin, Italy).

REFERENCES

1. A. Dix, J. Finlay, G. D. Abowd, and R. Beale. *Human–Computer Interaction*. Pearson Prentice-Hall, Harlow, UK, 3rd edition, 2004.
2. J. Nielsen and T. K. Landauer. A mathematical model of the finding of usability problems. In *Proceedings of the INTERACT and CHI conference on Human factors in computing systems*, pp. 206–213. ACM, 1993.
3. R. Langley. *Practical Statistics Simply Explained, Dover Books Explaining Science Series*. Dover Publications, Mineola, New York, 1971.
4. F. Heller and J. Borchers. PowerSocket: towards on-outlet power consumption visualization. In *CHI Extended Abstracts*, pp. 1981–1986. ACM, 2011.
5. T. Grace Holmes. Eco-visualization: combining art and technology to reduce energy consumption. In *Proceedings of C&C*, pp. 153–162. ACM, 2007.
6. D. Petersen, J. Steele, and J. Wilkerson. WattBot: a residential electricity monitoring and feedback system. In *CHI Extended Abstracts*, pp. 2847–2852. ACM, 2009.
7. M. Chetty, D. Tran, and R. E. Grinter. Getting to green: understanding resource consumption in the home. In *Proceedings of ACM UbiComp*, pp. 242–251. ACM, 2008.
8. T. Kim, H. Hong, and B. Magerko. Design requirements for ambient display that supports sustainable lifestyle. In *Proceedings of DIS*, pp. 103–112. ACM, 2010.
9. R. E. Grinter, W. K. Edwards, M. Chetty, E. S. Poole, J. Y. Sung, J. Yang, A. Crabtree, P. Tolmie, T. Rodden, C. Greenhalgh, and S. Benford. "The ins and outs of home networking: the case for useful and usable domestic networking," *Transactions on Computer-Human Interaction*, 16, 2009, 8.
10. E. S. Poole, M. Chetty, R. E. Grinter, and W. K. Edwards. More than meets the eye: transforming the user experience of home network management. In *Proceedings of ACM DIS*, pp. 455–464. ACM, 2008.
11. M. Chetty, R. Banks, R. Harper, T. Regan, A. Sellen, C. Gkantsidis, T. Karagiannis, and P. Key. Who's hogging the bandwidth? The consequences of revealing the invisible in the home. *Proceedings of ACM SIGCHI*, pp. 659–668. ACM, 2010.
12. J. Yang, W. K. Edwards, and D. Haslem. Eden: supporting home network management through interactive visual tools. In *Proceedings of ACM UIST*, pp. 109–118. ACM, 2010.
13. R. Mortier, T. Rodden, P. Tolmie, T. Lodge, R. Spencer, A. Crabtree, A. Koliousis, and J. Sventek. Homework: putting interaction into the infrastructure. In *Proceedings of ACM UIST*, pp. 197–206. ACM, 2012.
14. Comcast. About excessive use of data, Oct. 2012. Available at: http://customer.comcast.com/help-and-support/internet/data-usage-what-are-the-different-plans-launching.

15. FCC. Internet access services: status as of June 30, 2009, 2010.

16. M. Chetty, D. Haslem, A. Baird, U. Ofoha, B. Sumner, and R. Grinter. Why is my Internet slow? Making network speeds visible. In *Proceedings of ACM SIGCHI*, pp. 1889–1898. ACM, 2011.

17. M. Chetty, R. Banks, A. J. Bernheim Brush, J. Donner, and R. E. Grinter. While the meter is running: computing in a capped world. *Interactions*, 18(2), 2011, 72–75.

18. S. P Wyche, T. N. Smyth, M. Chetty, P. M. Aoki, and R. E Grinter. Deliberate interactions: characterizing technology use in Nairobi, Kenya. In *Proceedings of ACM SIGCHI*, pp. 2593–2602. ACM, 2010.

19. M. Chetty, R. Banks, A. J. Brush, J. Donner, and R. Grinter. You're capped: understanding the effects of bandwidth caps on broadband use in the home. In *Proceedings of ACM SIGCHI*, pp. 3021–3030. ACM, 2012.

20. H. Kim, S. Sundaresan, M. Chetty, N. Feamster, and W. K. Edwards. Communicating with caps: managing usage caps in home networks. In *Proceedings of ACM SIGCOMM (Posters and Demos)*, pp. 470–471. ACM, 2011.

21. H. Varian. The demand for bandwidth: evidence from the INDEX project, 2002. Available at: http://people.ischool.berkeley.edu/~hal/Papers/brookings.pdf.

22. R. Edell and P. Varaiya. Providing internet access: What we learn from INDEX, 1999. Keynote Speech, IEEE INFOCOM.

23. Y.-F. R. Chen and R. Jana. SpeedGate: a smart data pricing testbed based on speed tiers. In *Proceedings of the Workshop on Smart Data Pricing (SDP) at INFOCOM 2013*, pp. 3195–3200. IEEE, 2013.

24. J. M. Dyaberi, B. S. Parsons, V. S. Pai, K. Kannan, Y.-F. R. Chen, R. Jana, D. Stern, A. Varshavsky, and B. Wei. "Managing cellular congestion using incentives," *IEEE Communications Magazine*, 50(11), 2012, 100–107.

25. S. Sen, C. Joe-Wong, S. Ha, and M. Chiang. "Incentivizing time-shifting of data: a survey of time-dependent pricing for internet access," *IEEE Communications Magazine*, 50(11), 2012, 91–99.

26. S. Sen, C. Joe-Wong, S. Ha, and M. Chiang. "A survey of broadband data pricing: past proposals, current plans, and future trends," *ACM Computing Surveys*, 46(2), 2014, to appear.

27. The Economist. The Mother of Invention: Network Operators in the Poor World Are Cutting Costs and Increasing Access in Innovative Ways, Sept. 2009. Special Report.

28. The Economist J. Shih, R. Katz, and A. D. Joseph. Pricing experiments for a computer-telephony-service usage allocation. In *Proceedings of IEEE GLOBECOM*, pp. 2450–2454. IEEE, New York, 2001.

29. S. Ha, S. Sen, C. Joe-Wong, Y. Im, and M. Chiang. TUBE: time-dependent pricing for mobile data. In *Proceedings of ACM SIGCOMM*, pp. 247–258. ACM, 2012.

30. V. Glass, J. Prinzivalli, and S. Stefanova. Persistence of middle mile problems for rural exchanges local carriers. In *Smart Data Pricing Workshop*, July 2012. Available at: http://scenic.princeton.edu/SDP2012/Talks-VictorGlass.pdf.

31. A. Odlyzko, B. St. Arnaud, E. Stallman, and M. Weinberg. *Know Your Limits: Considering the Role of Data Caps and Usage Based Billing in Internet Access Service*. Public Knowledge, 2012. White Paper.

32. D. D. Clark. "Internet cost allocation and pricing" in L. W. McKnight and J. P. Bailey, eds., *Internet Economics*, MIT Press, Cambridge, Massachusetts, USA, 1997, pp. 215–252.

33. B. X. Chen. Shared mobile data plans: who benefits? New York Times, Bits Blog, July 2012.
34. S. Sen, C. Joe-Wong, S. Ha, J. Bawa, and M. Chiang. When the price is right: enabling time-dependent pricing of mobile data. In *Proceedings of ACM SIGCHI*, pp. 2477–2486. ACM, 2013.
35. N. Anderson. US rural broadband: you can get it, but you can't afford it. Ars Technica, 2008. Available at: http://arstechnica.com/uncategorized/2008/03/us-rural-broadband-you-can-get-it-but-you-cant-afford-it/.
36. J. Newman. Netflix has bandwidth cap sufferers covered. PC World, June 2011. Available at: http://www.pcworld.com/article/230982/Netflix_Has_Bandwidth_Cap_Sufferers_Covered.html.
37. Sergey Lossev. iOS 7 feature gem: cellular data usage statistics. For Techies Only, 2013. Available at: http://www.fortechiesonly.com/2013/07/ios-7-feature-gem-cellular-data-usage.html.
38. S. Perez. The best features of iOS 7. TechCrunch, 2013. Available at: http://techcrunch.com/2013/06/10/the-best-features-of-ios-7/.
39. C. S. Yoo. Network neutrality, consumers, and innovation. *University of Chicago Legal Forum*, 25, 2009, 179–262.
40. A. Schatz and S. E. Ante. "FCC chief backs usage-based broadband pricing," December 2 2010. Wall Street Journal.
41. J. Genachowski. New rules for an open internet. FCC, 2010.
42. C. Joe-Wong, S. Ha, and M. Chiang. Time-dependent broadband pricing: feasibility and benefits. In *Proceedings of IEEE ICDCS*, pp. 288–298. IEEE, 2011.

PART III
Usage-Based Pricing

7 Quantifying the Costs of Customers for Usage-Based Pricing

LÁSZLÓ GYARMATI, RADE STANOJEVIC, MICHAEL SIRIVIANOS, and NIKOLAOS LAOUTARIS

7.1 INTRODUCTION

The intense competition among telecom operators is a significant driving force behind the growing popularity of usage-based pricing in communication networks. The competition among the operators results in shrinking profit margins: the operators drop prices in order to attract or retain customers and face continuously increasing traffic volumes. In terms of reduction of capital expense (CAPEX) and operational expense (OPEX), the advancements in the technology are not able to keep up with the speed of the increase in traffic. As a consequence, prices offered to customers should take into account any induced expenses to sustain the profitability of the network operators.

Although operators have incentives to move toward usage-based pricing schemes, customers are also paying more attention to their expenses given the recent state of the global economy: they do not want to subsidize the costs of others. To meet the expectations toward usage-based pricing, operators have to understand the cost of their infrastructure and be aware as to how individual customers affect these costs. Understanding and optimizing the total cost of ownership (TCO) of a network, that is, both CAPEX and OPEX, is an important aspect of network operations, and therefore, it has received substantial attention from procurement, network development, and network planning departments of large telcos [1]. However, the impact of individual customers on the cost of the network is much less understood because of reasons such as difficulties of monitoring and nonlinearities of cost–capacity functions.

It is crucial to quantify how individual customers affect the TCO of a network as this can be used to design better tariff schemes that reflect the usage and costs of the

Smart Data Pricing, First Edition. Edited by Soumya Sen, Carlee Joe-Wong, Sangtae Ha, and Mung Chiang.

customers. For example, a customer could receive discounts if it inflicts low costs or charged more in the opposite case. In this chapter, we focus on the quantification of the customers' costs in communication networks, that is, how an operator should share the cost of the infrastructure among its customers. The structure of the chapter is as follows. First, we review the cost of the network infrastructure and also how this cost is affected by the aggregate traffic of the customers (Section 7.2). Afterwards, we introduce a metric, namely, discrepancy, that quantifies the differences of cost-sharing policies (Section 7.3). The forthcoming sections introduce several factors that impact the quantification of the customers' costs. In all the cases, we apply the introduced methodologies on real-world datasets: we derive the costs of the customers of a large backbone network. Section 7.8 discusses related work while the conclusion is arrived at in Section 7.9.

7.2 THE COST OF A CUSTOMER IN A NETWORK

We start with an example to show the pitfalls of the quantification of customers' cost in a communication network. In Figure 7.1, we depict a toy infrastructure used by two customers. The customers access web services using their mobile devices. Let us assume that the aggregate cost of the infrastructure is \$21k; we compute this by summing up all the expenditure of the parts of the network. Now we have to share this amount between the customers. A straightforward way is to look at the volume of the traffic they inject into the network and share the cost of the infrastructure proportional to this. As both customers have traffic of 10 Mbps, they both have a cost of \$10.5k. This is a simple cost-sharing policy; however, it has some limitations in terms of fairness. One may ask why should customer 1 partially cover the costs of the other customer? An alternate cost-sharing policy is when we consider exactly which parts of the network were used by which customer. In this way, customer 1 has only \$4.5k cost because it uses a link partially with the other customer and an additional one exclusively.

Additional factors impact the quantified cost of the customers including the precise cost of the infrastructure itself. The cost of a network consists of CAPEX and OPEX for all devices present in the network. The CAPEX is the one-time cost paid

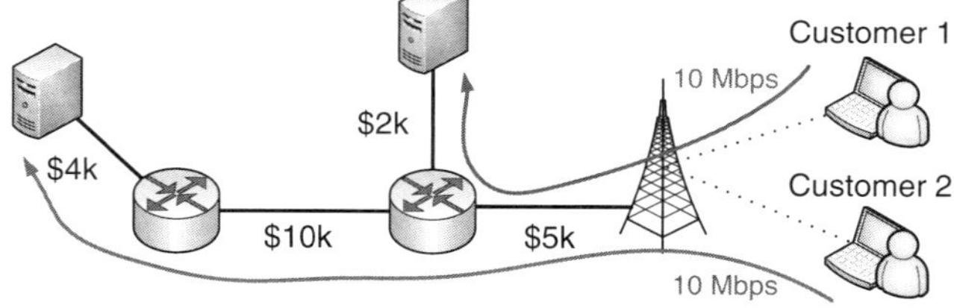

Figure 7.1 Quantifying the costs of the customers is challenging. The customers use different parts of the infrastructure and, hence, can be accounted for diverse costs even of their traffic is identical.

whenever equipment is bought and installed [1]. It depends on the amount of traffic the device must carry at a specific level of quality of service (QoS). A key observation is that *the capacity needed to guarantee a certain QoS depends on the peak traffic that needs to be carried.* This is because for a given capacity, QoS is minimized when the traffic peaks. The OPEX corresponds to operational costs such as real estate, energy, and personnel. It also depends on the amount of traffic and the QoS; however, that dependence is more elastic. The cost-sharing policies we discuss are generic enough to capture both CAPEX and OPEX with appropriate parameterization.

Next, we describe how we determine the aggregate cost that the sharing policies distribute. A network consists of various network devices, for example, routers and links. Let L denote the set of devices of the network. Let $x_i^l(t)$ denote the traffic volume of customer $i \in N$ on network device $l \in L$ during the time interval $t \in [1, T]$. Furthermore, let c^l denote the cost of network device $l \in L$. The cost of a specific device depends on the maximum amount of traffic that it has to carry during a certain time interval. Thus, we obtain c^l by examining the available capacity rates of the device (e.g., 1 and 10 Gbps) and then using the cost of the smallest device whose capacity satisfies the requested service-level agreement (SLA) for the given traffic demand. For example, the operator fulfills its SLA by upgrading its devices when utilization hits the 50% threshold. To this end, we assume that the costs follow a step function $C : \mathbb{R} \rightarrow \mathbb{R}$. Thus, the cost of device l is

$$c^l = C\left(\max_{t \in T} \sum_{i \in N} x_i^l(t) \right). \tag{7.1}$$

7.2.1 Datasets used in the Case Studies

In the remainder of the chapter, we introduce several factors that influence the quantified cost of the customers. We also present case studies to illustrate the applicability of the methodology in real-world scenarios. To this end, we use several datasets from a tier-1 backbone network, which interconnects with other Internet Service Providers (ISPs) that it serves. In our dataset, the network consists of tens of points of presence (PoPs) and each customer connects to the network at one or more PoPs through one or more interfaces. Overall, the datasets contain the per-link and per-customer traffic statistics of more than 1000 links over a period of 3 weeks, with a 2-h granularity (i.e., reporting volumes sent and received within 2 h). For additional details on the datasets, we refer to Reference 2.

The cost of a network link depends, on the one hand, on the capacity of the interface, that is, how much traffic it is capable of forwarding. On the other hand, the geographic location and the applied technology have an impact as well. Hardware costs, energy prices, deployment costs, and taxation, among others, contribute to the cost of a network device. Thereby, it is challenging to accurately quantify the cost of every single device. To estimate the cost of the network links, we use a wholesale point-to-point transport price database. In our empirical analysis, we apply the prices of network links with different bandwidth, ranging from E-1 (2 Mbps) throughout

STM-4 (622 Mbps) and 2.5G waves to 40G waves (40,000 Mbps). The costs of these links define a step function for the network expenditures.

Although we analyze a backbone network as a case study, we emphasize that the methodology we use to quantify the costs of the customers is general enough to be applied in access networks [3] or for the whole end-to-end traffic of the customers.

7.3 DISCREPANCY, THE METRIC OF COMPARING DIFFERENT COST-SHARING POLICIES

The operator can quantify the cost of the customers in numerous ways using cost-sharing policies. We will present several orthogonal factors that could be combined arbitrarily to create cost-sharing policies that are in line with the objectives of the operator. The quantification of the customers' costs is a multifacet problem covering the following questions:

- How do we compute the costs of the customers?
- Where do we meter the traffic?
- What is the impact of the diverse costs of the devices?
- Who is liable for the incurred costs?

Before addressing these questions and introducing the appropriate types of discrepancies, we first introduce the metric we use to quantify the discrepancies of a pair of cost-sharing procedures. Let N denote the set of customers who utilize resources in the network. Let A and B denote the sets of costs allocated to each customer using two different cost-sharing policies. It holds that $|A| = |B| = |N|$. Accordingly, $a_i \in A$ denotes the cost of customer $i \in N$ quantified based on the first cost-sharing policy while $b_i \in B$ represents customer i's cost based on the second policy. We define the discrepancy of the costs of customer i as

$$d(a_i, b_i) = \max \left\{ \frac{a_i}{b_i}, \frac{b_i}{a_i} \right\}. \tag{7.2}$$

We use this measure of discrepancy because it describes the relation of the costs with a simple, comprehensible value. In the case studies, we use several statistics of the customers' individual discrepancies to quantify the discrepancy of two cost-sharing policies including the 95th percentile and the median.

ILLUSTRATIVE EXAMPLE. Let us assume that three customers utilize the resources of the network with costs presented in Table 7.1. The operator quantified the costs of the customers using two cost-sharing policies. As we compute the discrepancy of the policies for each customer, we gain both individual and aggregate insights. For example, the two policies are identical in case of the third customer; however, the discrepancy of the policies is 4 in case of the second customer, that is, the cost of the customer is four times higher in case of policy A. For aggregate insights, we

TABLE 7.1 Illustrative Example for Computing Discrepancies

	Cost Based on Policy A	Cost Based on Policy B	Discrepancy
Customer 1	150	300	2
Customer 2	200	50	4
Customer 3	150	150	1

We quantify the costs of the customers using two cost-sharing policies and derive the discrepancies of these policies.

investigate the distribution of the discrepancies, for example, the median discrepancy of the policies is 2.

7.4 HOW DO WE COMPUTE THE COSTS OF THE CUSTOMERS?

The first source of discrepancies in some cost allocation methods is the *function* that the operator uses to compute the contribution of the customers to the aggregate cost (i.e., F-discrepancy). We next present four policies that strike different balances between precision and resource needs. We consider these methods because operators apply some of these policies (e.g., the 95Percentile-Customer and the Aggregate-Peak-Device) in practice to determine the costs a customer inflicts and consequently the price the customers pays. For example, one can easily map some of the tariffs (e.g., based on the purchased raw capacity or on the 95th percentile of the traffic) used in practice to the introduced policies (e.g., Volume-Customer and 95Percentile-Customer).

- *Volume-Customer.* We measure the amount of data that a single customer sends on a specific network device (e.g., on a single link) for the whole analyzed time period. Afterwards, we share the cost of the device proportionally to the traffic volumes of the customers using it. Hence, the cost of customer i for device l is

$$c_i^l = c^l \cdot \frac{\sum_{t \in T} x_i^l(t)}{\sum_{j \in N} \sum_{t \in T} x_j^l(t)}. \tag{7.3}$$

- *95Percentile-Customer.* We distribute the cost of the device proportional to the 95th percentile [4] of the customers' traffic that traverses the particular device:

$$c_i^l = c^l \cdot \frac{P_{95}\left(\dots, x_i^l(t), \dots\right)}{\sum_{j \in N} P_{95}\left(\dots, x_j^l(t), \dots\right)}, \tag{7.4}$$

  where $P_{95}()$ denotes the 95th percentile of the arguments.

- *Peak-Customer.* Under this policy, we share the expenditure of the network device proportional to the customers' maximum usage volumes for the given time interval:

$$c_i^l = c^l \cdot \frac{\max_{t \in T} x_i^l(t)}{\sum_{j \in N} \max_{t \in T} x_j^l(t)}. \tag{7.5}$$

- *Aggregate-Peak-Device.* Operators plan the capacity of the network based on the maximum utilization, for example, the 50% of the capacity of a device is larger than the expected maximum of the traffic that traverses it. Accordingly, we distribute the cost of the devices based on the contribution of individual customers to the peak utilization. Assuming that the peak utilization of device l happens at time step $t_m = \arg\max_t \sum_{j \in N} x_j^l(t)$, we allocate the following cost to customer i as

$$c_i^l = c^l \cdot \frac{x_i^l(t_m)}{\sum_{j \in N} x_j^l(t_m)}. \tag{7.6}$$

We evaluate F-discrepancies by comparing a policy with Aggregate-Peak-Device, which shares the costs in a fair way when it is guaranteed that new unallocated capacity from an upgrade will soon find a customer to amortize it. The F-discrepancies of the policies arise from the misalignment of traffic peaks: the peak of a customer's traffic may not coincide with the peak of the aggregate traffic the device carries. Before presenting a date-driven evaluation of F-discrepancies, we first illustrate the introduced policies.

ILLUSTRATIVE EXAMPLE. Let us assume that two customers utilize device L_1 with traffic volumes depicted in Figure 7.2. On the basis of the time series of the customers, we compute their costs, that is, what portion of the total cost of the device is related to each customer. The percentages of the cost that customer 1 covers are 69.9%, 56.2%, 60%, and 53.3% of the total cost of the device for the Volume-Customer, Peak-Customer, 95Percentile-Customer, and Aggregate-Peak-Device policies, respectively. In terms of discrepancies these are 1.31, 1.05, 1.13, and 1.0 having Aggregate-Peak-Device as reference. We use the introduced formulae of the policies to compute the shares of the customers. For example, in case of the 95Percentile-Customer policy, customer 1 has a traffic of 9 Mbps while the cumulative traffic is 15 Mbps resulting in 60% cost share. The main cause behind the discrepancies of the costs are the misalignment of the customers' peak, and that the different policies consider diverse parts of the time series to compute a value that describes the traffic of the customer.

The introduced policies quantify the cost of the customers using different functions on a per-device basis. The F-discrepancies of the customers emerge at two different levels.

- *Device-Level Discrepancies.* The discrepancy among different policies based on the customers' costs is computed separately for each network device. In this

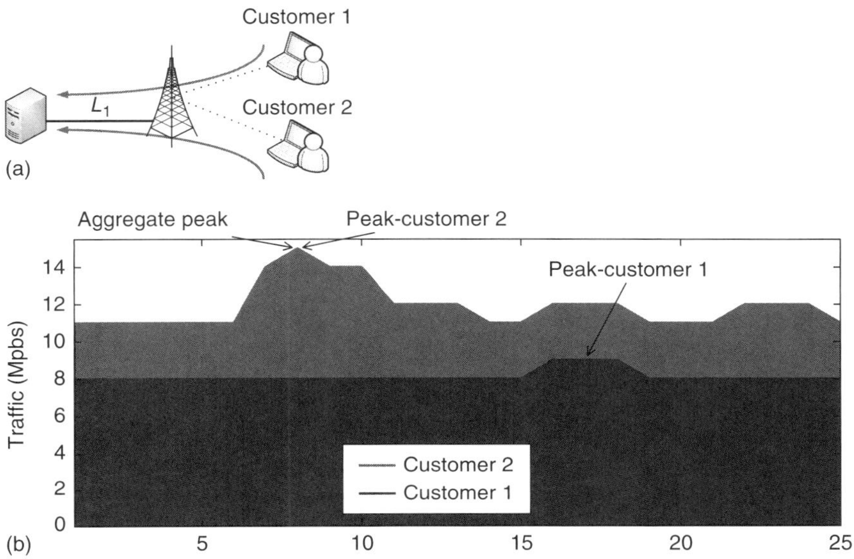

Figure 7.2 Illustrative example for F-discrepancies. The misalignment of the customers' peak causes cost differences across policies. (a) Toplogy and (b) traffic pattern of L_1.

case, the set of costs is

$$\left\{c_i^l \mid \forall i \in N\right\}, l \in L \tag{7.7}$$

and has cardinality $|N||L|$. For example, the F-discrepancy of customer i in case of policies a and b is $d(a_i^l, b_i^l)$.

- *Network-Level Discrepancies.* We first summarize the costs of a customer over all the devices of the network, that is, we compute the total cost of each customer. Afterwards, we compute the discrepancies of the policies. In this case, the set of the costs over which we compute the discrepancies is

$$\left\{\sum_{l \in L} c_i^l \mid \forall i \in N\right\} \tag{7.8}$$

and has cardinality $|N|$. For example, for policies a and b, the F-discrepancy of customer i is $d(\sum_{l \in L} a_i^l, \sum_{l \in L} b_i^l)$.

7.4.1 Case Study: F-Discrepancy in Backbone Networks

In the case studies, we use the datasets introduced in Section 7.2.1 to evaluate the various discrepancies of the cost-sharing policies. We use a uniform cost function for the network devices in this section to focus on the specific properties of the cost-sharing policies.

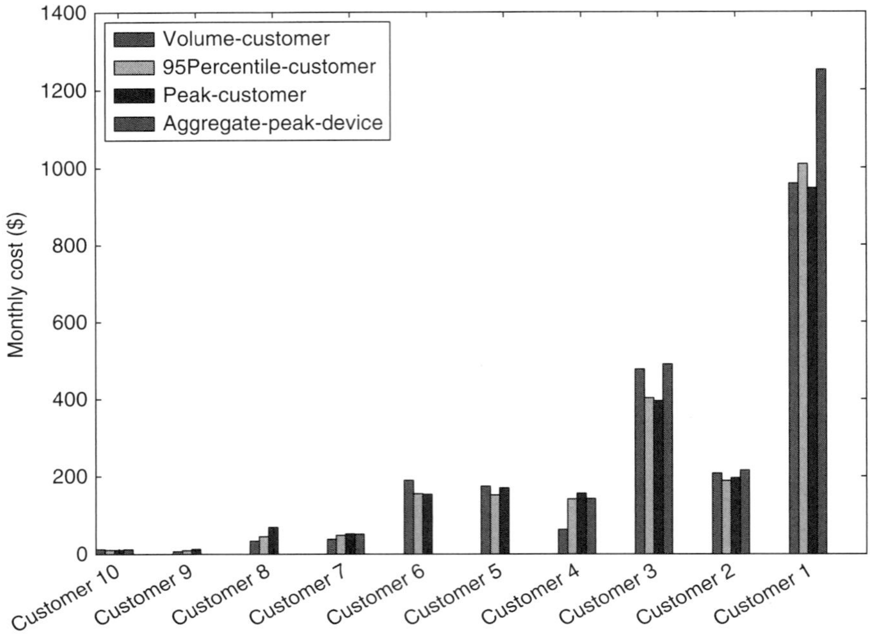

Figure 7.3 Device-level cost of customers for varying cost-sharing policies. Link between two European PoPs.

7.4.1.1 Device-Level F-Discrepancies To showcase the intricacies of F-discrepancies, we start with an example based on a backbone link between two major PoPs in Europe. The monthly cost of this link is \$2163. In Figure 7.3, we plot the amount of this cost attributed to each one of the 10 largest customers according to the four policies. The F-discrepancy, that is, the ratio of the cost computed by the Aggregate-Peak-Device policy and the cost computed by the simpler policy X∈ [Volume-Customer, 95Percentile-Customer, Peak-Customer] is as high as 2.36 for customer 4 in this example. This particular customer impacts the aggregate peak of the device disproportionally more than the other customers when we focus on the traffic volumes of the customers. For several other customers the F-discrepancies are much milder, that is, the different cost-sharing policies are more or less in agreement. In the case of customer 4, the Volume-Customer policy fails by 50% to approximate well the Aggregate-Peak-Device policy (thus, in this particular case, the F-discrepancy is 2). This means that the traffic of this customer is peaky instead of uniformly spread throughout the day.

We now look at F-discrepancies across all customers and all links in our dataset. In Figure 7.4, we plot the F-discrepancies for the three simpler policies and we summarize the main statistics in Table 7.2. The results show generally high F-discrepancies. For example, 60% of the customers are assigned 25% higher or lower cost than the real one they inflict according to Aggregate-Peak-Device. F-dis-

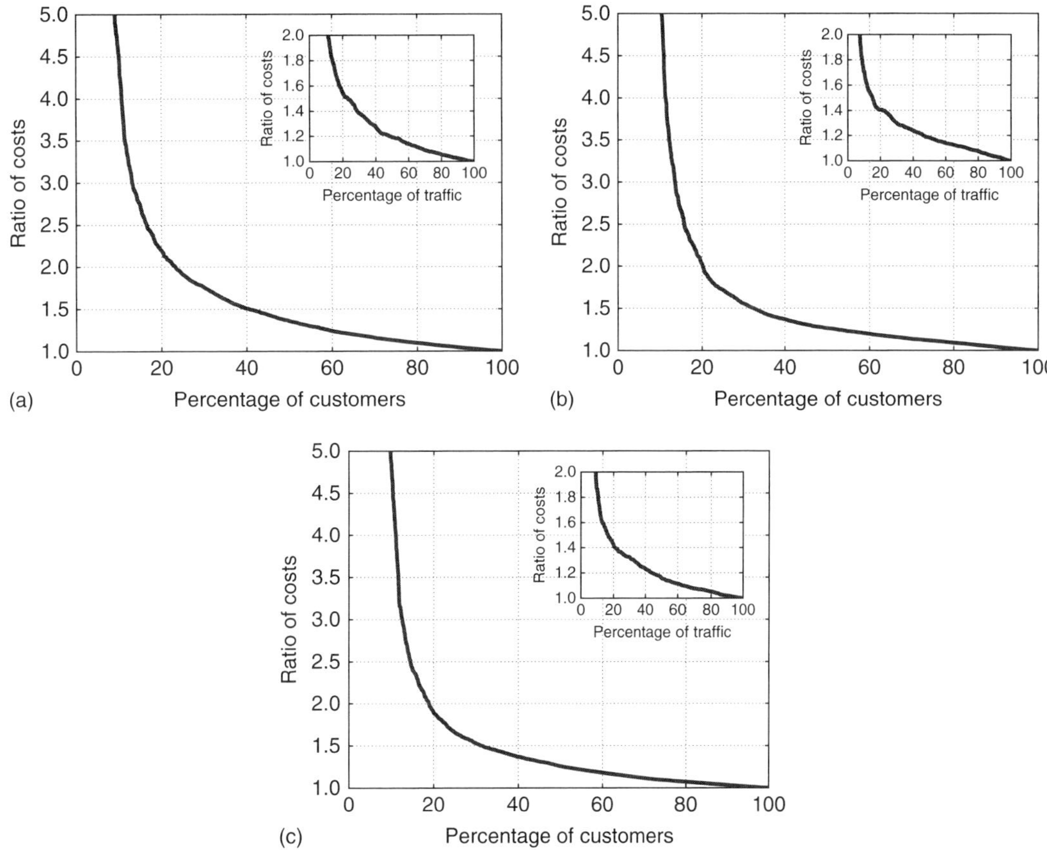

Figure 7.4 Distribution of the device-level F-discrepancies between the simpler cost-sharing policies and the Aggregate-Peak-Device policy. Distributions are based on the percentage of the customers and the traffic (insets). All the policies have high F-discrepancies, especially the F-discrepancies of the Volume-Customer policy are large. (a) Volume-Customer, (b) Peak-Customer, and (c) 95Percentile-Customer.

TABLE 7.2 Device-Level F-Discrepancies Compared to the Aggregate-Peak-Device Policy

Method	>25%	Median	95th Percentile
Volume-Customer	0.592	1.372	63.07
Peak-Customer	0.520	1.263	89.55
95Percentile-Customer	0.508	1.259	80.79

crepancies are particularly high for Volume-Customer and smaller for Peak-Customer and 95Percentile-Customer. The last two policies are sensitive to peaks, albeit those of particular customers instead of peaks of the aggregate traffic on the device. Volume-Customer is even less accurate because it is not looking at any peaks but only at aggregate volume over a longer time scale.

To illustrate the differences between the policies, Figure 7.5 depicts a portion of the time series of a link where a large F-discrepancy (3.07) exists between the Volume-Customer and the Aggregate-Peak-Device policies. The figure shows the traffic pattern of the customer with the large F-discrepancy and the aggregate traffic pattern of the other customers. The traffic of the customer is marginal compared to the traffic of the others, yielding a very low Volume-Customer cost. However, during the peak, the customer with the large discrepancy contributes a significant portion to the aggregate traffic, thereby inducing a 3.07 times higher Aggregate-Peak-Device than Volume-Customer cost.

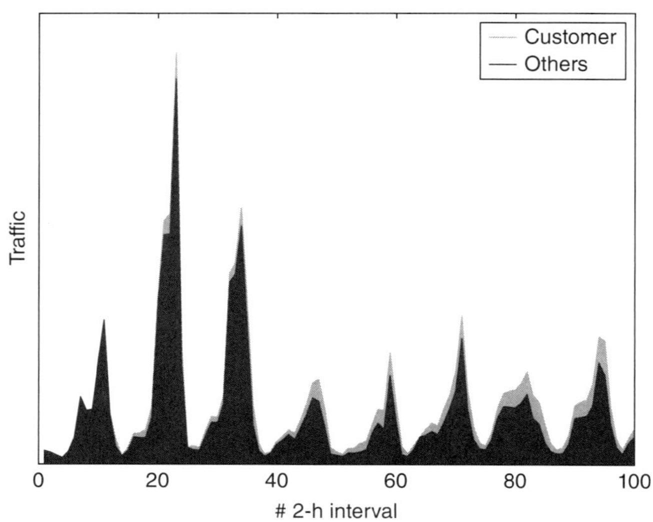

Figure 7.5 Time series of the traffic of a customer with a large (3.07) F-discrepancy on a single link. Volume-Customer versus Aggregate-Peak-Device policy.

SUMMARY AND IMPLICATIONS. In the case of device-level discrepancies, numerous and substantial F-discrepancies exist. This implies that operators should apply the Aggregate-Peak-Device policy for computing the costs in case of a single device instead of the simpler policies.

7.4.1.2 Network-Level F-Discrepancies We now examine F-discrepancies in the context of the entire backbone network. We do this by summing the costs of a customer over all the network's devices. We present the relative aggregate costs of the 10 largest customers in Figure 7.6; we consider the largest cost as the baseline. We present the F-discrepancies of the policies in Table 7.3. The results reveal that F-discrepancies at the network level are much smaller than that at the device level. For example, the network-level median F-discrepancies are approximately 40% less than the device-level ones. This is because in large networks, positive and negative cost differences at each device cancel each other out; thus, the cost predictions of the simpler policies become more aligned.

SUMMARY AND IMPLICATIONS. F-discrepancies although important for individual links or small networks tend to become less significant for larger networks.

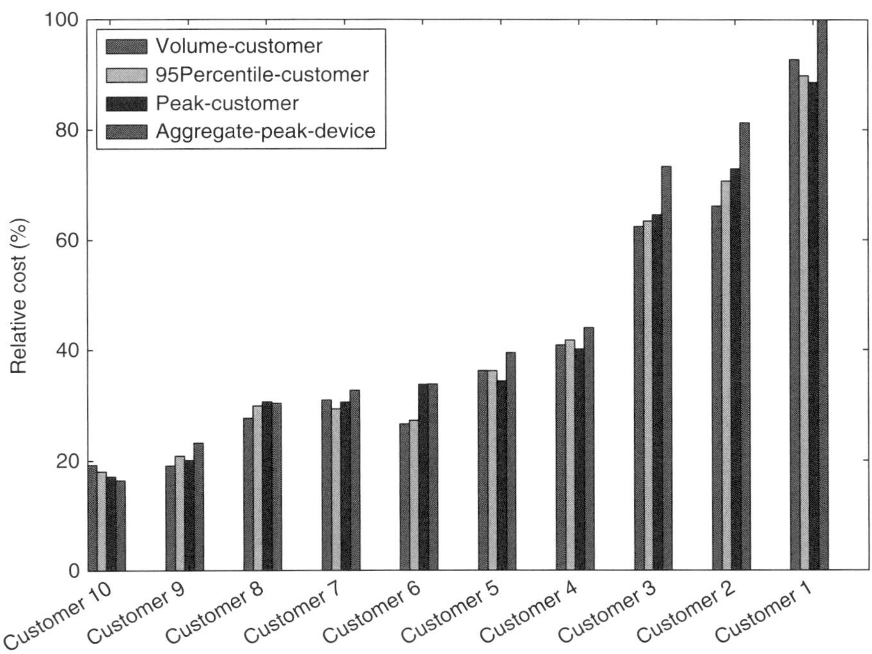

Figure 7.6 Aggregate relative costs of the 10 largest customers; comparison normalized by the largest cost.

TABLE 7.3 Network-Level F-Discrepancies Compared to the Aggregate-Peak-Device Policy

Method	> 25%	Median	95th Percentile
Volume-Customer	0.5	1.251	3.141
Peak-Customer	0.35	1.151	12.71
95Percentile-Customer	0.37	1.181	5.046

7.5 WHERE DO WE METER THE TRAFFIC?

The traffic *metering* method is the second source of the discrepancies (i.e., M-discrepancy). The resource requirements of the traffic monitoring tools depend on the resolution of metering. The main cause behind the M-discrepancies is the trade-off that operators face: increasing the precision of the metering improves the validity of the quantified cost; however, this comes with an elevated cost for traffic monitoring. We consider the two corner cases of traffic metering:

- *Customer-Ingress (CI).* Each customer can have several ingress devices through which it injects its traffic to the network. The operator keeps track of the customers' usage solely on the ingress devices. This is the least expensive metering method. The operator uses the ingress traffic time series to share the network-wide expenditures among the customers.
- *Customer-per-Device (CD).* If the operator deploys more advanced network monitoring tools, it can capture the time series of the customers not only on the ingress devices but also on all the devices located in the network. This is the most expensive metering method and is typically done using NetFlow technology, which comes at a high procurement and administration cost. Metering the actual traffic on each network device allows the operator to compute the costs of the customers based on the device-specific time series. Therefore, the operator faces a trade-off: more accurate expenditure sharing versus more cost efficient operation.

We define M-discrepancy as follows. First, we compute the cost of customer i on each device l using a given cost allocation function (e.g., based on the Volume-Customer policy of Section 7.4), and we compute the network-level cost of customer i as $\sum_{l\in L} c_i^l$. Second, we compute using the given cost allocation function the customer's share (c_i^*) of the network's total cost ($c = \sum_{l\in L} c^l$) using the ingress traffic time series of the customers. The total ingress traffic of customer i is $x_i^*(t) = \sum_{l\in I_i} x_i^l(t)$, where I_i denotes the set of ingress devices that customer i uses. Accordingly, the M-discrepancy of customer i is

$$d_i\left(\sum_{l\in L} c_i^l, c_i^*\right), \tag{7.9}$$

where d_i is our metric of discrepancy (Eq. 7.2).

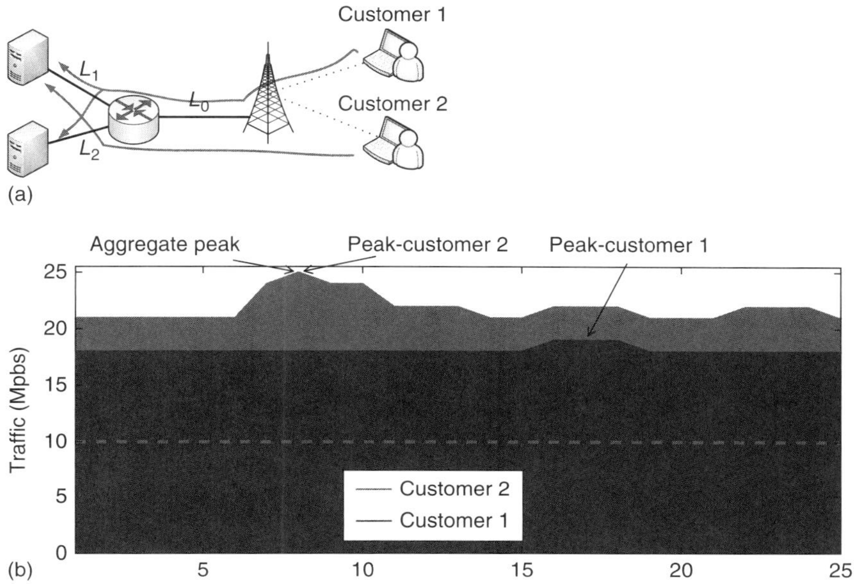

Figure 7.7 Illustrative example for M-discrepancies: metering traffic only at the ingress link (L_0) causes customer 1 to have a larger share of the cost than if we meter at all devices. (a) Topology and (b) traffic pattern of L_0.

ILLUSTRATIVE EXAMPLE. Let us now assume that the customers traffic on L_0 is as we show in Figure 7.7. The network consists of two additional devices: L_1 on which the traffic of the customers is as depicted in Figure 7.2 and L_2 which is solely utilized by customer 1 with a constant traffic of 10 Mbps. For illustration purposes, we separate the two flows of customer 1, one on L_1 and the other on L_2, with a dashed line in the Figure 7.7. Let us assume that the traffic on device L_2 can be transmitted on a device with modest capacity while the capacity of L_1 should be larger. The diverse device capacities imply diverse costs as well. In the case of ingress metering, that is, sharing the cost of the network just based on the traffic of L_0, the cost of customer 1 is 83.9%, 73.1%, 76%, and 72% of the cost of the whole network for the Volume-Customer, Peak-Customer, 95Percentile-Customer, and Aggregate-Peak-Device policies, respectively. However, if we measure the traffic of the customers on all the devices then shares of costs of customer 1 are 75.9%, 65%, 68%, and 62.7%. If we compare these cost fractions, we encounter large discrepancies caused by the level of the metering.

7.5.1 Case Study: M-Discrepancy in Backbone Networks

Similar to F-discrepancies, we use the datasets introduced in Section 7.2.1 and a uniform cost function to analyze M-discrepancies. We compute the discrepancy in the customer's network-level cost derived by (i) metering its traffic at its ingress links

TABLE 7.4 Network-Level M-discrepancies of the Cost-Sharing Policies

Method	>25%	Median	95th Percentile
Volume-Customer	0.695	1.543	34.53
Peak-Customer	0.752	1.738	32.34
95Percentile-Customer	0.750	1.630	19.10
Aggregate-Peak-Device	0.763	1.801	28.52

Comparison of the Customer-Ingress and the Customer-per-Device costs of the customers.

TABLE 7.5 Discrepancies with the Aggregate-Peak-Device Policy Using Customer-per-Device (Real Traffic) Metering

Method	>25%	Median	95th Percentile
Volume-Customer + CI	0.760	1.816	32.69
Aggregate-Peak-Device + CI	0.763	1.801	28.52
Volume-Customer + CD	0.500	1.251	3.141
Aggregate-Peak-Device + CD	0.0	1.0	1.0

Abbreviations: CI, Customer-Ingree; CD, Customer-per-Device.

(Customer-Ingress or CI) and (ii) metering its traffic on each device that the customer uses (Customer-per-Device or CD). All of the policies result in high M-discrepancies (ratios as high as 34) as summarized in Table 7.4.

Up to this point, we analyzed the impact of different discrepancies separately. Next, we quantify the joint effect of F-discrepancies and M-discrepancies, that is, how large can the difference be between the most and the least accurate combination of function and metering schemes. We do this by comparing the network-level costs of customers under the Volume-Customer + CI, Volume-Customer + CD, and Aggregate-Peak-Device + CI policies with the nominally accurate one, namely, the Aggregate-Peak-Device + CD policy. The results are summarized in Table 7.5. The Volume-Customer policy has the smallest M-discrepancy, that is, the median discrepancy of the Customer-Ingress and the Customer-per-Device costs is 1.5. On the contrary, the Aggregate-Peak-Device policy yields the largest M-discrepancies. The reason behind this is twofold. First, when metering traffic at the ingress links, traffic that results in peaks at individual links does not result in peaks of the aggregate ingress traffic. Second, under Customer-Ingress, the Aggregate-Peak-Device policy takes into account only the time interval with the largest aggregate traffic while the peaks of the internal devices may happen in other time intervals neglected by the Aggregate-Peak-Device + CI policy. We observe that under the Aggregate-Peak-Device + CI combination, the costs diverge by at least 25% for 76%

of the customers. In addition, we note that under the Volume-Customer + CI policy and metering, the discrepancy can be as high as 32.

SUMMARY AND IMPLICATIONS. The level at which the operator meters the traffic of the customers has a large impact on the quantified costs. Therefore, operators should apply sophisticated metering strategies (e.g., network-wide deployment of NetFlow-capable traffic monitoring devices) in order to accurately quantify the costs of the customers. Moreover, the simple methods are no longer aligned with the real cost of the customers (i.e., with the Aggregate-Peak-Customer policy) if the traffic is metered on the ingress links.

7.6 WHAT IS THE IMPACT OF THE DIVERSE COSTS OF THE DEVICES?

The third class of discrepancies is related to the *total cost of ownership (TCO)* of different devices of the network (i.e., TCO discrepancy). Owing to the heterogeneous nature of the network—caused by the geographic and technological differences of its parts—the same traffic patterns imply diverse expenditures for the operator on different devices. Therefore, additional discrepancies occur when we consider the TCO of the network in more detail. The following levels of TCO impact the costs and the discrepancies of the customers:

- *Pieces of Cost.* Even if the capacities of two particular equipment are equal, their costs can vary significantly because of the technology differences (newer vs older generation), location (cost of shipping), differences in purchase price, ad so on.
- *Location-Dependent Costs.* The network operator deals with diverse costs at each geographic location where it has a presence. The causes behind the varying costs include but are not limited to the following factors: energy (e.g., the energy price in Germany can be twice as much as that in the United Kingdom), facility costs (e.g., the rental cost of office space in Hong Kong can be four times higher than that in Germany [5]), taxation, and personnel costs.

Contrary to the former types of discrepancies, in the case of the TCO, only network-level discrepancies exist. Formally, we define the network-level TCO discrepancy of customer i as

$$d\left(\sum_{l\in L} c_i^l, \sum_{l\in L} c^l \cdot \frac{\sum_{l\in L} e_i^l}{\sum_{l\in L}\sum_{i\in N} e_i^l}\right). \qquad (7.10)$$

where the first term considers the diverse costs of the devices contrary to the second. e_i^l denotes the cost of customer i in case of device l assuming uniform cost across all the devices ($e^l = e^*, \forall l \in L$).

7.6.1 Case Study: TCO Discrepancy in Backbone Networks

Now we take into account that our dataset (Section 7.2.1) contains a geographically distributed set of links with diverse costs. We compute the TCO discrepancies by computing the ratio between the customers' costs given links with uniform and diverse costs. In Figure 7.8, we illustrate the TCO discrepancies under the Aggregate-Peak-Device policy. Each customer is affected by the TCO discrepancies. The difference between the two costs can be as high as 5% of the cost of the entire network.

We report the quantified TCO discrepancies of four policies in Table 7.6. The results show generally extreme TCO discrepancies; some customers have TCO discrepancies as high as 900. In addition, 80% of the customers are assigned 25% higher or lower cost when the diverse costs of the links is considered.

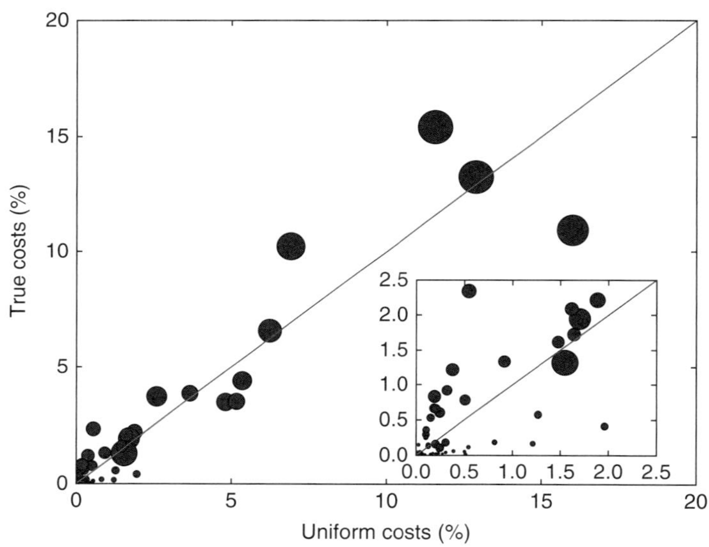

Figure 7.8 The customers' total costs for the uniform and true link costs using the Aggregate-Peak-Device cost-sharing policy; the size of the circles is proportional to the aggregate traffic volume of the customers.

TABLE 7.6 Network-Level TCO Discrepancies, That is, the Costs of the Customers Based on Uniform Versus Diverse Link Costs

Method	>25%	Median	95th Percentile
Volume-Customer	0.830	4.305	961.1
Peak-Customer	0.802	4.187	933.1
95Percentile-Customer	0.817	4.079	922.4
Aggregate-Peak-Device	0.840	4.019	862.1

SUMMARY AND IMPLICATIONS. TCO discrepancies have a very large impact on the costs of the customers. Cross-subsidization problems arise if the impact of TCO differences is neglected. Network operators are aware of the fact that different parts of their network have different TCOs. The implication of our results is that this diversity should also be reflected in the quantification of the customers' costs—and eventually in the tariffs too.

7.7 WHO IS LIABLE FOR THE INCURRED COSTS?

On the basis of the discussion of Section 7.2, one may conclude that splitting the cost among customers is straightforward: for each device of the network, each customer should pay in proportion to his/her contribution to the peak traffic carried by the device. Things, however, are not that simple owed to *liability* complications. If we were to build from scratch a new network for a fixed set of customers of known demand, then the cost attributed to each customer should be proportional to the sum of its contributions to the peaks of individual devices. Splitting costs based on the contribution to the peak is indeed exact but only for this "offline problem." However, in reality, networks are not deployed as a single event but grow organically with the addition of new customers and the ramping up of their traffic. Under this more realistic case, peak-based cost sharing is not guaranteed to be fair. Consider, for example, the case in which a network is already operating at the maximum utilization allowed by QoS constraints and a small new customer triggers an expensive upgrade that leads to a new network with plentiful unallocated capacity. In Figure 7.9, we illustrate the case when a new customer arrives to the network and pushes the aggregate traffic of the device above the upgrade threshold. Peak-based cost sharing would attribute only a small fraction of the overall cost to the new customer. Is that fair? The answer depends on what happens with the unallocated capacity. If the network can easily sell it to new or existing customers, then indeed it is fair. If, however, selling this leftover capacity is not guaranteed, then the new customer may have a larger liability for the upgrade costs.

The final type of discrepancies is caused by the different types of customer *liability* (i.e., L-discrepancy). We examine the following policies.

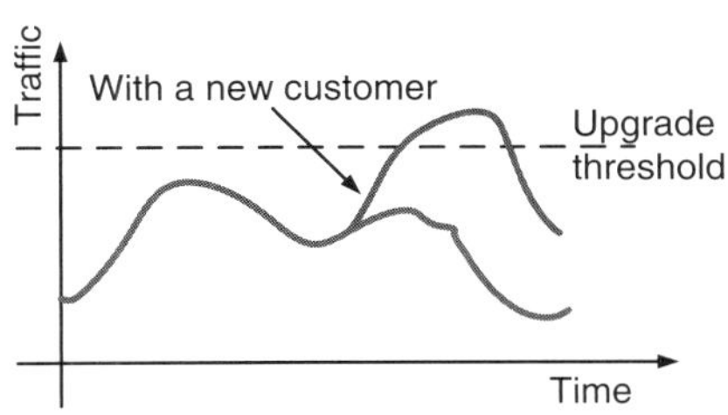

Figure 7.9 The traffic of a new customer may increase the aggregate traffic of a device to such an extent that the capacity of the device should be upgraded. Who should cover the elevated costs of the network in such a case?

- *Aggregate-Peak-Device.* This is the already introduced policy that is the measure of fairness when the customer liability is proportional to the aggregate peak of devices.
- *Trigger.* With this policy, the operator allocates the cost of the device exclusively to the customer that triggered the capacity upgrade. This policy is applied when the operator is not confident that it can sell the newly obtained but unallocated capacity. To this end, the operator utilizes the historical traffic patterns of the customers and their arriving order. For example, the cost of the first customer is equal to the cost of the device that is capable to transmit his traffic demand. We assume that the customers are numbered based on their arriving order while t_i denotes the time when the customer started to use the network. Accordingly, the cost of customer i in case of the Trigger policy is

$$c_i^l = C\left(\max_{t \le t_i} \sum_{j \in \{N | j \le i\}} x_j^l(t)\right) - C\left(\max_{t \le t_i} \sum_{j \in \{N | j < i\}} x_j^l(t_i)\right). \tag{7.11}$$

  The main drawbacks of Trigger are (i) it assigns cost only to the customer whose traffic triggers upgrades and 0 to everyone else; therefore, order of arrival can have a huge impact on the costs attributed to a customer; and (ii) it is difficult to compute Trigger because it requires extensive historical data on the order of customer arrival and traffic build up.
- *Shapley.* The Shapley cost-sharing policy lies between the two above presented extremes, the Aggregate-Peak-Device and Trigger policies. It assigns to customers partial liability for upgrades, thereby avoiding the all-or-nothing assignments of Triggers. Therefore, it is less strict than Trigger but more strict than Aggregate-Peak-Device because it assigns "averaged" liabilities rather than proportional liabilities based on a single time interval when a device peaks.

  The main advantage of the Shapley over the Aggregate-Peak-Device policy is that its allocations are more stable than that of the Aggregate-Peak-Device policy in view of customer churn. For example, let us imagine that the aggregate peak traffic of a device is P and appears at t_1 (see Fig. 7.10). We also assume

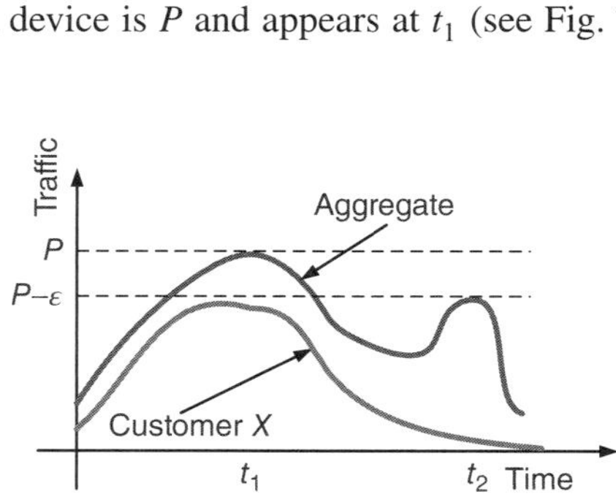

Figure 7.10 The Shapley policy is a robust cost-sharing policy as it takes into account all the local maxima of the aggregate traffic in quantifying the costs of the customers.

that the device at time t_2 has load $P - \varepsilon$. Now let us suppose that customer X is responsible for 90% of P at t_1 and 1% of $P - \varepsilon$ at t_2. Then if a small 2ε customer leaves from P at t_1, then the peak will move to t_2 and X will go from paying 90% to paying only 1% after a tiny 2ε perturbation of the aggregate traffic. On the contrary, the Shapley policy is aware of such situations as it takes into account all the local maxima of the aggregate traffic in quantifying the costs of the customers.

Under the Shapley policy, the cost of each customer is proportional to its average marginal contribution to the device's total cost. Particularly, let us consider all the possible $S \subset N$ subsets (coalitions) of the customers who utilize resources of the network device l. The cost of coalition S depends on the aggregate traffic volume of the participants, that is , it is equal to the cost of a network device that has sufficient capacity:

$$v^l(S) = C\left(\max_{t\in T} \sum_{j\in S} x_j^l(t)\right). \tag{7.12}$$

On the basis of the v cost function of the coalitions, we compute the Shapley value of customer i as

$$\phi_i(v^l) = \frac{1}{N!} \sum_{\Pi\in S_N} \left(v^l\left(S\left(\Pi, i\right)\right) - v^l\left(S\left(\Pi, i\right) \setminus i\right)\right), \tag{7.13}$$

where Π is a permutation of arrival order of the set N and $S(\Pi, i)$ denotes the set of players who arrived no later than i. The $(\phi_1(v), \ldots, \phi_N(v))$ Shapley values describe the fair distribution of costs in the case of the $S = N$ grand coalition. Fair in a way that it satisfies four intuitive fairness criteria [6–8]. We quantify the cost of customer i based on its Shapley value for the device l as

$$c_i^l = c^l \cdot \frac{\phi_i(v^l)}{\sum_{j\in N} \phi_j(v^l)}. \tag{7.14}$$

ILLUSTRATIVE EXAMPLE. We present the traffic patterns of two customers and the thresholds where the capacity of the device needs to be upgraded in Figure 7.11. Customer 1 is liable for 53.3% and 87.5% of the cost of the device in case of the Aggregate-Peak-Device and Shapley policies, respectively. Thus, the discrepancy is 1.64. The peak of the aggregate traffic happens in a time step where the customers' traffic volumes are balanced. Although there are local maxima where the traffic of customer 2 is small, it is not considered by the Aggregate-Peak-Device policy. From a Shapley policy viewpoint, the traffic peak of customer 1 is too large to be transmitted with a lower capacity device, that is , its traffic is mainly responsible for the total cost of the device. If we assume that customer 1 arrived first, it causes 100% of the costs according to the Trigger policy because its peak needs a larger-capacity device whose leftover capacity can be used by customer 2 afterwards.

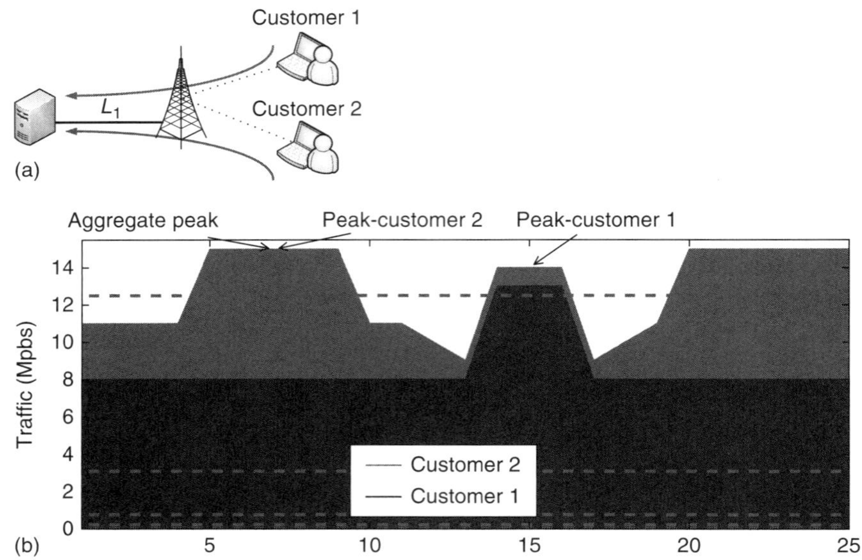

Figure 7.11 Illustrative example for L-discrepancies. The capacity thresholds of device L_1 are shown with dashed lines. (a) Topology and (b) traffic pattern of L_1.

Customers can have both *device-* or *network-level* L-discrepancies, depending on whether we consider the costs of the customers on particular devices (e.g., c_i^l) or on the aggregate (e.g., $\sum_{l \in L} c_i^l$).

7.7.1 Case Study: L-Discrepancy in Backbone Networks

Out of the three policies described earlier, one, the Trigger policy, requires historic information on customer arrival events as well as customer traffic information on long time scales that relate to network upgrade events. As we do not have full historic information on all the links, we omit the analysis of the Trigger policy in this case study.

7.7.1.1 Device-Level L-Discrepancies We present the L-discrepancies in Table 7.7 by computing the ratio between Shapley and the Aggregate-Peak-Device policies. L-discrepancies are quite high (ratios up to 472) pointing to the fact that liability can bias significantly the cost-sharing picture that a telco has. For example, the difference between the costs of the Shapley and the Aggregate-Peak-Device policies is substantial: the median ratio of the costs is 1.5; however, in some cases, the ratio can be larger than 400.

We present in Figure 7.12 a part of the time series of a customer with a large L-discrepancy (3.25) along with the aggregate time series of the other customers who utilize the same link. The dashed horizontal lines denote the traffic volumes where the capacity of the link needs to be upgraded. The traffic of the customer is

TABLE 7.7 Device-Level L-Discrepancies Compared to the Aggregate-Peak-Device Policy

Method	>25%	Median	95th Percentile
Shapley	0.674	1.497	472.4

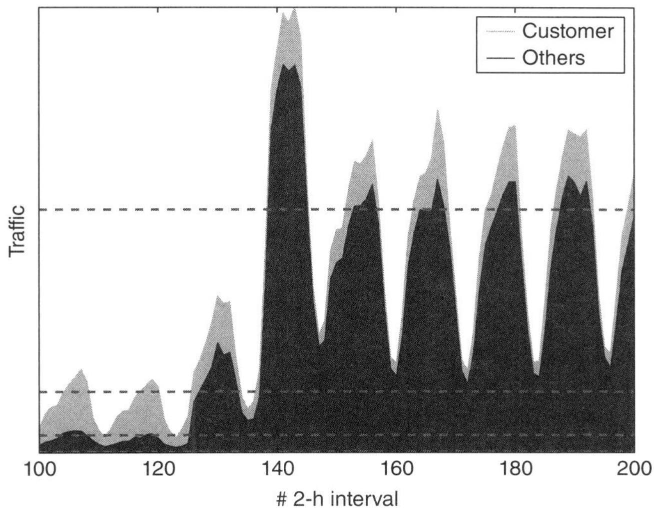

Figure 7.12 Time series of a customer with large (3.25) L-discrepancy (Shapley vs Aggregate-Peak-Device policy). The dashed lines represent the traffic volumes where the capacity of the link needs to be upgraded.

small enough to be transmitted over a link with lower capacity. However, the traffic of the other customers pushes the link to have larger capacity and thus larger cost. The Shapley policy considers this fact when it computes the average marginal contribution of the customer. As a result, the cost of the customer is less if we compute it based solely on time of the largest utilization of the device. On the contrary, the Aggregate-Peak-Customer policy focuses only on the time interval when the link has its aggregate peak. The particular customer has significant share of the aggregate peak and thus of the cost of the link according to the Aggregate-Peak-Customer. This, however, masks who is responsible for the link's larger capacity.

7.7.1.2 Network-Level L-Discrepancies We show the network-level L-discrepancies in Table 7.8. At the network level, the number and the magnitude of the L-discrepancies is smaller than that at the device level. Nevertheless, for more than 50% of the customers, the costs are off by at least 25%. The median L-discrepancies of the policies are notable too.

TABLE 7.8 Network-Level L-Discrepancies Compared to the Aggregate-Peak-Device Policy

Method	>25%	Median	95th Percentile
Shapley	0.54	1.316	179.3

SUMMARY AND IMPLICATIONS. The liability of network upgrades plays an important role in the quantification of the costs of customers. The implication of the results is that if the network is not built in one shot but is rather organically grown and upgraded, then the Aggregate-Peak-Customer policy may induce cross-subsidization problems: customers may be accounted for costs of upgrades for which they are not liable (or not in that degree). From a customer point of view, this cross-subsidization may not be tolerated in a long run, given the competitive environment of communication networks. That is, the customers may select other operator where they are not liable for the costs of others. From the operator point of view, the large L-discrepancies dictate that it needs to take them under serious consideration. If it is anticipated that the market for data services will be healthy, the operator should choose the Aggregate-Peak-Device policy. If, however, it expects difficulties in selling its capacity, our results indicate that Shapley should be the policy of choice.

7.8 RELATED WORK

We refer to the textbook of Courcoubetis and Weber [9] for a thorough treatment of pricing in communication networks. More detailed analyses of the challenge of cost sharing in backbone networks were carried out in References 2 and 10.

Several studies investigated how to reduce the transit costs including ISP peering [11–13], CDNs [14], P2P localization [15], and traffic smoothing [16]. Dimitropoulos et al. [4] presented a comprehensive analysis of the 95th percentile pricing. A proposal by Laoutaris et al. [17, 18] showed how traffic can be transferred in the network without increasing the 95th percentile of the customers. A recent proposal by Stanojevic et al. [19] proposes to the customers of transit providers to form a coalition to reduce their transit costs. Valancius et al. [20] show that a small number of pricing tiers are enough to extract close-to-optimal efficiency in the transit provider. Motiwala et al. [21] developed a cost model that operators can use to evaluate the costs of their routing and peering decisions. The net neutrality debate is in many ways related to the question of who is responsible for the costs in the network [22].

Owing to the desirable fairness properties [6–8] of the Shapley value [23], recent studies proposed pricing and cost-sharing mechanisms using Shapley values. Briscoe [24, 25] motivates the usage of mechanisms that share the costs of the users fairly as a way to reduce widely known cross-subsidization (the phenomenon in which a small set of customers is subsidized by a large fraction of other customers of the service) of the common infrastructure that often happens in the communication networks [26]. Stanojevic et al. [27] investigated the cross-subsidization of cellular subscribers from a service plan selection point of view. Cooperative approaches for cost sharing are investigated in case of interdomain routing [28, 29], and IP multicast [6, 7]. Ma et al. [30, 31] presented a fair revenue sharing method for ISPs that quantifies the importance of each ISP in the Internet ecosystem. The work of Stanojevic et al. [3] empirically investigated the temporal usage effects using the Shapley and the 95Percentile-Customer method in case of asymmetric digital subscriber line (ADSL) subscribers.

7.9 CONCLUSIONS

Network operators need to know accurately the costs of their customers to apply smart data pricing schemes in practice. Despite the widespread availability of big data infrastructures, the quantification of the costs of individual customers is a challenging task in communication networks. This chapter provided a thorough analysis of four nontrivial underlying mechanism impacting the quantification of the costs. The influencing factors include temporal/spatial characteristics of the customers, nonlinear cost–capacity relationships, measurement infrastructure issues, and high variability of the component costs. On the basis of the findings of our case studies, usage-based tariffs should include device-level expenditures and measurements to assure their accuracy and fairness.

REFERENCES

1. W. B. Norton. *The Internet Peering Playbook: Connecting to the Core of the Internet*. DrPeering Press, 2012.
2. L. Gyarmati, R. Stanojevic, M. Sirivianos, and N. Laoutaris. Sharing the cost of backbone networks: cui bono? In *Proceedings of the 2012 ACM Conference on Internet Measurement Conference, IMC '12*, pp. 509–522, New York, NY, USA, 2012. ACM.
3. R. Stanojevic, N. Laoutaris, and P. Rodriguez. On economic heavy hitters: shapley value analysis of 95th-percentile pricing. In *Proceedings of the 10th ACM SIGCOMM Conference on Internet Measurement, IMC '10*, pp. 75–80, New York, NY, USA, 2010. ACM.
4. X. Dimitropoulos, P. Hurley, A. Kind, and M. Stoecklin. "On the 95-percentile billing method,". In S. Moon, R. Teixeira, and S. Uhlig, eds., *Passive and Active Network Measurement*, *Lecture Notes in Computer Science*, vol. 5448, Springer Berlin, Heidelberg, 2009, pp. 207–216.
5. Cushman & Wakefield. Office Space Across the World, 2012.
6. A. Archer, J. Feigenbaum, A. Krishnamurthy, R. Sami, and S. Shenker. "Approximation and collusion in multicast cost sharing," *Games and Economic Behavior*, 47(1), 2004, 36–71.
7. J. Feigenbaum, C. H. Papadimitriou, and S. Shenker. "Sharing the cost of multicast transmissions," *Journal of Computer and System Sciences*, 63(1), 2001, 21–41.
8. H. Moulin and S. Shenker. "Strategyproof sharing of submodular costs: budget balance versus efficiency," *Economic Theory*, 18, 2001, 511–533. Doi: 10.1007/PL00004200.

9. C. Courcoubetis and R. Weber. *Pricing and Communications Networks*. John Wiley & Sons, Ltd, 2003, West Sussex, England.
10. L. Gyarmati, M. Sirivianos, and N. Laoutaris. Simplicity vs Precision: sharing the cost of backbone networks. In *NetEcon 2012 - Seventh Workshop on the Economics of Networks, Systems, and Computation*, 2012.
11. B. Augustin, B. Krishnamurthy, and W. Willinger. Ixps: mapped? In *Proceedings of the 9th ACM SIGCOMM Conference on Internet Measurement Conference, IMC '09*, pp. 336–349, New York, NY, USA, 2009. ACM.
12. A. Dhamdhere and C. Dovrolis. The internet is flat: modeling the transition from a transit hierarchy to a peering mesh. In *Proceedings of the 6th International Conference, Co-NEXT '10*, pp. 21:1–21:12, New York, NY, USA, 2010. ACM.
13. A. Dhamdhere, C. Dovrolis, and P. Francois. A value-based framework for internet peering agreements. In *Teletraffic Congress (ITC), 2010 22nd International*, 2010.
14. L. Qiu, V. N. Padmanabhan, and G. M. Voelker. On the placement of web server replicas. In *IEEE INFOCOM*, pp. 1587–1596, 2001.
15. D. R. Choffnes and F. E. Bustamante. Taming the torrent: a practical approach to reducing cross-ISP traffic in peer-to-peer systems. In *Proceedings of the ACM SIGCOMM 2008 Conference on Data Communication, SIGCOMM '08*, pp. 363–374, New York, NY, USA, 2008. ACM.
16. M. Marcon, M. Dischinger, K. P. Gummadi, and A. Vahdat. The local and global effects of traffic shaping in the internet. In *Third International Conference on Communication Systems and Networks (COMSNETS)*, 2011.
17. N. Laoutaris, G. Smaragdakis, P. Rodriguez, and R. Sundaram. Delay tolerant bulk data transfers on the internet. In *Proceedings of the Eleventh International Joint Conference on Measurement and Modeling of Computer Systems, SIGMETRICS '09*, pp. 229–238, New York, NY, USA, 2009. ACM.
18. N. Laoutaris, M. Sirivianos, X. Yang, and P. Rodriguez. Inter-data center bulk transfers with netstitcher. In *Proceedings of the ACM SIGCOMM 2011 Conference, SIGCOMM '11*, pp. 74–85, New York, NY, USA, 2011. ACM.
19. R. Stanojevic, I. Castro, and S. Gorinsky. Cipt: using tuangou to reduce ip transit costs. In *Proceedings of the Seventh Conference on Emerging Networking Experiments and Technologies, CoNEXT '11*, pp. 17:1–17:12, New York, NY, USA, 2011. ACM.
20. V. Valancius, C. Lumezanu, N. Feamster, R. Johari, and V. V. Vazirani. How many tiers? Pricing in the internet transit market. In *Proceedings of the ACM SIGCOMM 2011 Conference, SIGCOMM '11*, pp. 194–205, New York, NY, USA, 2011. ACM.
21. M. Motiwala, A. Dhamdhere, N. Feamster, and A. Lakhina. "Towards a cost model for network traffic," *SIGCOMM Computer Communication Review*, 42(1), 2012, 54–60.
22. K. C. Claffy. "Network neutrality": the meme, its cost, its future," *SIGCOMM Computer Communication Review*, 41(5), 2011, 44–45.
23. L. S. Shapley. "A Value for n-Person Games," In *Annals of Mathematical Studies*, H. W. Kuhn and A. W. Tucker (eds.), 1953, pp. 307–318.
24. B. Briscoe. "Flow rate fairness: dismantling a religion," *SIGCOMM Computer Communication Review*, 37(2), 2007, 63–74.
25. B. Briscoe. "A fairer, faster internet," *IEEE Spectrum*, 45(12), 2008, 42–47.
26. K. Cho, K. Fukuda, H. Esaki, and A. Kato. The impact and implications of the growth in residential user-to-user traffic. In *SIGCOMM '06*, pp. 207–218, New York, NY, USA, 2006. ACM.

27. R. Stanojevic, V. Erramilli, and K. Papagiannaki. Cognitive bias in network services. In *Proceedings of the 11th ACM Workshop on Hot Topics in Networks, HotNets-XI*, pp. 49–54, New York, NY, USA, 2012. ACM.
28. R. Mahajan, D. Wetherall, and T. Anderson. Negotiation-based routing between neighboring ISPs. In *Proceedings of the 2nd Conference on Symposium on Networked Systems Design & Implementation - Volume 2, NSDI'05*, pp. 29–42, Berkeley, CA, USA, 2005. USENIX Association.
29. G. Shrimali, A. Akella, and A. Mutapcic. "Cooperative interdomain traffic engineering using Nash bargaining and decomposition," *IEEE/ACM Transactions on Networking*, 18(2), 2010, 341–352.
30. R. T. B. Ma, D. M. Chiu, J. C. S. Lui, V. Misra, and D. Rubenstein. Internet economics: the use of shapley value for ISP settlement. In *Proceedings of the 2007 ACM CoNEXT Conference, CoNEXT '07*, pp. 6:1–6:12, New York, NY, USA, 2007. ACM.
31. R. T. B. Ma, D.-M. Chiu, J. C. S. Lui, V. Misra, and D. Rubenstein. On cooperative settlement between content, transit and eyeball internet service providers. In *Proceedings of the 2008 ACM CoNEXT Conference, CoNEXT '08*, pp. 7:1–7:12, New York, NY, USA, 2008. ACM.

8 Usage-Based Pricing Differentiation for Communication Networks: Incomplete Information and Limited Pricing Choices*

SHUQIN LI and JIANWEI HUANG

8.1 INTRODUCTION

Pricing is important for the design, operation, and management of communication networks. Pricing has been used with two different meanings in the area of communication networks. One is the "optimization-oriented" pricing for network resource allocation: it is made popular by Kelly's seminal work on network congestion control [2, 3]. For example, the Transmission Control Protocol (TCP) has been successfully reverse engineered as a congestion-pricing-based solution to a network optimization problem [4, 5]. A more general framework of network utility maximization (NUM) was subsequently developed to forward-engineer many new network protocols (see a recent survey in Reference [6]). In various NUM formulations, the "optimization-oriented" prices often represent the Lagrangian multipliers of various resource constraints and are used to coordinate different network entities to achieve the maximum system performance in a distributed manner. The other is the "economics-based" pricing, which is used by a network service provider to various objectives including revenue maximization. The proper design of such a pricing becomes particularly challenging today because of the exponential growth of data volume and applications in both wireline and wireless networks. In this

*This work is supported by the General Research Funds (Project Number CUHK 412710 and CUHK 412511) established under the University Grant Committee of the Hong Kong Special Administrative Region, China. Part of the results have appeared in IEEE GLOBECOM 2009 [1].

chapter, we focus on studying the "economics-based" pricing schemes for managing communication networks.

Economists have proposed many sophisticated pricing mechanisms to extract surpluses from the consumers and maximize revenue (or profits) for the providers. A typical example is the optimal nonlinear pricing (e.g., [7, 8]). In practice, however, we often observe simple pricing schemes deployed by the service providers. Typical examples include flat-fee pricing and (piecewise) linear usage-based pricing. One potential reason behind the gap between "theory" and "practice" is that the optimal pricing schemes derived in economics often has a high implementational complexity. Besides a higher maintenance cost, complex pricing schemes are not "customer friendly" and discourage customers from using the services [9, 10]. Furthermore, achieving the highest possible revenue with complicated pricing schemes requires knowing the information (identity and preference) of each customer, which can be challenging in large-scale communication networks. It is then natural to ask the following two questions:

1. How to design simple pricing schemes to achieve the best trade-off between complexity and performance?
2. How does the network information structure impact the design of pricing schemes?

This chapter tries to answer the above two questions with some stylized communication network models. Different from some previous work that considered a flat-fee pricing scheme in which the payment does not depend on the resource consumption (e.g., [9, 11, 12]), here we study the revenue maximization problem with the linear usage-based pricing schemes, where a user's total payment is linearly proportional to allocated resource. In wireless communication networks, the usage-based pricing scheme has become increasingly popular because of the rapid growth of wireless data traffic. In June 2010, AT&T in the United States switched from the flat-free-based pricing (i.e., unlimited data for a fixed fee) to the usage-based pricing schemes for 3G wireless data [13]. Verizon followed up with similar plans in July 2011. Similar usage-based pricing plans have been adopted by major Chinese wireless service providers including China Mobile and China UniCom. Thus, the research on the usage-based pricing is of great practical importance.

In this chapter, we consider the revenue maximization problem of a monopolist service provider facing multiple groups of users. Each user determines its optimal resource demand to maximize the surplus, which is the difference between its utility and payment. The service provider chooses the pricing schemes to maximize his/her revenue, subject to a limited resource. We consider both complete information and incomplete information scenarios and design different pricing schemes with different implementational complexity levels.

Our main contributions are as follows.

- *Complete Network Information.* We propose a polynomial algorithm to compute the optimal solution of the partial price differentiation problem, which includes the complete price differentiation scheme and the single pricing scheme as special cases. The optimal solution has a threshold structure, which allocates positive resources with priorities to users with high willingness to pay.

- *Revenue Gain under Complete Network Information.* Compared to the single pricing scheme, we identify the two important factors behind the revenue increase of the (complete and partial) price differentiation schemes: the differentiation gain and the effective market size. The revenue gain is the most significant when users with high willingness to pay are minority among the whole population and total resource is limited but not too small.
- *Incomplete Network Information.* We design an incentive-compatible scheme with the goal to achieve the same maximum revenue that can be achieved with the complete information. We find that if the differences of willingness to pays of users are larger than some thresholds, this incentive-compatible scheme can achieve the same maximum revenue. We further characterize the necessary and sufficient condition for the thresholds.

8.1.1 Related Work

It is often quite challenging to design a practical pricing schemes in communication networks. The main difficulties including dealing with the incomplete information structure and limiting the implementational complexity. Under incomplete network information, customers have the private information that is unknown by the service providers. The study of incomplete information is a significant part of microeconomics, especially in several important branches, such as incentive theory [14], information economics [15], organization theory [16], and contract theory [17]. In particular, there exists a rich body of literature on monopoly revenue maximization with incomplete information, for example, [7, 8, 18–26]. Mussa and Rosen in Reference [7] and Maskin and Riley in Reference [8] proposed the optimal price differentiation strategies based on product qualities and quantities, respectively. Armstrong in Reference [22], Bakos and Brynjolfsson in Reference [23], and Geng et al. in Reference [24] studied the optimal multiproduct bundling schemes. Stokey Reference [18], Baron and Besanko in [19], Hart and Tirole in Reference [20], and Acquisti and Varian in Reference [21] focused on multistage price differentiation. Cabral et al. in Reference [25] and Aoyagi in Reference [26] studied the pricing and revenue maximization problem with network externalities. Although these existing results on pricing provide theoretical optimal solutions under incomplete information, they are seldom directly applied in practice, mainly because of their high implementational complexity. In a practical system, a service provider often needs to constrain the number of pricing choices for the customers, either due to implementation complexity constraint or users' aversion to too many choices [27].

However, the issue of optimal pricing design subject to complexity constraint is less understood in the literature. One related analytical result is Reference [9], where the authors discussed complexity issue in flat-fee pricing. To the best of our knowledge, our study about partial price differentiation is the first result about the optimal design of limited price choices in usage-based pricing schemes. It is interesting to compare our results on partial price differentiation with results in References [9] and [28]. Shakkottai et al. in Reference [9] showed that the revenue gain of price differentiation is small with a flat entry-fee-based Paris Metro Pricing (e.g., [29]), and a complicated differentiation strategy may not be worthwhile. Chau et al. [28] further derived the sufficient conditions of congestion functions that guarantee the viability of these Paris Metro Pricing schemes. By contrast, our results show that the revenue gain of price differentiation can be substantial for a usage-based pricing system.

Some recent work of usage-based pricing and revenue management in communication network includes References [30–38]. Basar and Srikant in Reference [30] investigated the bandwidth allocation problem in a single-link network with the single pricing scheme. Shen and Basar in Reference [31] extended the study to a more general nonlinear pricing case with the incomplete network information scenario. They discussed the single pricing scheme under incomplete information with a continuum distribution of users' types. In contrast, our study on the incomplete information focuses on the linear pricing with a discrete setting of users' types. We also show that, besides the single pricing scheme, it is also possible to design differentiation pricing schemes under incomplete information. Daoud et al. [32] studied an uplink power allocation problem in a CDMA system, where the interference among users are the key constraint instead of the limited total resource considered in this chapter. Jiang et al. in Reference [33], Hande et al. in Reference [34], and Ha et al. in Reference [35] focused on the study of the time-dependent pricing. He and Walrand in Reference [36], Shakkottai and Srikant in Reference [37], and Gajic et al. in Reference [38] focused on the interaction between different service providers embodied in the pricing strategies, rather than the design of the pricing mechanism. Besides, none of the related work considered the partial differential pricing as in this chapter.

8.2 SYSTEM MODEL

We consider a network with a total amount of S limited resource (which can be in the form of rate, bandwidth, power, time slot, etc.). The resource is allocated by a monopolistic service provider to a set $\mathcal{I} = \{1, \ldots, I\}$ of user groups. Each group $i \in \mathcal{I}$ has N_i homogeneous users[1] with the same utility function

$$u_i(s_i) = \theta_i \ln(1 + s_i), \tag{8.1}$$

where s_i is the allocated resource to one user in group i and θ_i represents the willingness to pay of group i. The logarithmic utility function is commonly used to model the proportionally fair resource allocation in communication networks (see Reference [30] for detailed explanations). The analysis of the complete information case can also be extended to more general utility functions (see Appendix 8.A.1). Without the loss of generality, we assume that $\theta_1 > \theta_2 > \cdots > \theta_I$. In other words, group 1 contains users with the highest valuation, and group I contains users with the lowest valuation.

[1] A special case is N_i=1 for each group, that is, all users in the network are different.

We consider two types of information structures:

1. *Complete Information*. The service provider knows each user's utility function. Although the complete information is a very strong assumption, it is the most frequently studied scenario in the network pricing literature [30–34, 36–38]. The significance of studying the complete information is twofold. It serves as the benchmark of practical designs and provides important insights for the incomplete information analysis.
2. *Incomplete Information*. The service provider knows the total number of groups I, the number of users in each group $N_i, i \in \mathcal{I}$, and the utility function of each group $u_i, i \in \mathcal{I}$. It does not know which user belongs to which group. Such assumption in our discrete setting is analogous to that the service provider knows only the users' type distribution in a continuum case. Such statistical information can be obtained through long-term observations of a stationary user population.

The interaction between the service provider and users can be characterized as a two-stage Stackelberg model shown in Figure 8.1.

The service provider publishes the pricing scheme in Stage 1, and users respond with their demands in Stage 2. The users want to maximize their surpluses by optimizing their demands according to the pricing scheme. The service provider wants to maximize its revenue by setting the right pricing scheme to induce desirable demands from users. As the service provider has a limited total resource, it must guarantee that the total demand from users is no larger than what it can supply.

The details of pricing schemes depend on the information structure of the service provider. Under complete information, because the service provider can distinguish different groups of users, it announces the pricing and the admission control decisions to different groups of users. It can choose from the complete price differentiation scheme, the single pricing scheme, and the partial price differentiation scheme to realize a desired trade-off between the implementational complexity and the total

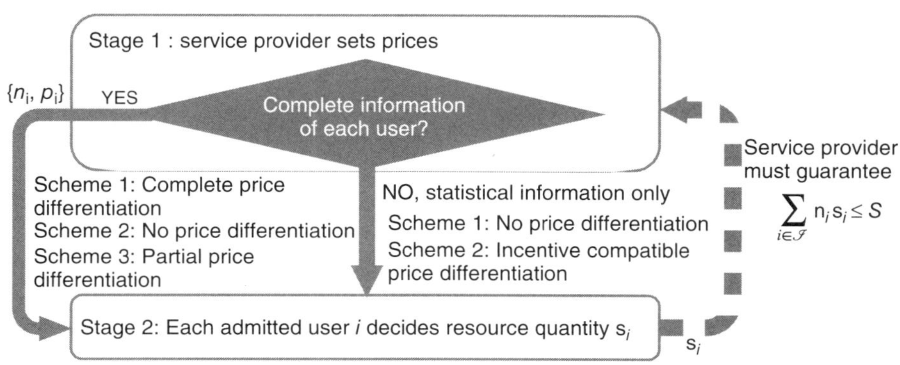

Figure 8.1 A two-stage Stackelberg model.

revenue. Under incomplete information, it publishes a common price *menu* to all users and allows users to freely choose a particular price option in this menu. All these pricing schemes are discussed one by one in the following sections.

8.3 COMPLETE PRICE DIFFERENTIATION UNDER COMPLETE INFORMATION

We first consider the complete information case. As the service provider knows the utility and the identity of each user, it is possible to maximize the revenue by charging a different price to each group of users. The analysis will be based on backward induction, starting from Stage 2 and then moving to Stage 1.

8.3.1 User's Surplus Maximization Problem in Stage 2

If a user in group i has been admitted into the network and offered a linear price p_i in Stage 1, then it solves the following surplus maximization problem,

$$\underset{s_i \geq 0}{\text{maximize}}\ u_i(s_i) - p_i s_i, \tag{8.2}$$

which leads to the following unique optimal demand

$$s_i(p_i) = \left(\frac{\theta_i}{p_i} - 1\right)^+, \quad \text{where } (\cdot)^+ \triangleq \max(\cdot, 0). \tag{8.3}$$

Remark 8.1 *The analysis of the Stage 2 user surplus maximization problem is the same for all pricing schemes. The result in Eq. (8.3) will be also used in Sections 8.4–8.6.*

8.3.2 Service Provider's Pricing and Admission Control Problem in Stage 1

In Stage 1, the service provider maximizes its revenue by choosing the price p_i and the number of admitted users n_i for each group i subject to the limited total resource S. The key idea is to perform a complete price differentiation (*CP*) scheme, that is, charging each group with a different price.

$$CP: \underset{\boldsymbol{p}\geq 0, \boldsymbol{s}\geq 0, \boldsymbol{n}}{\text{maximize}} \sum_{i\in\mathcal{I}} n_i p_i s_i \tag{8.4}$$

$$\text{subject to } s_i = \left(\frac{\theta_i}{p_i} - 1\right)^+, \quad i \in \mathcal{I}, \tag{8.5}$$

$$n_i \in \{0, \ldots, N_i\}, \quad i \in \mathcal{I}, \tag{8.6}$$

$$\sum_{i\in\mathcal{I}} n_i s_i \leq S, \tag{8.7}$$

where $\boldsymbol{p} \triangleq \{p_i, i \in \mathcal{I}\}$, $\boldsymbol{s} \triangleq \{s_i, i \in \mathcal{I}\}$, and $\boldsymbol{n} \triangleq \{n_i, i \in \mathcal{I}\}$. We use bold symbols to denote vectors in the sequel. Constraint (8.5) is the solution of the Stage 2 user surplus maximization problem in Eq. (8.3). Constraint (8.6) denotes the admission control decision, and constraint (8.7) represents the total limited resource in the network.

The *CP* problem is not straightforward to solve, because it is a nonconvex optimization problem with a nonconvex objective function (summation of products of n_i and p_i), a coupled constraint (8.7), and integer variables $\boldsymbol{n}$. However, it is possible to convert it into an equivalent convex formulation through a series of transformations, and thus the problem can be solved efficiently.

First, we can remove the $(\cdot)^+$ sign in constraint (8.5) by realizing the fact that there is no need to set p_i higher than θ_i for users in group i; users in group i already demand zero resource and generate zero revenue when $p_i = \theta_i$. This means that we can rewrite constraint (8.5) as

$$p_i = \frac{\theta_i}{s_i + 1} \text{ and } s_i \geq 0, i \in \mathcal{I}. \tag{8.8}$$

Plugging Eq. (8.8) into Eq. (8.4), the objective function becomes $\sum_{i \in \mathcal{I}} n_i \frac{\theta_i s_i}{s_i+1}$. We can further decompose the *CP* problem in the following two subproblems:

1. *Resource Allocation.* For a fixed admission control decision $\boldsymbol{n}$, solve for the optimal resource allocation $\boldsymbol{s}$.

$$\begin{aligned} CP_1 : \underset{s \geq 0}{\text{maximize}} & \sum_{i \in \mathcal{I}} n_i \frac{\theta_i s_i}{s_i + 1} \\ \text{subject to} & \sum_{i \in \mathcal{I}} n_i s_i \leq S. \end{aligned} \tag{8.9}$$

   Denote the solution of CP_1 as $\boldsymbol{s}^* = (s_i^*(\boldsymbol{n}), \forall i \in \mathcal{I})$. We further maximize the revenue of the integer admission control variables $\boldsymbol{n}$.

2. *Admission Control.*

$$\begin{aligned} CP_2 : \underset{\boldsymbol{n}}{\text{maximize}} & \sum_{i \in \mathcal{I}} n_i \frac{\theta_i s_i^*(\boldsymbol{n})}{s_i^*(\boldsymbol{n}) + 1} \\ \text{subject to } & n_i \in \{0, \ldots, N_i\} \ , \ i \in \mathcal{I}. \end{aligned} \tag{8.10}$$

Let us first solve the CP_1 subproblem in $\boldsymbol{s}$. Note that it is a convex optimization problem. By using Lagrange multiplier technique, we can get the first-order necessary and sufficient condition:

$$s_i^*(\lambda) = \left(\sqrt{\frac{\theta_i}{\lambda}} - 1\right)^+, \tag{8.11}$$

where λ is the Lagrange multiplier corresponding to the resource constraint (8.9).

Meanwhile, we note the resource constraint (8.9) must hold with equality, because the objective is a strictly increasing function in s_i. Thus, by plugging Eq. (8.11) into (8.9), we have

$$\sum_{i\in\mathcal{I}} n_i \left(\sqrt{\frac{\theta_i}{\lambda}} - 1\right)^+ = S. \tag{8.12}$$

This weighted water-filling problem (where $\frac{1}{\sqrt{\lambda}}$ can be viewed as the water level) in general has no closed-form solution for λ. However, we can efficiently determine the optimal solution λ^* by exploiting the special structure of our problem. Note that because $\theta_1 > \cdots > \theta_I$, then λ^* must satisfy the following condition:

$$\sum_{i=1}^{K^{cp}} n_i \left(\sqrt{\frac{\theta_i}{\lambda^*}} - 1\right) = S, \tag{8.13}$$

for a group index threshold value K^{cp} satisfying

$$\frac{\theta_{K^{cp}}}{\lambda^*} > 1 \text{ and } \frac{\theta_{K^{cp}+1}}{\lambda^*} \leq 1. \tag{8.14}$$

In other words, only groups with index no larger than K_{cp} will be allocated the positive resource. This property leads to the following simple Algorithm 8.1 to compute λ^* and group index threshold K^{cp}: we start by assuming $K^{cp} = I$ and compute λ. If Eq. (8.14) is not satisfied, we decrease K^{cp} by one and recompute λ until Eq. (8.14) is satisfied.

Algorithm 8.1

Resource-Allocation-CP$(\{n_i, \theta_i\}_{i\in\mathcal{I}}, S)$:

$k \leftarrow I,\ \lambda(k) \leftarrow \left(\frac{\sum_{i=1}^{k} n_i\sqrt{\theta_i}}{S+\sum_{i=1}^{k} n_i}\right)^2$

while $\theta_k \leq \lambda(k)$

$\quad k \leftarrow k-1,\ \lambda(k) \leftarrow \left(\frac{\sum_{i=1}^{k} n_i\sqrt{\theta_i}}{S+\sum_{i=1}^{k} n_i}\right)^2$

end while

$K^{cp} \leftarrow k,\ \lambda^* \leftarrow \lambda(k)$

return (K^{cp}, λ^*)

End pseudo-code.

As $\theta_1 > \lambda(1) = \left(\frac{n_1}{s+n_1}\right)^2 \theta_1$, Algorithm 8.1 always converges and returns the unique values of K^{cp} and λ^*. The complexity is $\mathcal{O}(I)$, that is, linear in the number of user groups (not the number of users).

With K^{cp} and λ^*, the solution of the resource allocation problem can be written as

$$s_i^* = \begin{cases} \sqrt{\dfrac{\theta_i}{\lambda^*}} - 1, & i = 1, \ldots, K^{cp}; \\ 0, & \text{otherwise,} \end{cases} \tag{8.15}$$

For the ease of discussion, we introduce a new notion of the *effective market*, which denotes all the groups allocated nonzero resource. For resource allocation subproblem CP_1, the threshold K^{cp} describes the size of the effective market. All groups with indices no larger than K^{cp} are *effective groups*, and users in these groups as *effective users*. An example of the effective market is illustrated in Figure 8.2.

Now let us solve the admission control subproblem CP_2. Denote the objective Eq. (8.10) as $R_{cp}(\boldsymbol{n})$, by Eq. (8.15), then $R_{cp}(\boldsymbol{n}) \overset{\Delta}{=} \sum_{i=1}^{K^{cp}} n_i \left(\sqrt{\frac{\theta_i}{\lambda^*(\boldsymbol{n})}} - 1 \right) \sqrt{\theta_i \lambda^*(\boldsymbol{n})}$. We first relax the integer domain constraint of n_i as $n_i \in [0, N_i]$. As Eq. (8.13), by taking the derivative of the objective function $R_{cp}(\boldsymbol{n})$ with respect to n_i, we have

$$\frac{\partial R_{cp}(\boldsymbol{n})}{\partial n_i} = n_i \left(\sqrt{\frac{\theta_i}{\lambda^*(\boldsymbol{n})}} - 1 \right) \frac{\partial \sqrt{\theta_i \lambda^*(\boldsymbol{n})}}{\partial n_i}, \tag{8.16}$$

Also from Eq (8.13), we have $\lambda^* = \left(\frac{\sum_{i=i}^{K^{cp}} n_i \sqrt{\theta_i}}{S + \sum_{i=1}^{K^{cp}} n_i} \right)^2$, thus $\frac{\partial \sqrt{\lambda^*(\boldsymbol{n})}}{\partial n_i} > 0$, for $i = 1, \ldots, K^{cp}$, and $\frac{\partial \sqrt{\lambda^*(\boldsymbol{n})}}{\partial n_i} = 0$, for $i = K^{cp} + 1, \ldots, I$. This means that the objective $R_{cp}(\boldsymbol{n})$ is strictly increasing in n_i for all $i = 1, \ldots, K^{cp}$, thus it is optimal to admit all users in the effective market. The admission decisions for groups not in the effective market are irrelevant to the optimization, because those users consume zero resource. Therefore, one of the optimal solutions of the CP_1 subproblem is $n_i^* = N_i$ for all $i \in \mathcal{I}$. Solving CP_1 and CP_2 subproblems leads to the optimal solution of the CP problem:

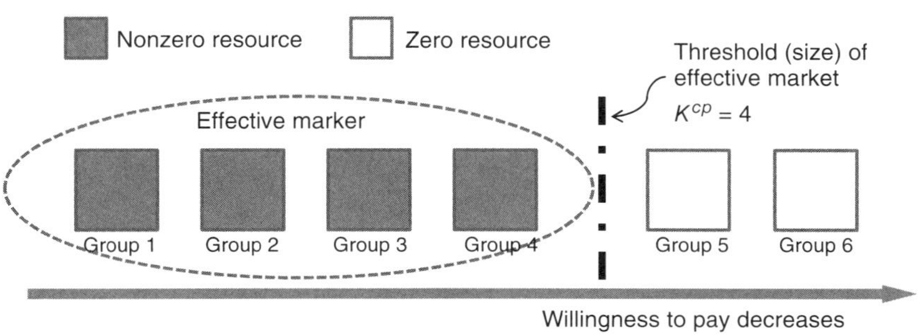

Figure 8.2 A six-group example for effective market: the willingness to pays decrease from group 1 to group 6. The effective market threshold can be obtained by Algorithm 8.1 and is four in this example.

Theorem 8.1 *There exists an optimal solution of the CP problem that satisfies the following conditions:*

- *All users are admitted:* $n_i^* = N_i$ *for all* $i \in \mathcal{I}$.
- *There exist a value* λ^* *and a group index threshold* $K^{cp} \leq I$, *such that only the top* K^{cp} *groups of users receive positive resource allocations,*

$$s_i^* = \begin{cases} \sqrt{\frac{\theta_i}{\lambda^*}} - 1, & i = 1, \ldots, K^{cp}; \\ 0, & \text{otherwise}, \end{cases}$$

with the prices

$$p_i^* = \begin{cases} \sqrt{\theta_i \lambda^*}, & i = 1, \ldots, K^{cp}; \\ \theta_i, & \text{otherwise}. \end{cases}$$

The values of λ^* *and* K^{cp} *can be computed as in Algorithm 8.1 by setting* $n_i = N_i$, *for all* $i \in \mathcal{I}$.

Theorem 8.1 provides the right economic intuition: the service provider maximizes its revenue by charging a higher price to users with a higher willingness to pay. It is easy to check that $p_i > p_j$ for any $i < j$. The users with low willingness to pay are excluded from the markets.

8.3.3 Properties

Here we summarize some interesting properties of the optimal complete price differentiation scheme:

8.3.3.1 Threshold Structure The threshold-based resource allocation means that higher willingness to pay groups have higher priories of obtaining the resource at the optimal solution.

To see this clearly, assume the effective market size is $K^{(1)}$ under parameters $\{\theta_i, N_i^{(1)}\}_{i\in\mathcal{I}}$ and S. Here the superscript (1) denotes the first parameter set. Now consider another set of parameters $\{\theta_i, N_i^{(2)}\}_{i\in\mathcal{I}}$ and S, where $N_i^{(2)} \geq N_i^{(1)}$ for each group i and the new effective market size is $K^{(2)}$. By Eq. (8.13), we can see that $K^{(2)} \leq K^{(1)}$. Furthermore, we can show that if some high willingness to pay group has many more users under the latter system parameters, that is, $N_i^{(2)}$ is much larger than $N_i^{(1)}$ for some $i < K^{(1)}$, then the effective size will be strictly decreased, that is, $K^{(2)} < K^{(1)}$. In other words, the increase of high willingness to pay users will drive the low willingness to pay users out of the effective market.

8.3.3.2 Admission Control with Pricing Theorem 8.1 shows the explicit admission control is not necessary at the optimal solution. Also from Theorem 8.1, we can see that when the number of users in any effective group increases, the price, p_i^*, for all $i \in \mathcal{I}$ increases and resource, s_i^*, for all $\forall\, i \leq K^{cp}$ decreases. The prices serve as the indications of the scarcity of the resources and will automatically prevent the low willingness to pay users to consume the network resource.

8.4 SINGLE PRICING SCHEME

In Section 8.3, we showed that the *CP* scheme is the optimal pricing scheme to maximize the revenue under complete information. However, such a complicated pricing scheme is of high implementational complexity. In this section, we study the single pricing scheme. It is clear that the scheme will in general suffer a revenue loss comparing with the *CP* scheme. We will try to characterize the impact of various system parameters on such revenue loss.

8.4.1 Problem Formulation and Solution

Let us first formulate the single pricing (*SP*) problem.

$$\begin{aligned} SP : \underset{p\geq 0,\ \boldsymbol{n}}{\text{maximize}}\ & p \sum_{i\in\mathcal{I}} n_i s_i \\ \text{subject to } & s_i = \left(\frac{\theta_i}{p} - 1\right)^+ , \quad i \in \mathcal{I} \\ & n_i \in \{0, \ldots, N_i\} \ , \quad i \in \mathcal{I} \\ & \sum_{i\in\mathcal{I}} n_i s_i \leq S. \end{aligned}$$

Comparing with the *CP* problem in Section 8.3, here the service provider charges a single price p to all groups of users. After a similar transformation as in Section 8.3, we can show that the optimal single price satisfies the following weighted water-filling condition

$$\sum_{i\in\mathcal{I}} N_i \left(\frac{\theta_i}{p} - 1\right)^+ = S.$$

Thus we can obtain the following solution that shares a similar structure as complete price differentiation.

Theorem 8.2 *There exists an optimal solution of the SP problem that satisfies the following conditions:*

- *All users are admitted:* $n_i^* = N_i$, *for all* $i \in \mathcal{I}$.
- *There exist a price* p^* *and a group index threshold* $K^{sp} \leq I$, *such that only the top* K^{sp} *groups of users receive positive resource allocations,*

$$s_i^* = \begin{cases} \dfrac{\theta_i}{p^*} - 1, & i = 1, 2, \ldots, K^{sp}, \\ 0, & \text{otherwise}, \end{cases}$$

with the price

$$p^* = p(K^{sp}) = \frac{\sum_{i=1}^{K^{sp}} N_i\theta_i}{S + \sum_{i=1}^{K^{sp}} N_i}.$$

The value of K^{sp} and p^* can be computed as in Algorithm 8.2.

Algorithm 8.2

Threshold-SP $(\{N_i, \theta_i\}_{i\in\mathcal{I}}, S)$:

$k \leftarrow I, p(k) \leftarrow \frac{\sum_{i=1}^{k} N_i\theta_i}{S+\sum_{i=1}^{k} N_i}$

while $\theta_k \le p(k)$

$\quad k \leftarrow k - 1, p(k) \leftarrow \frac{\sum_{i=1}^{k} N_i\theta_i}{S+\sum_{i=1}^{k} N_i}$

end while

$K^{sp} \leftarrow k, p^* \leftarrow p(k)$

return K^{sp}, p^*

End pseudo-code.

8.4.2 Properties

The *SP* scheme shares several similar properties as the *CP* scheme (Section 8.3.3), including the threshold structure and admission control with pricing. Similarly, we can define the effective market for the *SP* scheme.

It is more interesting to notice the differences between these two schemes. To distinguish solutions, we use the superscript "cp" for the *CP* scheme, and "sp" for the *SP* scheme.

Proposition 8.1 *Under same parameters $\{N_i, \theta_i\}_{i\in\mathcal{I}}$ and S:*

1. *the effective market of the SP scheme is no larger than the one of the CP scheme, that is, $K^{sp} \le K^{cp}$.*
2. *there exists a threshold $\bar{k} \in \{1, 2 \ldots, K^{sp}\}$, such that*
   - *groups with indices less than $\bar{k}$ (high willingness to pay users) are charged with higher prices and allocated less resources in the CP scheme, that is, $p_i^{cp} \ge p^*$ and $s_i^{cp} \le s_i^{sp}$, $\forall\, i \le \bar{k}$, where the equality holds if only if $i = \bar{k}$ and $\theta_{\bar{k}} = \frac{p^{*2}}{\lambda^*}$,*
   - *groups with indices greater than $\bar{k}$ (low willingness to pay users) are charged with lower prices and allocated more resources in the CP scheme, that is, $p_i^{cp} < p^*$ and $s_i^{cp} > s_i^{sp}$, $\forall\, i \ge \bar{k}$,*

   where p^* is the optimal single price.

The proof is given in Appendix 8.A.2. An illustrative example is shown in Figures 8.3 and 8.4.

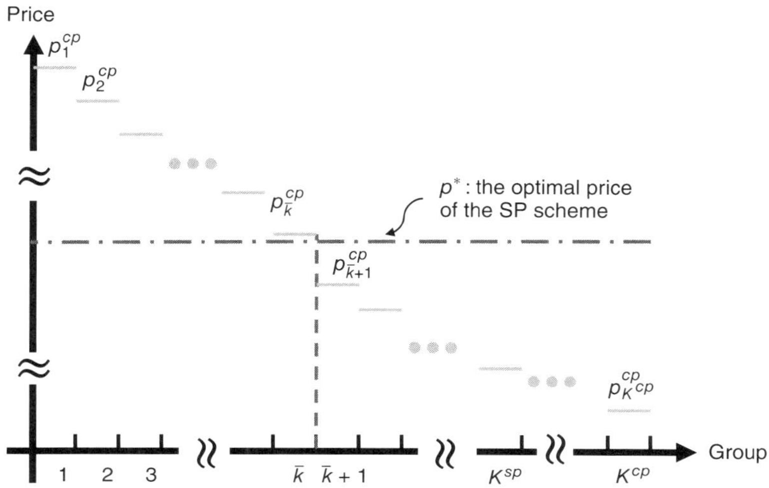

Figure 8.3 Comparison of prices between the *CP* scheme and the *SP* scheme.

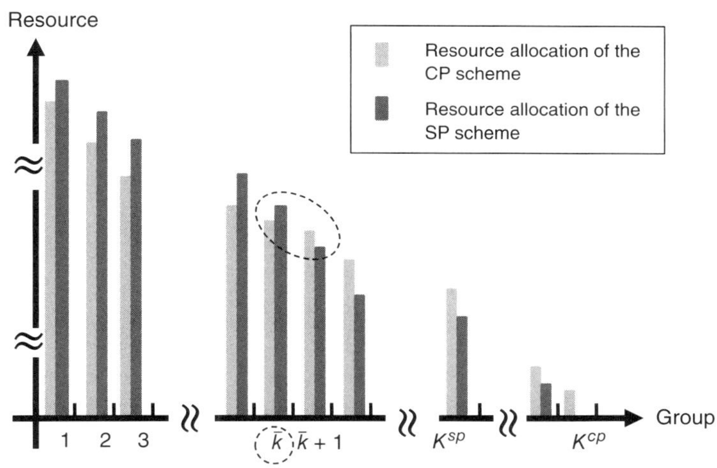

Figure 8.4 Comparison of resource allocation between the *CP* scheme and the *SP* scheme.

It is easy to understand that the *SP* scheme makes less revenue, because it is a feasible solution to the *CP* problem. A little bit more computation sheds more light on this comparison. We introduce the following notations to streamline the comparison:

- $N_{eff}(k) \triangleq \sum_{i=1}^{k} N_i$: the number of effective users, where k is the size of the effective market.
- $\gamma_i(k) \triangleq \frac{N_i}{N_{eff}(k)}$, $i = 1, 2, \ldots, k$: the fraction of group i's users in the effective market.

- $\bar{s}(k) \triangleq \frac{S}{N_{eff}(k)}$: the average resource per an effective user.
- $\bar{\theta}(k) \triangleq \sum_{i=1}^{k} \gamma_i \theta_i$: the average willingness to pay per an effective user.

On the basis of Theorem 8.1, the revenue of the *CP* scheme is

$$R^{cp}(K^{cp}) = N_{eff}(K^{cp})\left(\frac{\bar{s}(K^{cp})\bar{\theta}(K^{cp}) + g(K^{cp})}{\bar{s}(K^{cp}) + 1}\right), \tag{8.17}$$

where

$$g(K^{cp}) = \frac{1}{\lambda^*}\sum_{i=1}^{K^{cp}}\sum_{j>i}^{K^{cp}} \gamma_i\gamma_j(p_i^{cp} - p_j^{cp})^2. \tag{8.18}$$

On the basis of Theorem 8.2, the revenue of the *SP* scheme is

$$R^{sp}(K^{sp}) = N_{eff}(K^{sp})\left(\frac{\bar{s}(K^{sp})\bar{\theta}(K^{sp})}{\bar{s}(K^{sp}) + 1}\right). \tag{8.19}$$

From Eqs. (8.17) and (8.19), it is clear to see that $R^{cp} \geq R^{sp}$ because of two factors: one is the nonnegative term in Eq. (8.18) and the other is $K^{cp} \geq K^{sp}$: a higher level of differentiation implies a no smaller effective market. Let us further discuss them in the following two cases:

- If $K^{cp} = K^{sp}$, then the additional term of Eq. (8.18) in Eq. (8.17) means that $R^{cp} \geq R^{sp}$. The equality holds if and only if $K^{cp} = 1$, in which case $g(K^{cp}) = 0$. Note that in this case, the *CP* scheme degenerates to the *SP* scheme. We name the nonnegative term $g(K^{cp})$ in Eq. (8.18) as *price differentiation gain*, as it measures the average price difference between any effective users in the *CP* scheme. The larger the price difference, the larger the gain. When there is no differentiation in the degenerating case ($K^{cp} = 1$), the gain is zero.
- If $K^{cp} > K^{sp}$, because the common part of two revenue $N_{eff}(K)\left(\frac{\bar{s}(K)\bar{\theta}(K)}{\bar{s}(K)+1}\right) = \frac{S\bar{\theta}N_{eff}(K)}{S+N_{eff}(K)}$ is a strictly increasing function of $N_{eff}(K)$, price differentiation makes more revenue even if the positive differentiation gain $g(K^{cp})$ is not taken into consideration. The result that more consumers with purchasing power always mean more revenue in the service provider's pocket is intuitive.

Finally, we note that the *CP* scheme in Section 8.3 requires the complete network information. The *SP* scheme here, on the other hand, works in the incomplete information case as well. This distinction becomes important in Section 8.6.

8.5 PARTIAL PRICE DIFFERENTIATION UNDER COMPLETE INFORMATION

For a service provider facing thousands of user types, it is often impractical to design a price choice for each user type. The reasons behind this, as discussed in Reference [10], are mainly high system overheads and customers' aversion. However, as we have shown in Section 8.4, the single pricing scheme may suffer a considerable revenue loss. How to achieve a good trade-off between the implementational complexity and the total revenue? In reality, we usually see that the service provider offers only a few pricing plans for the entire users population; we term it as the *partial price differentiation* scheme. In this section, we answer the following question: if the service provider is constrained to maintain a limited number of prices, $p^1, \dots, p^J$, $J \leq I$, then what is the optimal pricing strategy and the maximum revenue? Concretely, the partial price differentiation (*PP*) problem is formulated as follows.

$$PP : \underset{n_i, p_i, s_i, p^j, a_i^j}{\text{maximize}} \sum_{i \in \mathcal{I}} n_i p_i s_i$$

$$\text{subject to } s_i = \left(\frac{\theta_i}{p_i} - 1\right)^+, \forall i \in \mathcal{I}, \tag{8.20}$$

$$n_i \in \{0, \dots, N_i\}, \ \forall i \in \mathcal{I}, \tag{8.21}$$

$$\sum_{i \in \mathcal{I}} n_i s_i \leq S, \tag{8.22}$$

$$p_i = \sum_{j \in \mathcal{J}} a_i^j p^j, \tag{8.23}$$

$$\sum_{j \in \mathcal{J}} a_i^j = 1, \ a_i^j \in \{0, 1\}, \forall i \in \mathcal{I}. \tag{8.24}$$

Here $\mathcal{J}$ denotes the set $\{1, 2, \dots, J\}$. As we consider the complete information scenario in this section, the service provider can choose the price charged to each group, thus constraints (8.20)–(8.22) are the same as in the *CP* problem. Constraints (8.23) and (8.24) mean that p_i charged to each group i is one of J choices from the set $\{p^j, j \in \mathcal{J}\}$. For convenience, we define *cluster* $\mathcal{C}^j \triangleq \{i \mid a_i^j = 1\}$, $j \in \mathcal{J}$, which is a set of groups charged with the same price p^j. We use superscript j to denote clusters and subscript i to denote groups through this section. We term the binary variables $\boldsymbol{a} \triangleq \{a_i^j, \ j \in \mathcal{J}, \ i \in \mathcal{I}\}$ as the *partition*, which determines which cluster each group belongs to.

The *PP* problem is a combinatorial optimization problem and is more difficult than the previous *CP* and *SP* problems. On the other hand, we notice that this *PP* problem formulation includes the *CP* scheme ($J = I$) and the *SP* scheme scenario ($J = 1$) as special cases. The insights we obtained from solving these two special cases in Sections 8.3 and 8.4 will be helpful to solve the general *PP* problem.

8.5.1 Three-Level Decomposition

To solve the *PP* problem, we decompose and tackle it in three levels. In the lowest level 3, we determine the pricing and resource allocation for each cluster, given a fixed partition and fixed resource allocation among clusters. In level 2, we compute the optimal resource allocation among clusters, given a fixed partition. In level 1, we optimize the partition among groups.

8.5.1.1 Level 3: Pricing and Resource Allocation in Each Cluster For a fix partition $\boldsymbol{a}$ and a cluster resource allocation $\boldsymbol{s} \overset{\Delta}{=} \{s^j\}_{j\in\mathcal{J}}$, we focus the pricing and resource allocation problems within each cluster $\mathcal{C}^j, j \in \mathcal{J}$:

$$\begin{aligned}
\text{Level-3: } \underset{n_i, s_i, p^j}{\text{maximize}} \quad & \sum_{i\in C^j} n_i p^j s_i \\
\text{subject to} \quad & s_i = \left(\frac{\theta_i}{p^j} - 1\right)^+, \quad \forall i \in \mathcal{C}^j, \\
& n_i \le N_i, \quad \forall i \in \mathcal{C}^j, \\
& \sum_{i\in\mathcal{C}^j} n_i s_i \le s^j.
\end{aligned}$$

The level-3 subproblem coincides with the *SP* scheme discussed in Section 8.4, because all groups within the same cluster $\mathcal{C}^j$ are charged with a single price p^j. We can then directly apply the results in Algorithm 8.2 to solve the level-3 problem. We denote the effective market threshold[2] for cluster $\mathcal{C}^j$ as K^j, which can be computed in Algorithm 8.2. An illustrative example is shown in Figure 8.5, where the cluster contains four groups (groups 4–7), and the effective market contains groups 4 and 5,

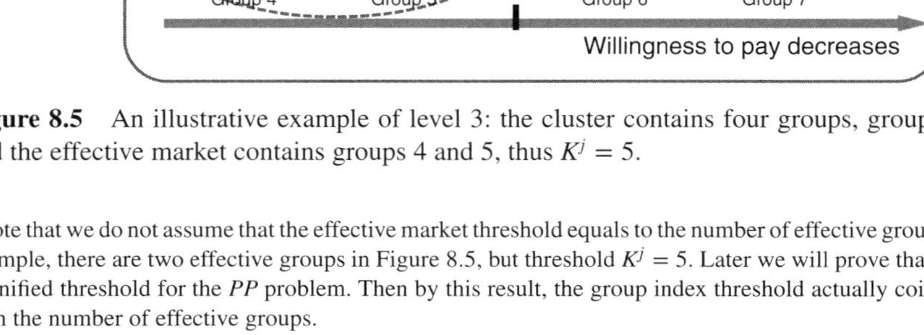

Figure 8.5 An illustrative example of level 3: the cluster contains four groups, group 4–7; and the effective market contains groups 4 and 5, thus $K^j = 5$.

[2]Note that we do not assume that the effective market threshold equals to the number of effective groups, for example, there are two effective groups in Figure 8.5, but threshold $K^j = 5$. Later we will prove that there is unified threshold for the *PP* problem. Then by this result, the group index threshold actually coincides with the number of effective groups.

thus $K^j = 5$. The service provider obtains the following maximum revenue obtained from cluster $\mathscr{C}^j$:

$$R^j(s^j, \boldsymbol{a}) = \frac{s^j \sum_{i \in C^j, i \leq K^j} N_i \theta_i}{s^j + \sum_{i \in C^j, i \leq K^j} N_i}. \tag{8.25}$$

8.5.1.2 Level 2: Resource Allocation among Clusters For a fix partition $\boldsymbol{a}$, we then consider the resource allocation among clusters.

$$\text{Level 2: } \underset{s^j \geq 0}{\text{maximize}} \sum_{j \in \mathcal{J}} R^j(s^j, \boldsymbol{a})$$

$$\text{subject to } \sum_{j \in \mathcal{J}} s^j \leq S.$$

We will show in Section 8.5.2 that subproblems in level 2 and level 3 can be transformed into a complete price differentiation problem under proper technique conditions. Let us denote its optimal value as $R_{pp}(\boldsymbol{a})$.

8.5.1.3 Level 1: Cluster Partition Finally, we solve the cluster partition problem.

$$\text{Level-1: } \underset{a_i^j \in \{0,1\}}{\text{maximize}} \, R_{pp}(\boldsymbol{a})$$

$$\text{subject to } \sum_{j \in \mathcal{J}} a_i^j = 1, \; i \in \mathcal{I}.$$

This partition problem is a combinatorial optimization problem. The size of its feasible set is $S(I, J) = \frac{1}{J!} \sum_{t=1}^{J} (-1)^{J+t} C(J, t) t^I$, *Stirling number of the second kind* [39, Chap.13], where $C(J, t)$ is the binomial coefficient. Some numerical examples are given in the third row of Table 8.1. If the number of prices J is given, the feasible set size is exponential in the total number of groups I. For our problem, however, it is possible to reduce the size of the feasible set by exploiting the special problem structure. More specifically, the group indices in each cluster should be consecutive at the optimum. This means that the size of the feasible set is $C(I-1, J-1)$ as shown in the last row of Table 8.1 and thus is much smaller than $S(I, J)$.

TABLE 8.1 Numerical Examples for Feasible Set Size of the Partition Problem in Level 1

Number of Groups	$I = 10$		$I = 100$		$I = 1000$
Number of Prices	$J = 2$	$J = 3$	$J = 2$	$J = 3$	$J = 2$
$S(I, J)$	511	9330	6.33825×10^{29}	8.58963×10^{46}	5.35754×10^{300}
$C(I-1, J-1)$	9	36	99	4851	999

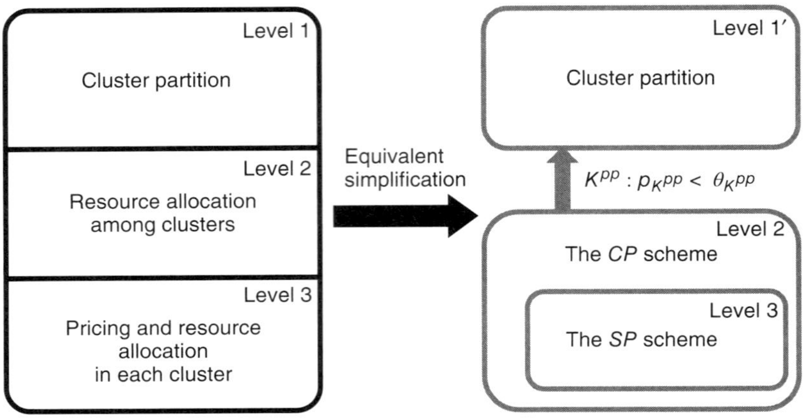

Figure 8.6 Decomposition and simplification of the general *PP* problem: The three-level decomposition structure of the *PP* problem is shown in the left-hand side (LHS). After simplifications in Sections 8.5.2 and 8.5.3, the problem will be reduced to structure in right-hand side (RHS).

Next we discuss how to solve the three-level subproblems. A route map for the whole solution process is given in Figure 8.6.

8.5.2 Solving Level 2 and Level 3

The optimal solution (8.25) of the level-3 problem can be equivalently written as

$$R^j(\boldsymbol{s}, \boldsymbol{a}) = \frac{s^j \sum_{i\in C^j, i\leq K^j} N_i\theta_i}{s^j + \sum_{i\in C^j, i\leq K^j} N_i} \stackrel{(a)}{=} \frac{s^j N^j \theta^j}{s^j + N^j}, \tag{8.26}$$

$$\text{where} \begin{cases} N^j = \sum\limits_{i\in C^j, i\leq K^j} N_i, \\ \theta^j = \sum\limits_{i\in C^j, i\leq K^j} \dfrac{N_i\theta_i}{N^j}. \end{cases} \tag{8.27}$$

The equality (a) in Eq. (8.26) means that each cluster $\mathscr{C}^j$ can be equivalently treated as a group with N^j homogeneous users with the same willings to pay θ^j. We name this equivalent group as a *super group* (SG). We summarize the above result as the following lemma.

Lemma 8.1 *For every cluster C^j and total resource s^j, $j \in \mathcal{J}$, we can find an equivalent SG that satisfies conditions in Eq. (8.27) and achieves the same revenue under the SP scheme.*

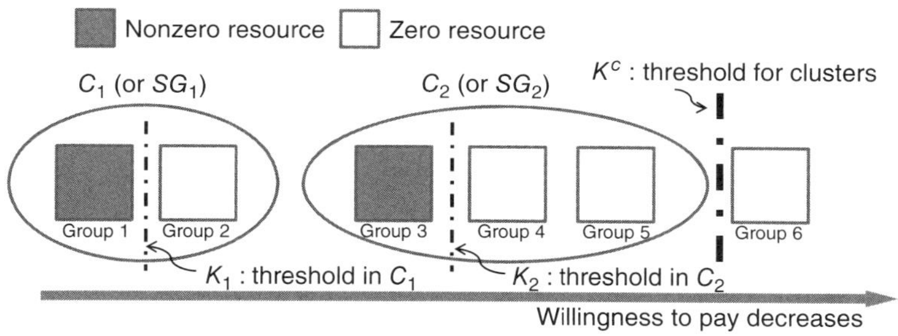

Figure 8.7 An example of coupling thresholds.

On the basis of Lemma 8.1, the level-2 and level-3 subproblems together can be viewed as the *CP* problem for SGs. As a cluster and its SG from a one-to-one mapping, we will use the two words interchangeably in the sequel.

However, simply combining Theorems 8.1 and 8.2 to solve level-2 and level-3 subproblems for a fixed partition $\boldsymbol{a}$ can result in a very high complexity. This is because the effective markets within each SG and between SGs are coupled together. An illustrative example of this coupling effective market is shown in Figure 8.7, where K^c is the threshold between clusters and has three possible positions (i.e., between group 2 and group 3, between group 5 and group 6, or after group 6); and K_1 and K_2 are thresholds within cluster $\mathscr{C}^1$ and $\mathscr{C}^2$, which have two or three possible positions, respectively. Thus, there are $(2 \times 3) \times 3 = 18$ possible thresholds possibilities in total.

The key idea resolving this coupling issue is to show that the situation in Figure 8.7 cannot be an optimal solution of the *PP* problem. The results in Sections 8.3 and 8.4 show that there is a unified threshold at the optimum in both the *CP* and *SP* cases, for example, Figure 8.2. Next we will show that a unified single threshold also exists in the *PP* case.

Lemma 8.2 *At any optimal solution of the PP scheme, the group indices of the effective market is consecutive.*

The proof of Lemma 8.2 can be found in Appendix 8.A.3. The intuition is that the resource should be always allocated to high willingness to pay users at the optimum. Thus, it is not possible to have Figure 8.7 at an optimal solution, where high willingness to pay users in group 2 are allocated zero resource while low willingness to pay users in group 3 are allocated positive resources.

On the basis of Lemma 8.2, we know that there is a unified effective market threshold for the *PP* problem, denoted as K^{pp}. As all groups with indices larger than K^{pp} make zero contribution to the revenue, we can ignore them and only consider the partition problem for the first K^{pp} groups. Given a partition that divides the K^{pp} groups into J clusters (SGs), we can apply the *CP* result in Section 8.3 to compute the optimal revenue in the level 2 based on Theorem 8.1.

$$R_{pp}(\boldsymbol{a}) = \sum_{j=1}^{J} N^j\theta^j - \frac{\left(\sum_{j=1}^{J} N^j\sqrt{\theta^j}\right)^2}{S + \sum_{j=1}^{J} N^j}$$

$$= \sum_{i=1}^{K^{pp}} N_i\theta_i - \frac{\left(\sum_{j=1}^{J} N^j\sqrt{\theta^j}\right)^2}{S + \sum_{i=1}^{K^{pp}} N_i}, \tag{8.28}$$

8.5.3 Solving Level 1

8.5.3.1 With a Given Effective Market Threshold K^{pp} On the basis of the previous results, we first simplify the level-1 subproblem and prove the following theorem.

Theorem 8.3 *For a given threshold K^{pp}, the optimal partition of the level-1 subproblem is the solution of the following optimization problem.*

$$\text{Level-1}' \underset{a_i^j, N^j, \theta^j}{\text{minimize}} \quad \sum_{j\in\mathcal{J}} N^j\sqrt{\theta^j}$$

$$\text{subject to} N^j = \sum_{i\in\mathcal{K}^{pp}} N_i a_i^j, \quad j \in \mathcal{J},$$

$$\theta^j = \sum_{i\in\mathcal{K}^{pp}} \frac{N_i a_i^j}{N^j}\theta_i \quad j \in \mathcal{J},$$

$$\sum_{j\in\mathcal{J}} a_i^j = 1, \; a_i^j \in \{0,1\}, i \in \mathcal{K}^{pp}\; j \in \mathcal{J},$$

$$\theta_{K^{pp}} > p^J = \sqrt{\theta^J(\boldsymbol{a})\lambda(\boldsymbol{a})}, \tag{8.29}$$

where $\mathcal{K}^{pp} \triangleq \{1, 2, \ldots, K^{pp}\}$, $\theta^J(\boldsymbol{a})$ is the value of average willingness to pay of the Jth group for the partition $\boldsymbol{a}$, and $\lambda(\boldsymbol{a}) = \left(\frac{\sum_{j\in\mathcal{J}} N^j\sqrt{\theta^j}}{S+\sum_{i=1}^{K^{pp}} N_i}\right)^2$.

Proof: The objective function and the first three constraints in the level-1 subproblem are easy to understand: if the effective market threshold K^{pp} is given, then the objective function of the level-1 subproblem, maximizing R_{pp} in Eq. (8.28) over $\boldsymbol{a}$, is as simple as minimizing $\sum_{j=1}^{J} N^j\sqrt{\theta^j}$ as the level-1 subproblem suggested; the first three constraints are given by the definition of the partition.

Constraint (8.29) is the threshold condition that supports Eq. (8.28), which means that the least willingness to pay users in the effective market has a positive demand. It ensures that calculating the revenue by Eq. (8.28) is valid. Remember that the solution of the *CP* problem of level 2 and level 3 is threshold based, and Lemma 8.2 indicates that Eq. (8.29) is sufficient for that all groups with willingness larger than group K^{pp} can have positive demands. Otherwise, we can construct another partition leading to a larger revenue (please refer to the proof of Lemma 8.2) or equivalently leading to a less objective value of the level-1 subproblem. This leads to a contradiction. ■

The level-1 subproblem is still a combinatorial optimization problem with a large feasible set of $\boldsymbol{a}$ (similar to the original level 1). The following result can help us to reduce the size of the feasible set.

Theorem 8.4 *For any effective market size K^{pp} and number of prices J, an optimal partition of the PP problem involves consecutive group indices within clusters.*

The proof of Theorem 8.4 is given in Appendix 8.A.4. We first prove that this result is true for the level-1 subproblem without constraint (8.29) and further show that this result will not be affected by Eq. (8.29). The intuition is that high willingness to pay users should be allocated positive resources with priority. It implies that groups with similar willingness to pays should be partitioned in the same cluster, instead of several far away clusters. Or equivalently, the group indices within each cluster should be consecutive.

We define $\mathscr{A}$ as the set of all partitions with consecutive group indices within each cluster, and $v(\boldsymbol{a}) = \sum_{j \in \mathcal{J}} N^j \sqrt{\theta^j}$ is the value of objective of the level-1 subproblem for a partition $\boldsymbol{a}$. Algorithm 8.3 finds the optimal solution of the level-1 subproblem. The main idea for this algorithm is to enumerate every possible partition in set $\mathscr{A}$, and then check whether the threshold condition (8.29) can be satisfied. The main part of this algorithm is to enumerate all partitions in set $\mathscr{A}$ of $C(K^{pp} - 1, J - 1)$ feasible partitions. Thus the complexity of Algorithm 8.3 is no more than $\mathcal{O}((K^{pp})^{J-1})$.

Algorithm 8.3

```
Level-1(K^pp, J):
k ← K^pp, v* ← sqrt(Σ_{i=1}^k N_i Σ_{i=1}^k N_i θ_i), a* ← 0
for a ∈ 𝒜
    if θ_k > sqrt(θ^J(a) λ(a))
        if v(a) < v*
            v* ← v(a), a* ← a
        end if
    end if
end for
return a*
End pseudo-code.
```

8.5.3.2 ***Search the Optimal Effective Market Threshold K^{pp}*** We know the optimal market threshold K^{pp} is upper bounded, that is, $K^{pp} \leq K^{cp} \leq I$. Thus we can first run Algorithm 8.1 to calculate the effective market size for the *CP* scheme K^{cp}. Then, we search the optimal K^{pp} iteratively using Algorithm 8.3 as an inner loop. We start by letting $K^{pp} = K^{cp}$ and run Algorithm 8.3. If there is no solution, we decrease K^{pp} by one and run Algorithm 8.3 again. The algorithm will terminate once we find an effective market threshold where Algorithm 8.3 has an optimal solution. Once the optimal threshold and the partition of the clusters are determined, we can further run Algorithm 8.1 to solve the joint optimal resource allocation and pricing scheme. The pseudo-code is given in Algorithm 8.4 as follows.

Algorithm 8.4

Solving PPD

$p_i \leftarrow \theta_i$
$(k, \lambda) \Leftarrow$ ***Resource-Allocation-CP***$(\{N_i, \theta_i\}_{i\in\mathcal{I}}, S)$,
$\boldsymbol{a}^* \Leftarrow$ ***Level-1***(k,J)
while $\boldsymbol{a}^* == \boldsymbol{0}$
  $k \leftarrow k - 1$
  $\boldsymbol{a}^* \Leftarrow$ ***Level-1***(k,J)
end while
for $j \leftarrow 1, \ldots J$
  $N^j \leftarrow \sum_{i=1}^{k} N_i a_i^j,\ \theta^j \leftarrow \sum_{i=1}^{k} \frac{N_i a_i^j}{N^j}\theta_i$
end for
$(k, \lambda) \Leftarrow$ ***Resource-Allocation-CP***$(\{N^j, \theta^j\}_{i\in\mathcal{J}}, S)$
for $j \leftarrow 1, , \ldots J$
  $p^j \leftarrow \sqrt{\theta^j \lambda},\ s^j \leftarrow \sqrt{\frac{\theta^j}{\lambda}} - 1$
end for
for $i \leftarrow 1, \ldots, k$
  $p_i \leftarrow \sum_{j=1}^{J} a_i^j \sqrt{\theta^j \lambda}$
end for
return $\{p_i\}_{i\in\mathcal{I}}$
End pseudo-code.

In Algorithm 8.4, it invokes two functions: ***Resource-Allocation-CP*** $(\{N_i \theta_i\}_{i\in\mathcal{I}}, S)$ as described in Algorithm 8.1 and ***Level-1*** (k, J) as in Algorithm 8.3.

The above analysis leads to the following theorem.

Theorem 8.5 *The solution obtained by Algorithm 8.4 is optimal for the PP problem.*

Proof: It is clear that Algorithm 8.4 enumerates every possible value of the effective market size for the *PP* problem K^{pp}, and for a given K^{pp} its inner loop Algorithm 8.3 enumerates every possible partition in set $\mathcal{A}$. Therefore, the result in Theorem 8.4 follows. ■

Next we discuss the complexity of Algorithm 8.4. The complexity of Algorithm 8.1 is $\mathcal{O}(I)$, and we run it twice in Algorithm 8.4. The worst case complexity of Algorithm 8.3 is $\mathcal{O}(I^{J-1})$, and we run it no more than $I - J$ times. Thus the whole complexity of Algorithm 8.4 is no more than $\mathcal{O}(I^J)$, which is polynomial of I.

8.6 PRICE DIFFERENTIATION UNDER INCOMPLETE INFORMATION

In Sections 8.3–8.5, we discuss various pricing schemes with different implementational complexity level under complete information, the revenues of which can be viewed as the benchmark of practical pricing designs. In this section, we further study the incomplete information scenario, where the service provider does not know the group association of each user. The challenge for pricing in this case is that the service provider needs to provide the right incentive so that a group i user does not want to pretend to be a user in a different group. It is clear that the *CP* scheme in Section 8.3 and the *PP* scheme in Section 8.5 cannot be directly applied here. The *SP* scheme in Section 8.4 is a special case, because it does not require the user-group association information in the first place and thus can be applied in the incomplete information scenario directly. On the other hand, we know that the *SP* scheme may suffer a considerable revenue loss compared with the *CP* scheme. Thus it is natural to ask whether it is possible to design an incentive-compatible differentiation scheme under incomplete information. In this section, we design a quantity-based price menu to incentivize users to make the right self-selection and to achieve the same maximum revenue of the *CP* scheme under complete information with proper technical conditions. We name it the incentive-compatible complete price (ICCP) differentiation scheme.

In the ICCP scheme, the service provider publishes a quantity-based price menu, which consists of several step functions of resource quantities. Users are allowed to freely choose their quantities. The aim of this price menu is to make the users *self-differentiated*, so that to mimic the same result (the same prices and resource allocations) of the *CP* scheme under complete information. On the basis of Theorem 8.1, there are only K (without confusion, we remove the superscript "cp" to simplify the notation) effective groups of users receiving nonzero resource allocations, thus there are K steps of unit prices, $p_1^* > p_2^* > \cdots > p_K^*$ in the price menu. These prices are exactly the same optimal prices that the service provider would charge for K effective groups as in Theorem 8.1. Note that for the $K+1, \ldots, I$ groups, all the prices in the menu are too high for them, then they will still demand zero resource. The quantity is divided into K intervals by $K-1$ thresholds, $s_{th}^1 > s_{th}^2 > \cdots > s_{th}^{K-1}$. The ICCP scheme can be specified as follows:

$$p(s) = \begin{cases} p_1^* & \text{when } s > s_{th}^1 \\ p_2^* & \text{when } s_{th}^1 \geq s > s_{th}^2 \\ \vdots & \\ p_K^* & \text{when } s_{th}^{K-1} \geq s > 0. \end{cases} \tag{8.30}$$

A four-group example is shown in Figure 8.8.

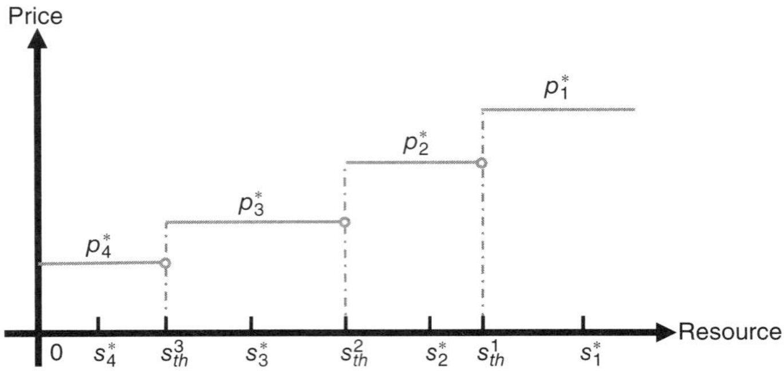

Figure 8.8 A four-group example of the *ICCP* scheme, where the prices $p_1^* > p_2^* > p_3^* > p_4^*$ are the same as the CP scheme. To mimic the same resource allocation as under the *CP* scheme, one necessary (but not sufficient) condition is $s_{th}^{j-1} \geq s_j^*$ for all j, where s_j^* is the optimal resource allocation of the *CP* scheme.

Note that in contrast to the usual "volume discount," here the price is nondecreasing in quantity. This is motivated by the resource allocation in Theorem 8.1, in which a user with a higher θ_i is charged a higher price for a larger resource allocation. Thus the observable quantity can be viewed as an indication of the unobservable users' willingness to pay and helps to realize price differentiation under incomplete information.

The key challenge in the ICCP scheme is to properly set the quantity thresholds so that users are perfectly segmented through self-differentiation. This is, however, not always possible. Next we derive the necessary and sufficient conditions to guarantee the perfect segmentation.

Let us first study the self-selection problem between two groups: group i and group q with $i < q$. Later on, we will generalize the results to multiple groups. Here group i has a higher willingness to pay and will be charged with a higher price p_i^* in the *CP* case. The incentive-compatible constraint is that a high willingness to pay user cannot get more surplus by pretending to be a low willingness to pay user, that is, $\max_s U_i(s; p_i^*) \geq \max_s U_i(s; p_q^*)$, where $U_i(s; p) = \theta_i \ln(1 + s) - ps$ is the surplus of a group i user when it is charged with price p.

Without confusion, we still use s_i^* to denote the optimal resource allocation under the optimal prices in Theorem 8.1, that is, $s_i^* = \arg\max_{s_i \geq 0} U_i(s_i; p_i^*)$. We define $s_{i \to q}$ as the quantity satisfying

$$\begin{cases} U_i(s_{i \to q}; p_q^*) = U_i(s_i^*; p_i^*) \\ s_{i \to q} < s_i^* \end{cases}. \tag{8.31}$$

In other words, when a group i user is charged with a lower price p_q^* and demands resource quantity $s_{i \to q}$, it achieves the same as the maximum surplus under the optimal price of the *CP* scheme p_i^*, as showed in Figure 8.9. As there are two solutions of the first equation of Eq. (8.31), we constraint $s_{i \to q}$ to be the one that is smaller than s_i^*.

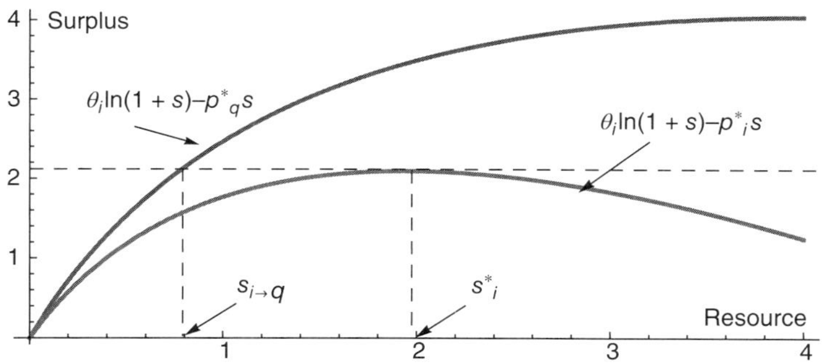

Figure 8.9 When the threshold $s_{th}^{q-1} < s_{i\to q}$, the group i user cannot obtain $U(s_i^*, p_i^*)$ if it chooses the lower price p_q at a quantity less than s_{th}^{q-1}. Therefore, it will automatically choose the high price p_i^* to maximize its surplus.

To maintain the group i users' incentive to choose the higher price p_i^* instead of p_q^*, we must have $s_{th}^{q-1} \le s_{i\to q}$, which means a group i user cannot obtain $U_i(s_i^*, p_i^*)$ if it chooses a quantity less than s_{th}^{q-1}. In other words, it will automatically choose the higher (and the desirable) price p_i^* to maximize its surplus. On the other hand, we must have $s_{th}^{q-1} \ge s_q^*$ in order to maintain the optimal resource allocation and allow a group q user to choose the right quantity-price combination (illustrated in Fig. 8.8).

Therefore, it is clear that the *necessary and sufficient* condition that the ICCP scheme under incomplete information achieves the same maximum revenue of the *CP* scheme under complete information is

$$s_q^* \le s_{i\to q}, \ \forall\, i < q, \forall\, q \in \{2, \dots, K\}. \tag{8.32}$$

By solving these inequalities, we can obtain the following theorem (detailed proof in Appendix 8.A.5).

Theorem 8.6 *There exist unique thresholds $\{t_1, \dots, t_{K-1}\}$, such that the ICCP scheme achieves the same maximum revenue as in the complete information case if*

$$\sqrt{\frac{\theta_q}{\theta_{q+1}}} \ge t_q \quad \textit{for } q = 1, \dots, K-1.$$

Moreover, t_q is the unique solution of the equation

$$t^2 \ln t - (t^2 - 1) + \frac{t\sum_{k=1}^{q} N_k + N_{q+1}}{S + \sum_{k=1}^{K^{cp}} N_k}(t-1) = 0$$

over the domain $t > 1$.

We want to mention that the condition in Theorem 8.6 is necessary and sufficient for the case of $K = 2$ effective groups.[3] For $K > 2$, Theorem 8.6 is sufficient but not necessary. The intuition of Theorem 8.6 is that users need to be sufficiently different to achieve the maximum revenue.

The following result immediately follows Theorem 8.6.

Corollary 8.1 *The t_qs in Theorem 8.6 satisfy $t_q < t_{root}$ for $q = 1, \ldots, K-1$, where $t_{root} \approx 2.21846$ is the larger root of equation $t^2 \ln t - (t^2 - 1) = 0$.*

The Corollary 8.1 means that the users do not need to be extremely different to achieve the maximum revenue.

When the conditions in Theorem 8.6 are not satisfied, there may be revenue loss by using the pricing menu in Eq. (8.30). As it is difficult to explicitly solve the parameterized transcend equation (8.31), we are not able to characterize the loss in a closed form yet.

8.6.1 Extensions to Partial Price Differentiation under Incomplete Information

For any given system parameters, we can numerically check whether a partial price differentiation scheme can achieve the same maximum revenue under both the complete and incomplete information scenarios. The idea is similar as we described in this section. As the *PP* problem can be viewed as the *CP* problem for all effective SGs, then we can check the ICCP bound in Theorem 8.6 for SGs (once the SG partition is determined by the searching using Algorithm 8.4). Deriving an analytical sufficient condition (as in Theorem 8.6) for an incentive-compatible partial price differentiation scheme, however, is highly nontrivial and is part of our future study.

8.7 CONNECTIONS WITH THE CLASSICAL PRICE DIFFERENTIATION TAXONOMY

In economics, price differentiation is often categorized by the first/second/third-degree price differentiation taxonomy [40]. This taxonomy is often used in the context of unlimited resources and general pricing functions. The proposed schemes in this chapter have several key differences from these standard concepts, mainly because of the assumption of limited total resources and the choice of linear usage-based pricing.

In the first-degree price differentiation, each user is charged a price based on its willingness to pay. Such a scheme is also called the *perfect price differentiation*, as it captures users' entire surpluses (i.e., leaving users with zero payoffs). For the complete price differentiation scheme under complete information in Section 8.3, the service provider does not extract all surpluses from users, mainly because of the choice of linear price functions. All effective users obtain positive payoffs.

[3] There might be other groups who are not allocated positive resource under the optimal pricing.

In the second-degree price differentiation, prices are set according to quantities sold (e.g., the volume discount). The pricing scheme under incomplete information in Section 8.6 has a similar flavor of quantity-based charging. However, our proposed pricing scheme charges a *higher* unit price for a *larger* quantity purchase, which is opposite to the usual practice of volume discount. This is due to our motivation of mimicking the optimal pricing differentiation scheme under the complete information. Our focus is to characterize the sufficient conditions, under which the revenue loss owing to incomplete information (also called *information rent* [14, 41]) is zero.

In the third-degree price differentiation, prices are set according to some customer segmentation. The segmentation is usually made based on users' certain attributes such as ages, occupations, and genders. The partial price differentiation scheme in Section 8.5 is analogous to the third-degree price differentiation, but here the user segmentation is still based on users' willingness to pay. The motivation of our scheme is to reduce the implementational complexity.

8.8 NUMERICAL RESULTS

We provide numerical examples to quantitatively study several key properties of price differentiation strategies in this section.

8.8.1 When is Price Differentiation Most Beneficial?

Definition 8.1 *(Revenue gain) We define the revenue gain G of one pricing scheme as the ratio of the revenue difference (between this pricing scheme and the single pricing scheme) normalized by the revenue of single pricing scheme.*

In this subsection, we study the revenue gain of the *CP* scheme, that is, $G(\boldsymbol{N}, \boldsymbol{\theta}, S) \stackrel{\Delta}{=} \frac{R_{cp}-R_{sp}}{R_{sp}}$, where $\boldsymbol{N} \stackrel{\Delta}{=} \{N_i, \forall i \in \mathcal{I}\}$ denotes the number of users in each groups, $\boldsymbol{\theta} \stackrel{\Delta}{=} \{\theta_i, \forall i \in \mathcal{I}\}$ denotes their willingness to pays, and S is the total resource. Notice that this gain is the maximum possible differentiation gain among all *PP* schemes.

We first study a simple two-group case. According to Theorems 8.1 and 8.2, the revenue under the *SP* scheme and the *CP* scheme can be calculated as follows:

$$R^{sp} = \begin{cases} \dfrac{S(N_1\theta_1 + N_2\theta_2)}{N_1 + N_2 + S}, & 1 \le t < \sqrt{\dfrac{S+N_1}{N_1}}; \\[2ex] \dfrac{SN_1\theta_1}{N_1 + S}, & t \ge \sqrt{\dfrac{S+N_1}{N_1}}; \end{cases}$$

and

$$R^{cp} = \begin{cases} \dfrac{S(N_1\theta_1 + N_2\theta_2) + N_1N_2(\sqrt{\theta_1} - \sqrt{\theta_2})^2}{N_1 + N_2 + S}, & 1 \leq t < \dfrac{S + N_1}{N_1}; \\ \dfrac{SN_1\theta_1}{N_1 + S}, & t \geq \dfrac{S + N_1}{N_1}; \end{cases}$$

where $t = \sqrt{\frac{\theta_1}{\theta_2}} > 1$.

The revenue gain will depend on five parameters, S, N_1, θ_1, N_2, and θ_2. To simplify notations, let $N = N_1 + N_2$ be the total number of the users, $\alpha = \frac{N_1}{N}$ the percentage of group 1 users, and $\bar{s} = \frac{S}{N}$ the level of normalized available resource. Thus the revenue gain can be expressed as

$$G(t, \alpha, \bar{s}) = \begin{cases} \dfrac{\alpha(1-\alpha)(t-1)^2}{\bar{s}(1+\alpha(t^2-1))}, & 1 < t < \sqrt{\dfrac{\bar{s}+\alpha}{\alpha}}; \\ \dfrac{(1-\alpha)(\bar{s}+\alpha-t\alpha)^2}{\alpha\bar{s}(1+\bar{s})t^2}, & \sqrt{\dfrac{\bar{s}+\alpha}{\alpha}} \leq t \leq \dfrac{\bar{s}+\alpha}{\alpha}. \end{cases} \tag{8.33}$$

Next we discuss the impact of each parameter.

Observation 8.1 *In terms of the parameter t, G monotonically increases in* $\left(1, \sqrt{\frac{\bar{s}+\alpha}{\alpha}}\right)$ *and decreases in* $\left[\sqrt{\frac{k+\alpha}{\alpha}}, \frac{k+\alpha}{\alpha}\right)$. *The maximum is obtained at* $t_{G-\max} = \sqrt{\frac{k+\alpha}{\alpha}}$, *when the resource allocated to the group 2 user just becomes zero in the SP scheme.*

One example is showed in Figure 8.10.

It is clear that the revenue gain is not monotonic in the willingness to pay ratio. Its behavior can be divided into three regions: the increasing Region (1) with $t \in \left(1, \sqrt{\frac{\bar{s}+\alpha}{\alpha}}\right)$, the decreasing Region (2) with $t \in \left[\sqrt{\frac{k+\alpha}{\alpha}}, \frac{k+\alpha}{\alpha}\right)$, and the zero Region (3) with $t \geq \frac{k+\alpha}{\alpha}$.

It is also interesting to note that three regions are closed related to the effective market sizes: $K^{sp} = K^{cp} = 2$ in Region (1); $K^{sp} = 1$ and $K^{cp} = 2$ in Region (2); and $K^{cp} = K^{sp} = 1$ in Region (3) where the *CP* scheme degenerates to the *SP* scheme. The peak point of the revenue gain correspond to the place where the effective market of the *SP* Scheme changes.

Intuitively, the *CP* scheme increases the revenue by charging the high willingness groups with high prices, thus the revenue gain increases first when the difference of willingness to pays increase. However, when the difference of willingness to pay is very large, the *CP* scheme obtains most revenue from the high willingness to pay

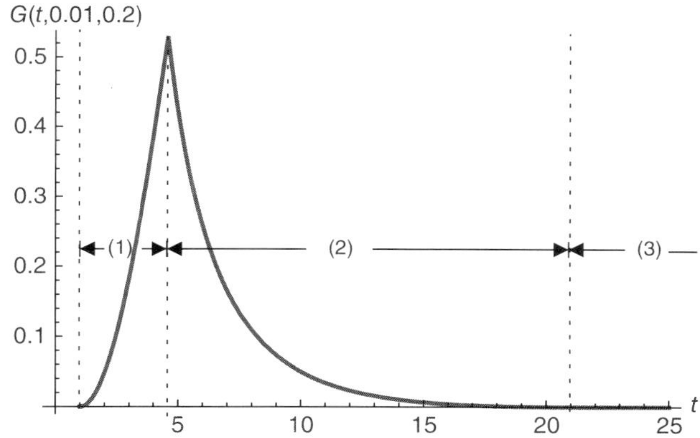

Figure 8.10 One example of the revenue gain $G(t, 0.01, 0.2)$ for the *CP* scheme. It is clear that the revenue gain can be divided into three regions. Region (1), increasing region, where $K^{cp} = K^{sp} = 2$ and the revenue gain comes from the differentiation gain. Region (2), decreasing region, where $K^{cp} = 2$, $K^{sp} = 1$, and the revenue gain comes from larger effective market and differentiation gain. Region (3), zero region, where $K^{cp} = K^{sp} = 1$ and is a degenerating case where two pricing scheme coincide.

users, while the *SP* scheme declines the low willingness to pay users but serves the high willingness to pays only. Both schemes lead to similar resource allocation in this region, and thus the revenue gain decreases as the difference of willingness to pays increases.

Figure 8.10 shows the revenue gain under usage-based pricing can be very high in some scenario, for example, over 50% in this example. We can define this peak revenue gain as

$$G_{\max}(\alpha, \bar{s}) = \max_{t \geq 1} G(t, \alpha, \bar{s}) = \frac{(\alpha - 1)(\sqrt{\bar{s}+\alpha} - \sqrt{\alpha})^2}{\bar{s}(1+\bar{s})}.$$

Figure 8.11 is shown how $G_{\max}$ changes in $\bar{s}$ with different parameters α.

Observation 8.2 *For a fixed* $\bar{s}$, $G_{\max}(\alpha, \bar{s})$ *monotonically decreases in* α.

When α is small, which means high willingness to pay users are minorities in the effective market, the advantage of price differentiation is very evident. As shown in Figure 8.11, when $\alpha = 0.1$, the maximum possible revenue gain can be over than 20%; and when $\alpha = 0.01$, this gain can be even higher than 50%. However, when high willingness to pay users are majority, the price differentiation gain is very limited, for example, the gain is no larger than 8% and 2% for $\alpha = 0.5$ and 0.9, respectively.

Intuitively, high willingness to pay users is the most profitable users in the market. Ignoring them is detrimental in terms of revenue even if they only occupy a small fraction of the population. As the *SP* scheme is set based on the average willingness to pay of the effective market, the high willingness to pay users will be ignored (in the

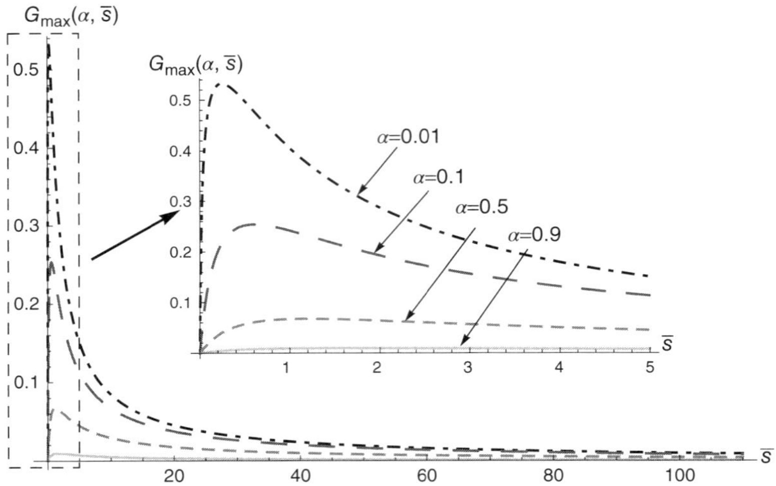

Figure 8.11 For a fixed $\bar{s}$, $G_{max}(\alpha, \bar{s})$ monotonically increases in α. For a fixed α, $G_{max}(\alpha, \bar{s})$ first increases in $\bar{s}$ and then decreases in $\bar{s}$.

sense of not charging the desirable high price) when α is small. In contrast, ignoring the low willingness to pay users when α is large is not a big issue.

Observation 8.3 *For parameter k, $G_{max}(\alpha, \bar{s})$ is not a monotonic function in $\bar{s}$. Its shape looks like a skewed bell. The gain is small when $\bar{s}$ is either very small or very large.*

Small $\bar{s}$ means that resource is very limited, and both schemes allocates the resource to high willingness to pay users (see the discussion of the threshold structure in Sections 8.3 and 8.4), and thus there is not much difference between two pricing schemes. While $\bar{s}$ is very large, that is, the resource is abundant, the prices and the resource allocation with or without differentiation become similar (which can be easily checked from formulations in Theorems 8.1 and 8.2). In these two scenarios, similar resource allocations lead to similar revenues. These explains the bell shape for parameter $\bar{s}$.

We find that the revenue gain can be very high under two conditions based on the above observations. First, the high willingness to pay users is minorities in the effective market. Second, the total resource is comparatively limited.

For cases with three or more groups, the analytical study becomes much more challenging because of many more parameters. Moreover, the complex threshold structure of the effective market makes the problem even complicated. We will present some numerical studies to illustrate some interesting insights.

For illustration convenience, we choose a three-group example and three different sets of parameters as shown in Table 8.2. To limit the dimension of the problem, we set the parameters such that the total number of users and the average willingness

TABLE 8.2 Parameter Settings of a Three-Group Example

	θ_1	N_1	θ_2	N_2	θ_3	N_3	$\overline{\theta}$
Case 1	9	10	3	10	1	80	2
Case 2	3	33	2	33	1	34	2
Case 3	2.2	80	1.5	10	1	10	2

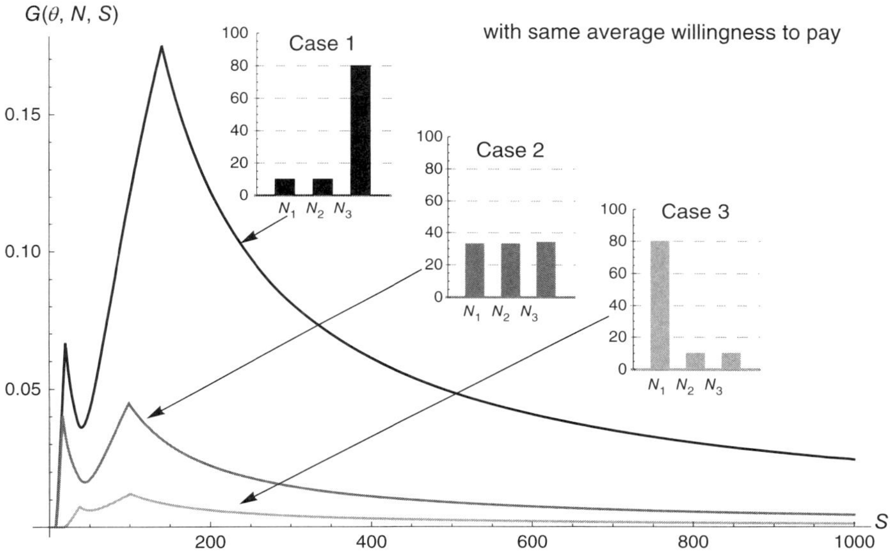

Figure 8.12 An example of the revenue gain of the three-group market with the same average willingness to pay.

to pay (i.e., $\overline{\theta} = \sum_{i=1}^{3} N_i\theta_i/(\sum_{i=1}^{3} N_i)$) of all users are the same across three different parameter settings. This ensures that the *SP* scheme achieves the same revenue in three different cases when resource is abundant. Figure 8.12 illustrates how the differentiation gain changing changes in resource S.

Similar to the analytical study of the two-group case, Figure 8.12 shows that the revenue gain is large only when the high willingness to pay users are minorities (e.g., case 1) in the effective market and the resource is limited but not too small ($100 \leq S \leq 150$ in all three cases). When resource S is large enough (e.g., ≥ 150), the gain will gradually diminish to zero as the resource increases. For each curve in Figure 8.12, there are two peak points. Each peak point represents a change of the effective market threshold in the *SP* scheme, that is, when the resource allocation to a group becomes zero. In numerical studies of networks with $I > 3$ groups (not shown in this chapter), we have observed the similar conditions for achieving a large differentiation gain and the phenomenon of $I - 1$ peak points.

8.8.2 What is the Best Trade-Off of Partial Price Differentiation?

In Section 8.5, we design Algorithm 8.4 that optimally solves the *PP* problem with a polynomial complexity. Here we study the trade-off between total revenue and implementational complexity.

To illustrate the trade-off, we consider a five-group example with parameters shown in Table 8.3. Note that high willingness to pay users is minorities here. Figure 8.13 shows the revenue gain G as a function of total resource S under different *PP* schemes (including *CP* scheme as a special case), and Figure 8.14 shows how the effective market thresholds change with the total resource.

We enlarge Figures 8.13 and 8.14 within the range of $S \in [0, 50]$, which is the most complex and interesting part because of several peak points. Similar to Figure 8.12, we observe $I - 1 = 4$ peak points for each curve in Figure 8.13. Each peak point again represents a change of effective market threshold of the single pricing scheme, as we can easily verify by comparing Figure 8.14 with Figure 8.13.

As the resource S increases from 0, all gains in Figure 8.13 first overlap with each other, then the two-price scheme separates from the others at $S = 3.41$, after that the three-price scheme separates at $S = 8.89$, and finally the four-price scheme separates at near $S = 20.84$. These phenomena are due to the threshold structure of the *PP*

TABLE 8.3 Parameter Setting of a Five-Group Example

Group Index i	1	2	3	4	5
θ_i	16	8	4	2	1
N_i	2	3	5	10	80

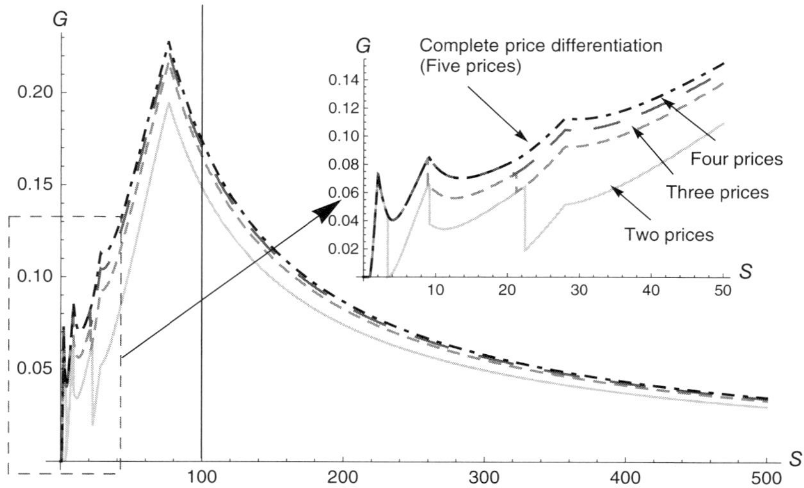

Figure 8.13 Revenue gain of a five-group example under different price differentiation schemes.

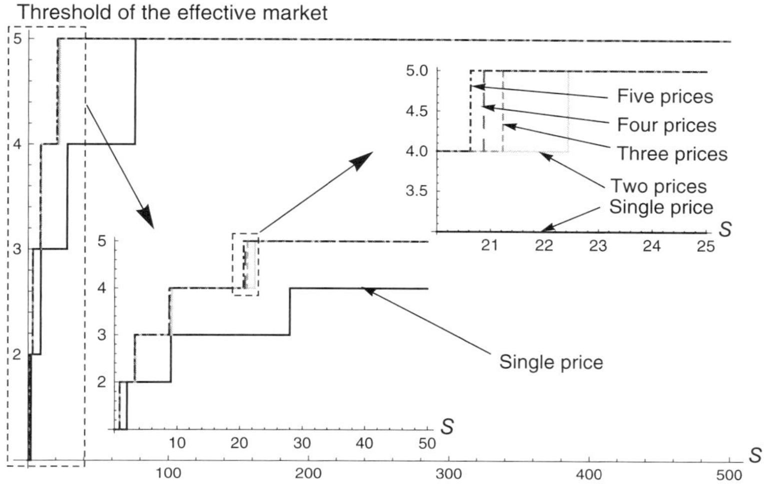

Figure 8.14 Corresponding thresholds of effective markets of Figure 8.13's Example.

scheme. When the resource is very limited, the effective markets under all pricing scheme include only one group with the highest willingness to pay and all pricing schemes coincide with the *SP* scheme. As the resource increases, the effective market enlarges from two groups to finally five groups. The change of the effective market threshold can be directly observed in Figure 8.14. Comparing across different curves in Figure 8.14, we find that the effective market size is nondecreasing with the number of prices for the same resource *S*. This agrees with our intuition in Section 8.4.2, which states that the size of effective market indicates the degree of differentiation.

Figure 8.13 provides the service provider a global picture of choosing the most proper pricing scheme according to achieve the desirable financial target under a certain parameter setting. For example, if the total resource $S = 100$, the two-price scheme seems to be a sweet spot, as it achieves a differential gain of 14.8% compared to the *SP* scheme and is only 2.4% worser than the *CP* scheme with five prices.

8.9 CONCLUSION

In this chapter, we study the revenue-maximizing problem for a monopoly service provider under both complete and incomplete network information. Under complete information, our focus is to investigate the trade-off between the total revenue and the implementational complexity (measured in the number of pricing choices available for users). Among the three pricing differentiation schemes we proposed (i.e., complete, single, and partial), the partial price differentiation is the most general one and includes the other two as special cases. By exploiting the unique problem structure, we designed an algorithm that computes the optimal partial pricing scheme in polynomial time and numerically quantizes the trade-off between implementational complexity and total revenue. Under incomplete information, designing an incentive-compatible differentiation pricing scheme is difficult in general. We show that when the users are significantly different, it is possible to design a quantity-based pricing scheme that achieves the same maximum revenue as under complete information.

APPENDIX 8.A

8.A.1 COMPLETE PRICE DIFFERENTIATION UNDER COMPLETE INFORMATION WITH GENERAL UTILITY FUNCTIONS

In this section, we extend the solution of complete price differentiation problem to general form of increasing and concave utility functions $u_i(s_i)$. We denote $R_i(s_i)$ as the revenue collected from one user in group i. On the basis of the Stackelberg model, the prices satisfy $p_i = u_i'(s_i)$, $s_i \geq 0$ $i \in \mathcal{I}$, thus

$$R_i(s_i) = u_i'(s_i)s_i, s_i \geq 0. \tag{8.A.1}$$

Therefore, we can rewrite the complete price differentiation problem with general utility function (CPG) as follows.

$$CPG : \underset{s \geq 0, \boldsymbol{n}}{\text{maximize}} \sum_{i \in \mathcal{I}} n_i R_i(s_i)$$

$$\text{subject to } n_i \in \{0, \ldots, N_i\} \quad , \quad i \in \mathcal{I}. \tag{8.A.2}$$

$$\sum_{i \in \mathcal{I}} n_i s_i \leq S \tag{8.A.3}$$

By similar solving technique in Section 8.3, we can solve CPG Problem by decomposing it into two subproblems: resource allocation subproblem CPG_1 and admission control subproblem CPG_2. In subproblem CPG_1, for given $\boldsymbol{n}$, we solve

$$CPG_1 : \underset{s \geq 0}{\text{maximize}} \sum_{i \in \mathcal{I}} n_i R_i(s_i)$$

$$\text{subject to} \sum_{i \in \mathcal{I}} n_i s_i \leq S.$$

After solving the optimal resource allocation $s_i^*(\boldsymbol{n})$, $i \in \mathcal{I}$, we further solve admission control subproblem:

$$CPG_2 : \underset{\boldsymbol{n}}{\text{maximize}} \sum_{i \in \mathcal{I}} n_i R_i(s_i^*(\boldsymbol{n}))$$

$$\text{subject to} : n_i \in \{0, \ldots, N_i\}.$$

We are especially interested in the case that constraint (8.A.3) is active in CPG Problem, which means the resource bound is tight in the considered problem;

otherwise, *CPG* problem degenerates to a revenue maximization without any bounded resource constraint. We can prove the following results.

Proposition 8.A.1 *If the resource constraint (8.A.3) is active in the optimal solution of the CPG problem (or CPG_1 Subproblem), then one of optimal solutions of CPG_2 Subproblem is*

$$n_i^* = N_i,\ i \in \mathcal{I}. \tag{8.A.4}$$

Proof: We first release the variable n_i to real number and calculate the first derivative as follows:

$$\frac{\partial R_i}{\partial n_i} = R_i(s_i^*) + n_i \frac{\partial R_i(s_i^*)}{\partial s_i} \frac{\partial s_i^*}{\partial n_i},\quad i \in \mathcal{I}. \tag{8.A.5}$$

Plugging Eq. (8.A.1), $R_i'(s_i) = u_i''(s_i)\, s_i + u_i'(s_i)$, and we have

$$\frac{\partial R_i}{\partial n_i} = u_i'(s_i^*) \left(s_i^* + n_i \frac{\partial s_i^*}{\partial n_i} \right) + n_i\, u_i''(s_i^*)\, s_i^* \frac{\partial s_i^*}{\partial n_i},\quad i \in \mathcal{I}. \tag{8.A.6}$$

As the resource constraint (8.A.3) is active in the optimal solution of CPG_1 subproblem, that is, $\sum_{i\in\mathcal{I}} n_i s_i = S$, by taking derivative of n_i in both sides of it, we have

$$s_i^* + n_i \frac{\partial s_i^*}{\partial n_i} = 0. \tag{8.A.7}$$

Substituting Eq. (8.A.7) into Eq. (8.A.6), because we assume the utility function $u_i(s_i)$ is increasing and concave function, then we have

$$\frac{\partial R_i}{\partial n_i} = -u_i''(s_i^*)\, s_i^{*2} \geq 0,\quad i \in \mathcal{K}. \tag{8.A.8}$$

Thus we can conclude that one of optimal solutions for CPG_2 subproblem is $n_i^* = N_i, i \in \mathcal{I}$. ■

Proposition 8.A.1 points out that when the resource constraint (8.A.3) is active, the *CPG* problem can be greatly simplified: its solution can be obtained by solving *CPG* subproblem with parameters $n_i = N_i, i = 1, \ldots, I$. The following proposition provides a sufficient condition that the resource constraint (8.A.3) is active.

Proposition 8.A.2 *If $u_i''(s_i)s_i + u_i(s_i) > 0,\ s_i \geq 0,\ i \in \mathcal{I}$, then the resource constraint is active at the optimal solution.*

Proof: Let λ and μ_i, $i \in \mathcal{I}$, be the Lagrange multiplier of constraint (8.A.3) and $s_i \geq 0, i \in \mathcal{I}$, respectively, thus the KKT conditions of CGP_1 subproblem is given as follows:

$$n_i \frac{\partial R_i(s_i^*)}{\partial s_i} - n_i \lambda^* + \mu_i^* = 0,\ i \in \mathcal{I};$$

$$\lambda^* \left(\sum_{i \in \mathcal{I}} n_i s_i^* - S \right) = 0;$$

$$\mu_i^* s_i^* = 0;$$

$$\lambda^* \geq 0;$$

$$\mu_i^* \geq 0,\ i \in \mathcal{I};$$

$$s_i^* \geq 0,\ i \in \mathcal{I}.$$

We denote $\mathcal{K} := \{i \mid s_i^* > 0\}$ and $\overline{\mathcal{K}} := \{i \mid s_i^* = 0\}$.
For $i \in \mathcal{K}$,

$$\frac{\partial R_i(s_i^*)}{\partial s_i} = \lambda^*,\ i \in \mathcal{I}; \tag{8.A.9}$$

$$\lambda^* \left(\sum_{i \in \mathcal{K}} n_i s_i^* - S \right) = 0. \tag{8.A.10}$$

For $i \in \overline{\mathcal{K}}$,

$$\frac{\partial R_i(0)}{\partial s_i} \leq \lambda^*,\ i \in \mathcal{I}; \tag{8.A.11}$$

As $u_i''(s_i)s_i + u_i(s_i) > 0$, $s_i \geq 0$, $i \in \mathcal{I}$ and Eq. (8.A.9), we have

$$\lambda^* = \frac{\partial R_i(s_i^*)}{\partial s_i} = u_i''(s_i^*)s_i^* + u_i(s_i^*) > 0.$$

By Eq. (8.A.10), we must have $\sum_{i \in \mathcal{I}} n_i s_i^* - S = 0$, that the resource constraint is active at the optimal solution. ■

Next, let us discuss how to calculate the optimal solution. To guarantee uniqueness resource allocation solution, we assume that the revenue is a strictly concave function of the demand,[4] that is, $\frac{\partial^2 R_i(s_i)}{\partial s_i^2} < 0$, $i \in \mathcal{I}$. Thus we have the following theorem.

Theorem 8.A.1 *If* $\frac{\partial^2 R_i(s_i)}{\partial s_i^2} < 0$, $i \in \mathcal{I}$, *then there exists an optimal solution of CGP problem as follows:*

- *All users are admitted:* $n_i^* = N_i$ *for all* $i \in \mathcal{I}$.
- *There exist a value* λ^* *and a group index threshold* $K^{cp} \leq I$, *such that only the top* K^{cp} *groups of users receive positive resource allocations,*

[4]This assumption has been frequently used in the revenue management literature [42].

$$s_i^* = \begin{cases} \dfrac{\partial R_i}{\partial s_i}^{-1}(\lambda^*), & i \in \mathcal{K}; \\ 0, & otherwise, \end{cases} \tag{8.A.12}$$

where values of λ^* and effective market $\mathcal{K}$ can be computed as in Algorithm 8.A.1.

In Algorithm 8.A.1, we use notation f^{-1} denotes its inverse function and rearrange the group index satisfying $\frac{\partial R_{(1)}}{\partial s_{(1)}}^{-1}(0) \geq \frac{\partial R_{(2)}}{\partial s_{(2)}}^{-1}(0) \geq \cdots \geq \frac{\partial R_{(I)}}{\partial s_{(I)}}^{-1}(0)$.

Algorithm 8.A.1

Threshold-General-Utility

$k \leftarrow I$, $\lambda \leftarrow \frac{\partial R_{(k)}}{\partial s_{(k)}}^{-1}(0)$

while $\sum_{i=1}^{k} n_{(i)} \left(\frac{\partial R_{(i)}}{\partial s_{(i)}}^{-1}(\lambda) \right)^+ \geq S$,

$k \leftarrow k - 1$

$\lambda \leftarrow \frac{\partial R_{(k)}}{\partial s_{(k)}}^{-1}(0)$

end while

return $\mathcal{K} = \{(1), (2), \ldots, (k)\}$

End pseudo-code.

Remark 8.A.1 *The complexity of Algorithm 8.A.1 is also $\mathcal{O}(I)$, that is, linear in the number of user groups (not the number of users).*

Remark 8.A.2 *There are several functions satisfying the technical conditions in Theorem 8.A.1, for example, the standard α-fairness functions*

$$u_i(s_i) = \begin{cases} (1-\alpha)^{-1} s_i^{1-\alpha}, & 0 \leq \alpha < 1; \\ \log s_i, & \alpha = 1. \end{cases}$$

8.A.2 PROOF OF PROPOSITION 8.1

Proof: We first focus on the key water-filling problems that we solve for the two pricing schemes (the *CP* scheme on the LHS and the *SP* scheme on the RHS):

$$\sum_{i\in\mathcal{I}} N_i\left(\sqrt{\frac{\theta_i}{\lambda^*}}-1\right)^+ = S = \sum_{i\in\mathcal{I}} N_i\left(\frac{\theta_i}{p^*}-1\right)^+. \tag{8.A.13}$$

Let $\theta = \frac{p^{*2}}{\lambda^*}$ be the solution of the equation of $\sqrt{\frac{\theta}{\lambda^*}} = \frac{\theta}{p^*}$. By comparing it with θ_i, $i \in \mathcal{I}$, there are three cases:

- *Case 1.*
  $\theta > \theta_1 \Rightarrow \sqrt{\frac{\theta_i}{\lambda^*}} = \frac{\sqrt{\theta_i}\sqrt{\theta}}{p^*} > \frac{\theta_i}{p^*}, \forall i \in \mathcal{I}$.
  This case cannot be possible. As if every term in the left summation is strictly larger than its counterpart in the right summation, then Eq. (8.A.13) cannot hold.
- *Case 2.* $\theta_I \geq \theta \Rightarrow \sqrt{\frac{\theta_i}{\lambda^*}} = \frac{\sqrt{\theta_i}\sqrt{\theta}}{p^*} \leq \frac{\theta_i}{p^*}, \forall i \in \mathcal{I}$. Similar to Case 1, it cannot hold, either.
- *Case 3.* $\exists k,\ s.t.\ 1 \leq k < I$ and $\theta_k \geq \theta \geq \theta_{k+1}$

$$\Rightarrow \begin{cases} \sqrt{\frac{\theta_i}{\lambda^*}} = \underbrace{\frac{\sqrt{\theta_i}\sqrt{\theta}}{p^*} \leq \frac{\theta_i}{p^*}, i = 1, 2, \ldots, k;}_{\text{The equality holds only when } \theta = \theta_k \text{ and } i = k.} \\ \sqrt{\frac{\theta_i}{\lambda^*}} = \underbrace{\frac{\sqrt{\theta_i}\sqrt{\theta}}{p^*} \geq \frac{\theta_i}{p^*},\ i = k+1, \ldots, I.}_{\text{The equality holds only when } \theta = \theta_{k+1} \text{ and } i = k+1.} \end{cases}$$

Similar argument as the above two case, we have $K^{cp} \geq k$ and $K^{sp} \geq k$, otherwise (8.A.13) can not hold. Further, $K^{cp} \geq K^{sp}$, since if $\frac{\theta_{K^{sp}}}{p^*} - 1 > 0$, then $\sqrt{\frac{\theta_{K^{cp}}}{\lambda^*}} - 1 > 0$.

By Theorems 8.1 and 8.2, we prove the proposition. ■

8.A.3 PROOF OF LEMMA 8.2

We can first prove the following lemma.

Lemma 8.A.1 *Suppose an effective market of the single pricing scheme is denoted as $\mathcal{K} = \{1, 2, \ldots, K\}$. If we add a new group v of N_v users with $\theta_v > \theta_K$, then the revenue strictly increases.*

Proof: We denote the single price before joining group v is p, the price after joining group v is p', the effective market become $\mathcal{K}'$. By Theorem 8.2, we have

$$p = \frac{\sum_{i=1}^{K} N_i \theta_i}{S + \sum_{i=1}^{K} N_i} \text{ with } \theta_K > p \text{ and } \theta_{K+1} \leq p.$$

As the optimal revenue is obtained by selling out the total resource S, thus to prove that the total revenue strictly increases if and only if, we can prove $p' > p$. We consider the following two cases.

- If after group v joining in, the new effective market satisfies $\mathscr{K}' = \mathscr{K} \cup \{v\}$, then we have

$$p' = \frac{\sum_{i=1}^{K} N_i \theta_i + N_v \theta_v}{S + \sum_{i=1}^{K} N_i + N_v}.$$

As $\theta_v > \theta_K > p$, we have $p' > p$, because of the following simple fact.

Fact 8.A.1 *For any a_1, b_1, a_2, $b_2 > 0$, the following two inequality are equivalent:*

$$\frac{a_1}{b_1} \geq \frac{a_2}{b_2} \Leftrightarrow \frac{a_1}{b_1} \geq \frac{a_1 + a_2}{b_1 + b_2} \geq \frac{a_2}{b_2}. \quad (8.A.14)$$

- If after group v joining in, the new effective market shrinks, namely, $\mathscr{K}' \subset \mathscr{K} \cup \{v\}$, $\mathscr{K}' \neq \mathscr{K} \cup \{v\}$, then we have $p' > \theta_K > p$.

■

By the above Lemma 8.A.1, we further prove Lemma 8.2.

Proof: We prove Lemma 8.2 by contradiction. Suppose that the group indices of the effective market under the optimal partition $\boldsymbol{a}$ is not consecutive. Suppose that group i is not an effective group, and there exists some group j, $j > i$, which is an effective group. We consider a new partition $\boldsymbol{a}'$ by putting group i into the cluster to which group j belongs and keeping other groups unchanged. According to Lemma 8.A.1, the revenue under partition $\boldsymbol{a}'$ is greater than that under partition $\boldsymbol{a}$, thus partition $\boldsymbol{a}$ is not optimal. This contradicts to our assumption and thus completes the proof. ■

8.A.4 PROOF OF THEOREM 8.4

For convenience, we use the notation $(\cdots \cup \cdots | \cdots \cup \cdots | \cdots)$ to denote a partition with the groups between bars connected with "$\cup$" representing a cluster, for example, three partitions for $J = 2$, $K^{pp} = 3$ are $(1|2 \cup 3)$, $(1 \cup 2 \mid 3)$, and $(1 \cup 3 \mid 2)$. In addition, we introduce the *compound group* to simplify the notation of complex clusters with multiple groups. A cluster containing group i can be simply represented as $Pre(i) \cup i \cup Post(i)$, where $Pre(i)$ (or $Post(i)$) refers as a compound group composing of all the groups with willingness to pay larger (or smaller) than that of group i in the cluster. Note that the compound groups can be empty in certain cases.

Before we prove the general case in Theorem 8.4, we first prove the results is true for the following two special cases in Lemmas 8.A.2 and 8.A.3.

Lemma 8.A.2 *For a three-group effective market with two prices, that is, $K^{pp} = 3$, $J = 2$, an optimal partition involves consecutive group indices within clusters.*

Proof: There are three partitions for $K^{pp} = 3$, $J = 2$, and only $(1 \cup 3 \mid 2)$ is with discontinuous group index within clusters. To show our result, we only need to prove one of partitions with group consecutive is better than $(1 \cup 3 \mid 2)$. We have two main steps in this proof; first, we prove this result is true for *PP* problem without consider constraint (8.29). Further, we show that constraint (8.29) will not affect the optimality of partitions with consecutive group indices within each cluster.

Step 1: **(Without Constraint (8.29))**

Without considering constraint (8.29), we want to show that $\boldsymbol{a}_1 = (1 \cup 2 \mid 3)$ is always better than $\boldsymbol{a}_2 = (1 \cup 3 \mid 2)$. Mathematically, what we try to prove is

$$v(\boldsymbol{a}_2) > v(\boldsymbol{a}_1), \tag{8.A.15}$$

where $v(\boldsymbol{a}_2) = (N_1 + N_3)\sqrt{\frac{N_1\theta_1+N_3\theta_3}{N_1+N_3}} + N_2\sqrt{\theta_2}$, and $v(\boldsymbol{a}_1) = (N_1 + N_2)\sqrt{\frac{N_1\theta_1+N_2\theta_2}{N_1+N_2}} + N_3\sqrt{\theta_3}$. With the new notation

$$\Delta V(i,j) := (N_i + N_j)\sqrt{\frac{N_i\theta_i + N_j\theta_j}{N_i + N_j}} - N_i\sqrt{\theta_i} - N_j\sqrt{\theta_j},$$

it is easy to see that (8.A.15) is equivalent to the following inequality:

$$\Delta V(1,3) > \Delta V(1,2). \tag{8.A.16}$$

We prove the inequality (8.A.16) by considering the following two cases.

(a) If $N_1 \le N_2$, we define a function of x as follows,

$$g(j;x) := (N_j + N_1)\sqrt{\frac{N_j\theta_j + N_1(\theta_j + x)}{N_j + N_1}} - N_1\sqrt{\theta_j + x} - N_j\sqrt{\theta_j}.$$

It is easy to check that

$$g(j;x)|_{x=\theta_1-\theta_j} = \Delta V(1,j), \text{ and } g(j;x)|_{x=0} = 0;$$

and if $x > 0$, then

$$g'(j;x) = \frac{\partial g(j;x)}{\partial x} = \frac{N_1}{2}\left(\frac{1}{\sqrt{\frac{N_j\theta_j+N_1(\theta_j+x)}{N_j+N_1}}} - \frac{1}{\sqrt{\theta_j + x}}\right) > 0$$

$$\text{and} \quad \frac{\partial g'(j;x)}{\partial \theta_j} = \frac{N_1}{4}\left(\frac{1}{(\theta_j + x)^{1.5}} - \frac{1}{\left(\frac{N_j\theta_j+N_1(\theta_j+x)}{N_j+N_1}\right)^{1.5}}\right) < 0.$$

As $\theta_2 > \theta_3$, it immediately follows that

$$g'(3;x) \geq \frac{N_1}{2}\left(\frac{1}{\sqrt{\frac{N_3\theta_2+N_1(\theta_2+x)}{N_1+N_3}}} - \frac{1}{\sqrt{\theta_2+x}}\right).$$

As $N_2 \geq N_1$, then we have

$$g'(3;x) \geq \frac{N_1}{2}\left(\frac{1}{\sqrt{\frac{N_3\theta_2+N_1(\theta_2+x)}{N_1+N_3}}} - \frac{1}{\sqrt{\theta_2+x}}\right) \geq g'(2;x).$$

Thus, it follows

$$\Delta V(1,3) = \int_0^{\theta_1-\theta_3} g'(3;x)dx > \int_0^{\theta_1-\theta_2} g'(2;x)dx = \Delta V(1,2),$$

that is, Eq. (8.A.16) is obtained.

Let us see a special case of Eq. (8.A.16). When $N_1 = N_2$, then

$$\Delta V(1,2) = (N_1+N_1)\sqrt{\frac{N_1\theta_2+N_1\theta_1}{N_1+N_1}} - N_1\sqrt{\theta_1} - N_1\sqrt{\theta_2},$$

then we have

$$\Delta V(1,3) > (N_1+N_1)\sqrt{\frac{N_1\theta_2+N_1\theta_1}{N_1+N_1}} - N_1\sqrt{\theta_1} - N_1\sqrt{\theta_2}. \tag{8.A.17}$$

Notice that although Eq. (8.A.17) is defined with the assumption that $N_1 \leq N_2$, it also holds for the case $N_1 > N_2$ as Eq. (8.A.17) does not contain the parameter N_2. This result will be used in the proof later.

(b) If $N_1 > N_2$, we define a function of m as

$$f(m) := (N_1+m)\sqrt{\frac{N_1\theta_1+m\theta_2}{N_1+m}} - N_1\sqrt{\theta_1} - m\sqrt{\theta_2}.$$

It is easy to obtain that

$$\frac{df(m)}{dm} = \frac{\left(\sqrt{\frac{N_1\theta_1+m\theta_2}{N_1+m}} - \sqrt{\theta_2}\right)^2}{2\sqrt{\frac{N_1\theta_1+m\theta_2}{N_1+m}}} > 0,$$

that is, the function f is an increasing function of m.

Thus it follows that

$$\Delta V(1,2) = f(N_2) < f(N_1) \overset{(a)}{<} \Delta V(1,3),$$

where (a) results from Eq. (8.A.17), the RHS of which is equal to $f(N_1)$.

***Step 2:* (Checking Constraint (8.29))**

We want to prove that $\boldsymbol{a}_1$ satisfying constraint (8.29) is the sufficient condition of $\boldsymbol{a}_2$ satisfying Eq. (8.29).

Consider if $\boldsymbol{a}_1$ does not satisfy Eq. (8.29), it means

$$\sqrt{\theta_3} \leq \sqrt{\lambda(\boldsymbol{a}_1)} = \sqrt{\frac{v(\boldsymbol{a}_1)}{S + \sum_{i=1}^{3} N_i}}.$$

By the result in Step 1, we know that $v(\boldsymbol{a}_1) < v(\boldsymbol{a}_2)$, then we have

$$\sqrt{\theta_3} < \sqrt{\lambda(\boldsymbol{a}_2)} = \sqrt{\frac{v(\boldsymbol{a}_2)}{S + \sum_{i=1}^{3} N_i}},$$

and further

$$\theta_3 < \sqrt{\theta_3 \lambda(\boldsymbol{a}_2)} < \sqrt{\frac{N_3\theta_3 + N_1\theta_1}{N_1 + N_3}\lambda(\boldsymbol{a}_2)} = \sqrt{\theta^1 \lambda(\boldsymbol{a}_2)}.$$

It means $\boldsymbol{a}_2$ cannot satisfy Eq. (8.29) either. Thus we see that constraint (8.29) actually does not affect the result in Step 1. In conclusion, we show that in a simple case with $K^{pp} = 3$, $J = 2$, an optimal partition involves consecutive group indices within clusters. ■

Further, based on Lemma 8.A.2, we prove another simple special case.

Lemma 8.A.3 *For a four-group effective market with two prices, that is, $K^{pp} = 4$, $J = 2$, an optimal partition involves consecutive group indices within clusters.*

Proof: For $K^{pp} = 4$ and $J = 2$ case, there are a total of seven possible partitions. Three among them are with consecutive group index, $(1 \mid 2 \cup 3 \cup 4)$, $(1 \cup 2 \mid 3 \cup 4)$, and $(1 \cup 2 \cup 3 \mid 4)$. We denote a set composed by these three partitions as Σ_c. We need to show the remaining four partitions are no better than some partition in Σ_c. To show this, we only need to transform them to some three-group case and apply the result of Lemma 8.A.2.

- *Case 1.* $(1 \cup 4,\ 2 \cup 3)$ is not optimal because we can prove $(1 \cup 2 \cup 3, 4) \in \Sigma_c$ is better. To show that, we take $2 \cup 3$ as a whole, then by Lemma 8.A.2, it follows that $\Delta V(1,4) > \Delta V(1,\ 2 \cup 3)$.

- *Case 2*. $(2,\ 1 \cup 3 \cup 4)$ is not optimal, because we can prove $(1 \cup 2, 3 \cup 4) \in \Sigma_c$ is better. To show that, we take $3 \cup 4$ as a whole, then by Lemma 8.A.2, it follows $\Delta V(1,\ 3 \cup 4) > \Delta V(1, 2)$.
- *Case 3*. $(3,\ 1 \cup 2 \cup 4)$ is not optimal, because we can prove $(1 \cup 2 \cup 3, 4) \in \Sigma_c$ is better. To show that, we take $1 \cup 2$ as a whole, then by Lemma 8.A.2, it follows that $\Delta V(1 \cup 2, 4) > \Delta V(1 \cup 2, 3)$.
- *Case 4*. $(1 \cup 3, 2 \cup 4)$ is not optimal, because we can prove $(1 \cup 2 \cup 3, 4) \in \Sigma_c$ is better. To show that, by Lemma 8.A.2, it follows that $\Delta V(2, 4) > \Delta V(2, 3)$ and that $\Delta V(1, 3) \overset{(b)}{>} \Delta V(1, 2 \cup 3)$. Here inequality (b) is also easily obtained, if we notice that $\theta_1 > \theta_{2\cup 3} > \theta_3$, thus group $2 \cup 3$ can be also treated as the role of group 2 in Lemma 8.A.2.

■

Now Let us prove Theorem 8.4. For convenience, we introduce the notation *Compound group*, such as *Pre*(i) or *Post*(i), which represents some part of a cluster with ordered group indices. For a group i in some cluster, *Pre*(i) (or *Post*(i)) refers to a compound group composing of all the groups with willingness to pay larger (or smaller) than that of group i. For example, in a cluster $1 \cup 2 \cup 3 \cup 5 \cup 7 \cup 8$, $Pre(3) = 1 \cup 2$, $Post(3) = 5 \cup 7 \cup 8$. Note that compound groups can be empty, denoted as $\emptyset$. In the last example, $Pre(1) = Post(8) = \emptyset$. As all the groups within the compound group belong to one cluster, we can apply Lemma 8.2. For example, with the previous cluster setting, $N_{Pre(3)} = N_1 + N_2$, and $\theta_{Pre(3)} = \frac{N_1\theta_1 + N_2\theta_2}{N_1 + N_2}$. By this equivalence rule, a compound group actually has not much difference with one original group. The conclusions of Lemmas 8.A.2 and 8.A.3 can be easily extended to compound groups.

Proof: Without the loss of generality, suppose that the group indices order within each cluster is increasing.

Now consider one partition with discontinuous group indices within some clusters. We can check the group indices continuity for every single group. For example, a group c belonging to a cluster $\mathscr{C}$, and its next neighbor in this cluster is group d, if $c - d = 1$, then the group indices until c are consecutive, and if $c - d > 1$, then the group indices are discontinuous and we find a gap between c and d.

Suppose that checking group indices continuity for each group following the increasing indices order (or equivalently decreasing willingness to pay order) from group 1 to group K^{pp}. We do not find any gap until group u_1 in cluster $\mathscr{U}$. We denote group u_1 next neighbor in cluster $\mathscr{U}$ is group u_2. As there is a gap between u_1 and u_2, there exists a group v in another cluster $\mathscr{V}$ and satisfying $v = u_1 + 1 < u_2$. Now we can construct a better partition by rearranging the two clusters $\mathscr{U}$ and $\mathscr{V}$, while keeping other clusters unchanged. We can view $\mathscr{U}$ as $(Pre(u_2) \cup Post(u_1))$, and $\mathscr{V}$ as $(v \cup Post(v))$, because there is no group before v in SG $\mathscr{V}$, otherwise it contradicts with the fact that we do not find any gap until group u_1. It is easy to show that there is some new partition better than the original one by Lemmas 8.A.2 and 8.A.3. There are two cases depending on whether *Post*(v) is empty or not. If $Post(v) = \emptyset$, according to Lemma 8.A.2, we find another partition with $\mathscr{U}' = Pre(u_2) \cup v$, $\mathscr{V}' = Post(u_1)$ better than the original $\mathscr{U}$ and $\mathscr{V}$. If $Post(v) \neq \emptyset$, no matter $\theta_{Post(v)}$ is larger than

$\theta_{Post(u_2)}$ or not, according to Lemma 8.A.3, it is easy to construct other partitions better than the original $\mathcal{U}$ and $\mathcal{V}$, because the compound groups in these original clusters ($\mathcal{U} = (Pre(u_2) \cup Post(u_1))$, and $\mathcal{V} = (v \cup Post(v)))$ does not satisfy the property of consecutive group indices within each cluster.

In conclusion, we show that for general cases, if there is any gap in the partition, then we can construct another partition that is better, which is equivalent to that the optimal partition must satisfy consecutive group indices within each cluster. ■

8.A.5 PROOF OF THEOREM 8.6

Proof: As $U_i(s, p_q)$ is a strictly increasing function in the interval $[0, s_i^*]$, then Eq. (8.32) holds, if and only if the following inequality holds:

$$U_i(s_q^*, p_q) \leq U_i(s_{i \to q}, p_q), \ \forall i < q. \tag{8.A.18}$$

As $t_{1q} > \cdots > t_{Kq}$, Eq. (8.A.18) can be simplified to

$$t_{q-1q}^2 \ln t_{q-1q} - (t_{q-1q}^2 - 1) + \frac{\sum_{k=1}^{K} N_k t_{kq}}{\sum_{k=1}^{K} N_k + S}(t_{q-1q} - 1) \geq 0, \tag{8.A.19}$$

where $t_{iq} = \sqrt{\frac{\theta_i}{\theta_q}}$. With a slight abuse of notation, we abbreviate t_{q-1q} as t_q, $(q = 2, \ldots, K)$ in the sequel. It is easy to see that the following inequality is the necessary and sufficient condition of Eq. (8.A.19) for $q = 2$, and sufficient condition of Eq. (8.A.19) for $q > 2$:

$$t_q^2 \ln t_q - (t_q^2 - 1) + \frac{t_q \sum_{k=1}^{q-1} N_k + N_q}{\sum_{k=1}^{K} N_k + S}(t_q - 1) \geq 0. \tag{8.A.20}$$

Let $g(t)$ be the LHS of the inequality (8.A.20). It is easy to check that $g(t)$ is a convex function, with $g(1) = 0$, $g(\infty) = \infty$ and $g'(1) < 0$. So there exists a root $t_q > 1$. When $t > t_q$, the inequality (8.A.20) holds, thus Eq. (8.A.18) holds, and the conclusion in Theorem 8.6 follows. ■

REFERENCES

1. S. Li, J. Huang, and S. Li. Revenue maximization for communication networks with usage-based pricing. In *Proceedings of the IEEE Global Telecommunications Conference*, pp. 1–6, 2009.
2. F. Kelly. Charging and rate control for elastic traffic. In *European Transactions on Telecommunications*, 1997.

3. F. Kelly, A. Maulloo, and D. Tan. "Rate control for communication networks: shadow prices, proportional fairness and stability," *Journal of the Operational Research Society*, 49, 1998, 237–252.

4. S. Low and D. Lapsley. "Optimization flow control: basic algorithm and convergence," *IEEE/ACM Transactions on Networking (TON)*, 7(6), 1999, 861–874.

5. S. Kunniyur and R. Srikant. "End-to-end congestion control schemes: utility functions, random losses and ecn marks," *IEEE/ACM Transactions on Networking (TON)*, 11(5), 2003, 702.

6. M. Chiang, S. Low, A. Calderbank, and J. Doyle. "Layering as optimization decomposition: a mathematical theory of network architectures," *Proceedings of the IEEE*, 95, 2007, 255–312.

7. M. Mussa and S. Rosen. "Monopoly and product quality," *Journal of Economic Theory*, 18(2), 1978, 301–317.

8. E. Maskin and J. Riley. "Monopoly with incomplete information," *RAND Journal of Economics*, 15, 1984, 171–196.

9. S. Shakkottai, R. Srikant, A. Ozdaglar, and D. Acemoglu. "The price of simplicity," *IEEE Journal on Selected Areas in Communications*, 26(7), 2008, 1269–1276.

10. V. Valancius, C. Lumezanu, N. Feamster, R. Johari, and V. Vazirani. "How many tiers? Pricing in the internet transit market," *SIGCOMM-Computer Communication Review*, 41(4), 2011, 194.

11. D. Acemoglu, A. Ozdaglar, and R. Srikant. The marginal user principle for resource allocation in wireless networks. In *Proceedings of IEEE Conference on Decision and Control*, vol. 2, 2004, pp. 1544–1549.

12. R. Gibbens, R. Mason, and R. Steinberg. "Internet service classes under competition," *IEEE Journal on Selected Areas in Communications*, 18(12), 2000, 2490–2498.

13. John Cox "AT&T shifts to usage-based wireless data plan," *Network World* Available online: http://www.networkworld.com/news/2010/060210-att-ends-unlimited-wireless-offering.html?source=nww rss.

14. J.-J. Laffont and D. Martmort. *The Theory of Incentives: The Principal Agent Modle*. Princeton University Press, Princeton, HJ, 2002.

15. J. Arrow. *Collected Papers of Kenneth J. Arrow, Volume 4: The Economics of Information*. Harvard University Press, Cambridge, MA, 1984.

16. J. Tirole. *The Theory of Industrial Organization*. MIT Press, Cambridge, MA, 1988.

17. P. Bolton and M. Dewatripont. *Contract Theory*. MIT Press, Cambridge, MA, 2005.

18. N. Stokey. "Intertemporal price discrimination," *Quarterly Journal of Economics*, 93(3), 1979, 355–371.

19. D. Baron and D. Besanko. "Regulation and information in a continuing relationship," *Information Economics and Policy*, 1(4), 1984, 267–302.

20. O. Hart and J. Tirole. "Contract renegotiation and coasian dynamics," *Review of Economic Studies*, 55(4), 1988, 509–540.

21. A. Acquisti and H. Varian. "Conditioning prices on purchase history," *Marketing Sciences*, 24(3), 2005, 367–381.

22. M. Armstrong. "Price discrimination by a many-product firm," *Review of Economic Studies*, 66(1), 1999, 151–168.

23. Y. Bakos and E. Brynjolfsson. "Bundling information goods: pricing, profits, and efficiency," *Management Science*, 45(12), 1999, 1613–1630.
24. X. Geng, M. Stinchcombe, and A. Winston. "Bundling information goods of decreasing value," *Magagement Science*, 51(4), 2005, 662–667.
25. L. Cabral, D. Salant, and G. Woroch. "Monopoly pricing with network externalities," *International Journal of Industrial Oranization*, 17, 1999, 199–214.
26. M. Aoyagi. "Coordinating adoption decisions under externalities and incomplete information," *Games and Economic Behavior*, 77(1), 2013, 77–89.
27. B. Schwartz. *The Paradox of Choice: Why More is Less*. Harper Perennial, New York, 2005.
28. C. Chau, Q. Wang, and D. Chiu. On the viability of paris metro pricing for communication and service networks. In *Proceedings of IEEE INFOCOM*, 2010, pp. 1–9.
29. A. Odlyzko. "Paris metro pricing for the internet. In *Proceedings of the 1st ACM Conference on Electronic Commerce*, 1999, pp. 140–147.
30. T. Basar and R. Srikant. Revenue-maximizing pricing and capacity expansion in a many-users regime. In *Proceedings of IEEE INFOCOM*, vol. 1, 2002, pp. 294–301.
31. H. Shen and T. Basar. "Optimal nonlinear pricing for a monopolistic network service provider with complete and incomplete information," *IEEE Journal on Selected Areas in Communications*, 25(6), 2007, 1216–1223.
32. A. Daoud, T. Alpcan, S. Agarwal, and M. Alanyali. A stackelberg game for pricing uplink power in wide-band cognitive radio networks. In *Proceedings of IEEE Conference on Decision and Control*, 2008, pp. 1422–1427.
33. L. Jiang, S. Parekh, and J. Walrand. Time-dependent network pricing and bandwidth trading. In *Proceedings of IEEE Network Operations and Management Symposium Workshops*, 2008, pp. 193–200.
34. P. Hande, M. Chiang, R. Calderbank, and J. Zhang. Pricing under constraints in access networks: Revenue maximization and congestion management. In *Proceedings of IEEE INFOCOM*, 2010.
35. S. Ha, S. Sen, C. Joe-Wong, Y. Im, and M. Chiang. Tube: time-dependent pricing for mobile data. In *Proceedings of ACM SIGCOMM, Helsinki, Finland*, 2012.
36. L. He and J. Walrand. Pricing and revenue sharing strategies for internet service providers. In *Proceedings IEEE INFOCOM*, vol. 1, 2005.
37. S. Shakkottai and R. Srikant. "Economics of network pricing with multiple ISPs," *IEEE/ACM Transactions on Networking (TON)*, 14(6), 2006, 1233–1245.
38. V. Gajic, J. Huang, and B. Rimoldi. Competition of wireless providers for atomic users: equilibrium and social optimality. In *Proceedings of 47th Allerton Conference on Communication, Control, and Computing, Urbana-Champaign, IL*, 2009.
39. J. van Lint and R. Wilson. *A Course in Combinatorics*. Cambridge University Press, 2001.
40. B. Pashigian. *Price Theory and Applications*. McGraw-Hill, New York, 1995.
41. A. Mas-Colell, M. Whinston, and J. Green. *Microeconomic Theory*. Oxford University Press, 1995.
42. K. Talluri and G. Van Ryzin. *The Theory and Practice of Revenue Management*. Springer-Verlag, 2005.

9 Telecommunication Pricing: Smart Versus Dumb Pipes*

ATANU LAHIRI

9.1 INTRODUCTION

Discriminatory pricing of telecommunication services today is reminiscent of practices that date back to the eighteenth century, when navigation tolls varied from one cargo type to another—the toll for a ton of sand was not the same as that for a ton of timber [3]. Similarly, consumers of wireless services are now paying different *per-byte* prices for traffic generated by different services, such as voice calls, SMS (short message service) text messages, picture mails, and mobile Internet [2]. For instance, if one assumes a data rate of 100 KB/min of voice, and 100 bytes per SMS text, a price of 10 cents/min of voice turns out to be equivalent to $1/MB, whereas a price of 20 cents per SMS text amounts to a whopping $2000/MB [1].

Such discriminatory pricing, particularly the pricing of text messages, has recently come under tremendous opposition from consumers—to the extent that even the US Senate felt compelled to examine its legality and economic impacts [4]. Experts testifying at a hearing at the US Senate Judiciary Committee argued that the average price of an SMS text message far exceeded the marginal cost of the wireless path [5], with no apparent economic justifications. Even the *Wall Street Journal*, which is known for its dislike for regulations on corporations, voiced opposition to such pricing, demanding a system where all bytes would be treated as just bytes. Mossberg [6], in his weekly column there,

> "We need a wireless mobile device ecosystem that mirrors the PC/Internet ecosystem, one where the consumers' purchase of network capacity is separate from their purchase of the hardware and software they use on that network. It will take government action, or some disruptive technology or business innovation, to get us there."

*This chapter is motivated by recent articles on wireless pricing by Lahiri et al. [1, 2]. For the sake of exposition, we use a different, but more elegant, model here.

Smart Data Pricing, First Edition. Edited by Soumya Sen, Carlee Joe-Wong, Sangtae Ha, and Mung Chiang.

Carriers, on the other hand, have been vocal at the Senate hearing about their support for discriminatory pricing. Some carriers even took additional steps towards protecting their smart-pipe strategy; for instance, fearing that open platforms such as Android—along with its user-developed services that could not be tagged or metered separately—will disrupt discriminatory pricing, AT&T stayed away from them until 2010 [7].

The common wisdom is thus that dumb pipes work in favor of consumers and smart pipes in favor of carriers. In this chapter, we put this wisdom to economic tests. We will do so by comparing the two pricing regimes in terms of their impacts on profits and welfare. In economic terms, the smart-pipe strategy means pricing different services differently. For example, consider the following pricing by a national carrier: for voice calls, one can either choose a plan that costs \$80 a month and includes 1000 min of call time or another that costs \$60 a month but includes only 700 min. Consumers can also purchase 300 text messages for a price of \$5 a month, or 1000 text messages for \$10 a month. Likewise, they can get limited access to the web for \$10 a month, or unlimited access for \$15 a month. For each service, consumers thus pick from a separate menu of plans.

In the case of dumb pipes, however, a carrier would simply price the traffic carried, giving each consumer the right to allocate the purchased traffic between various services according to his/her preferences. Thus, there is just one menu to choose from. The dumb-pipe strategy is best characterized as quasi-bundling [8]—it is a way of bundling several services into one common bundle, namely, "bytes," while leaving to each consumer the task of determining his/her actual bundle composition. An astute reader may immediately contrast this with conventional bundling that puts the seller in charge of deciding the bundle composition [9]. We develop a two-stage model of quasi-bundling in this chapter.

We will model the monopolist's tariff design problem assuming that it faces a heterogeneous market. The monopoly setting is relevant to telecommunications. Although the wireless market in the United States has several major players, each happens to possess substantial monopoly power. This is reflected in the fact that the price charged for carrying a text message is often several hundred times the marginal cost of carrying it. The close relationship of a carrier with handset manufacturers, along with various other consumer lock-in mechanisms, contributes to such pricing power. Further, while modeling the monopoly problem, we will ignore all over-limit usage and revenues resulting from such usage; over-limit usage occurs when, for example, a consumer who is on a 1000-min plan uses 1100 min in a month. As explained by Masuda and Whang [10], this simplification is reasonable when a consumer's monthly usage is deterministic.

For the purpose of exposition, we will only consider settings in which the monopolist faces discrete consumer types and sells multiple services. When it sells smart pipes, it offers a separate price schedule (or menu) for each service. Each schedule consists of a number of plans, one aimed at each consumer type, and each plan provides a fixed amount of usage for a price. On the other hand, when the carrier sells dumb pipes, it faces the task of designing only one price schedule to price traffic consumption; each plan then provides a certain level of traffic usage for a price.

We will examine two models, each with two types of consumers. In the first model, consumers differ primarily in their incomes and those with higher incomes are willing to pay more for every service. In other words, consumer preferences are uniformly ordered across services. In the second one, this ordering is not uniform, just as one would expect when preferences are driven primarily by consumers' idiosyncratic tastes: one segment of consumers is willing to pay more for certain services and the other segment is willing to do the same for the rest.

For the two models mentioned earlier, uniform ordering and nonuniform ordering, we will compare all three surpluses—producer, consumer, and social—under smart pipes with that under dumb pipes. We will show that the prevalent view regarding the welfare impacts of smart and dumb pipes hold in certain circumstances, but it does not in many others. In particular, we will identify circumstances in which smart pipes surprisingly lead to a higher consumer surplus. A related finding is that smart pipes can lead to a higher social surplus. Finally, we will discuss why the wisdom that smart pipes increase profits may not hold when consumer types are ordered nonuniformly.

9.2 UNIFORM ORDERING

Consider a carrier serving a market with two types of consumers, indistinguishable ex ante to the carrier. Let the consumer types be indexed by i, $i = 1$ or 2. Types 1 and 2 comprise fractions $1 - f$ and f, respectively, of the whole population. The firm offers n different wireless services, which are indexed by $j, j \in \{1, 2, ..., n\}$.

The value of a service to a consumer is taken to be of the form $\alpha \frac{t^{1-\rho}}{1-\rho}$, where α is a constant, t is the amount of the service used (in bytes), and $0 < \rho < 1$. Value functions of this form have the support of empiricists [11]. Besides, they have desirable properties—they are concave and also exhibit *constant relative risk aversion* [12]. In our context, the parameter ρ reflects the desire for diversity in use of services: a consumer with a higher ρ prefers allocating traffic more evenly between various services than does a consumer with a lower ρ. Following the paper on bundling by Armstrong [13], we will assume hereafter that the total value of a set of services to a consumer is the sum of their individual values:

Assumption 9.1 *The reservation price of a type-i consumer using t_j bytes on service j is given by $\sum_{j=1}^{n} v_{ij}(t_j)$, where $v_{1j}(t) = \alpha_j \frac{t^{1-\rho}}{1-\rho}$ and $v_{2j}(t) = \alpha_j \theta_j \frac{t^{1-\rho}}{1-\rho}$, with $\alpha_j > 0$ and $\theta_j > 0$, $\forall j$.*

When the consumer types are *uniformly ordered*, we can assume without loss of generality that $\theta_j \geq 1$, $\forall j$. Thus, one consumer type, namely, type 2, gets an equal or higher value from *every* service vis-á-vis the other, namely, type 1. Such a scenario is expected when type 2 has a substantially higher income vis-á-vis type 1 and, therefore, is willing to pay more for every telecommunication service. In the next section, we will examine the other scenario, in which consumer type 2 gets higher values from some services, but consumer type 1 gets so from the rest.

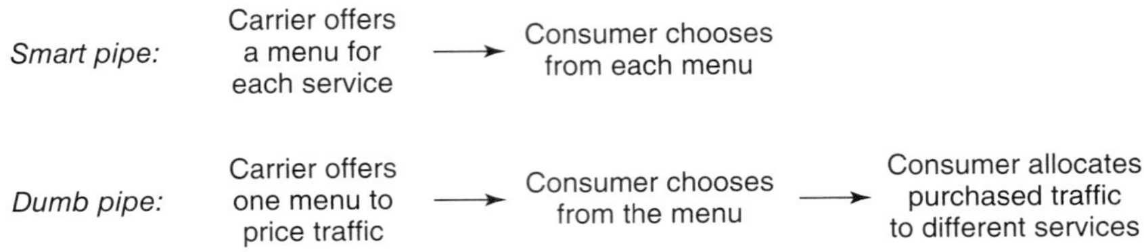

Figure 9.1 Timeline.

For the sake of clarity, we will assume a constant marginal cost throughout our analysis. We will also normalize all fixed costs to zero as they are sunk for our purposes.

Assumption 9.2 *The carrier's marginal cost is c dollars per byte of traffic.*

The firm (i.e., the carrier) offers nonlinear price schedules consisting of plans of the form (p, t), where p is the price charged for t bytes of usage. If the carrier employs the smart-pipe strategy, it announces a separate price schedule for each service. When using the dumb-pipe approach, it announces just one price schedule to price the total traffic consumption (see Fig. 9.1).

9.2.1 Dumb Pipe

In the case of dumb pipes, the carrier offers a single schedule to price traffic. The price schedule or menu consists of two plans here, (p_1, t_1) and (p_2, t_2), aimed at types 1 and 2, respectively. A consumer who purchases a plan has to decide on how much to use for each service. For a type-i consumer, this allocation problem—assuming that the consumer is allocating t bytes of traffic—is as follows:

$$\max_{t_{ij}\geq 0,\ \sum_{j=1}^{n} t_{ij}\leq t} \sum_{j=1}^{n} v_{ij}(t_{ij}).$$

Let the solution to the above problem be $(t^*_{ij}(t), j \in \{1, 2, ..., n\})$, that is, $t^*_{ij}(t)$ is the traffic that a type-i consumer will allocate to service j when dividing t units of traffic optimally between the n services. This solution can be found from

$$\begin{aligned} v'_{ij}(t^*_{ij}(t)) &= v'_{ik}(t^*_{ik}(t)),\ \forall j, k \in \{1, 2, ..., n\}, \text{ and} \\ \sum_{j=1}^{n} t^*_{ij}(t) &= t. \end{aligned} \tag{9.1}$$

Equation (9.1) is readily solved for the value functions described in Assumption 9.1. For type-1 consumers, they are equivalent to the following equations.

$$\begin{aligned}&\left(\frac{\alpha_j}{\alpha_k}\right)^{\frac{1}{\rho}} = \frac{t^*_{1j}(t)}{t^*_{1k}(t)}, \forall j, k \in \{1, 2, ..., n\}, \text{ and}\\&\sum_{j=1}^{n} t^*_{1j}(t) = t.\end{aligned} \tag{9.2}$$

The solution for type-1 consumers is, therefore,

$$t^*_{1j}(t) = \left(\frac{\alpha_j}{\alpha}\right)^{\frac{1}{\rho}} t, \text{ where } \alpha = \left(\sum_{j=1}^{n} \alpha_j^{\frac{1}{\rho}}\right)^{\rho}. \tag{9.3}$$

Figure 9.2, which assumes a two-service scenario, shows the fractions of total traffic that a type-1 consumer allocates to the two services.

When ρ is small, each consumer allocates a greater percentage to the service that he/she values more, which is the service with the higher α parameter—service 1 in the example. When ρ is large, his/her allocation becomes more even. Clearly, ρ reflects his/her affinity for diversity in use of services.

For type-2 consumers, Eq. (9.1) can be rewritten as

$$\begin{aligned}&\left(\frac{\alpha_j\theta_j}{\alpha_k\theta_k}\right)^{\frac{1}{\rho}} = \frac{t^*_{2j}(t)}{t^*_{2k}(t)}, \forall j, k \in \{1, 2, ..., n\}, \text{ and}\\&\sum_{j=1}^{n} t^*_{2j}(t) = t.\end{aligned} \tag{9.4}$$

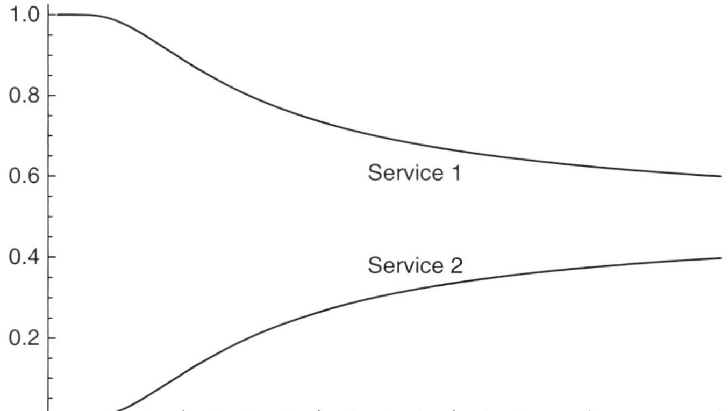

Figure 9.2 Type-1 consumer's traffic allocation for $\alpha_1 = 3$ and $\alpha_2 = 2$.

The solution for type-2 consumers is, therefore, similar.

$$t^*_{2j}(t) = \left(\frac{\alpha_j\theta_j}{\alpha\theta}\right)^{\frac{1}{\rho}} t, \text{ where } \theta = \left(\frac{\sum_{j=1}^{n} \alpha_j^{\frac{1}{\rho}} \theta_j^{\frac{1}{\rho}}}{\sum_{j=1}^{n} \alpha_j^{\frac{1}{\rho}}}\right)^{\rho}. \tag{9.5}$$

We can derive the maximum value or the reservation price for traffic of each consumer type by substituting the optimal $t^*_{ij}(t)$ values back into the respective value functions. Let $v_i(t)$ denote the maximum value that a consumer of type i can derive from optimally allocating t units of traffic to different services. Then

$$\begin{aligned} v_1(t) &= \sum_{j=1}^{n} v_{1j}(t^*_{1j}(t)) = \alpha\frac{t^{1-\rho}}{1-\rho}, \\ v_2(t) &= \sum_{j=1}^{n} v_{2j}(t^*_{2j}(t)) = \alpha\theta\frac{t^{1-\rho}}{1-\rho}. \end{aligned} \tag{9.6}$$

We now turn to the problem of designing a pricing menu consisting of two plans, namely, (p_1, t_1) and (p_2, t_2). Although this appears difficult at first sight, comparing Eq. (9.6) with Assumption 9.1, we can easily see that the problem of designing a menu for traffic is, in fact, structurally no different from the problem of pricing any one service. Further, $\theta_j \geq 1$, $\forall j$, implies that $\theta \geq 1$, which, in turn, means that type 2 is also the high type here just as it is in the case of any service.

Individual Rationality. A consumer will only purchase a plan that offers him/her a nonnegative utility. The individual rationality constraints for types 1 and 2 are

$$v_1(t_1) \geq p_1, \tag{IR1}$$

$$v_2(t_2) \geq p_2. \tag{IR2}$$

Incentive Compatibility. Each consumer picks the plan that generates the highest surplus for him/her. These constraints are

$$v_1(t_1) - p_1 \geq v_1(t_2) - p_2, \tag{IC1}$$

$$v_2(t_2) - p_2 \geq v_2(t_1) - p_1. \tag{IC2}$$

The carrier's problem is to maximize the profit subject to the individual rationality and incentive compatibility constraints above. Its profit, with the plans selected as posited, is

$$(1-f)\left(p_1 - ct_1\right) + f\left(p_2 - ct_2\right).$$

There are three possible outcomes [14]: no one is served, only high type consumers are served, or all are served. When only the high type (type 2) is served, the firm extracts the entire surplus—the price equals the reservation price of the high type. When both types are served, the carrier's profit is maximized when the low type is indifferent between purchasing and not purchasing, and the high type, between purchasing the plan aimed at the low type and its own plan. Hence, only the IR1 and IC2 constraints are binding. Using the IR1 and IC2 constraints to substitute for p_i in terms of v_i, $i = 1, 2$, we can immediately reduce the firm's problem to

$$\max_{t_1,t_2 \geq 0} \; (1-f)\left(v_1(t_1) - c\,t_1\right) + f\left(v_2(t_2) - v_2(t_1) + v_1(t_1) - c\,t_2\right).$$

Henceforth, we will refer to the optimal schedule—the solution to the second-degree discrimination problem—by $\{(p_1^{\mathrm{SB}}, t_1^{\mathrm{SB}}), (p_2^{\mathrm{SB}}, t_2^{\mathrm{SB}})\}$. The optimal usages, along with the corresponding optimal profit and the consumer surplus for each consumer type, are shown in Table 9.1.

The optimal prices can easily be obtained from the optimal usages and the fact that IR1 and IC2 become equalities.

As a benchmark, the table also provides the first-best (FB) solution, the one the carrier would use when it can identify each consumer's type and accordingly charge each a different price. Let us first compare usages for each consumer type. The usage of traffic by the high type (type 2) is unaffected by the degree of discrimination—it is always at a level where its marginal willingness to pay equals the marginal cost. The same is not the case for the low type (type 1). If low type consumers make up a small fraction of the population, that is, $f \geq 1/\theta$, the carrier will not serve them at all. Even when the carrier serves them, it only serves them a quantity that is below the FB level.

Lemma 9.1 *The solution to the second-degree discrimination problem (the optimal usage level for each consumer type) is continuous in f. The optimal profit and consumer surplus are also continuous in f.*

Proof: The optimal usage level for each consumer type, t_i^{SB} in Table 9.1, is continuous at $f = 1/\theta$ and, therefore, it is continuous everywhere in [0, 1]. Because $v_i(t)$ is continuous in t and t_i^{SB} is continuous in f, the result follows immediately. ∎

9.2.2 Smart Pipe

When using the smart-pipe strategy, the carrier offers a separate schedule for each of the n services. The schedule for a service j consists of two plans, $\{(p_{1j}, t_{1j}), (p_{2j}, t_{2j})\}$, one for each of the two consumer types. Note that this problem is separable across services—the firm can independently find the optimal nonlinear price schedule for each service. When $\theta_j \geq 1$, the solution for service j will be the same as the one in Table 9.1, except that α and θ will need to be replaced by α_j and θ_j, respectively. Henceforth, we are going to denote this optimal schedule for service j by $\{(p_{1j}^{\mathrm{SB}}, t_{1j}^{\mathrm{SB}}), (p_{2j}^{\mathrm{SB}}, t_{2j}^{\mathrm{SB}})\}$.

TABLE 9.1 Optimal Menu for Pricing Traffic; $\theta \geq 1$

Case	Usage	Profit	Consumer Surplus
First-Degree Discrimination			
$\forall f$,	$t_1^{\mathrm{FB}} = \left(\frac{\alpha}{c}\right)^{\frac{1}{\rho}}$	$\pi_1^{\mathrm{FB}} = \frac{(1-f)\rho c^{-\frac{1-\rho}{\rho}} \alpha^{\frac{1}{\rho}}}{1-\rho}$	0
all served	$t_2^{\mathrm{FB}} = \left(\frac{\alpha\theta}{c}\right)^{\frac{1}{\rho}}$	$\pi_2^{\mathrm{FB}} = \frac{f\rho c^{-\frac{1-\rho}{\rho}} (\alpha\theta)^{\frac{1}{\rho}}}{1-\rho}$	0
Second-Degree Discrimination			
$f < \frac{1}{\theta}$,	$t_1^{\mathrm{SB}} = t_1^{\mathrm{FB}} \left(\frac{1-f\theta}{1-f}\right)^{\frac{1}{\rho}}$	$\pi^{\mathrm{SB}} =$	0
all served	$t_2^{\mathrm{SB}} = t_2^{\mathrm{FB}}$	$\pi_1^{\mathrm{FB}} \left(\frac{1-f\theta}{1-f}\right)^{\frac{1}{\rho}} + \pi_2^{\mathrm{FB}}$	$\frac{f(\theta-1)\alpha^{\frac{1}{\rho}}}{(1-\rho)} \left(\frac{1-f\theta}{c(1-f)}\right)^{\frac{1-\rho}{\rho}}$
$\frac{1}{\theta} \leq f$, only	$t_1^{\mathrm{SB}} = 0$	$\pi^{\mathrm{SB}} = \pi_2^{\mathrm{FB}}$	0
type 2 served	$t_2^{\mathrm{SB}} = t_2^{\mathrm{FB}}$		0

9.2.3 Smart Pipe Versus Dumb Pipe

Without the loss of generality, we can assume that $\theta_n \geq \theta_{n-1} \geq \cdots \geq \theta_1 \geq 1$ so that $1/\theta_n \leq 1/\theta_{n-1} \leq \cdots \leq 1/\theta_1 \leq 1$. This ordering also implies that $1/\theta_n \leq 1/\theta \leq 1/\theta_1$. When not all θ_j parameters are equal,[1] the two pricing strategies or regimes are equivalent only if the proportion of the high type is at an extreme level, for example, when $f \geq 1/\theta_1$. When $f \geq 1/\theta_1$, under either regime only type-2 consumers are served and their usage is at a level where the marginal cost equals the marginal value. However, significant differences between the two pricing regimes exist at all other values of f. For example, when $0 < f < 1/\theta_n$, all consumers use all services regardless of the regime but their usage amounts differ; as a result, profits and consumer surpluses differ. When $1/\theta_n \leq f < 1/\theta$, the smart-pipe strategy restricts certain services to type-2 consumers, while the dumb-pipe strategy makes all services available to all consumers. When $1/\theta \leq f < 1/\theta_1$, type-1 consumers are not served at all in the case of dumb pipes; however, in the case of smart pipes, they are served all services whose θ_j parameters are strictly less than θ.

Theorem 9.1 *When the two consumer types are uniformly ordered, either smart pipes or dumb pipes can result in a higher consumer surplus. In particular,*

(a) *Smart pipes result in a strictly higher consumer surplus at values of f satisfying $1/\theta \leq f < 1/\theta_1$.*
(b) *There exists an f^a such that smart pipes lead to a strictly a higher consumer surplus at values of f satisfying $f^a < f < 1/\theta$. Therefore, smart pipes lead to a higher consumer surplus at all f satisfying $f^a < f < 1/\theta_1$.*
(c) *Smart pipes and dumb pipes are identical in terms of consumer welfare at values of f satisfying $f \geq 1/\theta_1$, as they both lead to a zero consumer surplus.*

Proof: We will prove the three parts in order.

(a) The consumer surplus under dumb pipes is 0 if $1/\theta \leq f < 1$. However, the consumer surplus under smart pipes is strictly positive if $1/\theta \leq f < 1/\theta_1$ (see Table 9.1). Therefore, smart pipes lead to a strictly higher surplus at values of f satisfying $1/\theta \leq f < 1/\theta_1$.
(b) If $f < 1/\theta_n$, the consumer surplus under dumb pipes is $\Gamma(\alpha, \theta, f)$, where

$$\Gamma(\alpha, \theta, f) = \alpha^{\frac{1}{\rho}} \frac{f(\theta - 1)}{(1-\rho)} \left(\frac{1 - f\theta}{c(1-f)} \right)^{\frac{1-\rho}{\rho}}.$$

It immediately follows that

$$\frac{\partial \Gamma(\alpha, \theta, 0)}{\partial f} = \alpha^{\frac{1}{\rho}} \frac{(\theta - 1)}{(1-\rho)} c^{-\left(\frac{1-\rho}{\rho}\right)}. \tag{P1}$$

[1]If $\theta_1 = \theta_2 = \cdots = \theta_n$, each θ_j also equals θ, making the smart-pipe and dumb-pipe strategies equivalent.

The consumer surplus under smart pipes is $\sum_{j=1}^{n}\Gamma(\alpha_j, \theta_j, f)$, and its derivative, too, can be similarly expressed.

Let $\Gamma^d(f)$ denote the difference between the two surpluses, that is, $\Gamma^d(f) = \Gamma(\alpha, \theta, f) - \sum_{j=1}^{n}\Gamma(\alpha_j, \theta_j, f)$. We can now use the Holder's Inequality to show that $\frac{\partial\Gamma^d(0)}{\partial f} > 0$.

$$\sum_{j=1}^{n} \alpha_j^{\frac{1}{\rho}}\theta_j = \sum_{j=1}^{n}(\alpha_j\theta_j)\left(\alpha_j^{\frac{1-\rho}{\rho}}\right)$$

$$< \left(\sum_{j=1}^{n} \alpha_j^{\frac{1}{\rho}}\theta_j^{\frac{1}{\rho}}\right)^{\rho}\left(\sum_{j=1}^{n} \alpha_j^{\frac{1}{\rho}}\right)^{1-\rho}$$

$$= \frac{\left(\sum_{j=1}^{n} \alpha_j^{\frac{1}{\rho}}\theta_j^{\frac{1}{\rho}}\right)^{\rho}}{\left(\sum_{j=1}^{n} \alpha_j^{\frac{1}{\rho}}\right)^{\rho}}\left(\sum_{j=1}^{n} \alpha_j^{\frac{1}{\rho}}\right) = \theta\alpha^{\frac{1}{\rho}}.$$

The inequality above is actually strict, because not all θ_j parameters are identical. It then immediately follows from (P1) that $\frac{\partial\Gamma^d(0)}{\partial f} > 0$. Further, $\Gamma^d(f)$ is 0 when $f = 0$, and by Lemma 9.1, it is also continuous in f. Therefore, there must exist an $f^t \in (0, 1/\theta)$ such that $\Gamma^d(f) > 0$ if $0 < f < f^t$, implying that dumb pipes lead to a higher consumer surplus at small values of f. However, according to part (a), $\Gamma^d(1/\theta) < 0$. Therefore, it follows from the Intermediate Value Theorem that there must be at least one $f \in (f^t, 1/\theta)$ such that $\Gamma^d(f) = 0$; let us denote the largest of these f values by f^a. This definition of f^a implies that smart pipes do lead to a higher consumer surplus at values of f satisfying $f^a < f < 1/\theta$.

(c) The proof follows from the fact that only type 2 is served under the stated condition.

■

One of the fascinating consequences of Theorem 9.1 is that the consumer surplus can be higher under smart pipes. To see this clearly, consider the scenario in which the monopolist sells two services, namely, 1 and 2, with $\theta_2 > \theta_1 > 1$. Depending on the value of f, the fraction of high type consumers, the equilibrium usages obtained from Table 9.1 is going to be in one of the four distinct regions, labeled U1, U2, U3, and U4 in Figure 9.3, with different equilibrium outcomes. In U1, where $0 < f < 1/\theta_2$, both types enjoy both services irrespective of the pricing strategy. However, in U2, where $1/\theta_2 \leq f < 1/\theta$, type 1 can no longer enjoy service 2 under smart pipes, but it still enjoys both services under dumb pipes. Moving on to U3, where $1/\theta \leq f < 1/\theta_1$, type 1 continues to enjoy service 1 if the carrier employs the smart-pipe strategy, although it gets fully shut out if the carrier implements dumb pipes. Finally, type 1 is fully shut out in U4, where $1/\theta_1 \leq f < 1$, regardless of the pricing strategy. Note that type 2, being the high type, enjoys all services in all regions.

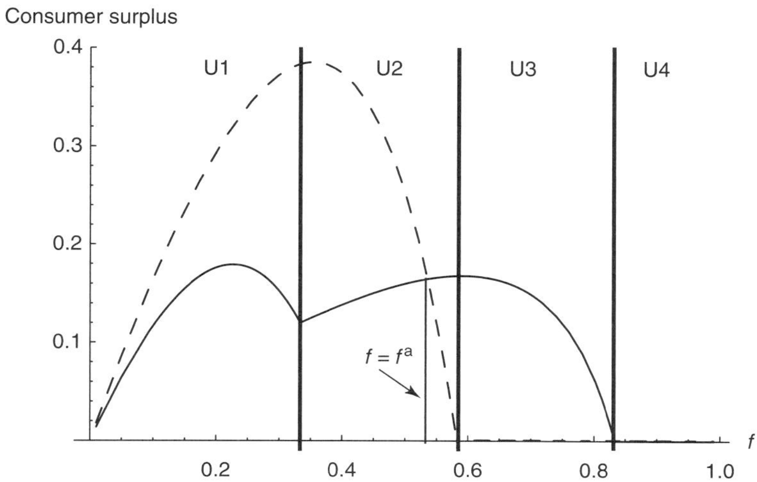

Figure 9.3 Consumer surplus comparison with uniformly ordered consumer types; smart pipe (solid) versus dumb pipe (dashed); $c = 1$, $\rho = 0.5$, $\alpha_1 = 1$, $\alpha_2 = 0.5$, $\theta_1 = 1.2$, and $\theta_2 = 3$.

Figure 9.3 also depicts how the consumer surplus varies with the fraction of type-2 consumers across these regions. In region U3, the dumb-pipe strategy shuts out the low type consumer, driving the consumer surplus to zero, but smart pipes still allow the low type to use one of the two services, leading to a strictly positive consumer surplus. More intriguingly, in parts of region U2, where both types are served both services under dumb pipes, smart pipes lead to a higher consumer surplus despite restricting the low type to only one service.

Let us now examine the intuition behind these apparently counterintuitive results. As expected, the dumb-pipe approach favorably affects the consumer surplus in U1, and in parts of U2, because it lets all consumers freely allocate the purchased traffic between the two services. However, in U3 and in parts of U2, such pricing significantly reduces the carrier's incentive to serve the low type. Because the carrier is not able to restrict the low type to only one service in these regions, serving the low type adequately there requires ceding too large an information rent (i.e., surplus) to the high type. On the other hand, the smart-pipe strategy allows the carrier to restrict service 2 to the high type, creating more flexibilities to curb the information rent. Viewed differently, when restricting one service to only one consumer type is no longer an option, the carrier is forced to severely underserve the low type in order to control the consumer surplus that goes to the high type. The net implication is that the prevalent view held by consumers against smart pipes becomes inapplicable when the fraction of the high type is moderate to large.

Let us now consider the impact of ρ. Figure 9.4 illustrates how f^a changes with ρ. Evidently, f^a is lower at lower values of ρ, that is, consumers find smart pipes preferable over a wider range of values of f.

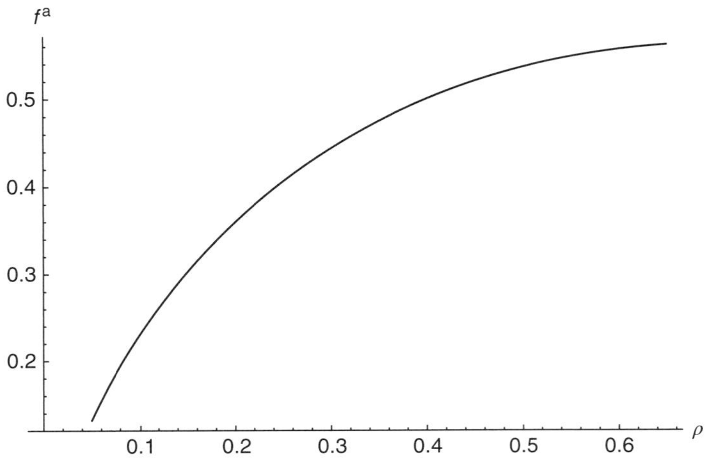

Figure 9.4 f^a versus ρ; $c = 1$, $\alpha_1 = 1$, $\alpha_2 = 0.5$, $\theta_1 = 1.2$, and $\theta_2 = 3$.

This phenomenon can be explained as follows. Recall that ρ indicates consumers' desire for diversity in their use of services. When ρ is close to 1, consumers want to allocate traffic more evenly between services. On the other hand, when ρ is close to 0, consumers benefit more by concentrating their usage on a few services. Hence, the primary benefit of dumb pipes—the freedom to allocate traffic between services—is less valuable to consumers at lower values of ρ.

Lemma 9.2 *When the two consumer types are uniformly ordered, smart pipes result in a strictly higher profit at values of f satisfying $0 < f < 1/\theta_1$. The two profits are equal at all other f.*

Proof: The two profits equal if $f \geq 1/\theta_1$, because only type-2 consumers are served under either regime and the usage under either regime is at a level where the marginal value equals the marginal cost. In other words, the two strategies lead to the same equilibrium.

Moving on to the scenario in which $0 < f < 1/\theta_1$, let t_1^{SB} and t_2^{SB} denote the equilibrium usages for the dumb-pipe strategy. Let the corresponding optimal prices be p_1^{SB} and p_2^{SB}. We have to allow the possibility that $t_1^{SB} = 0$.

We will now show that there exists a feasible solution under the smart-pipe regime, which leads to the same levels of usage. Let us first construct the smart-pipe plans, (p_{1j}, t_{1j}), for the low type consumer with the objective of completely extracting all its surplus: if $t_1^{SB} > 0$, we will set $t_{1j} = t_{1j}^*(t_1^{SB})$ and $p_{1j} = v_{1j}(t_{1j}^*(t_1^{SB}))$. Additionally, if $t_1^{SB} = 0$, we will set $t_{1j} = p_{1j} = 0$.

While designing the smart-pipe plans for the high type, (p_{2j}, t_{2j}), let us set $t_{2j} = t_{2j}^*(t_2^{SB})$ and fix p_{2j} so that the IC2 constraint becomes binding for all j, that is,

$$p_{2j} = p_{1j} + (v_{2j}(t_{2j}) - v_{2j}(t_{1j})).$$

Note that the revenue earned from the low type segment is the same under both strategies, that is, $\sum_{j=1}^{n} p_{1j} = p_1^{\mathrm{SB}}$. We can also show that the revenue earned from the high type segment is at least as large as, that is, $\sum_{j=1}^{n} p_{2j} \geq p_2^{\mathrm{SB}}$,

$$\begin{aligned}
\sum_{j=1}^{n} p_{2j} &= \sum_{j=1}^{n} p_{1j} + \sum_{j=1}^{n} (v_{2j}(t_{2j}) - v_{2j}(t_{1j})) \\
&= p_1^{\mathrm{SB}} + \sum_{j=1}^{n} (v_{2j}(t_{2j}^*(t_2^{\mathrm{SB}})) - v_{2j}(t_{1j}^*(t_1^{\mathrm{SB}}))) \\
&\geq p_1^{\mathrm{SB}} + \sum_{j=1}^{n} (v_{2j}(t_{2j}^*(t_2^{\mathrm{SB}})) - v_{2j}(t_{2j}^*(t_1^{\mathrm{SB}}))) \\
&= p_1^{\mathrm{SB}} + (v_2(t_2^{\mathrm{SB}}) - v_2(t_1^{\mathrm{SB}})) \\
&= p_2^{\mathrm{SB}}.
\end{aligned}$$

The last step in the derivation above involves exploiting the incentive compatibility constraint for type 2. The second last step (the inequality step) follows from the fact that consumers benefit more from allocating according to their own preferences than they do from allocating according to those of the other type, that is,

$$\sum_{j=1}^{n} v_{2j}(t_{2j}^*(t_1^{\mathrm{SB}})) \geq \sum_{j=1}^{n} v_{2j}(t_{1j}^*(t_1^{\mathrm{SB}})).$$

As a feasible solution results in an identical or higher profit for the carrier, from the carrier's perspective, smart pipes must be at least weakly superior to dumb pipes when $0 < f < 1/\theta_1$. This superiority of smart pipes is, in fact, strict, because the feasible solution designed above is clearly a suboptimal one. ■

We now turn to a comparison of the social surpluses. According to Lemma 9.2, when consumers are uniformly ordered, the smart-pipe strategy is superior from the carrier's viewpoint. More interestingly, it suggests that smart pipes accomplish a higher producer surplus even in regions in which it leads to a higher consumer surplus. Therefore, in contrast to what consumer advocates believe, the smart-pipe strategy is not a mechanism that merely transfers surplus from consumers to the carrier. As indicated by the numerical example in Figure 9.5, discrimination can also lead to a higher social surplus.

We now formally state the comparison of the social surpluses under the two strategies as our next theorem.

Theorem 9.2 *When the two consumer types are uniformly ordered, smart pipes lead to both a strictly higher profit and a strictly higher consumer surplus at values of* f *satisfying* $f^a < f < 1/\theta_1$*. Further, there exists an* $f^s < f^a$ *such that smart*

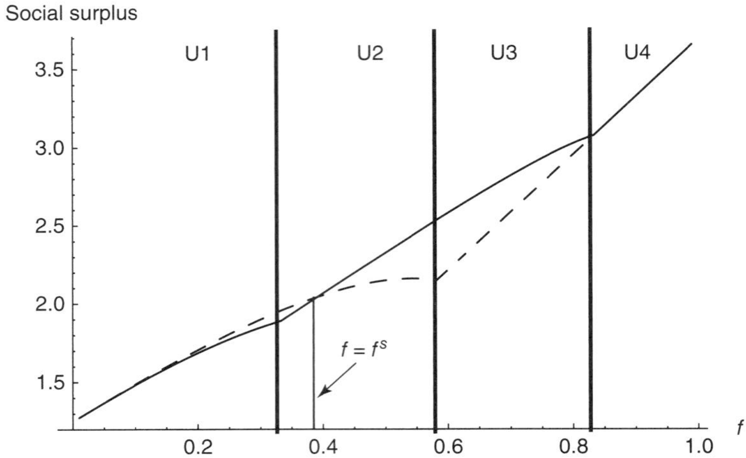

Figure 9.5 Social surplus comparison with uniformly ordered consumer types; smart pipe (solid) versus dumb pipe (dashed); $c = 1$, $\rho = 0.5$, $\alpha_1 = 1$, $\alpha_2 = 0.5$, $\theta_1 = 1.2$, and $\theta_2 = 3$.

pipes lead to a strictly higher social surplus at values of f satisfying $f^s < f \leq f^a$ and, consequently, at all f satisfying $f^s < f < 1/\theta_1$.

Proof: Note that f^a is as defined in Theorem 9.1. The first part follows immediately from Theorem 9.1 and Lemma 9.2. The rest follows from Lemma 9.1: because the difference between the social welfare under the two strategies is positive at $f = f^a$ and the difference is continuous by Lemma 9.1, there must exist an $f^s < f^a$ such that smart pipes lead to a strictly higher social surplus when $f^s < f \leq f^a$. ■

The implication is obvious. If we impose a ban on the smart-pipe strategy, it can quickly become a lose–lose proposition for both the carrier and consumers. Only in U1, which is characterized by a few high income consumers and a large number of low income consumers, the ban would work as expected, that is, it would lead to a higher consumer surplus as well as a higher profit.

The existing literature on second-degree price discrimination also identifies instances in which discrimination is preferable (to uniform pricing across consumer classes) from the social planner's viewpoint [15–17]. Similarly, the literature on third-degree discrimination also finds instances in which the discrimination across different consumer markets leads to a higher social surplus [18, 19]. In essence, we are here examining discrimination across services (calling, messaging, etc.) that use the same resource (traffic). Despite the difference in the nature of discrimination, the intuition behind Theorem 9.2 is still very similar to what one finds in the existing literature: when discrimination is not possible, the producer feels constrained; in response, it reduces its output and forgoes certain consumer segments, which, in turn, adversely affects social welfare. Clearly, our finding here has significant policy implications for governments that are currently debating how to regulate the telecommunication sector optimally from the societal perspective.

9.3 NONUNIFORM ORDERING

We now examine the other possible scenario, which is *nonuniform ordering* of consumer types: one consumer type is the high type for services $\{1, 2, ..., m-1\}$, while the other is the high type for services $\{m, m+1, ..., n\}$. One expects such nonuniform ordering when consumers' preferences depend more on their idiosyncratic tastes for different services and less on their income differences. Let us assume that type 1 is the high type for services $\{1, 2, ..., m-1\}$, that is, $\theta_j < 1$, $\forall j \in \{1, 2, ..., m-1\}$, and that type 2 is the high type for services $\{m, m+1, ..., n\}$, that is, $\theta_j \geq 1$, $\forall j \in \{m, m+1, ..., n\}$. As before, without the loss of generality, we can assume that $\theta_n \geq \theta_{n-1} \geq \cdots \geq \theta_m \geq 1 > \theta_{m-1} \geq \theta_{m-2} \geq \cdots \geq \theta_1$.

The pricing problem for the smart-pipe strategy is still identical to the problem discussed in Section 9.2 except that the roles of the two consumer types have now interchanged for services $j \in \{1, 2, ..., m-1\}$. Therefore, the solution for services $\{m, m+1, ..., n\}$ would still be as mentioned earlier. The solutions for services $\{1, 2, ..., m-1\}$ would, however, be as shown in Table 9.2.

A material implication is that, in the case of smart pipes, type 1 will end up with some surplus whenever type 2 is served any of the first $(m-1)$ services. Recall that, in the previous section, type 1 could never enjoy any surplus.

Note that, in the case of dumb pipes, either type-1 or type-2 consumers can be the high type depending on the relative sizes of the α_j parameters. Without the loss of generality, we will assume that type 2 is the high type in the case of dumb pipes, that is, $\theta \geq 1$. Therefore, the solution in the case of dumb pipes is also the same as that discussed in the previous section.

9.3.1 Smart Pipe Versus Dumb Pipe Revisited

What is perhaps most interesting about nonuniform ordering is that discrimination no longer guarantees a higher profit. Before we can discuss this outcome, the following result is necessary:

Lemma 9.3 *When the two consumer types are not uniformly ordered, smart pipes lead to a strictly higher profit at values of f satisfying $1/\theta \leq f < 1$.*

Proof: Recall that α and θ are as given by Eqs. (9.3) and (9.5), respectively. When $f \geq \frac{1}{\theta}$, under dumb pipes, only the high type (type 2) is served. Therefore, the profit is

$$\frac{f\rho c^{-\frac{1-\rho}{\rho}}(\alpha\theta)^{\frac{1}{\rho}}}{1-\rho} = \frac{f\rho c^{-\frac{1-\rho}{\rho}}\sum_{j=1}^{n}(\alpha_j\theta_j)^{\frac{1}{\rho}}}{1-\rho}.$$

Henceforth, we will refer to the jth term of the sum above as the dumb-pipe profit attributable to service j; thus the dumb-pipe profit attributable to service j is

TABLE 9.2 Optimal Menu for Service j; $\theta_j < 1$

Case	Usage	Profit	Consumer Surplus
First-Degree Discrimination			
$\forall f$,	$t_{1j}^{\mathrm{FB}} = \left(\frac{\alpha_j}{c}\right)^{\frac{1}{\rho}}$	$\pi_{1j}^{\mathrm{FB}} = \frac{(1-f)\rho c^{-\frac{1-\rho}{\rho}} \alpha_j^{\frac{1}{\rho}}}{1-\rho}$	0
all served	$t_{2j}^{\mathrm{FB}} = \left(\frac{\alpha_j \theta_j}{c}\right)^{\frac{1}{\rho}}$	$\pi_{2j}^{\mathrm{FB}} = \frac{f \rho c^{-\frac{1-\rho}{\rho}} (\alpha_j \theta_j)^{\frac{1}{\rho}}}{1-\rho}$	0
Second-Degree Discrimination			
$1-\theta_j < f$,	$t_{1j}^{\mathrm{SB}} = t_{1j}^{\mathrm{FB}}$	$\pi_j^{\mathrm{SB}} = \pi_{1j}^{\mathrm{FB}} +$	$\frac{(1-f)(1-\theta_j)\alpha_j^{\frac{1}{\rho}}}{1-\rho} \left(\frac{f-1+\theta_j}{cf}\right)^{\frac{1-\rho}{\rho}}$
all served	$t_{2j}^{\mathrm{SB}} = t_{2j}^{\mathrm{FB}} \left(\frac{1-\left(\frac{1-f}{\theta_j}\right)}{f}\right)^{\frac{1}{\rho}}$	$\pi_{2j}^{\mathrm{FB}} \left(\frac{1-\left(\frac{1-f}{\theta_j}\right)}{f}\right)^{\frac{1}{\rho}}$	0
$f \leq 1-\theta_j$, only	$t_{1j}^{\mathrm{SB}} = t_{1j}^{\mathrm{FB}}$	$\pi_j^{\mathrm{SB}} = \pi_{1j}^{\mathrm{FB}}$	0
type 1 served	$t_{2j}^{\mathrm{SB}} = 0$		0

$$\frac{f\rho c^{-\frac{1-\rho}{\rho}}(\alpha_j\theta_j)^{\frac{1}{\rho}}}{1-\rho}.$$

Under smart pipes, the total profit from services $\{m, m+1, ..., n\}$ is at least what the carrier gets by serving the high type alone. As type 2 is the high type for $\{m, m+1, ..., n\}$, the profit from these services is at least

$$\frac{f\rho c^{-\frac{1-\rho}{\rho}}\sum_{j=m}^{n}(\alpha_j\theta_j)^{\frac{1}{\rho}}}{1-\rho}.$$

This last expression is identical to the sum of the profits attributable to services $\{m, m+1, ..., n\}$ under dumb pipes.

Now consider services $\{1, 2, ..., m-1\}$, that is, the ones for which type 1 is the high type. We are going to partition this set of services into two subsets and show that, for each set, smart pipes lead to a strictly higher profit. First, consider the subset of these services that are served to type 1 only. Let us call this subset J_1; note that $f \leq 1-\theta_j$, $\forall j \in J_1$. The total profit under smart pipes from these services is

$$\frac{(1-f)\rho c^{-\frac{1-\rho}{\rho}}\sum_{j\in J_1}\alpha_j^{\frac{1}{\rho}}}{1-\rho}.$$

For each $j \in J_1, f\theta_j^{\frac{1}{\rho}} < \theta_j^{\frac{1}{\rho}} < \theta_j \leq (1-f)$, as $0 < \theta_j, \rho < 1$, and $f \leq 1-\theta_j$. Rearranging the terms, we get $(1-f) > f\theta_j^{\frac{1}{\rho}}$, implying that the smart-pipe profit from service j is indeed higher than that attributable to it under dumb pipes.

Now, consider the subset J_2 of $\{1, 2, ..., m-1\}$, which consists of services offered to both types. Note that $f > 1-\theta_j$, $\forall j \in J_2$. The total profit under smart pipes attributable to J_2 is

$$\sum_{j\in J_2}\frac{\rho c^{-\frac{1-\rho}{\rho}}\alpha_j^{\frac{1}{\rho}}}{1-\rho}\left((1-f)+f\left(\frac{f-1+\theta_j}{f}\right)^{\frac{1}{\rho}}\right).$$

When $f \geq 1/\theta$, the difference between this profit and that under dumb pipes from J_2 is

$$\sum_{j\in J_2}\frac{\rho c^{-\frac{1-\rho}{\rho}}\alpha_j^{\frac{1}{\rho}}}{1-\rho}\left((1-f)+f\left(\left(\frac{f-1+\theta_j}{f}\right)^{\frac{1}{\rho}}-\theta_j^{\frac{1}{\rho}}\right)\right).$$

Each term in the sum above is 0 when $f = 1$. Further, when $1-\theta_j \leq f < 1$, the derivative of the jth term with respect to f is

$$\frac{\rho c^{-\frac{1-\rho}{\rho}} \alpha_j^{\frac{1}{\rho}}}{1-\rho} \left(-(1+\theta_j^{\frac{1}{\rho}}) + \frac{\left(\frac{f-1+\theta_j}{f}\right)^{\frac{1-\rho}{\rho}} (1-\theta_j+(f+\theta_j-1)\rho)}{f\rho} \right).$$

The second derivative with respect to f is

$$\frac{\rho c^{-\frac{1-\rho}{\rho}} \alpha_j^{\frac{1}{\rho}}}{1-\rho} \left(\frac{(1-\theta_j)^2 \left(\frac{f-1+\theta_j}{f}\right)^{\frac{1+\rho}{\rho}} (1-\rho)}{(f+\theta_j-1)^3\rho^2} \right).$$

Because this second derivative is positive when $f > 1-\theta_j$, the first derivative is maximized at $f = 1$, and the maximum value is

$$\frac{\rho c^{-\frac{1-\rho}{\rho}} \alpha_j^{\frac{1}{\rho}}}{1-\rho} \left(-1 + \theta_j^{\frac{1-\rho}{\rho}} \left(\frac{1-\theta_j}{\rho} \right) \right).$$

By differentiating the expression above with respect to θ_j, we can show that its maximum value is, in fact, the following negative number:

$$\frac{\rho c^{-\frac{1-\rho}{\rho}} \alpha_j^{\frac{1}{\rho}}}{1-\rho} \left(-1 + (1-\rho)^{\frac{1-\rho}{\rho}} \right).$$

The first derivative of each term, therefore, is negative. Furthermore, as noted earlier, each term in the sum equals 0 at $f = 1$. Hence, the sum must be positive if $f < 1$, that is, the smart-pipe profit must be higher.

The three comparisons presented above, for the sets $\{m, m+1, ..., n\}$, J_1, and J_2, respectively, together imply that smart pipes result in a strictly higher profit at values of f satisfying $f \geq 1/\theta$. ■

We are now ready for our next result, which shows that dumb pipes are superior from the carrier's viewpoint when θ is below a threshold and f is not too large. Is it not interesting that a pricing strategy, which apparently constrains the carrier and offers consumers greater freedom, can actually lead to a higher profit?

Theorem 9.3 *When the two consumer types are not uniformly ordered, either smart pipes or dumb pipes may lead to a higher profit. In particular, if* $\theta < \theta^k$, *where*

$$\theta^k = \frac{\sum_{j=1}^{m-1} \alpha_j^{\frac{1}{\rho}} + \sum_{j=m}^{n} \alpha_j^{\frac{1}{\rho}} \theta_j + \rho \sum_{j=1}^{m-1} (\alpha_j \theta_j)^{\frac{1}{\rho}}}{\alpha^{\frac{1}{\rho}}},$$

there exists an $f^k \in (0, 1/\theta)$ *such that dumb pipes result in a strictly higher profit at values of* f *satisfying* $0 < f < f^k$.

Proof: Before proceeding with the proof, let us note that $\theta^k > 1$ by definition. In other words, we aim to show that, when θ is between this threshold and 1, dumb pipes lead to a higher profit unless f is too large.

When $f < \min(1/\theta, 1 - \theta_{m-1}, 1/\theta_n)$, under dumb pipes, all services are served to both consumer types. So, the dumb-pipe profit equals

$$\frac{\rho c^{-\frac{1-\rho}{\rho}}}{1-\rho}\left(f(\alpha\theta)^{\frac{1}{\rho}} + (1-f)\left(\frac{\alpha(1-f\theta)}{1-f}\right)^{\frac{1}{\rho}}\right).$$

And, its derivative at $f = 0$ is

$$\frac{\rho c^{-\frac{1-\rho}{\rho}}}{1-\rho}\alpha^{\frac{1}{\rho}}\left(\frac{1-\theta-\rho}{\rho} + \theta^{\frac{1}{\rho}}\right) = \frac{\rho c^{-\frac{1-\rho}{\rho}}}{1-\rho}\left(\alpha^{\frac{1}{\rho}}\left(\frac{1-\theta}{\rho} - 1\right) + \sum_{j=1}^{n}\alpha_j^{\frac{1}{\rho}}\theta_j^{\frac{1}{\rho}}\right).$$

Under smart pipes, when $f < \min(1/\theta, 1 - \theta_{m-1}, 1/\theta_n)$, services $\{1, 2, ..., m-1\}$ are offered only to type 1. Hence, the profit is

$$\sum_{j=m}^{n}\left(\frac{\rho c^{-\frac{1-\rho}{\rho}}}{1-\rho}\left(f(\alpha_j\theta_j)^{\frac{1}{\rho}} + (1-f)\left(\frac{\alpha_j(1-f\theta_j)}{1-f}\right)^{\frac{1}{\rho}}\right)\right) + \sum_{j=1}^{m-1}\left(\frac{(1-f)\rho c^{-\frac{1-\rho}{\rho}}\alpha_j^{\frac{1}{\rho}}}{1-\rho}\right),$$

and its derivative at $f = 0$ is

$$\begin{aligned}
&\frac{\rho c^{-\frac{1-\rho}{\rho}}}{1-\rho}\left(\sum_{j=m}^{n}\alpha_j^{\frac{1}{\rho}}\left(\frac{1-\theta_j-\rho}{\rho} + \theta_j^{\frac{1}{\rho}}\right) - \sum_{j=1}^{m-1}\alpha_j^{\frac{1}{\rho}}\right)\\
&= \frac{\rho c^{-\frac{1-\rho}{\rho}}}{1-\rho}\left(\sum_{j=m}^{n}\alpha_j^{\frac{1}{\rho}}\left(\frac{1-\theta_j}{\rho} + \theta_j^{\frac{1}{\rho}}\right) - \sum_{j=1}^{n}\alpha_j^{\frac{1}{\rho}}\right)\\
&= \frac{\rho c^{-\frac{1-\rho}{\rho}}}{1-\rho}\left(\sum_{j=m}^{n}\alpha_j^{\frac{1}{\rho}}\left(\frac{1-\theta_j}{\rho} + \theta_j^{\frac{1}{\rho}}\right) - \alpha^{\frac{1}{\rho}}\right).
\end{aligned}$$

Therefore, the derivative of the difference between the two profits—the dumb-pipe profit minus the smart-pipe profit—is positive at $f = 0$ when

$$\theta < \frac{\sum_{j=m}^{n}\alpha_j^{\frac{1}{\rho}}\theta_j + \alpha^{\frac{1}{\rho}} - \sum_{j=m}^{n}\alpha_j^{\frac{1}{\rho}} + \rho\sum_{j=1}^{m-1}(\alpha_j\theta_j)^{\frac{1}{\rho}}}{\alpha^{\frac{1}{\rho}}}$$

$$= \frac{\sum_{j=m}^{n} \alpha_j^{\frac{1}{\rho}} \theta_j + \sum_{j=1}^{m-1} \alpha_j^{\frac{1}{\rho}} + \rho \sum_{j=1}^{m-1} (\alpha_j \theta_j)^{\frac{1}{\rho}}}{\alpha^{\frac{1}{\rho}}} = \theta^k.$$

Now, note that the difference between the two profits is 0 at $f = 0$, and by Lemma 9.1, the difference is continuous. Hence, when $\theta < \theta^k$, Lemma 9.3 implies the existence of a threshold $f^k \in (0, 1/\theta)$ such that the difference is positive at values of f satisfying $0 < f < f^k$. ■

Theorem 9.3 differs from the results found in the existing literature, which show that discrimination typically results in a higher profit when compared to the case of uniform or nondiscriminatory pricing [15–19]. The reason for this difference is that the dumb-pipe strategy essentially amounts to quasi-bundling—it is, in fact, a means of bundling all services into one bundle, even though the seller does not play any role in determining the bundle composition.

Let us now carefully examine the case of two services, namely, 1 and 2. Let $\theta_2 > 1 > \theta_1$ and $\theta > 1$. It follows that $1/\theta_2 < 1/\theta < 1$. This time, however, Tables 9.1 and 9.2 together imply that we need to examine three scenarios, and not just one, to get the complete picture. They are as follows:

$$\text{Case X: } 1 - \theta_1 < \frac{1}{\theta_2} < \frac{1}{\theta};$$

$$\text{Case Y: } \frac{1}{\theta_2} \leq 1 - \theta_1 < \frac{1}{\theta};$$

$$\text{Case Z: } \frac{1}{\theta_2} < \frac{1}{\theta} \leq 1 - \theta_1.$$

For each scenario, we again end up with four distinct regions, with different equilibrium outcomes. Consider the profit plot shown in Figure 9.6, which provides us with an example of Case X. In this case, if the carrier uses the smart-pipe approach, type 1 always enjoys service 1, for which it is the high type. Similarly, type 2 always enjoys service 2. However, type 1 can enjoy service 2 only in X1 and X2, where $f < 1/\theta_2$, and type 2 can enjoy service 1 only in X2, X3, and X4, where $f > 1 - \theta_1$. Under dumb pipes, type 2 is always served, because $\theta > 1$ for this example. Type 1 is served only if $f < 1/\theta$, that is, in regions X1, X2, and X3.

Further, for the example in Figure 9.6, $\theta^k = 1.4341$ and $\theta = 1.2411$. According to Theorem 9.3, quasi-bundling should be preferable when θ is small, specifically, when it is below θ^k. As can be seen from Figure 9.6, dumb pipes indeed lead to a higher profit in both X1 and X2. It further shows that dumb pipes perform better even in parts of X3. For the sake of brevity, we will skip similar discussions for Cases Y and Z.

Note that a small θ, that is, a θ close to 1, means that the reservation prices for traffic of the two consumer types are very similar. In other words, a small θ implies a strong inverse relationship between the preferences of the two consumer types. Prior

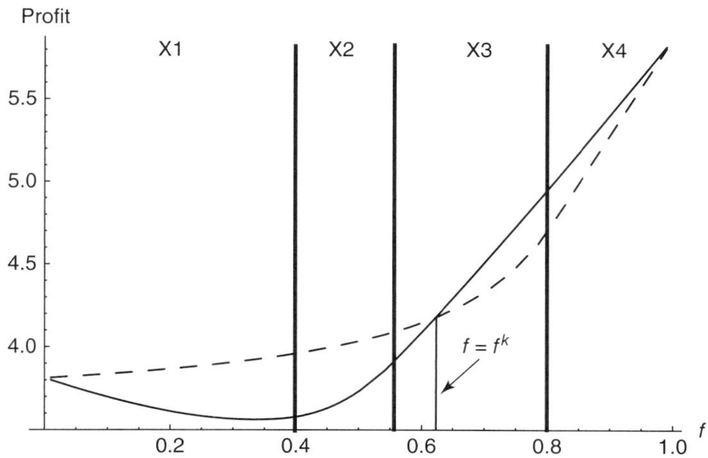

Figure 9.6 Profit comparison with nonuniformly ordered consumer types; smart pipe (solid) versus dumb pipe (dashed); $c = 1$, $\rho = 0.5$, $\alpha_1 = 1.5$, $\alpha_2 = 1.25$, $\theta_1 = 0.6$ and $\theta_2 = 1.8$.

research has shown that such strong inverse relationships increase the effectiveness of bundling [9]. Theorem 9.3 extends the same insight to the case of quasi-bundling. Also, Lemma 9.2 and Theorem 9.3 collectively establish that a necessary condition for quasi-bundling to be effective from a producer's viewpoint is that the primary source of consumer heterogeneity is taste and not income, and a necessary condition for it to be highly effective is that the reservation price for traffic be similar across different segments.

We now turn our attention to consumer welfare. The following theorem describes certain sufficient conditions under which the smart-pipe strategy unexpectedly leads to a higher consumer surplus.

Theorem 9.4 *When the two consumer types are not uniformly ordered, either smart pipes or dumb pipes can lead to a higher consumer surplus. In particular, the following holds:*

(a) *Smart pipes lead to at least as much in consumer surplus as do dumb pipes at values of f satisfying $f \geq 1/\theta$ and a strictly higher consumer surplus at values of f satisfying $f > \max(1/\theta, 1 - \theta_{m-1})$.*

(b) *If $1/\theta > 1 - \theta_{m-1}$, there exists an f^b in $(0, 1/\theta)$ such that smart pipes result in a strictly higher consumer surplus at values of f satisfying $f^b < f < 1/\theta$; consequently, under smart pipes, the consumer surplus is higher at all f satisfying $f^b < f < 1$.*

(c) *If $1/\theta < 1/\theta_m$, there exists an f^c in $(0, 1/\theta)$ such that smart pipes result in a strictly higher consumer surplus at values of f satisfying $f^c < f < 1/\theta_m$.*

Proof:

We will prove the three parts in order.

(a) The first part follows directly from the fact that the consumer surplus under dumb pipes is 0 if $f \geq 1/\theta$. Further, if $f > \max(1/\theta, 1 - \theta_{m-1})$, under smart pipes, the surplus is positive—type 1 must get a positive surplus, because it is profitable for the carrier to offer service $(m - 1)$ to type 2 there.

(b) Let $\Gamma^d(f)$ denote the difference between the two consumer surpluses (i.e., the consumer surplus under dumb pipes minus that under smart pipes) at a given f. It follows from part (a) that $\Gamma^d(f)$ is negative for $f \geq 1/\theta$. Also, by Lemma 9.1, $\Gamma^d(f)$ is continuous in f. Therefore, there must exist an f^b in $(0, 1/\theta)$ such that $\Gamma^d(f)$ is negative at values of f satisfying $f^b < f < 1/\theta$.

(c) If $1/\theta < 1/\theta_m$, $\Gamma^d(f)$ is again negative for any f satisfying $1/\theta \leq f < 1/\theta_m$. This is because, under smart pipes, the consumer surplus is positive—type 2 must get a positive surplus, because it is profitable for the carrier to offer service m to type 1 there. Moreover, $\Gamma^d(f)$ is continuous in f. Therefore, there ought to exist an f^c in $(0, 1/\theta)$ such that $\Gamma^d(f)$ is negative for values of f satisfying $f^c < f < 1/\theta_m$. ∎

The conditions described in parts (b) and (c) of Theorem 9.4 are very mild, in the sense that our assumptions regarding the θ_j parameters imply that $\min_{j \leq m-1}(1 - \theta_j) = (1 - \theta_{m-1})$ and that $\max_{j \geq m}(1/\theta_j) = 1/\theta_m$. Figure 9.7, which uses the same parameter values as Figure 9.6, illustrates Theorem 9.4. In Figure 9.7, as predicted by the theorem, smart pipes indeed lead to a higher consumer surplus in X4 and a part of X3.

Moving on to social welfare, Figure 9.8 indicates that smart pipes lead to a higher social surplus in X4 and a part of X3.

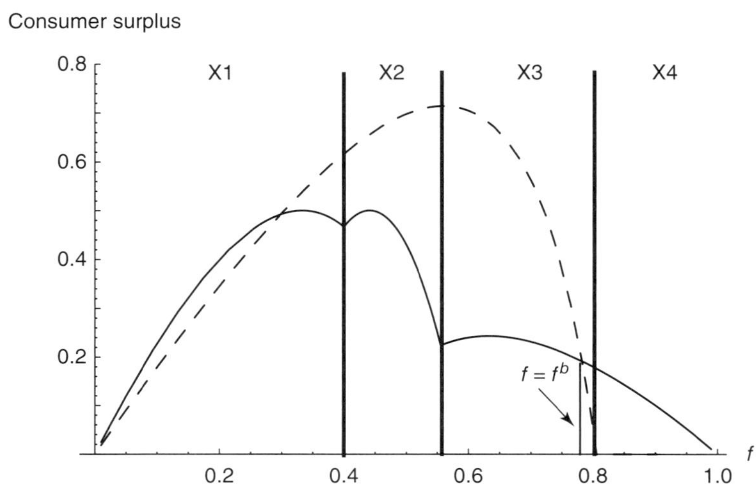

Figure 9.7 Consumer surplus comparison with nonuniformly ordered consumer types; smart pipe (solid) versus dumb pipe (dashed); $c = 1$, $\rho = 0.5$, $\alpha_1 = 1.5$, $\alpha_2 = 1.25$, $\theta_1 = 0.6$ and $\theta_2 = 1.8$.

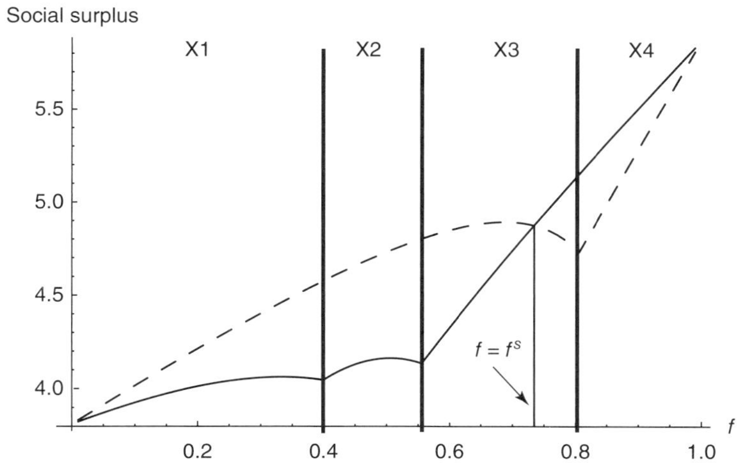

Figure 9.8 Social surplus comparison with nonuniformly ordered consumer types; smart pipe (solid) versus dumb pipe (dashed); $c = 1$, $\rho = 0.5$, $\alpha_1 = 1.5$, $\alpha_2 = 1.25$, $\theta_1 = 0.6$ and $\theta_2 = 1.8$.

This figure also shows that dumb pipes can result in a higher social surplus at relatively moderate values of f, for example, in X2. Evidently, the common wisdom regarding which strategy is better from the point of view of the social planner is not applicable in this context as well—depending on the sizes of different market segments, either strategy might result in a higher social surplus.

Theorem 9.5 *When the two consumer types are not uniformly ordered, smart pipes lead to a strictly higher profit while leading to at least as much in consumer surplus at values of f satisfying $f \geq 1/\theta$. Further, if $1/\theta > 1 - \theta_{m-1}$, the smart-pipe strategy leads to strictly higher surpluses for both the carrier and consumers at values of f satisfying $f \geq 1/\theta$; on the other hand, if $1/\theta < 1/\theta_m$, it leads to strictly higher surpluses at values of f satisfying $1/\theta \leq f < 1/\theta_m$. Finally, there exists an $f^s < 1/\theta$ such that smart pipes lead to a strictly higher social surplus at values of f satisfying $f^s < f < 1/\theta$ and, consequently, at all f satisfying $f^s < f < 1$.*

Proof: All parts except the very last one follow immediately from Lemma 9.3 and Theorem 9.4. The last part follows from the continuity of these surpluses with respect to f: because the difference between the social surpluses under the two regimes—the one under dumb pipes minus that under smart pipes—is negative at $f = 1/\theta$ and the difference is also continuous by Lemma 9.1, there must exist an $f^s < 1/\theta$ such that smart pipes also lead to a strictly higher social surplus at all f satisfying $f^s < f < 1/\theta$. ∎

Theorems 9.4 and 9.5 extend the counterintuitive findings described in the context of uniform ordering to the case of nonuniform ordering. The intuition is still the same. While using smart pipes, the carrier selectively serves the two consumer types; for example, in X4, it restricts service 2 to type-2 consumers only. However, when discrimination across services is not possible, the carrier has to choose between serving type 1 both services there and serving them none. Serving both adequately requires ceding too large a surplus to type 2. The carrier, therefore, finds it preferable to shut out type-1 consumers in X4 while significantly raising the price on type 2 consumers.

9.4 CONCLUSION

In this chapter, we have examined the economic impacts of service-based discrimination on the market for telecommunication services. We have analyzed a market with discrete consumer types, who may or may not be uniformly ordered with regards to their preferences for various services. We have closely examined—analytically and numerically—how various surpluses behave depending on the pricing strategy used by the carrier. When a carrier employs the smart-pipe strategy, it announces a separate schedule for each service. When it does not discriminate across services, that is, uses the dumb-pipe strategy, it offers just one schedule to price all traffic consumption.

Service-based discrimination is commonly expected to lower consumer and social welfare. It is also expected to lead to a higher producer surplus. We have established—through rigorous economic arguments—that these expectations are often untrue. Specifically, the impact of discrimination is far more complex than often argued, or even imagined. As unambiguously demonstrated in this work, the nature of consumer heterogeneity and relative sizes of different market segments play a pivotal role in determining which strategy would lead to a higher surplus.

Intriguingly, when consumers differ mainly in their incomes and their ordering is thus uniform across services, the expectation that smart pipes harm consumers is true only in situations that involve a few high type consumers and a large number of low type consumers. In fact, despite their natural appeal to consumers, dumb pipes can be detrimental to consumer welfare—in situations, unable to prevent certain consumers from using a service, the carrier may decide to completely forgo them while raising the price on the rest. A similar phenomenon also occurs when consumers differ mainly in their tastes for different services and their ordering is consequently nonuniform. The main lesson, therefore, is that discrimination allows a producer greater flexibility with regards to how it serves different market segments, which may even help it expand coverage, leading to higher surpluses for consumers. This ability of discriminatory pricing to lead to higher consumer welfare gets more pronounced when consumers' affinity for diversity in use, captured by the parameter ρ, becomes lower. This is because, when ρ is low, the benefits derived from dumb pipes, which primarily arise from the freedom to allocate traffic between services, are also low.

Another insight that emerges from this chapter is that the perception that discrimination is always beneficial for the producer holds only in the case of uniform ordering. It does not hold in the case of nonuniform ordering. The reason is that an inverse ordering of consumer preferences can turn dumb pipes into a very effective bundling strategy. Equally noteworthy is the insight that service-based discrimination is not always detrimental to social welfare as often believed. The smart-pipe strategy is, in fact, win–win for the carrier and consumers in many situations. The implication is intriguing. The usual arguments about how smart pipes serve as a mechanism for reallocating surplus from consumers to the carrier have little merit. These usual "zero-sum" arguments fail to recognize that smart pipes can indeed drive the social surplus higher, creating the possibility of a higher surplus for everyone.

In summary, the salient contribution of this chapter is that it explains why banning smart pipes may not benefit consumers as expected. Further, it explains why dumb pipes do not always lead to a lower producer surplus or a higher social surplus. All these findings have critical implications for everyone, including carriers, consumers, and policymakers. Consumers need to understand that service-based discrimination can work in their favor before they extend unqualified support for a ban on discrimination, or they embrace open platforms with the hope of rendering such discrimination ineffective. Likewise, policymakers should consider the negative impacts a ban on discrimination may produce, including the most unexpected impact of a lower surplus for everyone. A carrier should also note that smart pipes are not necessarily more profitable—they can indeed be suboptimal when consumers are nonuniformly ordered.

This chapter also contributes towards the broader debate on net neutrality. In particular, it sheds light on the issue of neutrality with respect to different services that use the same communication network. Contrary to what supporters of neutrality may claim, the results above all point to the fact that when assessing economic impacts of neutrality, relying on the prevalent wisdom can be futile. Analytical models that rest on well-tested economic theories are desired for proper evaluation of likely consequences.

A word on extending this work to competitive settings is in order now. One way to model a firm, which is not a monopoly but has substantial pricing power, could be to use slightly different participation (Individual Rationality) constraints. So far we have assumed that a consumer must get a nonnegative utility from any plan he/she purchases. Equivalently, we have implicitly assumed the so-called "outside" utility to be zero. It is possible to imagine a setting where the outside utility is positive, as will be the case if consumers have the ability to switch to alternatives offered by the competing firms. Each consumer will then seek a plan that provides them with a utility at least as large as that obtained from the best of all such alternatives. However, extending the work this way is not at all trival, because many more mathematical possibilities will emerge. For example, if the outside utility for one service is high and that for another is low, under dumb pipes, a consumer may allocate all of his/her purchased traffic to the service with the lower outside utility. Thus, the pricing problem may not have interior solutions, implying that analyses of corner solutions may also become necessary. Interested readers may refer to Reference 20 for additional useful guidelines.

REFERENCES

1. A. Lahiri, R. M. Dewan, and M. Freimer. "The disruptive effect of open platforms on the market for wireless services," *Journal of Management Information Systems*, 29(3), 2011, 81–109.
2. A. Lahiri, R. M. Dewan, and M. Freimer. "Pricing of wireless services: service pricing vs. traffic pricing," *Information Systems Research*, 24(2), 2013, 418–435.
3. A. Odlyzko. "The evolution of price discrimination in transportation and its implications for the internet," *Review of Network Economics*, 3(3), 2004, 323–346.
4. ABC News. Cell phone execs grilled by Senate over texting charges, June 18, 2009. Available at: http://abcnews.go.com/video/playerindex?id=7856703.
5. S. Keshav. Testimony of Professor Srinivasan Keshav at the hearing on June 16 at the United States Senate Judiciary Committee, 2009. Available at: http://www.judiciary.senate.gov/hearings/testimony.cfm?id=e655f9e2809e5476862f735da14b66b2/&wit_id=e655f9e2809e5476862f735da14b66b2-0-3.
6. W. S. Mossberg. Free my phone. *The Wall Street Journal*, Oct. 22, 2007. Available at: http://online.wsj.com/article/SB119264941158362317.html.
7. R. Jackson. AT&T Android phones from Dell, HTC and Motorola coming soon! Jan. 6, 2010, Available at: http://phandroid.com/2010/01/06/att-android-phones-from-dell-htc-and-motorola-coming -soon/.
8. K. Brown and P. J. Alexander. "Bundling in cable television: a pedagogical note with a policy option," *International Journal on Media Management*, 6(3), 2004, 162–167.
9. W. J. Adam and J. L. Yellen. "Commodity bundling and the burden of monopoly," *Quarterly Journal of Economics*, 90(3), 1976, 475–498.
10. Y. Masuda and S. Whang. "On the optimality of fixed-up-to tariff for telecommunication service," *Information Systems Research*, 17(3), 2006, 247–253.
11. I. Friend and M. E. Blume. "The demand for risky assets," *American Economic Review*, 65(5), 1975, 900–922.
12. P. S. Armington. "A theory of demand for products distinguished by place of production," *IMF Staff Papers*, 16(1), 1969, 159–178.
13. M. Armstrong. "Price discrimination by a many-product firm," *Review of Economic Studies*, 66(1), 1999, 151–168.
14. M. Mussa and S. Rosen. "Monopoly and product quality," *Journal of Economic Theory*, 18(2), 1978, 301–317.
15. M. Katz. "Nonuniform pricing, output and welfare under monopoly," *Review of Economic Studies*, 50(1), 1983, 37–56.
16. K. W. S. Roberts. "Welfare considerations of nonlinear prices," *Economic Journal*, 89(353), 1979, 66–83.
17. A. Spence. "Nonlinear prices and welfare," *Journal of Public Economics*, 8(1), 1977, 1–18.
18. R. Schmalensee. "Output and welfare implications of monopolistic third-degree price discrimination," *American Economic Review*, 71(1), 1981, 242–247.
19. H. Varian. "Price discrimination and social welfare," *American Economic Review*, 75(4), 1985, 870–875.
20. M. Armstrong and J. Vickers. "Competitive price discrimination," *RAND Journal of Economics*, 32(4), 2001, 1–27.

PART IV
Content-Based Pricing

10 Economic Models of Sponsored Content in Wireless Networks with Uncertain Demand

MATTHEW ANDREWS, ULAS OZEN, MARTIN I. REIMAN, and QIONG WANG

10.1 INTRODUCTION

One persistent trend in wireless communications in recent years is that demand for bandwidth is exploding, whereas revenue per subscriber is either flat or growing at a much slower rate. Hence, there is a mismatch between the amount of revenue coming in from the end users (EUs) of the network and the amount of investment that is required for wireless capacity to keep pace with demand.

Variable pricing and bundling are two traditional ways for revenue enhancement. In the first case (e.g. [1]), the service provider charges different fees for the same service provided at different times of day or different locations, effectively using pricing as a control device to fit demand within capacity while discriminating between users with different needs. In the second case, the service provider offers a single package that contains both high revenue-per-bit services and low revenue-per-bit services, and uses the gains from the former to cover the loss of the latter. The potential of either method ultimately rests on the pockets of EUs and, hence, is limited by the latter's budget.

In this chapter, we investigate an approach whereby the service provider can tap into an alternative source of revenue, originating from sales of advertisements or products and channelled by the content provider in the form of sponsorship of viewing. The gain comes from removing inefficiency of the current arrangement under which the content provider derives profit from its content (e.g., by showing advertisements) while EUs pay for the cost of viewing it. To maximize its profit, the provider naturally wants to increase the number of views of its content but is unlikely to get help from EUs who are wary about wasting their precious bandwidth quotas on content such as embedded advertisements. The reluctance is strong not only because the use

Smart Data Pricing, First Edition. Edited by Soumya Sen, Carlee Joe-Wong, Sangtae Ha, and Mung Chiang.

of bandwidth can be heavy (in the case when an advertisement includes rich video) but also because EUs are uncertain and thus cannot control such use. Unlike voice service, which is charged by minutes, data service is sold in units of bytes and it is harder for the EUs to interpret how many bytes will be consumed when they perform a web action. Typically, when an EU clicks on a link, the only indication they would get that the resulting web page is large is if it takes a long time to load.

We consider a solution that allows the content provider to "sponsor" its content so that it does not get charged to the EUs' monthly quotas. The arrangement removes EUs' concern about paying an uncertain amount of bandwidth cost for content that is of little immediate value to them. As a consequence, more content will be accessed, not only because some of it is free but also because users are effectively given more quota. The content provider's profit increases as long as the cost of sponsoring stays below the new revenue from increased viewing of its content. Moreover, the provider's image also improves as fewer users will think of it as an irresponsible party that pushes costly and worthless materials to them. Content sponsoring also benefits the service provider by giving it the opportunity to charge content providers who, as the "over-the-top" companies, have a greater willingness to pay than EUs. Income from this new source enables the service provider to recover the value of some of the mobile services that it is enabling and use that revenue to finance capacity expansion.

10.1.1 Research Questions

The concept of sponsored content introduces the following set of research questions. In this chapter, we shall mostly focus on the first one.

- What is the proper way to design a contract between the service provider and the content provider for sponsored content? The service provider needs to design the contract in such a way that induces the content provider to participate in sponsoring and make decisions that improve the service provider's profit. The main contribution of this chapter is the design of a contract that not only does this but also maximizes the gain to the entire system, while at the same time transferring a significant share of the gain to the service provider itself. Moreover, we will show that the contract is win–win–win for the service provider, the content provider and the EUs in the sense that if sponsoring occurs, then all participants in the system will be better off than if there were no sponsoring.
- What are the main types of content for which the sponsored content concept is most applicable? Our model will shed some light on this problem because it will generate a set of conditions regarding when the content provider will decide to sponsor. At a crude level, the concept makes most sense for content that generates significant economic value to the content provider on each view. Some examples include advertising (especially advertising for luxury items), shopping sites (where the content provider cost of sponsoring could be wrapped into the purchase), large content such as movies (especially for sites where content must be purchased because again the cost of sponsoring could be wrapped into

the cost of the purchase) and applications such as games that include "in-app purchases."

- To what extent should consumers be informed that sponsored content is free to them and if so, how should this information be communicated? We believe that the sponsored content concept has value even if the EU is unaware of which content is sponsored because it would still prevent a big depletion of quota for large content that the EU would prefer to not pay for. However, it is likely that the dynamics of the system would be significantly transformed if the users do know what content is free to them.
- How can sponsored content be implemented at scale? Realizing the concept at a small scale is simple because all we need to do is inspect packets within the network to decide whether the EU will be charged or not. [This could for example be done on the basis of Internet Protocol (IP) address.] In addition, we would need a way to manage the payments from the content provider to the service. Achieving all this on the scale of a large service provider without bringing significant extra signaling into the network is an important challenge.
- What are the "net-neutrality" implications of sponsored content? The topic is relevant because if a content provider sponsors its content, then the service provider will treat that content differently than the content of other content providers. Hence, the service provider would not be treating the content of all content providers in an identical manner which is required under some definitions of network neutrality.

10.1.2 Previous Work

Networks that allow the option of content provider pricing were studied in References 2 and 3. The first paper considers a network utility maximization (NUM) setting where both the content provider and the EU have a utility and we wish to set prices at both end points and route traffic so that the aggregate utility minus the cost of the network is maximized. This setup was considered for both an Internet Service Provider (ISP) in a competitive market where the prices are determined by the market as well as a monopolistic ISP that has complete freedom to set prices. The second paper [3] considers a model where there is one local ISP for the content provider and one local ISP for the EUs. Two mechanisms are studied in this setting. In the first, there is no collusion and each ISP tries to selfishly maximize profit. In the second, a Nash bargaining solution is used to determine the profit division between the two ISPs.

There have also been a number of studies motivated by the question of network neutrality and some of these are related to the concept of sponsored content. In Reference 4, Economides and Hermalin consider a situation where the service provider partitions bandwidth and charges content providers for access to a given bandwidth partition. Conditions are presented for which a single partition is welfare maximizing. Mitra and Wang [5] study a model where the service provider maintains two pipes that can be accessed by content providers: a best-effort pipe and a managed bandwidth pipe that provides improved quality for an additional fee. In this setting, the service provider optimizes over the amount of best-effort bandwidth and the price of managed bandwidth. A key feature of this model is that when new applications enter the market, they typically rely on the best-effort pipe and so the rate of new service generation critically depends on the amount of best-effort bandwidth that is available. Lastly, in Reference 6, Njoroge et al. consider a model where each service provider controls access to a certain subset of EUs. Moreover, each service provider can charge a content provider for access to "its" consumers. The paper describes a setting in which this nonneutral regime can drive higher investment and, hence, maximize social welfare.

As already discussed, much of the focus of this chapter relates to how contracts should be designed between a content provider and a service provider. Similar issues of contract design have been discussed in a different context in the marketing discipline under the banner of "channel coordination" [7] and have been widely addressed in the supply chain literature (e.g., see Reference 8 for review). More recent work in References 9 and 3 has started the consideration of coordinating contracts in the network setting. Our paper fits closely with this stream of thoughts. In particular, our contracting arrangement is equivalent to the stylized buyback contract discussed in both References 7 and 8.

10.1.3 Designing Contracts Under Uncertain Demand

As already discussed, the focus of our presentation is on designing contracts between the service provider and content provider so that there is a win–win–win for all parties in the system (including the EUs). By this, we mean that all parties are better off than if the content was not sponsored. Our model differs from the prior models discussed earlier in that the underlying demand from the end users is uncertain. It is not simply a function of price and user/content provider utility. (In addition, the NUM model of Reference 2 assumes strictly concave utility functions as a consequence of demand being elastic with respect to price and, hence, does not capture a natural situation where the content provider is paid a fixed price per view by advertisers.) Lastly, our model extends beyond simple per-byte pricing and attempts to capture the notion of EU quota dynamics.

The fact that we treat underlying demand as a random variable has two effects on our analysis. First, the problem faced by the content provider is similar to the "Newsvendor" problem that is common in supply chain analysis. (In a Newsvendor problem, a retailer must purchase inventory to cover uncertain demand over a fixed time period. When the time period is over, the inventory only has a salvage value that is well below the purchasing price.) In our setting, the Newsvendor problem arises because the content provider must decide how much content to sponsor in a time period without knowing what the EU demand is. Second, the uncertain demand coupled with a reservation fee paid in advance will allow the service provider to control how much content is sponsored by the content provider, even when the latter's revenue grows in proportion to the number of views of its content.

We model the interaction between the service and the content providers as a Stackelberg game in which the service provider offers a contract parameterized by two fixed fees: a reservation fee proportional to the maximum number of views to be sponsored and a usage fee for each sponsored view that actually takes place. By accepting this contract, the content provider determines the maximum number of sponsored views and pays the corresponding reservation fee in advance, and assumes the payment of the usage fee for each view of its content by EUs, up to the aforementioned maximum.

We divide our discussions into two parts that reflect different ways of modeling EU payments.

- In Section 10.2, we present a simple model in which service provider congestion costs, EU bandwidth costs, and the price that the content provider must pay for sponsoring content are all determined on a per-byte basis. A key feature of this model is that the underlying demand from the EUs is a random variable but the service provider would like to control the amount of bandwidth it has to provide.

  We focus on the relationship between the service provider and a single content provider. We show that the aforementioned two-fee contract is incentive compatible: by charging a proper reservation fee, the content provider will be induced to choose the maximum number of sponsored views to optimize the total expected profit of both parties. It is also in the service provider's best interest to charge such a fee to bring about this outcome, because it can then use the per-use fee to transfer the profit to itself. In Section 10.2.5, we present a numerical example to demonstrate how the optimization might work in practice.
- In Section 10.3, we indicate how the results can be adapted for the case of EU quotas. In most current wireless data plans, the EUs pay a certain fee for a fixed quota of data. They do not pay on a byte-by-byte basis. This has a significant effect on the model because it is now much less explicit how much EU revenue the service provider is giving up when content is sponsored. In particular, we model an EU quota via a Markov Chain and describe the dynamics of the process according to whether the content is sponsored.

10.1.4 The Models

In this paper, we consider a contractual relationship between a single service provider (SP) and a single content provider (CP) in offering sponsored views of content (e.g., a webpage, an advertisement, or an online video). The situation is formally modeled as a Stackelberg game in which the SP is the leader who sets price parameters of the contract and the CP is the follower who responds by determining the maximum number of views it is willing to sponsor within a fixed period, for example, monthly. The purpose of sponsoring is to raise revenue by increasing the number of views of the said content. To model this effect, we assume EU will always access the said content if it is sponsored and with a smaller probability if it is not.

This problem can be naturally extended to the case of multiple CPs competing for the attentions of the EUs by sponsoring their own content. That situation leads to interesting competitive dynamics between the content providers and we leave its analysis for future work.

We define two basic models. In the first, the EUs pay for bandwidth on a per-byte "pay-as-you-go" basis. In the second model, we aim to capture the more common situation in which EUs pay for bandwidth via monthly quotas.

10.1.4.1 Model of Sponsored Content with Per-Byte End-User Costs The EUs generate N (a random variable) potential views in a period for content items that for ease of exposition are all assumed to have the same size θ. (It would not be difficult to extend to a situation where θ is the mean size of heterogeneous content.) Let F be the cumulative distribution function of N and let $\overline{F} = 1 - F$. If the content is sponsored, it is viewed with probability 1. If content is not sponsored, it is viewed with probability $q < 1$. (The parameter q here captures the strategic behavior of the EUs.) Let $\text{bin}(m, q)$ denote a binomial random variable with $m \in \mathbf{N}^0$ trials and success probability q. Then $E[\text{bin}(m, q)] = mq$.

The SP charges only the CP for sponsored content and the EUs for nonsponsored content, all on a per-byte basis. As we assume constant size for content, we denote EUs' payment per view for nonsponsored content by r. (The fact that an EU does not have to pay this amount for sponsored content is the main reason why the probability (=1) of viewing sponsored content is more than the probability (=q) of viewing nonsponsored content. The CP's decision is denoted by B, defined as the maximum number of views the CP is willing to sponsor. The actual number of sponsored views is, therefore, $\min\{N, B\}$ and the total number of views is $\min\{N, B\} + \text{bin}([N - B]^+, q)$. As mentioned earlier, the payment from the CP to the SP is structured as an ex ante reservation fee and an ex post usage fee. We define c to be the reservation fee per view and b to be the usage fee per view. Hence, the total revenue that the SP collects from the CP in a given period is $cB + b\min\{N, B\}$.

It remains to define the advertising revenue earned by CP and the bandwidth cost incurred by SP. We assume that CP earns revenue a for each view and, hence, its total revenue is $a(\min\{N, B\} + \text{bin}([N - B]^+, q))$. (We assume that all parameters are known by the SP and leave the interesting case where a and q are private information of the CP for future work.) The cost for the SP is dependent on the total congestion on its network, which is given by a nondecreasing function $C(\cdot)$. We let $\underline{B}$ be the total load on the network excluding the EU's views of the CP's content. The total load without sponsoring is

$$B_0 = \underline{B} + \theta\text{bin}(N, q)$$

and the total load with a sponsoring level of B is

$$\underline{B} + \theta\min[N, B] + \theta\text{bin}\left([N - B]^+, q\right).$$

So the expected congestion cost paid by the SP is

$$E\left[C\left(\underline{B}+\theta\min[N,B]+\theta\text{bin}\left([N-B]^{+},q\right)\right)\right]. \tag{10.1}$$

Our goal is to determine the maximum number of views that the CP should sponsor to maximize the total profit of both CP and SP. We also study the fees charged by SP that can induce this outcome.

10.1.4.2 Model of Sponsored Content with EU Quotas We now describe a more refined model in which EUs do not pay for bandwidth on a per-byte basis. Each EU instead pays periodically for a base data quota and has the ability to buy additional quota in case the base quota is exhausted in a period. More formally, we assume there is a homogeneous population of K EUs, all of whom are served by a single SP who periodically charges a fixed subscription fee. At the beginning of each billing cycle, every EU gets a bandwidth quota that she can use anytime within the period. The starting point of the first cycle of EUs is uniformly distributed over a period length. When an EU has exhausted her quota before the end of a period, she can wait until she gets new quota at the beginning of the next period or refill her quota immediately by paying an additional amount d. The choice between waiting and refilling is assumed to be independent of the number of times that the EU has refilled before.

An EU's opportunity to access content within a unit time period is a Poisson distributed random variable with mean λ. The likelihood of an EU taking the opportunity to view the content depends on the amount of the quota she has for the remainder of the period (this is a strong assumption because it does not take into account that an EU may use her quota more aggressively when it is about to expire). We model EU's decision by a discrete-state Markov chain. States are indexed by $i = 0, ..., S$, where EUs in the states of smaller index have more available quota left. EUs in state S have exhausted their quotas and are waiting for the next period to arrive. An EU in state i views unsponsored content with probability q_i ($i = 0, ..., S$), where $q_S = 0$. We remark that by using the Markov model, the periods will not have an equal length. However, we focus on this model as an approximation to a regular billing cycle because of its tractability.

10.1.4.3 Implementation Issues We conclude this section with a brief discussion of what the SP needs to track in order to implement a sponsored content offering. As already mentioned, the interaction between the SP and CP would happen on a periodic basis, for example, monthly. In order to perform the correct optimization, the SP needs to know a number of parameters, for example, q, a, and the distribution of N. The SP could estimate q and the distribution of N by monitoring EU behavior. However, the correct value of a (the value of a view to the CP) has to come from the CP itself. (The case in which the CP can try to "cheat" by giving an incorrect value of a generates a whole new set of interesting research questions that we will address in future work.)

Once SP has decided on the prices b, c and CP has decided on the sponsoring level B, the system is then operated by the SP. It has to identify which traffic is associated with the CP and charge the CP or EU appropriately. This depends on whether or not the particular view is sponsored which in turn depends on whether or not we have reached the sponsoring level B.

Of all these implementation issues, probably the most challenging is identifying the traffic associated with the CP because that involves monitoring the traffic at line rate and determining the content provider from which the content is being requested.

10.2 ANALYZING SPONSORED CONTENT WHEN EUs PAY PER BYTE

10.2.1 Content Provider's Problem

We start with the simplest situation in which the reservation fee $c = 0$. Recall that B denotes the maximum number of sponsored views, $a > 0$ denotes the (advertising) revenue to the CP of each view, and $b \geq 0$ denotes the usage fee per sponsored view paid by the CP to the SP. The revenue received by the CP is $E[a(\min(N, B) + \text{bin}([N - B]^+, q)]$ and the cost paid by the CP to the SP is $E[b\min(N, B)]$. The net revenue to the CP is

$$E\left[(a - b)\min(N, B) + a\text{bin}\left([N - B]^+, q\right)\right]$$
$$= aqE[N] + \left(a\overline{q} - b\right)E[\min(N, B)], \tag{10.2}$$

where $\overline{q} = 1 - q$, and we have used $[x - y]^+ = x - \min(x, y)$. Let $B^*(b)$ denote the maximizing value of B for given b. Then $B^*(b) = \infty$ if $a\overline{q} > b$ and $B^*(b) = 0$ if $a\overline{q} < b$. If $a\overline{q} = b$, the CP's net revenue function becomes a constant, and hence, the CP is indifferent between any choice of $B^*(b) \in [0, \infty)$. In other words, if the CP does not need to pay a reservation fee in advance for sponsoring but price b is paid for each view of sponsored content, then the CP's optimal choice of $B^*(b)$ is either zero or infinity with a transition point where the CP is indifferent between sponsoring any content or not.

Now, consider the case with a per-unit reservation fee $c > 0$. Then CP's revenue function becomes

$$aqE[N] + (a\overline{q} - b)E[\min(N, B)] - cB.$$

This is a standard Newsvendor model. If $c \geq a\overline{q} - b$, then $B^* = 0$, that is, the CP will not sponsor any content viewing if the combined reservation and usage fees exceed the additional revenue from advertisement. If N has a continuous distribution function (F has no jumps), then

$$B^* = \overline{F}^{-1}\left(\frac{c}{a\overline{q} - b}\right). \tag{10.3}$$

In our setting, N is a discrete random variable, so F has jumps, and there may not be a B^* such that Eq. (10.3) holds exactly. On the other hand, N is likely to be an extremely large integer (in our numerical Example, $E[N]$ is on the order of 10^7), so Eq. (10.3) will hold *almost* exactly. In particular, because B^* is defined by

$$\overline{F}(B^* - 1) > \frac{c}{a\overline{q} - b}, \quad \overline{F}(B^*) \leq \frac{c}{a\overline{q} - b},$$

the error we make in assuming that Eq. (10.3) holds is miniscule and will henceforth be ignored.

10.2.2 Service Provider's Problem

First, consider the case with no contract cost (i.e., $c = 0$) and no revenue from EUs (i.e., $r = 0$). The SP's revenue from the CP is $bE[\min(N, B)]$. The SP pays a congestion cost, given by a function C. Recall that $\underline{B}$ is the "baseline" congestion without the EU, and the congestion cost is given by Eq. (10.1). We remark that congestion cost may not be a convex function of B even if C is linear because $\min[N, B]$ is concave.

The SP wants to choose b to maximize

$$\begin{aligned} &bE\left[\min\left(N, B^*(b)\right)\right] \\ &\quad - E\left[C\left(\underline{B} + \theta\min\left[N, B^*(b)\right] + \theta\text{bin}\left(\left[N - B^*(b)\right]^+, q\right)\right)\right]. \end{aligned}$$

Recall that with $b < a\overline{q}$, $B^*(b) = \infty$, and with $b > a\overline{q}$, $B^*(b) = 0$. Thus the SP wants to choose either b as large as possible subject to $b < a\overline{q}$, in which case the SP's profit is

$$\begin{aligned} &\lim_{b\uparrow a\overline{q}} bE\left[\min\left(N, B^*(b)\right)\right] \\ &\quad - E\left[C\left(\underline{B} + \theta\min\left[N, B^*(b)\right] + \theta\text{bin}\left(\left[N - B^*(b)\right]^+, q\right)\right)\right] \\ &= a\overline{q}E[N] - E\left[C\left(\underline{B} + \theta N\right)\right], \end{aligned}$$

or $b > a\overline{q}$, in which case the SP's profit is $-C(\underline{B} + \theta\text{bin}(N, q))$. The SP will choose the alternative yielding the higher profit.

Now consider the case with fixed contract cost $c > 0$. The SP's revenue from the CP is $bE[\min(N, B)] + cB$. So the SP wants to choose c and b to maximize

$$\begin{aligned} &bE\left[\min(N, B^*(b, c))\right] + cB^*(b, c) \\ &\quad - E\left[C\left(\underline{B} + \theta\min\left[N, B^*(b, c)\right]\right.\right. \\ &\quad \left.\left. + \theta\text{bin}\left(\left[N - B^*(b, c)\right]^+, q\right)\right)\right]. \end{aligned}$$

Given the relationship $\overline{F}(B^*(b, c)) = \frac{c}{a\overline{q} - b}$ from Eq. (10.3), the SP's problem is equivalent to choosing $B \geq 0$ and $b \in [0, a\overline{q})$ to maximize

$$\begin{aligned} &bE[\min(N, B)] + \overline{F}(B)\left(a\overline{q} - b\right)B \\ &\quad - E\left[C\left(\underline{B} + \theta\min[N, B] + \theta\text{bin}\left([N - B]^+, q\right)\right)\right]. \end{aligned}$$

We now consider the optimal b for a given B. Looking at the first-order derivative of the profit function

$$E\left[\min\left(N,B\right)\right]-\overline{F}\left(B\right)B=\int_{0}^{B}\overline{F}\left(x\right)\mathrm{dx}-\overline{F}\left(B\right)B\geq 0,$$

we conclude that SP wants to set b as high as possible such that $b < a\overline{q}$. Thus with B fixed, the SP's profit is

$$\begin{aligned}&\lim_{b\uparrow a\overline{q}} bE\left[\min\left(N,B\right)\right]+\overline{F}\left(B\right)\left(a\overline{q}-b\right)B\\&\quad -E\left[C\left(\underline{B}+\theta\min\left[N,B\right]+\theta\mathrm{bin}\left(\left[N-B\right]^{+},q\right)\right)\right]\\&\quad = a\overline{q}E\left[\min\left(N,B\right)\right]\\&\quad -E\left[C\left(\underline{B}+\theta\min\left[N,B\right]+\theta\mathrm{bin}\left(\left[N-B\right]^{+},q\right)\right)\right],\end{aligned}$$

so that the optimal profit is attained using B^* given by

$$\begin{aligned}B^* &= \arg\max_{B}\left\{a\overline{q}E\left[\min\left(N,B\right)\right]\right.\\&\quad\left. -E\left[C\left(\underline{B}+\theta\min\left[N,B\right]+\theta\mathrm{bin}\left(\left[N-B\right]^{+},q\right)\right)\right]\right\}.\end{aligned}$$

Of course, in practice the limit cannot be attained: the SP needs to keep $c > 0$ to induce the CP to choose the SP's desired B^*. Thus there is some small ε such that $b = a\overline{q} - \varepsilon$ and $c = \overline{F}(B^*)\varepsilon$, that is, while a positive reservation fee is necessary to induce optimal B, the SP is better off to keep it as low as possible and derive all its profit by setting b as high as possible.

We also remark that because of Eq. (10.3), when finding B^* we must optimize over the support of N. In reality, we would typically wish to restrict the optimization further to between (say) $F(0.02)$ and $F(0.98)$ because otherwise the system would be overly sensitive to the exact values of b and c.

10.2.2.1 Revenue from EUs Suppose that SP earns revenue from the EUs, that is, $rE[\mathrm{bin}((N-B)^{+},q)]$ where r is the revenue rate. So the SP wants to choose c and b to maximize

$$\begin{aligned}&bE\left[\min\left(N,B^*\left(b,c\right)\right)\right]+cB^*\left(b,c\right)\\&\quad -E\left[C\left(\underline{B}+\theta\min\left[N,B^*\left(b,c\right)\right]+\theta\mathrm{bin}\left(\left[N-B^*\left(b,c\right)\right]^{+},q\right)\right)\right]\\&\quad +rE\left[\mathrm{bin}\left(\left(N-B^*\left(b,c\right)\right)^{+},q\right)\right].\end{aligned}$$

If $r \geq \frac{a\overline{q}}{q}$, it is not beneficial for the SP to offer sponsored content option to CP because the additional revenue from CP is not high enough to compensate for the loss in revenues from the EUs . This is the case if content revenues are low (i.e., for low a values) and/or the content is popular (i.e., for high q values).

Now, consider the case with $r < \frac{a\overline{q}}{q}$. Similarly as above, the SP's problem is equivalent to choosing B and b to maximize

$$\begin{aligned} &bE\left[\min\left(N,B\right)\right]+\overline{F}\left(B\right)\left(a\overline{q}-b\right)B \\ &-E\left[C\left(\underline{B}+\theta\min\left[N,B\right]+\theta\text{bin}\left(\left[N-B\right]^{+},q\right)\right)\right] \\ &+rE\left[\text{bin}\left(\left(N-B\right)^{+},q\right)\right]. \end{aligned}$$

Checking the derivative with respect to b, we conclude that SP wants to set b as high as possible such that $b < a\overline{q}$ as above. We can again optimize over B, the only difference being the additional term $rE[\text{bin}((N-B)^{+},q)]$.

10.2.3 A Pareto Analysis of the Two-Parameter Contract

The system performance, that is, the aggregate profit achieved by the SP and the CP, is given by

$$\begin{aligned} \pi^{S}\left(B\right)=E&\left[a\min\left(N,B\right)+\left(a+r\right)\text{bin}\left(\left[N-B\right]^{+},q\right)\right] \\ &-E\left[C\left(\underline{B}+\theta\min\left[N,B\right]+\theta\text{bin}\left(\left[N-B\right]^{+},q\right)\right)\right], \end{aligned}$$

and the SP takes the following share

$$\begin{aligned} &bE\left[\min\left(N,B\right)\right]+\overline{F}\left(B\right)\left(a\overline{q}-b\right)B \\ &-E\left[C\left(\underline{B}+\theta\min\left[N,B\right]+\theta\text{bin}\left(\left[N-B\right]^{+},q\right)\right)\right] \\ &+rE\left[\text{bin}\left(\left(N-B\right)^{+},q\right)\right]. \end{aligned}$$

Let B^{S} denote the system optimal number of sponsored views, that is, $B^{S}=\arg\max_{B}\{\pi^{S}(B)\}$. The increase of the total expected profit from sponsoring is

$$\pi^{S}\left(B^{S}\right)-\left(r+a\right)qE\left[N\right]-E\left[C\left(\underline{B}+\theta\text{bin}\left(N,q\right)\right)\right],$$

and from the system's perspective, sponsoring only makes sense if

$$\pi^{S}\left(B^{S}\right)>\left(r+a\right)qE\left[N\right]-E\left[C\left(\underline{B}+\theta\text{bin}\left(N,q\right)\right)\right].$$

From the earlier discussion, we know that the inequality is not satisfied if $r \geq \frac{a\overline{q}}{q}$. However, we remark that even if $r \leq \frac{a\overline{q}}{q}$, sponsored content might not generate sufficient advertising revenue to offset the combined effect of losing EU revenue and increasing congestion cost. Such situations will be identified by the optimization of

$\pi^S(B)$ when the optimal solution $B^S = 0$. This will happen if the congestion function $C(\cdot)$ increases steeply, for example, if,

$$\left(a\overline{q} - rq\right) \overline{F}(0) < G'(0),$$

where,

$$G(B) \equiv E\left[C\left(\underline{B} + \theta \min(N, B) + \theta \text{bin}\left((N-B)^+, q\right)\right)\right]$$

is an increasing and convex function of B, in which case $d\pi^S/dB < 0$ for all $B \geq 0$.

Assume now that $\pi^S(B^S) \geq (r+a)qE[N] - C(\underline{B} + \theta\text{bin}(N,q))$. As the system profit is maximized at B^S and the SP's share is increasing with b for any given B (and, hence, CP's share is decreasing with b), a contract (b, c) is Pareto efficient if and only if $B^*(b,c) = B^S$. Therefore, Pareto efficient contracts can be characterized by the single parameter $b < a\overline{q}$ where $c = \overline{F}(B^S)(a\overline{q} - b)$. Under the set of Pareto efficient contracts, any allocation of additional system profit can be possible. The SP's share of profit will have a range of $[rqE[N] - C(\underline{B} + \theta\text{bin}(N,q)), \pi^S(B^S) - aqE[N])$, whereas CP's profit will have a range of $[aqE[N], \pi^S(B^S) - rqE[N] + C(\underline{B} + \theta\text{bin}(N,q))]$. We remark that for CP to achieve profit of $\pi^S(B^S) - rqE[N] + C(\underline{B} + \theta\text{bin}(N,q))$, CP needs to negotiate from SP the most favorable per-unit price b such that

$$\begin{aligned}
&rqE[N] - C\left(\underline{B} + \theta \text{bin}(N,q)\right) \\
&= bE\left[\min\left(N, B^S\right)\right] + \overline{F}\left(B^S\right)\left(a\overline{q} - b\right)B^S \\
&- E\left[C\left(\underline{B} + \theta \min\left[N, B^S\right] + \theta \text{bin}\left(\left(N - B^S\right)^+, q\right)\right)\right] \\
&+ rE\left[\text{bin}\left(\left(N - B^S\right)^+, q\right)\right].
\end{aligned}$$

When r is small, b that satisfies this equality might be negative.

10.2.4 Summary of the Analysis with a Contract Price c and Additional Revenue from End Users

The findings from the above analysis can be summarized as follows.

- The system performance, that is, the total expected profit for both the SP and the CP, is given by

$$\begin{aligned}
\pi^S(B) = &E\left[a\min(N,B) + (a+r)\,\text{bin}\left([N-B]^+, q\right)\right] \\
&- E\left[C\left(\underline{B} + \theta \min[N,B] + \theta \text{bin}\left([N-B]^+, q\right)\right)\right].
\end{aligned}$$

  Optimizing the above determines whether sponsoring should take place.
- Although, as noted above, the system profit function $\pi^S(B)$ is not necessarily a concave function, finding the system optimal B is not a hard problem given that $\pi^S(B)$ is defined by one variable.

- Charging a single usage fee b is not enough to enforce a sponsored view to the CP. SP needs to charge a reservation fee c to induce the optimal limit on the sponsored views. By keeping c as low as possible and b as high as possible, the SP transfers all the expected gains from sponsoring to itself.
- Under any coordinating contract, the system profit is maximized (i.e., achieves Pareto optimum of the CP and the SP profits). Moreover, any allocation of additional profits is possible. Such a contract is definitely a win–win–win contract for CP, SP, and EUs.

10.2.5 Numerical Example

We now present a numerical example to illustrate the above concepts. Consider the case of a large CP for which N has a truncated normal distribution with mean $\overline{N} = 5 \times 10^7$ views per month. (The distribution is truncated to two standard deviations on each side.) The size of the content θ is 7.416 Mbit and the CP receives \$0.0125 profit for each view (before paying any sponsoring charge to the SP). We assume EUs pay at a rate \$10/GB for nonsponsored content and so this translates to a cost per view of $r = 10\theta/(8 \times 10^9)$. For the SP congestion cost, we set the baseline congestion $\underline{B} = 0$ and use a piecewise linear function given by

$$C(x) = \begin{cases} \dfrac{3rx}{5\theta} & \text{if } x \leq \dfrac{19\overline{N}}{20} \\[2ex] 10r\left(x - \dfrac{893\overline{N}}{1000}\right)\Big/\theta & \text{otherwise.} \end{cases}$$

(This stylized cost function reflects, in a simple manner, the additional costs, such as lost customer good will, of exceeding the nominal system capacity.) We set $q = 0.2$, that is, an EU is five times as likely to view the content when it is free to them than they are when they have to pay for the bandwidth. In Figures 10.1 and 10.2, we show system profit as a function of B when the standard deviation of the underlying normal distribution is $2\overline{N}/5$ and $\overline{N}/10$, respectively. We can see that as the uncertainty in N increases (i.e., the standard deviation increases), the optimal amount of content to sponsor decreases because there is more likelihood that the realization of N will correspond to the steep part of the SP congestion cost curve. We can also see that although the system profit is not concave, it is simple to identify the optimal value of B.

In Figure 10.3, we fix B to its optimal value (in this case, 4.5×10^7 views) for the case that the standard deviation is $\overline{N}/10$. We then plot both SP profit and CP profit as a function of b. (Recall that c is then determined from b and B via Eq. (10.3). In particular, as b increases from 0 to $a\overline{q} = 0.01$, c decreases from 0.0086 to 0.) As b increases, more of the excess system profit generated from the sponsored content is transferred from CP to SP. As a comparison, we also plot the baseline profit for SP and CP that would occur in the case that no content is sponsored (i.e., $B = 0$).

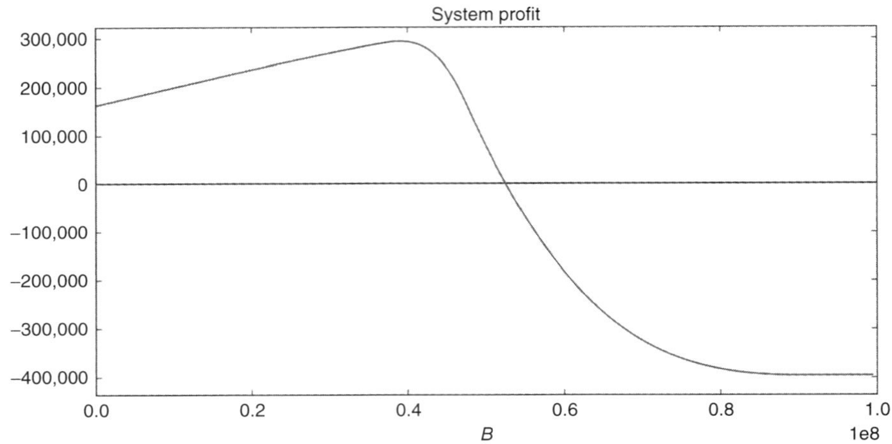

Figure 10.1 System profit when standard deviation is $2\overline{N}/5$.

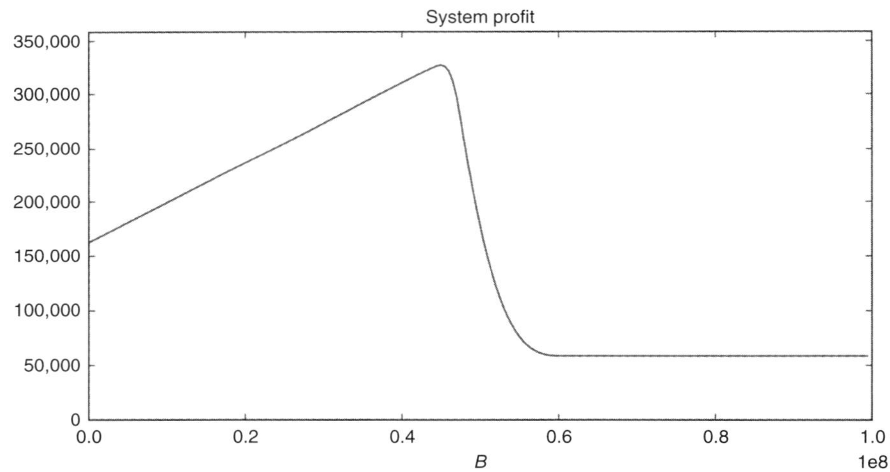

Figure 10.2 System profit when standard deviation is $\overline{N}/10$.

10.3 ANALYZING SPONSORED CONTENT IN THE CASE OF EU QUOTAS

Recall that we model quota usage via a discrete-state Markov Chain with state space $\{0, \ldots, S\}$. An EU in state i views the content with probability q_i. EUs in state S have exhausted their quota and are waiting for the next period to arrive. (Hence, $q_S = 0$.) When a user exhausts its quota it can pay to immediately renew it via an additional charge. An EU's opportunity to access content within a unit time period is a Poisson distributed random variable with mean λ. We let K denote the number of EUs and so

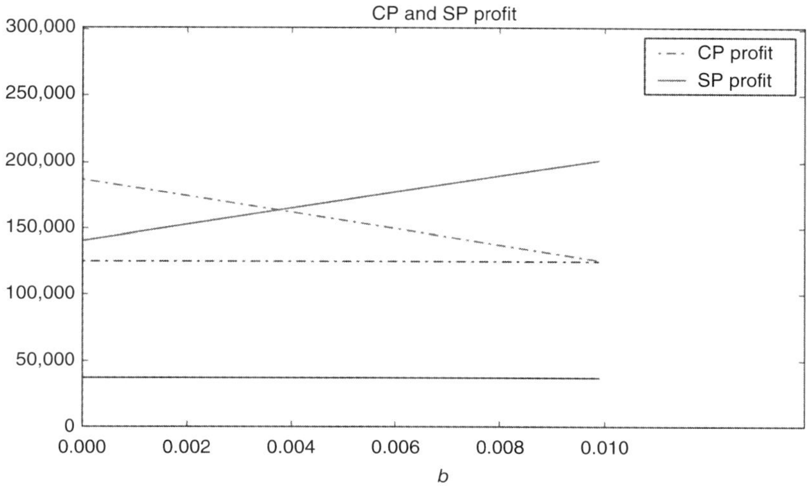

Figure 10.3 The sloped lines represent SP and CP profits as a function of b when the standard deviation is $\overline{N}/10$ and B is optimized for system profit. The horizontal lines represent the baseline profit for SP and CP in the case of no sponsoring.

the total number of potential views is given by the random variable Λ^K, which has a Poisson distribution with mean λK.

Let p_i be the transition rate from state i to state $i+1$ $(i = 0, \dots, S-2)$; p'_{S-1} and p''_{S-1} be the rates from state $S-1$ to state S and state 0, respectively; and p_S be the rate from state S to state 0. Transition rates between all other states are zero. Rates p_i $(i = 0, ..., S-2)$ reflect how fast EUs run down their quota. Define

$$\chi = \frac{p'_{S-1}}{p'_{S-1} + p''_{S-1}}$$

as the fraction of EUs who choose to wait for the next period after exhausting their quotas and denote p_{S-1} as the rate of EUs in state $S-1$ using up their quotas. Then

$$p'_{S-1} = \chi p_{S-1} \text{ and } p''_{S-1} = (1-\chi) p_{S-1}.$$

The rate p_S is inversely related to the residual time until the next billing cycle.

The transition rates imply the following steady-state probability for an EU to be in state i:

$$P_i = \frac{\rho_s}{1 + \rho_1 + \cdots + \rho_{S-1} + \chi\rho_S} \quad (i = 0, ..., S-1)$$

$$P_S = \frac{\chi\rho_S}{1 + \rho_1 + \cdots + \rho_{S-1} + \chi\rho_S}, \tag{10.4}$$

where $\rho_i = \dfrac{p_0}{p_i}$ $(i = 1, ..., S)$.

Although, as in the previous model, there are again two possible outcomes for each unsponsored potential view—viewed or not viewed, the probability of viewing is a bit more complicated. In particular, a potential view is associated with a user in state i with probability $P_i, 0 \le i \le S$, and such a user views content with probability q_i. Thus the number of actual views when there are m potential views is a binomial random variable bin(m, Q), where $Q = \sum_{i=0}^{S-1} q_i P_i$. Let $\overline{Q} = 1 - Q$, and note that $Q < 1$.

Let $\underline{D}$ be the base revenue that SP receives from the EUs for their regular monthly quotas and let τ be the rate at which EUs refill their quota "early." From the Markov Chain transition probabilities, we have

$$\tau = K(1 - \chi)p_{S-1}P_{S-1}.$$

Hence, the SP revenue from the EUs is $\underline{D} + d\tau = \underline{D} + dK(1 - \chi)p_{S-1}P_{S-1}$.

10.3.1 Case 1: Sponsorship-Insensitive Transition Probabilities

We begin with the case in which the Markov Chain transition probabilities do not change when CP's content is sponsored (i.e., the users simply switch their viewing from another content provider). We can obtain similar conclusions as before with Q playing the role of q. In particular, the CP's expected profit is given by

$$\begin{aligned} &E\left[(a-b)\min(\Lambda^K, B) + a\text{bin}\left(\left[\Lambda^K - B\right]^+, Q\right)\right] - cB \\ &= \left(a\overline{Q} - b\right)E\left[\min\left(\Lambda^K, B\right)\right] + aQE\left[\Lambda^K\right] - cB. \end{aligned}$$

As in Section 10.2.1, this is again a standard Newsvendor model and so the CP decision leads to a relationship of the form

$$B^*(b, c) = \begin{cases} \overline{F}^{-1}\left(\dfrac{c}{a\overline{Q} - b}\right) & \text{if } c < a\overline{Q} - b \\ 0 & \text{otherwise.} \end{cases} \tag{10.5}$$

The SP choice of b and c is, therefore, equivalent to a choice of b and B so long as $c > 0$. (Once again, therefore, we need $c > 0$ in order for the SP to be able to control the system.)

Similar to the per-byte cases, the system profit for a fixed value of B is

$$\begin{aligned} &a\overline{Q}E\left[\min\left(\Lambda^K, B\right)\right] + aQE\left[\Lambda^K\right] + \left(\underline{D} + d\tau\right) \\ &- E\left[C\left(\underline{B} + \theta\min\left[\Lambda^K, B\right] + \theta\text{bin}\left(\left(\Lambda^K - B\right)^+, Q\right)\right)\right]. \end{aligned}$$

Hence, as in the previous model, we can optimize system profit via a univariate optimization over B. Once the optimal value of the B has been obtained, the split between the SP and CP can be controlled by an appropriate choice of b.

10.3.2 Case 2: Sponsorship-Sensitive Transition Probabilities

The above analysis assumed that sponsoring content does not have a material affect on how fast the EUs consume their quota. It just causes the EUs to consume additional bandwidth corresponding to the sponsored content. In reality, of course, the knowledge that the CP's content is sponsored may affect the dynamics of quota usage. In particular, sponsoring may slow the rate at which quota is consumed, thereby lowering the probability that a user has to refill. In this case, the sponsoring of the content is *cannibalizing* the revenue that the SP obtains from the EUs.

A general analysis of this case is beyond the scope of this chapter. Here we take an initial step in the following by using a simple case to highlight issues involved. In particular, we consider a two-state model as in Figure 10.4. EUs are in state 0 if they have available quota to use and in state 1 if they do not. We define $p_0(B)$ as EUs' transition rate out of state 0 and assume it is a decreasing function of B, that is, EUs exhaust their quotas more slowly if they get more sponsored bandwidth. Let χ be the (constant) fraction of EUs who do not refill their quotas, so $\chi p_0(B)$ is EUs' transition rate into state 1. We also define $p_1(B)$ as the transition rate at which EUs move from state 1 back to state 0 as a result of monthly replenishment of quotas. We assume $p_1(B)$ to be an increasing function of B. As more content is sponsored, those who run out of their quotas will do so later in their monthly cycles, hence, get replenishment sooner and move back to state 0 faster.

Following the above definitions, the steady-state probability for EUs to be in states 0 and 1 are

$$P_0(B) = \frac{p_1(B)}{\chi p_0(B) + p_1(B)} \text{ and } P_1(B) = \frac{\chi p_0(B)}{\chi p_0(B) + p_1(B)}. \tag{10.6}$$

Obviously, $p_i(0)$ and $P_i(0)$ ($i = 0, 1$) are transition rates and steady-state probabilities, respectively, for the case without sponsorship. Observe that $p_1(B)P_1(B)$ is the rate at which an EU moves from state 1 back to state 0. As more sponsored content results in fewer EUs in state 1, it is natural to assume that $p_1(B)$ and $p_0(B)$ should be such that $p_1(B)P_1(B)$ decreases in B.

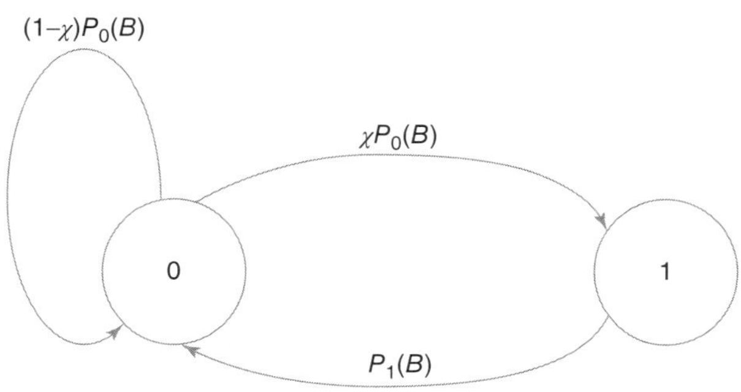

Figure 10.4 Transition of EU states.

As before, we define $\underline{D}$ as the subscription revenue that SP receives from the EUs to pay for their monthly quotas. Define

$$\tau(B) = K(1-\chi)p_0(B)P_0(B)$$

to be the rate at which EUs refill their quota "early." For the convenience of discussion, we denote $Q(B) \equiv q_0 P_0(B)$ as the probability that EUs access nonsponsored content. Let $\overline{Q}(B) = 1 - Q(B)$. Given B, the expected profit of the system is

$$a\overline{Q}(B)E\left[\min\left(\Lambda^K, B\right)\right] + aQ(B)E\left[\Lambda^K\right] + \left(\underline{D} + d\tau(B)\right) \\ - E\left[C\left(\underline{B} + \theta\min\left(\Lambda^K, B\right) + \theta\mathrm{bin}\left(\left(\Lambda^K - B\right)^+, Q(B)\right)\right)\right].$$

Compare the above with the case without sponsorship, the difference in profit is

$$a\overline{Q}(B)E\left[\min\left(\Lambda^K, B\right)\right] + aQ(B)E\left[\Lambda^K\right] - aQ(0)E\left[\Lambda^K\right] \\ + d\tau(B) - d\tau(0) \\ - E\left[C\left(\underline{B} + \theta\min\left(\Lambda^K, B\right) + \theta\mathrm{bin}\left(\left(\Lambda^K - B\right)^+, Q(B)\right)\right)\right] \\ + E\left[C\left(\underline{B} + \theta\mathrm{bin}\left(\Lambda^K, Q(0)\right)\right)\right].$$

Examining each component in the above,

$$a\overline{Q}(B)E\left[\min(\Lambda^K, B)\right] + aQ(B)E\left[\Lambda^K\right] - aQ(0)E\left[\Lambda^K\right] \tag{10.7}$$

is the increase of advertising revenue because of sponsoring, which increases in B, that is, more sponsoring leads to higher advertisement revenue. The incremental cost from sponsoring contents

$$E\left[C\left(\underline{B} + \theta\min\left(\Lambda^K, B\right) + \theta\mathrm{bin}\left(\left(\Lambda^K - B\right)^+, Q(B)\right)\right)\right] \\ - E\left[C\left(\underline{B} + \theta\mathrm{bin}\left(\Lambda^K, Q(0)\right)\right)\right]$$

always increases in B. The change of the refill revenue

$$d\tau(B) - d\tau(0) = dK(1-\chi)\left[p_0(B)P_0(B) - p_0(0)P_0(0)\right] \tag{10.8}$$

is always negative because from Eq. (10.6),

$$p_0(B)P_0(B) = p_1(B)P_1(B)/\chi,$$

and the right-hand side decreases in B.

In comparison with Case 1, the additional advertisement revenue in Eq. (10.7) is higher here because $Q(B)(\equiv q_0 P_0(B))$ increases in B instead of being fixed. This extra reward is accrued by the CP. On the other hand, the last component Eq. (10.8) shows an additional negative profit impact of content sponsoring, the cannibalization of the SP's refill revenue because of slower use of EUs' quotas. Like the bandwidth cost, this loss of revenue is assumed by the SP. To recoup its loss and share the extra gain, it is in the SP's best interest to require a more demanding transfer payment than that in Case 1 from the CP.

10.4 SUMMARY

In this chapter, we have introduced some of the natural research questions that arise if content providers in a wireless network are allowed to sponsor their content and thereby make it free to EUs. We considered the case of a single content provider, a single service provider, and a pool of EUs and showed that it is possible to design contracts that are win–win–win for all participants in system. This was first done in a setting where EUs pay for nonsponsored content on a per-byte basis. We then extended the model to incorporate EU data quotas. A key feature of all the models that we considered is that the content has uncertain demand, that is, the number of potential views is a random variable. This naturally led to a two-component price structure with a reservation fee and a usage fee.

We believe that this work can be extended in a number of natural ways. First, many interesting questions arise when we consider multiple content providers interacting with a single SP. For example, how should the SP decide which CPs should be allowed to sponsor content? Moreover, if different content has a different value of q (i.e., the probability that an EU views nonsponsored content is dependent on the identity of that content), how should that affect the cost of sponsoring?

Other potential variations include making the cost of sponsoring dependent on the user location, the time of day, or the current congestion in the network. We can also envisage a situation where sponsored content is provided its own quality-of-service guarantees by the network. Lastly, our entire framework is predicated on the knowledge of a number of different parameters, for example, a, q, and the distribution of N. One interesting network measurement task would be to monitor current network traffic and then estimate each of these parameters.

REFERENCES

1. S. Ha, S. Sen, C. Joe-Wong, Y. Im, and M. Chiang. Tube: time-dependent pricing for mobile data. In *SIGCOMM*, pp. 247–258, 2012.
2. P. Hande, M. Chiang, A. Calderbank, and S. Rangan. Network pricing and rate allocation with content provider participation. In *INFOCOM*, pp. 990–998, 2009.
3. Y. Wu, H. Kim, P. H. Hande, M. Chiang, and D. H. K. Tsang. Revenue sharing among ISPs in two-sided markets. In *INFOCOM*, pp. 596–600, 2011.
4. N. Economides and B. E. Hermalin. "The economics of network neutrality," *RAND Journal of Economics*, 43(4), 2012, 602–629.

5. D. Mitra and Q. Wang. A model-based study of the impact of managed services and the spawning of applications in broadband networks. *Workshop on Telecom Economics, Engineering and Policy, 24th International Teletraffic Congress (ITC 2012)*, 2012.
6. P. Njoroge, A. Ozdaglar, N. Stier-Moses, and G. Weintraub. Investment in two sided markets and the net neutrality debate. In *Columbia Working Paper # DRO-2010-05*, 2010.
7. B. A. Pasternack. "Optimal pricing and return policies for perishable commodities," *Marketing Science*, 4(2), 1985, 166–176.
8. G. P. Cachon. "Supply chain coordination with contracts," *Handbooks in Operations Research and Management Science*, 11, 2003, 229–340.
9. L. He and J. Walrand. Pricing and revenue sharing strategies for internet service providers. In *INFOCOM 2005*, pp. 205–216. IEEE, 2005.

11 CDN Pricing and Investment Strategies under Competition

YANG SONG, LIXIN GAO, and ARUN VENKATARAMANI

11.1 INTRODUCTION

A content delivery network (CDN) is a large distributed system that caches content at multiple locations on the Internet. When an end user makes a request, a CDN chooses the best server (usually the nearest one) to serve the content. CDNs enhance the user-perceived experience by reducing the delay and improving the availability of the content. By aggregating traffic across different content producers, CDNs also save individual investments in infrastructure for peak demand.

Owing to the technological and economical advantages, the CDN market has developed rapidly since its conception. The tremendous growth of Internet content further boosts the market. Today, 35–45% of the backbone traffic is from CDNs [1]. The majority of CDN traffic is delivered by three CDNs: Akamai technologies (Akamai for short), Level 3 Communications (Level 3 for short), and Limelight networks (Limelight for short). Their respective shares of the total CDN traffic are 48%, 25%, and 18% [2]. The rest of the CDN traffic is carried by relatively smaller CDNs such as Edgecast networks and Amazon CloudFront.

With CDNs competing with each other to attract business from content producers, the impact of this competition on shaping the CDN market is poorly understood. The unit price of CDN service has been dropping by more than 15% for at least five consecutive years [3]. It is unclear whether the dropping prices may result in a price war wherein no CDN can remain profitable. Recent events such as Level 3 charging a lower price than Akamai's to attract Netflix's video content drew a lot of attention [4, 5]. In light of such events, it is important to establish a rigorous technical foundation to understand and analyze pricing strategies practiced by CDNs and maintain a healthy market to benefit all participants.

Not only pricing competition is heated among CDNs but CDN market structure is also forced to change because of dramatic traffic increase (video traffic in particular). The growing traffic attracts potential new players to enter the market and

---

Smart Data Pricing, First Edition. Edited by Soumya Sen, Carlee Joe-Wong, Sangtae Ha, and Mung Chiang.

motivates incumbent players to expand and improve. For example, several Internet service providers (ISPs) attempt to build their own CDN service [6], and multiple Tier-1 ISPs and regional ISPs have already done so [7]. The incumbent players such as Akamai and Level 3 keep expanding their footprints [8, 9]. The changing structure not only makes the market highly competitive but also introduces complex dynamics. As a result, it is hard for both newcomers and incumbents to make wise decisions. For example, is it economically beneficial for a new CDN to enter the market? How much investment should one make? We need a theoretical model to guide decisions under a market structure that is changing constantly.

CDNs form a market with unique features. CDNs deploy servers close to end users, including computers or devices that request the content. Depending on the servers' locations, end users may receive various service qualities. For example, end users in China are likely to see little improvement in service quality by going through a CDN that has no cache presence in Asia. We refer to the set of end users who can receive improved service quality through a CDN as the *coverage* of the CDN. In other words, the set of end users is covered by the CDN. CDNs get paid by content providers for delivering their content to end users. Ideally, a content producer will select a CDN that covers all of its end users. However, in most cases, even the biggest CDN may not be able to include all the end users of a content producer. Content producers thus may select different CDNs for different end users. As a result, a content producer can subscribe to multiple CDNs at the same time [10]. We refer to the market where a customer can simultaneously select multiple providers as *multiple-choice market*. The multiple-choice market is different from the classical *single-choice market* where only one provider is chosen from a set of mutually exclusive alternatives. The single-choice market assumes that the alternatives can perfectly substitute for each other. For example, a cell phone user only needs to subscribe to one service provider for all the phone calls, no matter where the callee is located. We demonstrate that the multiple-choice markets can be prone to price wars. We provide sufficient conditions for price wars in two-CDN pricing games and propose an incentive-compatible pricing strategy to avoid them.

Another unique feature of the CDN market is that CDNs, especially smaller CDNs, are open for cooperation. Large CDNs are more appealing to large content producers, which usually require a broad coverage and a big capacity. Because the majority of Internet traffic is generated by large content producers [11], if small CDNs federate with each other to become a big CDN, they can attract more traffic than the total traffic they attract as individual CDNs. Not surprisingly, several CDNs have already formed a federation in order to compete more directly with the big CDNs in the market [12]. However, we show that CDN federation is not always beneficial for small CDNs. It is important to derive conditions of a profitable federation and instruct small CDNs to act accordingly.

With all the unique features of the CDN market, it is interesting to know how efficient this type of market is and how selfish behavior can degrade the efficiency. The results will be valuable for any multiple-choice market.

In this chapter, we address the challenges described earlier. We first focus on CDN pricing. We introduce a game-theoretic model to analyze the price competition among

CDNs. We derive sufficient conditions for price wars in a two-CDN pricing game. Then we prove that if CDNs incorporate future impact into current decision making, then price wars can be avoided. We proceed to introduce a predictive pricing strategy that is both incentive-compatible and efficient. That is, either CDN that deviates from the strategy will get a lower utility, and the total utility under the strategy is at least two-thirds of the social optimal utility. Then, we discuss a model to analyze the price competition for n-CDN games. We provide pricing strategies under various market structures through empirical analysis. The result shows that CDNs with bigger coverage always attract more traffic and charge a higher price.

Next, we focus on CDNs' investment strategies. We formulate a dynamic CDN game where the market structure, for example, the number of CDNs and their corresponding coverage, can be changed. We first analyze market changes caused by CDN federation and provide the conditions for small CDNs to benefit from a CDN federation. Then, we consider market changes because of new CDNs' entry and incumbent CDNs' investment and exit. The game is complicated by the fact that a CDN must consider the actions of its rivals while determining the best reaction scheme. Using the Markov perfect equilibrium approach, we derive the best investment strategy in the dynamic market. More importantly, we show that the investment strategy can lead to an equilibrium state that achieves up to 90% of the social optimal utility. The result indicates that the competition in multiple-choice markets can be efficient, and it may not be necessary for policy makers to interfere with a multiple-choice market to improve its social utility. The conclusion is applicable for any market with similar features.

The rest of the chapter is organized as follows. We begin by providing an overview of previous works and background in CDN economics (Section 11.2 and Section 11.3). Section 11.4 discusses the content producers' CDN selection problem. We then present the pricing and development strategy in Sections 11.5 and 11.6. Lastly, we conclude this chapter in Section 11.7.

11.2 RELATED WORKS

We will introduce three research areas that are closely related to CDN economics under competition. We first show the studies on the pricing of a monopoly CDN. Then we talk about the economics of CDNs as a part of the content delivery chain. Lastly, we introduce the difference between CDN market and several well-known multiple-choice markets.

11.2.1 The Pricing of a Monopoly CDN

We first introduce the pricing models used in the CDN industry. The most common model is to charge based on traffic volume delivered by a CDN over a period of time, for example, a month. This pricing model usually offers a volume discount, the larger the volume, the more the deduction [16]. The disadvantage of this volume-based model is that it fails to consider the bursty traffic [17], which can demand a costly investment from CDNs. The second pricing model enables CDNs to charge based on the bursty traffic. With the second charging model, a CDN periodically monitors the bandwidth usage and charges customers according to the 95-percentile usage. In this case, customers commit to a maximum bandwidth usage in advance. If the actual usage goes over the commitment by more than 5%, a penalty may be charged. The last pricing model is based on request. A customer is charged by the number of requests served by a CDN. This model is not widely used. A CDN may provide all types of pricing models, and customers can select the one that is best for their business.

Previous works on CDN pricing mainly focus on the best pricing scheme of a monopoly CDN when facing multiple heterogeneous content producers [18–20]. Content producers are given the option of either self-provisioning or subscribing to a CDN. Content producers' decisions are a function of the CDN's pricing scheme. The studies [19, 20] show that the optimal pricing scheme is determined by the patterns of the traffic. When traffic is Possion, CDNs should provide volume discounts to content providers. While facing bursty traffic and the burst varies across content providers, volume discounts can be suboptimal. In this case, a percentile-based pricing is more profitable than volume-based pricing.

Because of the rapid development of the CDN market, the monopoly CDN model may not be suitable for the current situation any more. As we have mentioned in Section 11.1, a content producer can subscribe to multiple CDNs at the same time. Under this situation, a CDN's price is mainly driven by the heated competition among CDNs rather than traffic patterns. In this chapter, we study the pricing schemes under competition.

11.2.2 CDNs in Content Delivery Supply Chain

Before CDNs and big content providers became big players, the supply chain of the Internet was clear and simple. Essentially, small networks (including content producers, CDNs and access networks) purchased transit from backbone networks, that is, they establish transit relationships, and both backbone networks and small networks peered among themselves to exchange traffic for free. However, as CDNs and content producers emerge as the biggest traffic contributors in the Internet [11], it becomes unclear which side gains more benefit from the connections, the content or the transit. Thus, it is hard to judge who should get paid in the supply chain. Transit networks argue that delivering the high volume content involves huge cost, and thus content producers or CDNs should pay for the service. On the other hand, content producers and CDNs insist that their content and service are valuable and help transit networks attract end users. There are many papers addressing the network neutrality debate [21, 22], and in this chapter, we focus on the debate that involves CDNs.

The debate is highlighted by a recent event between Level 3 (a backbone network and a CDN) and Comcast (an access network) [23, 24]. Originally, the relationship between Level 3 and Comcast is peering. In late 2010, Level 3 started to carry the traffic from Netflix, which significantly increased the traffic sent from Level 3 to Comcast. Level 3 then informed Comcast that they needed to add additional interconnect capacity. Comcast rejected the request, citing that their relationship was no longer peering because the traffic exchanged was unbalanced and requested Level 3 to pay for the connection. Level 3 eventually paid Comcast for the extra traffic. The result is supported by a study of Clark and Bauer [5]. Their paper points out that it is economically efficient for a CDN to pay access networks.

A number of studies formulate models to describe the position of CDNs in the Internet supply chain. One popular model involves two-sided markets [25]. It combines the access networks and the CDNs together as one platform and analyzes the best policy for the platform facing end users on one side and content producers on the other. In another related paper, Chuang [4] argues that the network structure is better measured under bilateral oligopoly relationship as shown in Figure 11.1. Compared with the two-sided market model, bilateral oligopoly preserves the market power for both CDNs and access networks, and it is capable of analyzing the competition within both CDN and access networks. According to the author, the competition is an indispensable factor to maintain economical sustainability in the Internet. Although the bilateral oligopoly relationship is of great interest, as far as we know, it has not been used by any analytical model. We try to take the first step toward the goal by focusing on the competition in the locus of CDN market (marked by the dashed line in Fig. 11.1). Because the relationship between CDNs and content producers has several unique properties that are ubiquitous among entities in the Internet, for example, customers can subscribe to several CDNs at the same time, and CDNs have both overlapped and unique service ranges, analyzing the competition between CDNs and content producers can provide useful insight for other competition in the Internet as well.

11.2.3 Compare CDN and Other Multiple-Choice Markets

There are many markets where customers face multiple choices. For example, a firm or a company may purchase multiple units of computers from multiple brands to meet their needs; a household may have multiple vehicles for different usages; and on one trip to a grocery store, customers regularly purchase multiple products on carbonated soft drinks, ready-to-eat cereals, canned soups, and so on. The previous research in this area focuses on modeling customers' multiple-choice behavior as firms' policies, customers' preference and demographics, and other related parameters change. Dube [26] proposed a model to analyze customers' multiple choices on carbonated

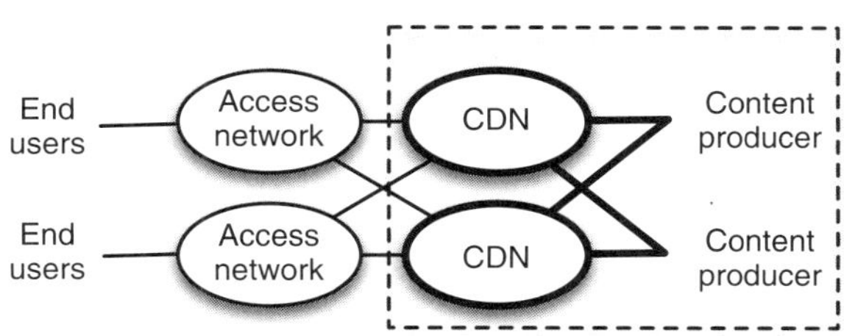

Figure 11.1 Bilateral oligopoly model.

soft drinks, and used the model to predict customers' future purchase. Hendel [27] developed a model to characterize firms' computer purchase behavior. Chandra Bhat and Sen [28] designed a model not only to estimate vehicle holdings in a family but also to predict the miles of travel for each vehicle. All these previous models focus on one-time multiple-choice purchase rather than a series of purchases that happen over time. There are two reasons that these models do not need (and never intend) to consider the market and customer behavior change over time. First, the markets addressed in these models are relatively stable. The number of competitors and their efficiency tend to stay consistent, and prices are also steady. Second, goods sold in these markets are durable and usually last a while after purchase. However, the CDN market is a different type of multiple-choice market. First, the CDN market sells service and charges content producers based on usage. It is easy for customers to switch among CDNs. Second, the CDN market exhibits high dynamics. New CDNs enter the market constantly, and incumbent CDNs keep expending their coverages, and the pricing of the CDN market changes dramatically. In the dynamic CDN market, we are not only interested in how customers make multiple-CDN selection as the market state changes, but more importantly, we also want to understand how CDNs survive and compete with customers' multiple-choice behavior and whether the competition is efficient and sustainable over the long term.

11.3 BACKGROUND

11.3.1 Static Analysis

In economics, competition is the rivalry among firms or providers that aim to achieve higher utilities, market share, and so on by adjusting the price, investment, innovation, or other marketing strategies [29]. Competition in economics has a long history stretching back to the nineteenth century. The original work was introduced in Reference 14 by Cournot. Cournot's model assumes that the demand in the market is a decreasing function of price and the competing firms choose quantities of a good to produce in order to maximize their own utility. The conclusion is that every firm eventually chooses the same quantity to produce and earns positive utility. As the number of firms increases, the price approaches marginal cost. Following the work of Cournot, Bertrand [13] took a different analysis approach. In Bertrand's model, the quantity is determined solely by the demand in the market and firms control prices rather than quantity. As a result, customers only purchase the good from the firm with the minimum price. The final outcome is an equilibrium where any firm in the market has zero utility. Since the two original papers, there have been a variety of models that discuss the competition among firms. Some of them produce more realistic results with the assumption that the firms are willing to supply only up to a quantity that maximizes their utility [30, 31]. Others focus on differentiated goods. For example, the Hotelling model [32] assumes that the goods associate with a cost depending on the location of the firms.

The previous models we introduced assume that an action has no long-term effect, for example, when a firm needs to make a decision, it ignores what other firms would do to react to its action in the future. The models also assume the state of the market, for example, the number of firms or the efficiency of each firm, is static. We refer to the analysis on games with only immediate impact and static market state as the *static analysis*. In the following subsections, we relax the constraints and introduce more realistic models to analyze the competition.

11.3.2 Predictive Analysis

One of the important factors that the static analysis rules out is the impact of competitors' current actions on the future. If actions are made with respect to the future, the same competitors may behave very differently. Repeated games capture the idea that a player takes into account the impact of his/her current action on the future. In a repeated game, a base game (or called a *stage game*) is played in finite or infinite repetitions with prediction. We name the analysis used in repeated games the *predictive analysis*.

We take the prisoners' dilemma as an example. In the classical version of the game, two prisoners are arrested and under investigation. If neither of them confesses, both of them will be sentenced to 1 year. If one of them confesses and the other keeps silence, then the one who confesses will be set free, and the other one will be sentenced to 3 years. If both of them confess, then both of them will get 2 years. Table 11.1 shows the utility of the prisoners with different actions. Note that, the longer the sentence, the lower the utility.

Suppose the game is played iteratively with the same pair of prisoners. We first assume no prisoner is concerned about the future impact. Then any rational prisoner would confess, because if they only consider the gain of the current iteration, confession always rewards more than silence. Thus, the only possible outcome in every iteration is that both prisoners confess. The interesting part of this result is that pursuing individual reward logically leads the prisoners to confess, but they would get a better reward if they both keep silence. The utility gained by each prisoner over time will be

$$\begin{aligned} U_s &= 0 + \delta \cdot 0 + \delta^2 \cdot 0 + \delta^3 \cdot 0 + \cdots \\ &= 0, \end{aligned} \tag{11.1}$$

where $\delta < 1$ is a discount factor. The discount factor indicates the value reduction for the future utility because the future utility cannot be obtained immediately.

Next, we assume that every prisoner makes decisions knowing that, in the next iteration, the other prisoner will act based on his/her current action. In this case, the

TABLE 11.1 Utility of Prisoners

	Prisoner 1: Silence	Prisoner 1: Confession
Prisoner 2: Silence	Both get 2	Prisoner 1: 3, prisoner 2: −1
Prisoner 2: Confession	Prisoner 1: −1, prisoner 2: 3	Both get 0

outcome will be changed. The strategy of prisoners is (i) always keep silence in the first iteration and (ii) if the other prisoner also keeps silence, then play silence in the next iteration; otherwise, always confess in all the future iterations. We will show in the following text that prisoners always have incentive to keep silence if the discount factor is small. If a prisoner keeps silence in the current iteration, the utility U_s he/she gets is

$$\begin{aligned} U_s &= 2 + \delta \cdot 2 + \delta^2 \cdot 2 + \delta^3 \cdot 2 + \cdots \\ &= \frac{2}{1-\delta}. \end{aligned} \tag{11.2}$$

In this case, both prisoners will play silence in every iteration and earn a utility of 2. If a prisoner confesses, then the utility U_c he/she gets is

$$\begin{aligned} U_s &= 3 + \delta \cdot 0 + \delta^2 \cdot 0 + \delta^3 \cdot 0 + \cdots \\ &= 3. \end{aligned} \tag{11.3}$$

Although the confessed prisoner earns a high utility in the first iteration, he/she will get 0 for all the future iterations. Note that, as long as the discount factor is larger than $\frac{1}{3}$, silence is always a better strategy than confession.

We can see that the future impact may dramatically change players' strategy. This is because the promise of future reward or the threat of future punishment provides incentive for both players to cooperate in the current iteration.

Although predictive analysis models the impact of the current actions on the future, it fails to model the state change of the game. For example, in the prisoners' dilemma game, we assume the utility gained under the same condition is constant. However, the utility may change over time in reality. If silence no long gives a utility of 2 but 0, then the prisoners' strategy should be changed accordingly. In general, the state change can be driven by both internal or external forces. An internal force can be a firm's investment or improvement in its technology, and an external force can be the outside market shock that may change the demand in the market we are interested in. The next model we introduce helps us to handle the case where the game state can change.

11.3.3 Dynamic Analysis

Both static and predictive analysis assume that the state of the game, for example, the efficiency of the firms or the number of competitors, stays the same over time. However, this assumption is not always true in reality. For example, players or firms can actively impact their efficiency by reducing their cost or adopting a new technology, and they can also change the total number of players by entering or exiting the game. The state of the game can also be changed by external causes. For example, the demand of customers can decrease if an outside alternative emerges. Introducing the state change can make the analysis more complicated, but it can provide useful insight for more realistic scenarios.

11.3.3.1 One-Player Dynamic Analysis We start with the simplest case, where there is only one player. We assume the base game is played iteratively. The utility that the player gains at time t is

$$U^t = \pi(\omega^t, x^t), \tag{11.4}$$

where ω^t is the state of the game and x^t is the action taken by the player at time t. The utility over time is thus

$$U = \sum_{t=0}^{\infty} \pi(\omega^t, x^t). \tag{11.5}$$

The state of the game can change over time. The state at time $t + 1$ is determined by the previous state, the action taken, and the outside market change. That is, $\omega^{t+1} = T(\omega^t, x^t, o^t)$, where o^t represents the change from the outside.

The goal of the game is to determine the best action to maximize the player's utility over time. Maximizing utility over time is equivalent to solving the following maximization problem.

$$\max\{U\} = \max_x \left\{ \sum_{t=0}^{\infty} \pi(\omega^t, x^t) \right\}. \tag{11.6}$$

Define a new function $V = \max\{U\}$ and assume the initial state is ω^0, then Eq. (11.6) will be expanded as follows.

$$\begin{aligned} V(\omega^0) = \max\{U\} &= \max_x \left\{ \sum_{t=0}^{\infty} \pi(\omega^t, x^t) \right\} \\ &= \max_{x^0} \left\{ \pi\left(\omega^0, x^0\right) + \delta \left[\max_x \sum_{t=1}^{\infty} \delta^{(t-1)} \pi\left(\omega^t, x^t\right) \right] \right\} \\ &= \max_{x^0} \left\{ \pi\left(\omega^0, x^0\right) + \delta V\left(\omega^1\right) \right\} \\ &= \max_{x^0} \left\{ \pi\left(\omega^0, x^0\right) + \delta V\left(T\left(\omega^0, x^0, o^0\right)\right) \right\}. \end{aligned} \tag{11.7}$$

Drop the time subscripts, we get

$$V(\omega) = \max_x \{\pi(\omega, x) + \delta V(T(\omega, x, o))\}. \tag{11.8}$$

We refer to Eq. (11.8) as the *value equation* of the one-player dynamic game. Similar to solving the Bellman equations [33], we can solve Eq. (11.8). The solution is a reaction function $x = R(\omega)$, which gives the best reaction x under state ω. We name the reaction function the *optimal action or strategy*. We will briefly talk about a numerical method to solve Eq. (11.8). Please refer to References 33–35 for other methods to solve the equation.

Rewrite Eq. (11.8) in the iterative form

$$V^k(\omega) = \max_x \left\{ \pi(\omega, x) + \delta \left(V^{k-1}\left(T(\omega, x, o)\right)\right) \right\}, \tag{11.9}$$

where $V^k(\omega)$ represents the result derived from the k-th iteration. Start with $k = 1$. Let $V^0(\omega)$ be any bounded function that maps ω to a real number. Solve the maximization function on the right side of the equation by letting the derivative of x equal to 0. Replace x with the solution of the maximization function, and then we get the function of $V^1(\omega)$. Next, let $k = 2$, and solve the right side with $V^1(\omega)$ and get $V^2(\omega)$. Repeat the process until $||V^k(\omega) - V^{k-1}(\omega)|| \leq \varepsilon$, where ε is the desired degree of accuracy.

11.3.3.2 Multiple-Player Dynamic Analysis In the previous subsection, we focus on the case where there is only one player in the game. Now, we discuss the case where n players are involved. Player i, where $i = 1, \ldots, n$, solves the following maximization function to maximize his/her utility over time:

$$\max \left\{U_i(\omega)\right\} = \max \left\{ \sum_{t=0}^{\infty} \pi_i \left(\omega^t, x_i^t\right) \right\}, \tag{11.10}$$

where π_i and x_i are player i's utility and action. Note that, for a n-player game, the state of the game depends on all players' actions. That is, $\omega^{t+1} = T(\omega^t, x_1^t, \ldots, x_i^t, \ldots, x_n^t, o^t)$. Define a new function $V_i(\omega)$ to be "max $\left\{U_i(\omega)\right\}$," and then player i's value equation can be written as

$$V_i(\omega) = \max_{x_i} \left\{\pi_i\left(\omega, x_i\right) + \delta\left(V_i\left(T\left(\omega, x_1, \ldots, x_i, \ldots, x_n, o\right)\right)\right)\right\}. \tag{11.11}$$

The solution of the n-player game is n reaction functions $R_1, \ldots, R_i, \ldots R_n$, where $x_i = R_i(\omega)$, and we name them the *optimal action or strategy*. Pakes and McGuire [34] and Ericson and Pakes [35] propose a method to solve Eq. (11.11). We will briefly introduce this method.

The iterative forms of the n players' value equations are as follows:

$$V_1^{k+1}(\omega) = \max_{x_1^k} \left\{\pi\left(\omega, x_1^k\right) + \delta V_1^k\left(T\left(\omega, x_1^k, \ldots, x_i^k, \ldots, x_n^k, o\right)\right)\right\} \tag{11.12}$$

$$V_2^{k+1}(\omega) = \max_{x_2^k} \left\{\pi\left(\omega, x_2^k\right) + \delta V_2^k\left(T\left(\omega, x_1^k, \ldots, x_i^k, \ldots, x_n^k, o\right)\right)\right\} \tag{11.13}$$

$$\cdots$$

$$V_n^{k+1}(\omega) = \max_{x_n^k} \left\{\pi\left(\omega, x_n^k\right) + \delta V_n^k\left(T\left(\omega, x_1^k, \ldots, x_i^k, \ldots, x_n^k, o\right)\right)\right\}, \tag{11.14}$$

where V_i^k is the function derived from the k-th iteration. To derive V_i^{k+1}, we let the derivative of the function on the right side equal to 0. That is,

$$\left\{\pi\left(\omega, x_1^k\right) + \delta V_1^k\left(T\left(\omega, x_1^k, \ldots, x_i^k, \ldots, x_n^k, o\right)\right)\right\}'\Big|_{x_1^k} = 0 \tag{11.15}$$

$$\left\{\pi\left(\omega, x_2^k\right) + \delta V_2^k\left(T\left(\omega, x_1^k, \ldots, x_i^k, \ldots, x_n^k, o\right)\right)\right\}'\Big|_{x_2^k} = 0 \tag{11.16}$$

$$\cdots$$

$$\left\{\pi\left(\omega, x_n^k\right) + \delta V_n^k\left(T\left(\omega, x_1^k, \ldots, x_i^k, \ldots, x_n^k, o\right)\right)\right\}'\Big|_{x_n^k} = 0. \tag{11.17}$$

By solving the system of nonlinear equations, we can get x_i^k as a function of ω, denoted by $x_i^k = f_i^k(\omega)$. Note that, the system of equations may not be easy to solve, because they include n variables. We can use the methods in References 34 and 35 to simplify the process. In the i-th equation of the system, replace x_j^k by the function $f_j^{k-1}(\omega)$ derived from the previous iteration, where $j = 1, \ldots, n$ and $j \neq i$. Then, the equations become

$$\left\{\pi\left(\omega, x_1^k\right) + \delta V_1^k\left(T\left(\omega, x_1^k, f_2^{k-1}, \ldots, f_n^{k-1}, o\right)\right)\right\}'\Big|_{x_1^k} = 0 \tag{11.18}$$

$$\left\{\pi\left(\omega, x_2^k\right) + \delta V_2^k\left(T\left(\omega, f_1^{k-1}, x_2^k, \ldots, f_n^{k-1}, o\right)\right)\right\}'\Big|_{x_2^k} = 0 \tag{11.19}$$

$$\cdots$$

$$\left\{\pi\left(\omega, x_n^k\right) + \delta V_n^k\left(T\left(\omega, f_1^{k-1}, \ldots, f_{n-1}^{k-1}, x_n^k, o\right)\right)\right\}'\Big|_{x_n^k} = 0. \tag{11.20}$$

As a result, we reduce the n-variable system of nonlinear equations to n one-variable equations. This significantly reduces the computational cost.

Compared to static analysis and predictive analysis, dynamic analysis allows the state of games to change, and thus it can provide a solution suitable for more realistic scenarios. However, we can also see that deriving the optimal strategy using dynamic analysis can be hard and, in most cases, we can only use numerical methods to solve the problem. Dynamic analysis is explicitly summarized in the original works [34, 35]: "we give up on analytic elegance in favor of an ability to numerically analyze the more complex situations that might better approximate what is observed in real data or situations."

11.3.3.3 Applications of Dynamic Analysis Dynamic analysis has been widely used to study a variety of markets under a dynamic state [15]. A number of papers apply dynamic analysis to show how policies can change the market state. Gowrisankaran and Town [36] use it to examine the hospital industry. It formulates a model where patients choose admission to a single hospital, while hospitals choose investment, entry, exit, and pricing strategies. They demonstrate the effect of policies such as universal health care or medicare reimbursement on hospitals' development decisions. Ryan [37] uses dynamic analysis to study the impact of environmental regulations on the cement industry where each customer selects one firm for cement supplement. The author shows that 1990 Amendments to the Clean Air Act on the US Portland cement industry has significantly increased the sunk cost of entry. The author also shows that static analysis can miss the welfare penalty on customers and draw a wrong conclusion on the amendments' effect on incumbent firms. Dynamic analysis is also used to explain industry structure and dynamics. For example, Lainz [38] uses the model to show that cost increase can result in the increase of firms' sizes in the market. Dynamic analysis has also proved to be useful to study mergers [39] and technology adoption [40]. In this chapter, we use dynamic analysis to study multiple-choice markets (CDN market in particular). The customers in multiple-choice markets make provider selection in a fine-grained manner, and the feature distinguishes it from all the markets mentioned earlier. We explore the best development strategy in this type of market and demonstrate what the market structure is at the equilibrium state and how efficient the market is in terms of price of anarchy.

11.3.4 Summary

In this section, we introduce three analytical models to characterize competition in economics. Static analysis ignores the impact of a current action on the future and the market state change. Predictive analysis considers the impact but it ignores the market state change. Dynamic analysis can model the market state change but the computation cost can be high. In the following sections, we use these methods to characterize the competition among CDNs.

11.4 CONTENT PRODUCERS' CDN SELECTION PROBLEM

We introduce two methods to model content producers' CDN selection. The first one models the coverages of CDNs with a Venn diagram, and it is capable of explicitly representing coverage overlaps among CDNs. We refer to the model as the *precise-coverage model.* It is obvious that the precise-coverage model does not scale well, because the regions in the Venn diagram increase exponentially as the number of CDNs increases. The second method is more suitable for large-scale analysis. It ignores the overlaps and uses the size to quantify CDNs' coverages approximately. The simplification makes it easy to scale. We refer to the model as *approximate-coverage model.*

In both models, we assume that content producers make decisions independently. This allows us to focus on one content producer, named S. S may select one or multiple CDNs from N available ones, represented by $D_1, \dots, D_i, \dots, D_N$. The selection depends on their performance and prices.

In reality, a CDN's performance is determined by many factors, including coverage, cache size at the CDN servers, congestion level, server selection algorithms, load balancing algorithms, traffic demand patterns, and network conditions. However, one of the most dominant factors is the coverage. In this chapter, we select the coverage as the single metric to measure CDNs' performance.

11.4.1 Precise-Coverage Model

The precise-coverage model describes the universe of S's end users in a Venn diagram as shown in Figure 11.2. Each circle represents the perceived coverage of a CDN. The circles divide the universe into subsets. We use G_I to represent the subset that is exclusively covered by a group of CDNs whose indexes are in the set, I. For example, $G_{\{1\}}$ represents the subset of end users who are covered by D_1 *only* and $G_{\{1,2\}}$ represents the end users who are covered by both D_1 and D_2. Note that, $G_{\{1\}}$ and $G_{\{1,2\}}$ are subsets that do not overlap. Figure 11.2 shows the universe of end users

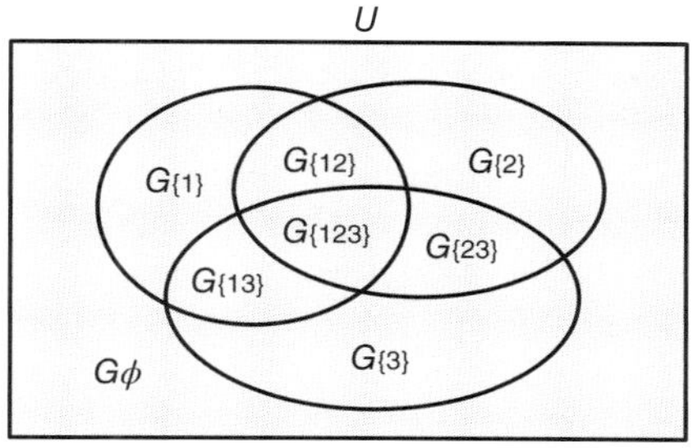

Figure 11.2 Divide end users into eight groups according to CDNs' perceived coverage.

who are divided into eight subsets by three CDNs. The traffic volume that S sends to G_I is represented by v_{G_I}.

If a subset is served by a CDN that covers the region, then each unit of traffic gets benefit η_1. If a subset is served by a CDN that does not cover it, the benefit is η_2. Apparently, $\eta_1 > \eta_2$. Suppose p_i is the price charged by D_i. Then, S chooses D_i for G_I if and only if S prefers it over all the alternatives. That is,

$$\eta_i^{G_I} - p_i > \eta_q^{G_I} - p_q, \tag{11.21}$$

where $q = 1, 2, \ldots, N$ and $q \neq i$. If D_i covers G_I, $\eta_i^{G_I} = \eta_1$. Otherwise, $\eta_i^{G_I} = \eta_2$. The CDN selection problem is equivalent to the following maximization problem. Select the best CDNs so that

$$\pi_s = \sum_{G_I} \max_{D_i} \left\{ \eta_i^{G_I} - p_i \right\} v_{G_I}. \tag{11.22}$$

If several CDNs provide the same utility, then S divides traffic evenly among them.

We use an $M \times K$ matrix named E to represent S's CDN selection. The rows of E represent CDNs, and the columns represent end users' subsets divided by CDNs' perceived coverages. Therefore, $K = 2^M$. E_{ij} is the percentage of subset j's traffic that is delivered by D_i. As a result, the utility of S can be written as

$$\pi_s = \sum_{j=1}^{K} \sum_{i=1}^{N} \left[E_{ij} \left(\eta_i^{G_j} - p_i \right) v_{G_j} \right]. \tag{11.23}$$

11.4.2 Approximate-Coverage Model

The precise-coverage model is able to provide fine-grained details about coverage. But the model does not scale well. The next model we will introduce is more suitable for a large number of CDNs.

We use parameter α_i to represent content producer S's perceived coverage of CDN D_i, where $0 \leq \alpha_i \leq 1$. The parameter reflects how well D_i can serve S's need. Thus, the parameter changes as D_i's coverage or S's need changes. The larger D_i's coverage, the larger α_i, and the higher S's traffic volume, the smaller α_i.

To enable S to make multiple-CDN selection, we assume S's benefit from D_i is equal to $\xi(v_i + \gamma)^{\alpha_i}$, where v_i is the traffic volume served by D_i, and ξ and γ are constants whose values stay the same across different CDNs. This benefit function is increasing and concave. The increasing property means that the more traffic volume S sends to D_i, the more benefit. The concave property indicates that the increase in the benefit generated by a one-unit volume increase becomes smaller as the total volume increases. The "diminishing return" as volume scales enables S to split traffic to multiple CDNs rather than use one CDN for all the traffic. This model has been widely used by multiple discrete choice analysis (refer to Reference 41 for details). S's utility gained from D_i is the benefit minus price: $u_i = \xi(v_i + \gamma)^{\alpha_i} - p_i v_i$. The total utility of S then can be written as $U = \sum_{i=1}^{N} u_i = \sum_{i=1}^{N}(\xi(v_i + \gamma)^{\alpha_i} - p_i v_i)$. Suppose the total traffic of S is V_s, then S's CDN selection problem is equivalent to the following optimization problem. Choose the optimal v_i so that

$$\max U = \sum_{i=1}^{N}(\xi(v_i + \gamma)^{\alpha_i} - p_i v_i) \tag{11.24}$$

$$\text{s.t.} \sum_{i=1}^{N} v_i \leq V_s. \tag{11.25}$$

Note that $\sum_{i=1}^{N} v_i$ can be smaller than V_s, because S's traffic can be delivered by ISPs directly.

In order to make the CDN selection problem scale, the approximate-coverage model simplifies the way to describe the service quality that S receives from CDNs. The model uses a single parameter α_i to represent S's perceived coverage size for D_i and implicitly assumes that the perceived coverage size is the only parameter to determine D_i's service. An important factor that the approximate-coverage model ignores is the overlap among CDNs. If a CDN's coverage is already covered by other CDNs, no matter how big its coverage is, using the CDN may provide little improvement for S's end users' experience. The precise-coverage model is able to describe the coverage overlaps and adjust service quality accordingly, but it does not scale well. While deciding which model to use for the CDN selection problem, we have to make the trade-off between the scalability and the precision.

11.5 CDN PRICING GAME UNDER COMPETITION

First, we analytically determine the optimal pricing strategy for two-CDN games. Then, we numerically characterize the strategy for n-CDN games.

11.5.1 Two-CDN Pricing Games

A two-CDN pricing game is defined as a repeated game where price decisions are made iteratively by two CDNs, D_1 and D_2. D_1 selects a price at the beginning of a period, and that price is stable for the rest of the period. D_2 cannot change its price during the period when D_1 is about to move and vice versa. After a CDN decides its price, S makes CDN selection.

We use v_1 and v_2 to represent the volume of S's traffic that is exclusively covered by D_1 and D_2, respectively, and v_{12} is the volume that is covered by both CDNs, and v_ϕ is the rest of the traffic, for example, the traffic that is covered by neither of the CDNs.

Assume the utility of D_i during period t is $\pi_i = p_i v_{D_i} - c$, where p_i is the price, v_{D_i} is the traffic served by D_i, and c is the cost. According to the precise-coverage model in Section 11.4.1, content producer S can make four possible CDN selections: case 1: if $p_2 < p_1 - (\eta_1 - \eta_2)$, S uses D_2 for all the traffic, and $\pi_1 = -c$, $\pi_2 = p_2(v_1 + v_2 + v_{12} + v_\phi) - c$; case 2: if $p_1 - (\eta_1 - \eta_2) < p_2 < p_1$, S uses D_1 for v_1 and D_2 for the rest, and $\pi_1 = p_1 v_1 - c$, $\pi_2 = p_2(v_2 + v_{12} + v_\phi) - c$; case 3: if $p_1 < p_2 < p_1 + (\eta_1 - \eta_2)$, S uses D_2 for v_2 and D_1 for the rest, and $\pi_1 = p_1(v_1 + v_{12} + v_\phi) - c$, $\pi_2 = p_2 v_2 - c$; case 4: if $p_2 > p_1 + (\eta_1 - \eta_2)$, S uses D_1 for all the traffic, and $\pi_1 = p_1(v_1 + v_2 + v_{12} + v_\phi) - c$, $\pi_2 = -c$.

If both CDNs ignore the impact of the current action on the future and maximize their *current* utility, then the competition can lead to price wars. Fortunately, we also show that if both CDNs plan ahead and maximize utility over time, then a price war can be avoided. We name the former strategy the *nonpredictive strategy* (Section 11.5.1.1) and the latter the *predictive strategy* (Section 11.5.1.2).

11.5.1.1 Nonpredictive Strategy A price war is commercial competition characterized by repeated cutting of prices below those of competitors, and it may eventually force firms to close because of profit reduction. In this chapter, we say a price war occurs if a series of price reduction leads to a state where at least one CDN cannot avoid zero or negative utility by changing its price.

Any CDN that applies the nonpredictive strategy aims to maximize its current utility. Assume D_2 is about to move and the current price of D_1 is p_1. It is straightforward that D_2 will select a price that makes S choose one of the first three CDN selection cases. Within each case, D_2 can reach the highest possible utility if it charges the highest price in corresponding range. Note that, only in case 3, D_2 selects a price that is higher than p_1, and in case 1 and case 2, D_2 will choose an action that is prone to price war (charge a price lower than D_1). Only if case 3 yields more utility than case 1 and case 2, then D_2 will charge a higher price than D_1. It is easy to verify that the condition is

$$p_1 < \min\left\{\frac{2\Delta\eta v_2}{v_1 + v_{12} + v_\phi} + \Delta\eta + \varepsilon, \frac{\Delta\eta v_2}{v_{12} + v_\phi} + \varepsilon\right\}, \tag{11.26}$$

where $\Delta\eta = \eta_1 - \eta_2$ and ε is the smallest price unit. We define the right side of the inequality as the *turning price* of D_2, denoted by p_2^T. Similarly, we can derive the same condition for D_1 to charge a higher price than the current price of D_2, which is

$$p_2 < \min\left\{\frac{2\Delta\eta v_1}{v_2 + v_{12} + v_\phi} + \Delta\eta + \varepsilon, \frac{\Delta\eta v_1}{v_{12} + v_\phi} + \varepsilon\right\}. \tag{11.27}$$

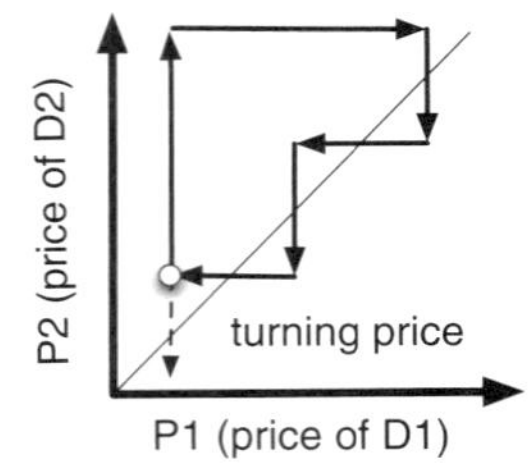

Figure 11.3 Prices oscillate in a cycle.

We define the right side of the inequality as the *turning price* of D_1, denoted by p_1^T. We use p^T to represent the larger one between p_1^T and p_2^T. Note that, the price will keep decreasing till it hits p^T and bounce back to a high price, where a new round of decreasing starts. The price oscillates in a cycle as shown in Figure 11.3. This figure shows a unique property of the multiple-choice market, where one of the CDNs will voluntarily increase its price in order to get a higher utility. The price increase can happen because the multiple-choice market allows a CDN to take a full advantage of its unique coverage. Whenever it is not profitable to compete for the overlap coverage, a CDN can focus on attracting end users in its exclusive coverage and charge a high price for the service. This will never happen to a single-choice market, where providers have to always compete for the same service, and increasing price will only make a provider less competitive. Thus, the multiple-choice market is less prone to price wars compared to the single-choice market.

However, price wars can still happen in multiple-choice markets. As the price decreases, the utility decreases as well. As the lowest price can be reached is p^T, the lowest income gained by a CDN is $\max\{p^T(v_1 + v_2 + v_{12} + v_\phi), (p^T - \Delta\eta)(v_1 + v_{12} + v_\phi), (p^T - \Delta\eta)(v_2 + v_{12} + v_\phi)\}$. We name the value the *bottom value*. If the bottom value is smaller than c, then one CDN will find itself having negative utility before reaching p^T, and the CDN cannot make positive utility by changing its price. The result is a price war. When a price war happens, the best nonpredictive strategy is to exit the game. We have the following theorem.

Theorem 11.1 *In the two-CDN pricing game, if CDNs apply the nonpredictive strategy and the bottom value is smaller than c, then the pricing game must lead to price wars.*

The overlapped coverage is linked to the likelihood of price wars. If two CDNs are fully overlapped ($v_1 = v_2 = 0$), then $p^T = 0$ and the bottom price is 0. Then a price war must occur. The game becomes less prone to price wars as the overlapped coverage decreases. Let us consider a game where two CDNs have partially overlapped coverage and assume the game leads to a price war with the nonpredictive strategy. Without the loss of generality, suppose $p_1^T > p_2^T$. Let D_1 keep its coverage stable, and the other CDN D_2 reduce its coverage by decreasing the overlapped coverage. Thus, v_1 increases and v_{12} decreases. According to the definitions, both p^T and the bottom value will increase. If the increase makes the bottom value larger than c, then p^T will be reached before the utility goes below zero, and the price bounces back before a price war can happen. That is, if the total covered region stays the same, the less the overlap, the less prone a game to a price war.

11.5.1.2 Predictive Strategies In this subsection, we assume any CDN that is about to move knows that its rival CDN will select a price based on its current move in the next period. The price strategy can be denoted by a pair of reaction functions $R_1(\cdot)$ and $R_2(\cdot)$ as follows:

$$p_1^t = R_1\left(p_2^{t-1}\right), \quad p_2^{t+1} = R_2\left(p_1^t\right), \tag{11.28}$$

where p_i^t is D_i's price at t. Because the reaction functions consider the consequences of the current action on the future, we name them the predictive strategy.

D_1's utility over time is $U_1 = \pi_1(p_1^t, p_2^{t-1}) + \delta\pi_1(p_1^t, p_2^{t+1}) + \delta^2\pi_1(p_1^{t+2}, p_2^{t+1}) + \delta^3\pi_1(p_1^{t+2}, p_2^{t+3}) + \cdots$. Substitute D_2's price using its reaction function, and then the utility over time can be written as $U_1 = \pi_1(p_1^t, p_2^{t-1}) + \delta\pi_1(p_1^t, R_2(p_1^t)) + \delta^2\pi_1(p_1^{t+2}, R_2(p_1^t)) + \delta^2\pi_1(p_1^{t+2}, R_2(p_1^{t+2})) + \cdots$. Note that at period $t+2$, the game will repeat the same decision process as it has at period t. We define a new function $V_1 = \max\left\{U_1\right\}$. Then, to maximize D_1's utility over time is equivalent to solving the following maximization function:

$$V_1\left(p_2^{t-1}\right) = \max_{p_1^t}\left\{\pi_1\left(p_1^t, p_2^{t-1}\right) + \delta\pi_1\left(p_1^t, R_2\left(p_1^t\right)\right) + \delta^2 V_1\left(R_2\left(p_1^t\right)\right)\right\}. \tag{11.29}$$

The solution of Eq. (11.29) is the optimal reaction function R_1. Similarly, we can derive the maximization function of D_2:

$$V_2\left(p_1^{t-1}\right) = \max_{p_2^t}\left\{\pi_2\left(p_1^{t-1}, p_2^t\right) + \delta\pi_2\left(R_1\left(p_2^t\right), p_2^t\right) + \delta^2 V_2\left(R_1\left(p_2^t\right)\right)\right\}. \tag{11.30}$$

The solution of Eq. (11.30) is the optimal reaction function R_2. We can solve the pair of equations using the Markov perfect equilibrium theorem [42, 43]. In the following content, we introduce the results for two different two-CDN pricing games, one with the same-size coverage ($v_1 = v_2$) and the other with the subsumed coverage ($v_2 = 0$).

For the CDN pricing game with the same-size coverage, the optimal predictive strategy is given by

$$R_i(p) = \begin{cases} p_f & \text{for } p \geq p_f, \\ p_l & \text{for } p_f > p > p_l, \\ p_f & \text{for } p \leq p_l, \end{cases} \tag{11.31}$$

where $i = 1, 2$ and p_f and p_l satisfy two functions (details can be found in Reference 44): $(\eta_1(2v_1 + v_{12}) + \eta_2 v_\phi)/(2v_1 + (5/4)(v_{12} + v_\phi)) < p_f < (\eta_1(2v_1 + v_{12}) + \eta_2 v_\phi)/(2v_1 + v_{12} + v_\phi)$ and $(1 + (3/2\beta + 1))(p_l - \varepsilon) < p_f < (1 + (3/2\beta + 1))p_l$, and

$\beta = \frac{2v_1}{v_{12}+v_\phi}$. Note that, at the equilibrium state, both CDNs will charge a price equal to p_f. According to the strategy, if the rival CDN charges a price that is greater than or equal to p_f, or smaller than p_l, then a CDN needs to charge p_f. If the rival CDN's price is between p_f and p_l, then a CDN needs to charge p_l. The optimal predictive strategy of the same-size coverage game is motivated by incentives, and it strategically forces the two CDNs to stay at p_f instead of competing to price wars. When any CDN deviates from p_f, the strategy makes sure that the other CDN will choose a price so that the deviating CDN can get the maximal utility only if its price goes back to p_f.

For the subsumed coverage CDN pricing game, the optimal predictive strategy is given by

$$R_1(p_2) = \begin{cases} p_2 - \varepsilon & \text{for } p_2 > p_2^f \\ p_1^f & \text{for } p_2^f \geq p_2 > p_1^f - (\eta_1 - \eta_2) \\ p_2 + (\eta_1 - \eta_2) - \varepsilon & \text{for } p_2 \leq p_1^f - (\eta_1 - \eta_2) \end{cases}$$

$$R_2(p_1) = \begin{cases} p_1 - (\eta_1 - \eta_2) - \varepsilon & \text{for } p_1 > p_1^f \\ p_2^f & \text{for } p_1^f \geq p_1 > p_2^f \\ p_1 - \varepsilon & \text{for } p_2^f \geq p_1 > p_l \\ p_2^f & \text{for } p_l \geq p_1 \end{cases} \tag{11.32}$$

where p_1^f, p_2^f, and p_l satisfy $(p_2^f - \varepsilon)(1+\beta) < p_1^f\beta < p_2^f(1+\beta)$, $(p_1^f - \varepsilon - (\eta_1 - \eta_2))((\beta/2)+2) < 2p_2^f < (p_1^f - (\eta_1 - \eta_2))((\beta/2)+2)$, $p_2^f(v_1 + v_{12} + v_\phi) > \eta_1 v_{12} + \eta_2(v_1 + v_\phi)$, and $2(p_l - \varepsilon) < p_2^f < 2p_l$. Note that, at the equilibrium state, D_1 and D_2 will charge p_1^f and p_2^f, respectively. The strategy adopts a similar method as the same-size case to force the two CDNs to stay at p_1^f and p_2^f. It is interesting to see that, under the optimal strategy, the larger CDN always charges a higher price.

Not all the CDN pricing games have an optimal strategy. The sufficient condition for its existence in the same-size game is that $0 \leq \beta < 2$ and $(2v_{12} + 5v_\phi)\eta_1 > (39v_{12} + 46v_\phi)\eta_2$. The first condition requires the overlapped coverage of D_1 and D_2 to be large enough, and the second condition requires η_1 to be larger than η_2. Note that a game that satisfies the first condition can be prone to a price war, and the optimal predictive strategy ensures that the price war can be avoided. More details about the sufficient conditions can be found in References 44 and 45.

We define the social optimal utility as the utility obtained by CDNs if they maximize their total utility instead of individual utilities. The social optimal utility is thus the highest utility that can be reached if all CDNs act selflessly. The closer the actual utility to the social optimal utility, the more efficient the strategy. Because the predictive strategy strategically forces CDNs to stay at a high price rather than compete to price wars, it is efficient in terms of price of anarchy. We have the following theorem.

Theorem 11.2 *The utility obtained under the optimal predictive strategies in Eqs. (11.31) and (11.32) is at least two-thirds of the social optimal utility.*

Please refer to Reference 44 for proof. The predictive strategy is apparently beneficial for CDNs, because the total utility is close to the social optimal utility (defined as the highest utility the CDN market can get). But the benefit for CDN market does not necessarily mean the loss for content producers, because content producers will never pay more than what CDN service can provide. That is because content producers have two options to deliver traffic to end users: going through an ISP or a CDN. If CDNs charge more than what their service worths, content producers can always switch to ISPs. The predictive strategy is designed to help CDNs to compete more efficiently rather than squeeze the revenue from content producers.

11.5.2 The n-CDN Pricing Games

In Section 11.5.1, we focus on two-CDN pricing games. Now, we extend the game to n CDNs. In a n-CDN pricing game, CDN D_i decides prices to maximize its own utility $U_{D_i} = p_i v_{D_i} - c$, knowing that content producer S makes CDN selection to maximize Eqs. (11.24) and (11.25). We say the game is under the equilibrium state if (i) content producer S cannot increase its utility by changing traffic distribution and (ii) CDNs cannot increase their utility by adjusting prices.

We will only demonstrate how to solve this problem with S as a big content producer. We omit the CDN pricing game that competes for small content producers because, unlike the big content producers who normally select multiple CDNs, small content producers usually select one CDN only, and any single-choice model [46] can solve the problem.[1]

Big content producers are the most important customers of the CDN market, because they possess huge traffic volume and pay great amount of money for traffic delivery [10]. To simplify the discussion but still preserve meaningful insights, we assume that a big content producer will only select among big CDNs (This assumption is consistent with the decision made by big content producers in reality [10]). Further, from a big content producer's perspective, the big CDNs have either a high coverage α_1 or a lower coverage α_2. We call them type 1 and type 2 big CDN, respectively. A type 1 CDN is intended to model Akamai, and a type 2 CDN resembles Level 3 or Limelight. We further assume CDNs with the same coverage have the same price. p_1 and p_2 represent the prices of the two types of big CDNs, and n_1 and n_2 represent their numbers.

[1]As small content producers make little impact on the traffic distribution of the CDN market, henceforth, we simplify the computation by assuming that small content producers choose one CDN randomly from all CDNs.

Unfortunately, the solution of the CDN selection problem in Eqs. (11.24) and (11.25) has no closed form, so we apply a numerical method to solve it. Because of the way we solve the CDN selection problem, the CDN pricing game has to be solved numerically as well. We enumerate all possible prices that CDNs can charge (with a step equal to 0.1) and solve the CDN selection problem. Then let CDNs choose the best price that can maximize their own utility.

We play eight CDN pricing games where the number of type 1 CDNs are either one or two and the number of type 2 CDNs ranges from one to four. We set $V_s = 10, \gamma = 1, \xi = 10, \alpha_1 = 0.8, \alpha_2 = 0.7$. We choose these values so that when there are one type 1 CDN and two type 2 CDNs in the market, their traffic shares are consistent with the percentages in the real market (introduced in Section 11.1). Note that, an arbitrary unit can be attached to V_s, for example, MB,GB,TB, and PB. For all the runs, there is always an equilibrium state. The result is shown in Table 11.2. We compare different games' price, traffic volume, and utility at the equilibrium state. We can see that type 1 CDNs always charge a higher price than type 2 CDNs do. But as the number of type 2 CDN increases, type 1 CDNs' price will decrease slightly. Moreover, type 1 CDNs always attract more traffic, and thus they always obtain a higher utility. An interesting observation is as the number of type 2 CDNs increases, the total utility gained by all CDNs can reduce. The reduction can be caused by the heated competition among CDNs. If there is one type 1 CDN, then having two type 2 CDNs can result in the highest utility. If there are two type 2 CDNs, then having one type 2 CDN can result in the highest utility.

Note that, if we play the n-CDN pricing game repeatedly as we did in the two-CDN pricing game, the outcome of the game will not change. That is because, with a stable market state, the base game will produce the same result in each iteration.

11.6 CDN COMPETITION UNDER MARKET STRUCTURE CHANGE

The CDN pricing game shows that a CDN with a larger coverage can obtain more utility than the CDNs with a smaller coverage. An interesting question now arises as to why not small CDNs invest to become big CDNs. In reality, CDNs indeed manage

TABLE 11.2 The Equilibrium State with Different Numbers of CDNs

Number of CDNs		Price		Volume		Utility		
Type 1	Type 2	Type 1	Type 2	Type 1	Type 2	Type 1	Type 2	Total
1	1	3.0	2.6	3.2	1.7	3.2	1.0	4.2
1	2	3.0	2.8	2.9	1.0	2.9	0.8	4.5
1	3	2.9	2.8	2.7	0.7	2.5	0.6	4.3
1	4	2.8	2.7	2.4	0.6	1.9	0.4	3.5
2	0	2.8	–	2.5	–	2.0	–	4
2	1	3.2	2.6	1.7	1.5	2.1	0.8	5.0
2	2	3.1	2.7	1.6	0.8	1.7	0.6	4.6
2	3	2.9	2.6	1.5	0.6	1.3	0.4	3.8
2	4	2.8	2.6	1.5	0.4	1.2	0.3	3.6

to expand their coverages through a variety of approaches including federation or investment. Nevertheless, the models presented in the previous section are no longer suitable to analyze this case because they all assume the market state, for example, the number of CDNs and their coverages, is fixed. In the following subsections, we use dynamic analysis to solve the CDN competition game. To simulate the situation in reality, we allow coverage change, as well as entry and exit. With these additional features, we address the following questions. How to make the federation or investment decision so that a CDN can get the highest utility? What is the equilibrium state in the dynamic game? Is the equilibrium state efficient compared to social optimal utility? We first discuss the case where market state is changed by federation, and then we focus on the case where market state is changed by investment, entry, exit, and market shocks.

11.6.1 Assumptions

Suppose there are M content producers, represented by $S_1, \ldots, S_j, \ldots, S_M$. The end users are evenly distributed over the network and request the same amount of traffic from the same content producer. Depending on the traffic volume generated, we divide content producers into K levels from low to high. There are s_k content producers at the kth level, and their volume is v_k, where $k = 1, \ldots, K$. We assume both s_k and v_k follow power law distribution over k, because the majority of the Internet traffic is generated by a small number of content producers [11]. More precisely, $v_k = v_0 k^{\beta_v}$, where $\beta_v > 0$, and $s_k = s_0 k^{\beta_s}$, where $\beta_s < 0$. In the following subsections, we suppose content producers are categorized into three levels, corresponding to small, medium, and big content producers, and $v_k = v_0 k^6$, and $s_k = s_0 k^{-5}$. Content producers in different levels have different perceived coverage as shown in Table 11.3.

11.6.2 Market State Change Through CDN Federation

CDN federation is a collection of CDNs that are operated autonomously by different entities but are interconnected through open interfaces, so that they can work collectively as a logical content delivery infrastructure [47]. CDN federation is an efficient and economic way for small CDNs to expand their coverage and compete more directly against large CDNs, such as Akamai. It has been under fast development recently. In June 2011, a group of small CDNs make one step forward to federation by founding an Operator Carrier Exchange (OCX) to interconnect their networks [12]. Edgecast networks built a platform where small CDNs can buy and sell capacity with an easy-to-use online control panel. In this section, we will not discuss the technical

TABLE 11.3 Perceived Coverage α_i of Content Producers in Different Levels

Content Producers	Type 1 CDN	Type 2 CDN	Small CDN
Small	1	1	1
Medium	0.8	0.75	–
Big	0.75	0.7	–

problems and feasibility of CDN federation [48]. Instead, we will focus on the economic incentives of CDN federation. We will answer the questions of whether it is profitable for small CDNs to join a federation, and to what extent.

Federation impacts the market state by changing the number of CDNs and their coverage. If k small CDNs form a federation, the number of small CDNs decreases by k, and the number of big CDNs increases by 1, the new big CDNs' coverage is the combination of the small CDNs' coverage. To determine whether a federation is profitable, we compute the utility with and without a federation.

On the basis of the current status of the CDN market, we assume the numbers of type 1, type 2, and small CDNs are $n_1 = 1$, $n_2 = 2$, and $n_3 = 20$, respectively, and small CDNs can federate to become a type 2 big CDN. CDNs conduct the n-CDN pricing game for every content producer independently (introduced in Section 11.5.2). We compare the utility gained by CDNs with and without a federation. Note that, a CDN's utility is equal to the summation of the utility gained from every content producer. The result is shown in Figure 11.4. We demonstrate four cases in the figure: no federation, three CDNs forming a federation, four CDNs forming a federation, and five CDNs forming a federation. For each case, we plot the utility of a type 1 CDN, a type 2 CDN, a small CDN, and a member of the CDN federation. We can see that a federation with three or four CDNs can be beneficial, because the members of the federation can gain more utility than individual small CDNs. Not surprisingly, as the number of members grows, the utility of each member decreases. If more than four CDNs are needed in order to form a federation, small CDNs may not have the motivation to join in, because acting as independent small CDNs can gain more utility than the small CDNs in the federation. The reason is twofold. First, the more members in the federation, the smaller share of the utility each of them will get. Second, as more small CDNs join the federation to compete with big CDNs, the competition among small CDNs becomes moderate and small CDNs' utility increases.

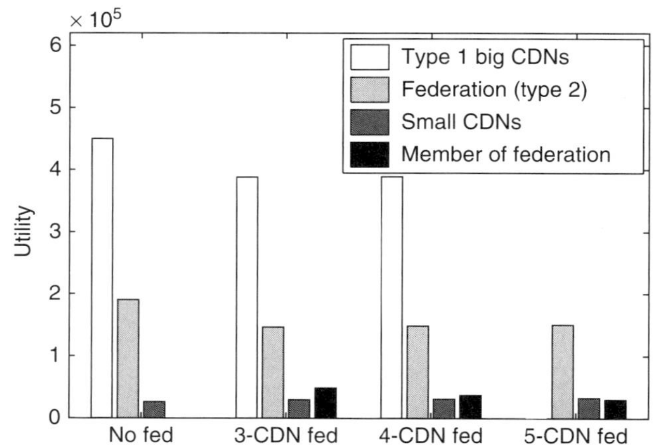

Figure 11.4 Utility comparison before and after federation.

For the current CDN market, a beneficial CDN federation has four or less member CDNs. If the CDN market has more big CDNs, the threshold will become smaller. But if two small CDNs can form a type 2 CDN, then forming a federation is always a better choice for these small CDNs, no matter how many big CDNs are in the market. That is because the federation can attract traffic from big content producers which contribute more traffic than small content producers.

11.6.3 The Dynamic CDN Game

In this subsection, we discuss the dynamic CDN game where the market state can be changed through the voluntary investment, entry and exit, as well as the involuntary market shocks. We model the dynamic CDN game as an infinitely repeated game where, at the beginning of each period, every CDN chooses an action (invest, enter, or exit) to maximize its utility over time. If a CDN decides to exit, it will get a scrap value in the current period. If it chooses to enter or invest, the market state change will take effect at the beginning of the next period. Within each period, the market state is stable and CDNs conduct the n-CDN pricing game introduced in Section 11.5.2. As CDNs' utility is totally determined by the market state, in the dynamic CDN game, each CDN aims to choose an action that can result in a market state that can benefit itself most. We use a vector $\omega = [l_1, l_2, \ldots, l_i, \ldots, l_n]$ to represent the market state, where l_i is the type of incumbent CDN D_i. We start with the case where the market state is changed by voluntary actions. Then we consider the case with market shocks.

11.6.3.1 Voluntary Dynamic CDN Game We first introduce a game where the market state can be changed by voluntary actions only (no market shocks). The game is referred to as the *voluntary dynamic CDN game*. We apply the Markov perfect equilibrium theory [34, 43] to derive the optimal reaction scheme in the game.

Any CDN, represented by D_i, aims to select proper actions to maximize its utility over time. D_i is allowed to invest, enter or exit the market. For simple explanation, we first consider the case where D_i is allowed to invest only. Later, we will add the other actions. We use x_i^t to represent D_i's investment at time t, then D_i's utility over time is equal to

$$I_i\left(\omega^0\right) = \sum_{t=0}^{\infty} \delta^t \left(\pi_i\left(\omega^t\right) - x_i^t\right), \tag{11.33}$$

where ω^t is the market state at t and $\pi_i(\omega^t)$ represents D_i's utility from the n-CDN pricing game under ω^t.

According to the dynamic analysis in Section 11.3.3, a CDN D_i should solve the following value equation.

$$\begin{aligned} V_i\left(\omega^0\right) &= \max \left\{ \sum_{t=0}^{\infty} \delta^t \left(\pi_i\left(\omega^t\right) - x_i^t\right) \right\} \\ &= \max_{x_i^0} \left\{ \pi_i\left(\omega^0\right) - x_i^0 + \delta V_i\left(\omega^1\right) \right\}. \end{aligned} \tag{11.34}$$

Note that at time $t = 1$, D_i faces the same maximization problem as the one it faces at $t = 0$, but the market state changes to ω^1.

Assume investment x_i either enhances D_i by one level in the type rank or fails to do so. We use υ_i to represent the result of investment x_i. υ_i is a random variable whose value depends on x_i.

$$\Pr(\upsilon_i|x_i) = \begin{cases} \dfrac{\lambda x_i}{1+\lambda x_i} & \upsilon_i = 1 \text{ (level increment)}, \\ \dfrac{1}{1+\lambda x_i} & \upsilon_i = 0 \text{ (no level increment)}. \end{cases} \tag{11.35}$$

Note that the probability to have a level increment is a monotonically increasing concave function of x_i. If $x_i = 0$, the probability is 0. We define a vector $\upsilon = [\upsilon_1, \ldots, \upsilon_i, \ldots, \upsilon_n]$ to represent the market change after the investment $x_1^t, \ldots, x_n^t$. The market state at time $t+1$, ω^{t+1}, depends on the market state at time t and the investments of all n incumbent CDNs. That is, $\omega^{t+1} = \omega^t + \upsilon$. Thus, the expected value of $V_i(\omega^1)$ is

$$V_i\left(\omega^1\right) = \sum_{\upsilon_1} \cdots \sum_{\upsilon_i} \cdots \sum_{\upsilon_n} V_i\left(\omega^0 + \upsilon\right) \Pr\left(\upsilon_1|x_1^0\right) \cdots \Pr\left(\upsilon_i|x_i^0\right) \cdots \Pr\left(\upsilon_n|x_n^0\right).$$

Then, D_i's value equation can be written as

$$V_i\left(\omega^0\right) = \max_{x_i^0}\Big\{\pi_i\left(\omega^0\right) - x_i^0 + \delta\Big(\sum_{\upsilon_1} \cdots \sum_{\upsilon_i} \cdots \sum_{\upsilon_n} V_i\left(\omega^0 + \upsilon\right) \Pr\left(\upsilon_1|x_1^0\right) \cdots \Pr\left(\upsilon_i|x_i^0\right) \cdots \Pr\left(\upsilon_n|x_n^0\right)\Big\}.$$

Next, we expand the value equation to include exit strategy. We assume D_i obtains a scrap value Φ when it exits the market. Thus, if the expected utility over time is less than Φ, D_i will quit. D_i always prefers an action that can result in a higher value between exit and staying. After including exit, D_i's value equation becomes

$$V_i\left(\omega^0\right) = \max\Big\{\Phi,\ \max_{x_i^0}\Big\{\pi_i\left(\omega^0\right) - x_i^0 + \delta\Big(\sum_{\upsilon_1} \cdots \sum_{\upsilon_i} \cdots \sum_{\upsilon_n} V_i\left(\omega^0 + \upsilon\right) \Pr\left(\upsilon_1|x_1^0\right) \cdots \Pr\left(\upsilon_i|x_i^0\right) \cdots \Pr\left(\upsilon_n|x_n^0\right)\Big\}\Big\}.$$

Dropping the time subscripts, we get D_i's value equation with investment and exit.

$$V_i(\omega) = \max\Big\{\Phi,\ \max_{x_i}\Big\{\pi_i(\omega) - x_i + \delta\Big(\sum_{\upsilon_1} \cdots \sum_{\upsilon_i} \cdots \sum_{\upsilon_n} V_i(\omega + \upsilon)\Pr(\upsilon_1|x_1) \cdots \Pr(\upsilon_i|x_i) \cdots \Pr(\upsilon_n|x_n)\Big\}\Big\}.$$

Next, we introduce entry to the value equation. We assume a new CDN enters the market as a small CDN and it pays a sunk cost x_e. If the expected utility over time is higher than x_e, then a CDN will enter. Otherwise, it stays inactive. A new CDN's expected utility over time is

$$V_e(\omega) = \delta\Big(\sum_{\upsilon_1} \cdots \sum_{\upsilon_i} \cdots \sum_{\upsilon_n} V_e(\omega + e + \upsilon) \\ \Pr(\upsilon_1|x_1) \cdots \Pr(\upsilon_i|x_i) \cdots \Pr(\upsilon_n|x_n), \tag{11.36}$$

where e represents market state change after the entry. Note that, if a CDN enters, the size of ω and υ will increase by one in order to match the total number of CDNs in the new market state.

The incumbent CDNs need to take entries into account while estimating their utility over time. Assume a new CDN's sunk cost x_e is a random variable that is uniformly distributed on $[x_{el}, x_{eu}]$. We use $E(\omega)$ to represent the probability that $x_e < V_e(\omega)$, hat is, an entry occurs, and $(1 - E(\omega))$ to represent the probability that there is no entry. After including entries, D_i's value equation becomes

$$V_i(\omega) = \max\Big\{ \Phi,\ \max_{x_i} \Big\{ \pi_i(\omega) - x_i + \delta E(\omega)\Big(\sum_{\upsilon_1} \cdots \sum_{\upsilon_i} \\ \cdots \sum_{\upsilon_n} V_i(\omega + e + \upsilon) \Pr(\upsilon_1|x_1) \cdots \Pr(\upsilon_i|x_i) \cdots \\ \Pr(\upsilon_n|x_n) + \delta(1 - E(\omega))\Big(\sum_{\upsilon_1} \cdots \sum_{\upsilon_i} \cdots \sum_{\upsilon_n} \\ V_i(\omega + \upsilon) \Pr(\upsilon_1|x_1) \cdots \Pr(\upsilon_i|x_i) \cdots \Pr(\upsilon_n|x_n) \Big\} \Big\}.$$

We refer to the equation as D_i's value equation of the voluntary dynamic CDN game.

In the following subsections, we make the same assumptions as we did in Section 11.6.1. In order to resemble the reality, for example, a new CDN is willing to enter when one type 1 and two type 2 CDNs are in the market, and there are less than five big CDNs at the equilibrium state, we assume that the scrap value Φ is 0.8 and x_{el} and x_{eu} are 1 and 2, respectively. The discount factor δ is set to be 0.925, a value commonly used in dynamic analysis [34, 35]. Note that, we do not intend to precisely predict what will happen in the real CDN market. We aim to show how CDNs' strategy may change as the market state changes, and whether the dynamics can produce an efficient market.

The best investment strategy is shown in Table 11.4. In this table, the first two columns show the number of type 1 and type 2 CDNs at the current market state and the rest of the columns show type 2 and small CDNs' best investment under corresponding market states (The best strategy of type 1 big CDNs is to avoid investment in any of the states). In general, if there are four or more big CDNs in the market, then small CDNs do not invest. Otherwise, they always invest. The investment increases as more small CDNs are competing in the market (the number of small CDNs is specified in the third row). Type 2 CDNs' best strategy is to invest only if there is no type 1

TABLE 11.4 Best Investment Strategy for Voluntary Dynamic CDN Games

Current state		Best investment					
		Small CDNs					Type 2 CDNs
Type 1 #	Type 2 #	1	2	3	4	5	
0	0	2.3	4.3	6.1	7.6	9.0	1.1
0	1	3.6	5.1	6.3	7.8	8.3	0.85
0	2	4.0	4.9	5.6	6.3	6.9	0.73
0	3	3.1	3.7	4.1	4.4	5.0	0.58
0	≥4	0	0	0	0	0	0.35
1	0	3.0	4.1	5.0	5.8	6.5	0
1	1	2.8	3.4	4.0	4.5	5.3	0
1	2	2.1	2.5	2.8	3.0	3.3	0
1	≥3	0	0	0	0	0	0
2	0	4.2	4.9	5.6	6.1	6.5	0
2	1	2.5	2.8	3.1	3.2	3.3	0
2	≥2	0	0	0	0	0	0
3	0	7.3	7.9	8.4	9.0	9.2	0
3	≥1	0	0	0	0	0	0

CDN in the market. The best entry strategy is as follows. If there are more than seven CDNs in the market, then no CDN should enter. Otherwise, CDNs can enter.

If all CDNs apply the best strategy, the voluntary dynamic CDN game will converge to an equilibrium state. Because the level increment v and the sunk cost x_e are random variables, different games may have different equilibria. We simulate 10,000 games with an initial state where there are one type 1 CDN and two type 2 CDNs. At the equilibriums state, there are 1.0 type 1 CDN, 3.0 type 2 CDNs, and 2.9 small CDNs on an average (Fig. 11.5a), and the type 1 CDN attracts 38.2% of all the traffic, and each type 2 and small CDN attract 17.6% and 2.7%, respectively. The utilities gained by the three types of CDNs are 1.16, 0.55, and 0.08 per period.

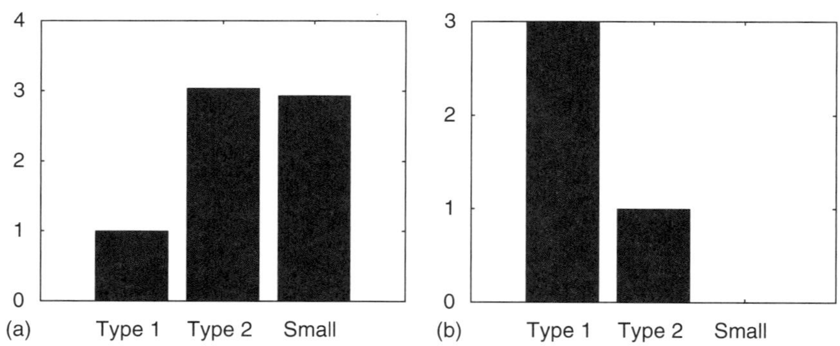

Figure 11.5 Average number of CDNs at the equilibrium state. (a) The voluntary dynamic CDN game and (b) Perfect cartel game.

To understand how efficient the equilibrium is, we compare the utility at the equilibrium state with the social optimal utility. We define a perfect cartel game to compute the social optimal utility. Assume that a perfect cartel can control the investment of all the CDNs and it aims to maximize the social utility of the CDN market. The value equation of the perfect cartel game is as follows.

$$V(\omega) = \max\Big\{\Phi, \max_x\Big\{\sum_{i=1}^{n}(\pi_i(\omega) - x_i) + \delta E(\omega)\Big(\sum_{\upsilon_1}\cdots \sum_{\upsilon_i}\cdots\sum_{\upsilon_n} V(\omega + e + \upsilon)\Pr(\upsilon_1|x_1)\cdots\Pr(\upsilon_i|x_i) \cdots \Pr(\upsilon_n|x_n) + \delta(1 - E(\omega))\Big(\sum_{\upsilon_1}\cdots\sum_{\upsilon_i}\cdots\sum_{\upsilon_n} V(\omega + \upsilon)\Pr(\upsilon_1|x_1)\cdots \Pr(\upsilon_i|x_i)\cdots \Pr(\upsilon_n|x_n)\Big\}\Big\}$$

Assume the conditions stay the same as the voluntary dynamic CDN game. The best strategy of the perfect cartel game is to promote a type 2 CDN to be a type 1 CDN if there are less than three type 1 CDNs in the market. The equilibrium state of the perfect cartel game has three type 1 CDNs, one type 2 CDNs, and no small CDN (Fig. 11.5b). The utilities of each type 1 CDN and type 2 CDN are 0.92 and 0.80, respectively, and thus the social optimal utility is 3.56. As a result, the total utility of the voluntary dynamic CDN game is 85.4% of the social optimal utility.

11.6.3.2 Dynamic CDN Game with Market Shocks In the dynamic CDN game, we allow both voluntary market state change and market shocks. We say a market shock happens when content producers downgrade the perceived type of the incumbents CDNs. The downgrade can be caused by several reasons, including the increase of content producers' traffic or the improvement of outside competitors such as ISPs. We assume that when a downgrade happens, it happens to all the incumbent CDNs and the probability of its occurrence is $\Pr(\tau)$. With market shocks, D_i's value equation becomes

$$V_i(\omega) = \max\Big\{\Phi, \ \max_{x_i}\Big\{\pi_i(\omega) - x_i + \delta E(\omega)\Big(\sum_{\tau}\sum_{\upsilon_1}\cdots \sum_{\upsilon_i}\cdots\sum_{\upsilon_n} V_i(\omega + e + \upsilon)\Pr(\upsilon_1|x_1, \tau)\cdots \Pr(\upsilon_i|x_i, \tau)\cdots \Pr(\upsilon_n|x_n, \tau)\Pr(\tau) + \delta(1 - E(\omega)) \Big(\sum_{\tau}\sum_{\upsilon_1}\cdots\sum_{\upsilon_i}\cdots\sum_{\upsilon_n} V_i(\omega + \upsilon)\Pr(\upsilon_1|x_1, \tau)\cdots \Pr(\upsilon_i|x_i, \tau)\cdots \Pr(\upsilon_n|x_n, \tau)\Pr(\tau)\Big\}\Big\}.$$

The best investment strategy under different $\Pr(\tau)$ is shown in Table 11.5. The first two columns show the numbers of type 1 and type 2 CDNs in the current market

TABLE 11.5 Best Investment Strategy with Market Shocks

Current State		Best Investment Strategy					
		$p = 0.1$		$p = 0.05$		$p = 0.01$	
T-1	T-2	Small	$T-2$	Small	$T-2$	Small	$T-2$
0	0	0.5–0.7		0.7–1.0		0.9–1.3	
0	1	0	1.1	0.5–0.7	1.1	0.7–1.0	1.2
0	2	0	0.7	0.1–0.2	0.9	0.5–0.7	0.9
0	3	0	0.6	0	0.6	0.2–0.2	0.8
0	4	0	0.6	0	0.4	0	0.6
0	5	0	0.5	0	0.5	0	0.4
1	0	0		0.3–0.5		0.5–0.8	
1	1	0	0	0	0	0.3–0.4	0
1	2	0	0.1	0	0.2	0.1–0.1	0
1	3	0	0	0	0.1	0	0
1	≥4	0	0	0	0	0	0
2	0	0		0.1–0.2		0.5–0.8	
2	1	0	0	0	0	0.1–0.1	0
2	≥2	0	0	0	0	0	0
3	0	0		0		0.5–0.7	
3	≥1	0	0	0	0	0	0

state, and the rest of the columns show the best investment strategy of small and type 2 CDNs (The best strategy of type 1 big CDNs is to avoid investment in any of the states). If the probability of market shocks is high, it is hard for small CDNs to survive. Thus, small CDNs only invest when there is no big CDN in the market. The best strategy of a type 2 CDN is to invest if there is no type 1 CDN. If there is already one type 1 CDN, then invest if two or three type 2 CDNs are in the market and the probability of market shock is high. The best entry strategy is as follows. When $\Pr(\tau) = 0.1$, a CDN should never enter if there are four or more CDNs in the market. The threshold is five and seven CDNs for $\Pr(\tau) = 0.05$ and $\Pr(\tau) = 0.01$, respectively.

To demonstrate the features at the equilibrium state, we simulate 10,000 periods with all CDNs applying their best strategy. The result is shown in Table 11.6. The first

TABLE 11.6 Average Number of CDNs at Equilibrium State

	Dynamic CDN Game			Perfect Cartel Game		
p	Type 1	Type 2	Small	Type 1	Type 2	Small
0.1	0.5	0.4	2.0	0	1.7	1.8
0.05	0.8	1.0	2.6	0	2.7	2.7
0.01	1.1	2.6	2.9	0	3.4	2.0

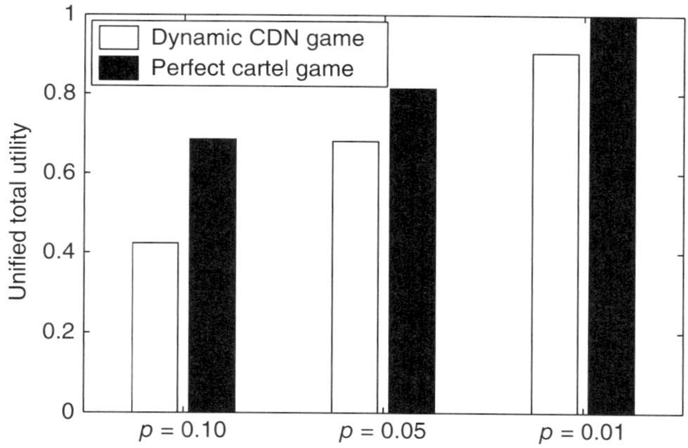

Figure 11.6 Compare total utility of the dynamic CDN game and perfect cartel game.

column represents the probability of market shocks. The next three columns show the average numbers of different types of CDNs at the equilibrium state. We can see that there are always more small CDNs than type 1 and type 2 big CDNs. As the probability of market shock decreases, the number of CDNs at the equilibrium state increases. We also show the equilibrium state of the perfect cartel game with market shocks in the last three columns. There is no type 1 CDN, but there are more type 2 CDNs compared to the dynamic CDN game.

We compare the total utility of the dynamic CDN game with the social optimal utility (the perfect cartel game). The result is shown in Figure 11.6. The first observation is that as the probability of market shocks decreases, the utility of both games increases. Further, as the probability decreases, the utility of the dynamic game gets closer to the social optimal utility. With the probability ranges from 0.1 to 0.01, the dynamic game can achieve 63–90% of the social optimal utility. That is, the dynamic CDN game becomes more efficient as the probability of market shocks decreases.

11.7 CONCLUSION

In this chapter, we study the competition in CDN market. CDN market is a unique market because content producers can select multiple CDNs at the same time, and small CDNs are open for cooperation. We propose the best pricing and investment strategies in the competitive and dynamic market. We demonstrate that by implementing the proposed strategies, CDN competition can be efficient in terms of price of anarchy. Because of the complexity of the CDN competition problem, we are only able to provide analytical result for small-scale CDN pricing games. For the large-scale pricing and investment games, we give up the analytic elegance for a better approximation for realistic scenarios. The results in this chapter provide an insightful analysis for the CDN market, and it is also generally applicable to any multiple-choice market with similar features.

ACKNOWLEDGMENTS

This work is supported by US NSF grants CNS-0917078.

REFERENCES

1. C. Labovitz. CDN and Over-The-Top Traffic Data, 2012.
2. C. Labovitz. First Data on Changing Netflix and CDN Market Share, 2012.
3. The State Of The CDN Market: Video, Pricing, Contract, Volume and Market Sizing Trends. Available at: www.cdnpricing.com.
4. J. Chuang. "Loci of competition for future Internet architectures," *Communications Magazine*, 49(7), 2011, 38–43.
5. D. Clark and S. Bauer. Interconnection in the Internet: the policy challenge. In *The 39th Research Conference on Communication, Information and Internet Policy*, 2011.
6. Dan Rayburn. Content Delivery Summit. Understand the CDN economics: build vs buy. Available at: http://www.contentdeliverysummit.com/2012/.
7. List of Vendors in the Content Delivery Ecosystem. Available at: http://blog.streamingmedia.com/the_business_of_online_vi/2011/07/updated-list-of-vendors-in-the-content-delivery-ecosystem.html.
8. BusinessTech. Level 3 enters South Africa. Available at: http://businesstech.co.za/news/internet/30472/level-3-enters-south-africa/.
9. Justin Lee. CDN Provider Akamai Expands Services to Costa Rica, Names Asia-Pacific Executives. Available at: http://www.thewhir.com/web-hosting-news/cdn-provider-akamai-expands-services-to-costa-rica-names-asia-pacific-executives.
10. D. Rayburn. CDN Pricing Stable: Survey Data Shows Pricing Down 15% This year. Available at: http://blog.streamingmedia.com/the_business_of_online_vi/2012/09/cdn-pricing-stable-survey-data-shows-pricing-down-15-this-year.html.
11. C. Labovitz, S. Iekel-Johnson, D. McPherson, J. Oberheide, and F. Jahanian. "Internet inter-domain traffic," *SIGCOMM Computer Communication Review*, 41(4), 75–86, 2010.
12. Dan Rayburn Federated CDN. Available at: http://blog.streamingmedia.com/the_business_of_online_vi/2011/06/telco-and-carriers-forming-new-federated-cdn-group-called-ocx-operator-carrier-exchange.html.
13. J. Bertrand. Revue de la Theorie Mathematique de la Richesse Sociale et des Recherches sur les Principes Mathhatiques de la Theorie des Richesses. Journal des Savants, 1883.
14. A. A. Cournot. Recherches sur les principes mathématiques de la théorie des richesses. L. Hachette, 1838.
15. U. Doraszelski and A. Pakes. *A Framework for Applied Dynamic Analysis in IO*, vol. 3. Elsevier, New York, 2007. Draft Date: September, 2006. Chapter 33, pp. 2183–2162.
16. Pricing chart of Amazon CloudFront. Available at: http://aws.amazon.com/cloudfront/pricing/.
17. Choosing a CDN. Available at: http://www.streamingmedia.com/Articles/Editorial/Featured-Articles/Choosing-a-CDN-65114.aspx.
18. R. Buyya, M. Pathan, and A. Vakali. *Content Delivery Networks*. Springer Publishing Company, Incorporated, 1st edition, 2008, New York.

19. K. Hosanagar, J. Chuang, R. Krishnan, and M. D. Smith. "Service adoption and pricing of content delivery network (CDN) services," *Management Science*, 54(9), 2008, 1579–1593.

20. K. Hosanagar, R. Krishnan, M. Smith, and J. Chuang. Optimal pricing of content delivery network (CDN) services. In *Proceedings of the Proceedings of the 37th Annual Hawaii International Conference on System Sciences (HICSS'04) - Track 7 - Volume 7*, HICSS '04, pages 70205.1–, Washington, DC, USA, 2004. IEEE Computer Society.

21. V. Kocsis and P. Bijl. "Network neutrality and the nature of competition between network operators," *International Economics and Economic Policy*, 4(2), 2007, 159–184.

22. R. Mason. "Simple competitive Internet pricing," *European Economic Review*, 44(4-6), 2000, 1045–1056.

23. D. Goldman. Netflix is a bandwidth hog. Who will pay? (Hint: You.). Available at: http://money.cnn.com/2010/11/30/technology/netflix_level3_comcast_traffic/index.htm.

24. W. Norton. The Netflix, Comcast and Level 3 Story. Available at: http://drpeering.net/AskDrPeering/blog/articles/Ask_DrPeering/Entries/2011/9/6_Access_Power_Peering.html.

25. N. S. Economides and J. Tg. "Network neutrality on the Internet: a two-sided market analysis," *Information Economics and Policy*, 24(2), 2012, 91–104.

26. J.-P. Dube. "Multiple discreteness and product differentiation: Demand for carbonated soft drinks,". *Marketing Science*, 23(1), 2004, 66–81.

27. I. Hendel. Estimating Multiple-Discrete Choice Models: An Application to Computeri-Zzation Returns. Working Paper 168, National Bureau of Economic Research, October 1994.

28. C. R. Bhat and S. Sen. "Household vehicle type holdings and usage: an application of the multiple discrete-continuous extreme value (mdcev) model," *Transportation Research Part B: Methodological*, 40(1), 2006, 35–53.

29. Competition in economics. Available at: https://en.wikipedia.org/wiki/Competition_(economics).

30. M. Shubik. *Strategy and Market Structure*. John Wiley & Sons, Inc., New York, 1959.

31. X. Vives. "Edgeworth and modern oligopoly theory," *European Economic Review*, 37(2-3), 1993, 463–476.

32. H. Hotelling. "Stability in competition,". *Economic Journal*, 39(1), 1929, 41–57.

33. Bellman equations. Available at: http://en.wikipedia.org/wiki/Bellman_equation.

34. A. Pakes and P. McGuire. "Computing Markov-perfect Nash equilibria: Numerical implications of a dynamic differentiated product model," *RAND Journal of Economics*, 25(4), 1994, 555–589.

35. R. Ericson and A. Pakes. "Markov-perfect industry dynamics: a framework for empirical work," *Review of Economic Studies*, 62(1), 1995, 53–82.

36. G. Gowrisankaran and R. J. Town. "Dynamic equilibrium in the hospital industry," *Journal of Economics & Management Strategy*, 6(1), 1997, 45–74.

37. S. P. Ryan. The Costs of Environmental Regulation in a Concentrated Industry. 2006 Meeting Papers 9, Society for Economic Dynamics, 2006.

38. C. A. Laincz. "Market structure and endogenous productivity growth: how do R&D subsidies affect market structure?" *Journal of Economic Dynamics and Control*, 29(1-2), 2005, 187–223.

39. S. Berry and A. Pakes. "Some applications and limitations of recent advances in empirical industrial organization: merger analysis," *American Economic Review*, 83(2), 1993–, 247–52.
40. F. Schivardi and M. Schneider. Strategic Experimentation and Disruptive Technological Change. CEPR Discussion Papers 4925, C.E.P.R. Discussion Papers, 2005.
41. J. Kim, G. M. Allenby, and P. E. Rossi. "Modeling consumer demand for variety," *Marketing Science*, 21(3), 2002, 229–250.
42. E. Maskin and J. Tirole. Markov Perfect Equilibrium, I: Observable Actions. Harvard Institute of Economic Research Working Papers 1799, Harvard–Institute of Economic Research, 1997.
43. A. Pakes. IO Class Notes: Multiple Agent Dynamics; An Intro to Markov Perfect Equilibria, 2009.
44. Y. Song, A. Venkataramani, and L. Gao. On CDN pricing game. Available at: https://sites.google.com/site/cdngametechreport/cdn.
45. Y. Song, A. Venkataramani, and L. Gao. On CDN pricing game. In *The 2nd IEEE International Workshop on Smart Data Pricing*, 2013.
46. M. E. B. Akiva and S. R. Lerman. *Discrete Choice Analysis: Theory and Application to Predict Travel Demand*, Mit Press Series in Transportation Studies, 9. Mit Press, 1985, Boston.
47. Dan Rayburn Cisco: the need for CDN federation. Available at: http://www.streamingmedia.com/Articles/Editorial/Featured-Articles/Cisco-The-Need-for-CDN-Federation-83223.aspx.
48. Dan Rayburn CDN Federation: A Badly-Defined Solution in Search of a Real Problem? Available at: http://www.streamingmedia.com/Articles/Editorial/Featured-Articles/CDN-Federation-A-Badly-Defined-Solution-in-Search-of-a-Real-Problem-80757.aspx.

12 Smart Pricing and Market Formation in Hybrid Networks

ARIS M. OUKSEL, DOUG LUNDQUIST, and
SID BHATTACHARYYA

12.1 SPECTRUM SHORTAGE

Over the past decade, the total demand in the United States for wireless services has approximately doubled annually. Although that trend is slowing (with 60% growth forecast for 2014), growth is still expected to be strong for the next several years [1]. This growing demand for wireless services is creating the so-called "spectrum shortage," which the United States is starting to address through its National Broadband Plan. In short, the demand for information too often exceeds the capacity of wireless networks to deliver it. In major markets, this phenomenon is already affecting service quality, as users experience reduced data rates because of peak activity.

The growth in demand for wireless services has three main sources. First, the number of network subscribers has been steadily increasing; this is largely due to the popularity of smartphones and tablets, which are continually becoming more powerful and more useful. Second, individual applications are becoming increasingly data intensive; streaming video is the most familiar example and, as video quality improves, becomes more data intensive. Third, the peak density of network subscribers is increasing. In the near future, applications requiring machine-to-machine communication (such as vehicular safety) will materially add to demand. Together, these growth factors are making demand outpace expected service availability in many areas.

The Federal Communication Commission's forecasts project 275 MHz of additional spectrum will be needed to meet demand by 2014 [2]. The shortage will not, of course, be experienced uniformly. While rural areas may not experience substantial

Smart Data Pricing, First Edition. Edited by Soumya Sen, Carlee Joe-Wong, Sangtae Ha, and Mung Chiang.

quality-of-service (QoS) degradation, urban areas (where the densities of both users and infrastructure are highest) certainly will, especially during times of peak demand. Even now, networks must often cap consumption of wireless services, whether by imposing steep fees for exceeding data plans or simply limiting users' download speeds. A thorough and current survey of extant methods for reducing congestion is provided by Sen et al. [3]. Networks that can maintain or improve their QoS in the face of this growing demand will have a clear competitive advantage.

Numerous solutions have been proposed and implemented to increase network capacity and address the imminent spectrum shortage. One class of solutions is technological, which includes developing wireless standards with faster data rates, such as the recent IEEE 802.11ac. Another involves more efficient ownership. Many usage rights are owned by firms with no near-term prospects of actually using them; networks with an urgent need for additional usage rights might buy them to expand capacity. Researchers have also proposed methods to flatten peak demand by incentivizing delayed network activity [4, 5]. Our solution is a novel network architecture and incentive program to encourage use of peer-to-peer (P2P) networking to expand the capacity of cellular networks.

Our model, in essence, provides a relief valve for congestion in cellular networks. When cellular networks are congested, they can subcontract a set of P2P networks to manage some of their load. Each P2P network will consist of multiple WiFi base stations, each governing network traffic in its vicinity. We envision that not all nodes will be within direct communication range of a base station, so the P2P network will require participants to act cooperatively—forwarding network traffic for others. Participation incentives will be offered to join the P2P network rather than remaining in the congested cellular network and may be adjusted to reflect differing participation, increasing for nodes that relay network traffic for other nodes and decreasing for nodes that consume a larger fraction of the P2P network's capacity. Additional incentives may be offered to encourage more continuous participation (and longer-term data caching), as opposed to nodes only remaining in the P2P network while they are actively requesting information. The roles and incentives of this hybrid network architecture are given in Table 12.1.

TABLE 12.1 Network Roles and Incentives

Party	Role	Incentive
Cellular network	Pays P2P network Provides aggregate usage data	Improves response time Can delay adding capacity Can support more subscribers
P2P network	Provides network services Distributes incentives to participants	Paid by cellular network Optimizes responsiveness
Network subscribers	Opts into P2P network	Paid by P2P network Potentially paid for forwarding other users' traffic

Currently, cellular network subscription plans typically allow users to switch between 3G/4G and WiFi networks, the incentive for users being that using the WiFi network does not deplete the subscriber's minutes or data plan, as applicable. The differences between our proposed hybrid network and extant commercial models are

- cellular network selection of P2P participants,
- incentive distribution for P2P network participation,
- multi-hop access to WiFi base stations (rather than only direct communication),
- automatic activity-driven incentive adjustments in the P2P network.

Managing this hybrid network to maximize profitability will entail solving multiple optimization problems, which we address in the following sections of this chapter. What participation level should the P2P network seek? Given the need for multi-hop communication between nodes and the base station, the P2P network will need sufficient density to enable continuous routing paths. However, although some additional density will be desirable as a safety margin (for when nodes exit the P2P network), additional participation may provide little value. Thus, the optimal network density is slightly higher than the minimum sufficient density. How should participants be incentivized to achieve the optimal density, especially given that willingness to participate will differ according to location, time, and other factors? And, finally, what contextual information needs to be gathered to support these decisions in dynamic environments?

We present our model from a bottom-up perspective, introducing its components and underlying concepts before explaining how they operate together. Before introducing our optimization model, the next three sections will cover

- P2P networking,
- commercial viability,
- Self-Balancing Supply/Demand (SBSD)protocol.

12.2 PEER-TO-PEER NETWORKING

The defining characteristic of P2P networking is the equality of network participants. In client–server models, all communication progresses up and down the network hierarchy. For example, a message sent from one client to another is not transmitted directly but must be sent up to a server and then down to the destination client. In contrast, the network architecture and node roles are more flexible in P2P models; nodes typically act as routers for their *peers* (other nodes in the network) and forward network traffic for them. In fact, any node in a P2P network may function as a client or server in regard to a particular query or response—provided, of course, that it has the requested information and that the nodes have common protocols to regulate their interactions. Compared to traditional client–server models, P2P offers some important advantages:

- *Load Sharing.* The decentralized structure of P2P networks enables better load sharing, because network traffic does not have to always be routed through a centralized server. Better predictive caching models (placing information near likely query sources) support improved load sharing.
- *Faster Response Time.* By decentralizing network traffic, each peer server may only need to respond to relatively few queries even under heavy total network traffic. Consequently, response time can be faster. This is especially important for multi-hop routing in wireless ad hoc networks, where the communication medium must be shared among potentially large node groups.
- *Minimal Oversight.* Typically, P2P applications require minimal oversight, because they do not rely on the continuous operation and availability of centralized servers. The loss of a server node will not disable the network.

Although our concern is wireless networking, the principles of P2P models apply to both wired and wireless networks. For wired networks, the music sharing application Napster is a well-known early P2P model, although in fact it was supported by centralized servers. In Napster's P2P model, users would register their available files at one of Napster's servers, where other users were able to search for content. The actual file transfer, however, occurred directly between the two users in question. Other more recent P2P applications, such as Freenet (www.freenet.net), operate entirely without such centralized server support.

As far as wireless networks, P2P models have been proposed and implemented for numerous uses; although somewhat dated, Akyildiz and Wang [6] provides a good introduction to their basic practical issues. Of particular interest are situational awareness applications, where groups of nodes share their knowledge about contextual variables, such as the location, mobility, density, and activity of groups of nodes. Many such applications (e.g., vehicular safety by sharing real-time mobility data) rely on all nodes gathering and forwarding continuous updates on variables of interest. These applications naturally conform to P2P models when contextual data is not sent to or obtained from centralized locations. Situational awareness applications can rely on networks of static or mobile nodes, with somewhat different characteristics in practice for each.

When nodes are static, data is typically gathered continuously by sensor nodes and directed to one or more data sinks. Most often, the data sinks may simply store the data (e.g., in remotely deployed environmental monitoring networks) or intermittently forward batches of data outside the sensor network (to respond to queries or provide real-time updates to network users). In some cases, however, the information might be used by the sensor nodes themselves (e.g., cross-checking observed variables as a means to detect observation errors or defective sensors).

For networks of mobile nodes, similar concerns arise. Mobile node groups may elect data sinks or interface with fixed infrastructure points. Compared to static topologies, the fluidity of such networks tends to make reliance on direct communication between nodes more advisable. Routing data to sinks can impose delays that degrade the value of continuous updates. For some applications, such as suggesting travel routes, moderate delays may be acceptable. Other applications, such as vehicular safety, may be extremely time sensitive. For those, pure P2P models are a natural and promising approach. Although P2P computing has numerous uses, it also presents a number of challenges not present in client–server models, especially for multi-hop wireless networking.

- *Free Riding*. P2P networking for multi-hop wireless networks requires that users forward traffic for others. Policies must be implemented to make free riding (using the P2P service without forwarding traffic for others) infeasible, essentially by making free access contingent on providing services to other nodes. Other nodes will either have their network access restricted or be charged for it.
- *Weaker Network Security*. Wireless networks are inherently less secure than wired networks, as any transmission can be overheard by nearby nodes. Depending on the nature of the communication protocol and messages, security concerns may include not only eavesdropping but even the proper operation of the communication model.
- *User Density Requirements*. In order to route packets quickly, multi-hop wireless routing requires sufficient node density to build complete end-to-end routes without waiting for nodes to move into mutual communication range. The variety of potential P2P transmission ranges can, of course, allow for varying densities over a route. However, as the maximum data rate is generally higher for shorter transmission ranges, higher density should allow greater P2P network throughput.
- *Encouraging Participation*. User density provides an upper bound on the number of potential P2P network participants. Not all of them are likely to be willing to forward network traffic for other users. Yet the presence of more willing users will tend to drive down the cost of establishing a given P2P network density.

12.3 COMMERCIAL VIABILITY

In order to apply P2P as a support mechanism for cellular networks, the system must be commercially viable. That is, the network company must recoup any expenses associated with providing and monitoring the P2P service. Although a great deal of research has been directed toward technical issues of P2P schemes, the relevant economic considerations have been largely ignored. The basic economic issues have been briefly covered in Reference [7], although little beyond recognizing that they exist as open research problems.

We advocate a voluntary participation model, where the network compensates cooperative P2P behaviors in the form of additional usage rights or reduced subscription charges. Extant research, such as Reference [8], suggests that wireless network subscribers are generally willing to modify their network usage in accordance with incentive programs, even if the incentives are small. The core economic question is what performance improvements can be obtained at what cost to the network provider. The P2P network can provide additional network capacity, but the end result of this extra capacity may take different forms depending on the network provider's strategic goals, such as the following.

- *More Data per User*. The most direct use of the additional capacity is to simply increase monthly data plans across the board for all users.
- *Increased Subscription Revenues*. A network might keep its existing subscription plans in place and use the additional capacity for promotions to attract new customers.
- *Driving Down Operating Costs*. Network providers will need to continuously add capacity in order to keep up with growing demand for wireless services. However, adopting P2P models can delay the need for such expansions, with consequent lower operating costs.

These basic principles of P2P system goals are easily understood but do not address the crucial question of participation pricing. That is, what compensation should be offered to attract the needed P2P participation density? Not so low as to fail to form a useful P2P network but not so high as to bring in many more nodes than those required. This efficiency problem becomes more difficult in dynamic wireless environments, where nodes may move rapidly, participate intermittently, and so on. Ideally, the network would also monitor the popularity of queries to drive data caching decisions. To achieve that, network providers need to continuously monitor node density, the popularity of queries, and the persistence of nodes in the network.

While the cellular network is congested, the P2P network will operate continuously in parallel with it. To ensure its continuous operation, incentives must be paid out to encourage cooperative behaviors among nodes. This gives rise to two fundamental questions that must be answered. First, what total incentives will be distributed to the P2P network participants; that is, how much will it cost to create and maintain it? Second, how will those incentives be distributed to the individual participants?

In general, we would assume an increasing asymptotic function for the total value of the set of participating nodes over any given time period. This implies a decreasing and approximately hyperbolic function for the value of each individual node's participation. Further, assuming nodes have variable willingness to participate, increasing the node density will reduce the cost of obtaining a given number of participants. The exact function parameters will, of course, vary across different networks but its general shape will be known. Importantly, however, *when node density is high in the cellular network, the costs of creating the P2P network will be relatively low*.

The next question is devising functions that describe this model. Here, there are analogues with congestion pricing in transportation. In those models, such as Nie [9], the goal is inducing a sufficient number of drivers to leave the roads to achieve a system optimal set of travel times. In a real-time environment, this can be achieved by requiring payment of tolls and, in principle, accruing additional payments to drivers that are not on the road. Applying this type of model to our proposed P2P network requires a slightly different framework.

Instead of charging drivers to remain on the roads, we would approach the problem strictly from an incentivization standpoint. That is, because network users already pay subscription fees, there is no need to levy additional charges. Instead, the P2P network functions to support the cellular network, and so participation (to enable gathering state information as well as actually forwarding packets for other nodes) would be rewarded. The incentives distributed to create the P2P network can be obtained as a prorated share of subscription fees for the congested periods. Because these incentives will be distributed according to individual nodes' assistance in creating the network, the P2P network can function entirely on a compensated voluntary basis.

We can formulate the costs of maintaining the P2P network as follows. Assume that n users wish to use the cellular network that can only support c users at the desired QoS. Assume also that each of the n users has paid k_i to use the cellular network during interval i. Then, the users above the cellular network's capacity can provide $(n - c)\, k_i$ over interval i. This quantity must be shared with the P2P network participants, whose share in interval i we indicate by $\omega_i, 0 < \omega_i < 1$. There will remain the quantity $(n - c)\, k_i \left(1 - \omega_i\right)$ to be kept by the cellular and P2P network providers.

Although the above model serves to illustrate how the P2P network will be funded, multiple practical refinements suggest themselves. First, the $(n - c)$ users may be insufficient (or indeed too numerous) to form a P2P network with adequate QoS. Second, the cellular network provider will wish to minimize ω_i within each interval in order to maximize profitability, while maintaining adequate QoS in both the P2P and cellular networks. Lastly, the size of the interval i is important; finer granularity will enable faster response to changing network conditions, potentially enhancing both profitability and QoS.

In order to set appropriate incentives in dynamic environments, the network provider must continuously gather smart pricing information (SPI). This information can then be leveraged to enhance routing performance (in terms of packet delivery ratio and latency). The key difference between the cellular network and the P2P network is that the cellular network can reasonably ensure access to any requested information, while the P2P network only allows access to information cached at participating nodes and then only probabilistically. So, it is imperative for the P2P network to not only provide complete routing paths but also to cache information likely to be requested by nearby nodes.

The P2P network requires estimates of the following.

- *Demand for Information*. It is a well-known fact that information is often requested by multiple users. In fact, this is essential to Google's caching functions, where more popular pages are kept more accessible. Applications with a geographic component (such as travel assistance) are especially likely to have coocated demand. On the other hand, it is also true that unique or highly infrequent requests should not be retained in the P2P network, as their likelihood of reuse is small.
- *Required Incentives*. We expect that users will largely rely on pricing assistance software to set their participation threshold. That is because few users would wish to constantly monitor such microtransactions. These automated assistants

will inform the network provider, at the very least, about which price offers are accepted and which are rejected. Combined with data about time and participation density, this information will support estimating participation as a function of incentive price.

- *Predictive Caching*. To place information near likely queries, it is necessary to gather information about the locations of query sources. Then, participating nodes can cache information in accordance with estimated future demand. This will both reduce the communication costs of searching for information (i.e., a smaller set of nodes can be searched) and of obtaining it (because routing paths will be shorter).

12.4 SELF-BALANCING SUPPLY/DEMAND

Our SBSD protocol is a paradigm for demand-driven information search and discovery. Unlike point-to-point routing protocols [such as Ad Hoc On-Demand Distance Vector (AODV) and Dynamic Source Routing (DSR)], which focus narrowly on constructing routing paths between known pairs of nodes, SBSD combines dynamic controlled flooding with predictive caching [10]. Our SBSD's predictive caching model and medium access control functions encompass the SPI requirements for our proposed P2P network model.

Flooding-based routing performs well in high mobility environments, where message destinations are very often unknown and any routes between them are ephemeral. Further, when multiple nodes request the same information, flooding can enable substantially faster responses to network users by widely caching popular information. This is a sensible approach when query sources do not know the query's destination and when multiple nodes often request the same information.

The basic SBSD model strives to flood query packets (and any matching responses found) in the vicinity of their sources. The depth of a given query's flooding depends on its popularity (how many other nodes have posted the same query) relative to others nearby. All else equal, the size of the flooding area is approximately a linear function of the frequency of demand. SBSD also controls the persistence of information according to a spatiotemporal utility function (given in the next paragraph). Packets closer in time and distance to their sources will be retained longer.

Each node will retain a queue of queries and responses to be retained until expiration (or perhaps discarded in the unlikely event of memory storage being overloaded). The items in this queue are ranked according to the utility function:

$$u = \frac{-a\sqrt{h}}{\sqrt[3]{f}},$$

where u is the item's *utility*, a is its *age* (the time since the item was created or most recently requested), h is the *hop count* (the number of hops traversed *en route* to its location), and f is its *frequency* (the number of unique users interested in the item, either posting the same query or receiving the same response).

The utility-based ranking is used to enable autonomous node decisions to create predictable group behaviors. Each node individually estimates the set of high utility items it will be able to forward during their remaining times to live. This estimation creates a forwarding threshold. Items with sufficiently high utility are forwarded until their utility declines below the threshold (primarily due to aging without compensating demand increases). The low utility items are simply cached until expiration, unless new demand for them boosts their frequency or resets their age. As most nearby copies of an item will have similar utility values, an item's flooding area tends to expand to a predictable flooding depth given a locally known arrival rate for new items.

Our utility function provides the important result of creating a linear relationship between an item's popular and its total contribution to network congestion during its lifetime [11]. That is, more popular items are granted a proportionally greater consumption of network resources. The above utility function approximates the negative ratio between the network congestion an item will ultimately be allowed to create and how much it has already created. That is, as an item's flooding proceeds, the transmissions of its copies add more to network congestion, which lowers the utility of creating new copies in the network. Further, this ratio's approximation is made without incurring the communication overhead to calculate either item separately.

We have developed a number of extensions to our SBSD protocol enabling more efficient delivery than provided through flooding. In a variety of ways, these extensions reduce the prevalence of redundant transmissions (where recipients have already received a copy of a transmitted packet). They can be applied to increase search depth for queries, propagation depth for response data, additional transmissions for more reliable delivery, or combinations thereof. The most recent of these extensions, Reference [12], considers geographically constrained flooding areas for vehicular networking in order to boost network throughput.

The SBSD model provides a means to obtain the following essential data, which may be delivered intermittently in batches or via sampling to the network provider.

- *What Information Is Popular?* Highly skewed demands for information are seen in many domains. Although somewhat dated, Reference [13] for example, analyzed a large search engine log (of a billion searches) and found that the most popular 25 queries accounted for 1.5% of search activity. Streaming video is another case, where the most popular videos can accumulate hundreds of millions of views while many videos struggle to reach a few thousand.
- *Where Is It Popular?* In terms of predictive caching for the P2P network, it is vital to know the corelation between popularity and location. If, for example, two users want the same information but are miles apart, multi-hop routing is probably not a good approach. However, many situational awareness applications tend to naturally provide users with information that is closely correlated with location. Vehicular safety and travel information are good examples, as many cars in an area will need the same information about nearby vehicles and travel routes.

- *Node Density, Topology, and Mobility*. In order to estimate the participation requirements for the P2P network, the network provider must know an approximate distribution of the node population. The alternative would be to learn this on the fly, which naturally adds an extra layer of delay to the P2P network. That is, if node density is insufficient, that fact has to be learned before incentives can be offered to boost density.
- *Willingness to Participate*. The set of nodes in a given area will be known to the cellular network. From their past participation history, the cellular network will be able to estimate what incentives are necessary to ensure sufficient cooperation to achieve performance targets. Such estimates may be further refined by considering node density and cellular network congestion

12.5 HYBRID NETWORK MODEL OVERVIEW

Our proposed model [14] will regulate load sharing in the hybrid cellular–P2P network. When the cellular network becomes congested, it will release a portion of its customers and assign them to the P2P network. The goal is that all users will receive better QoS than in the congested cellular network. The customers who remain in the cellular network will benefit from the reduced congestion level, while those added to the P2P network will receive QoS at least equal to what they would have obtained if they remained in the congested cellular network. As customers enter and leave the networks, they will need to be assigned to one network or the other. Ultimately, when the cellular network's congestion is sufficiently low, all customers in the P2P network will resume using the cellular network.

Our intention is to implement our model without disrupting the cellular network's existing methods of operation. Instead, our model will seamlessly add capacity to the cellular network through a system of voluntary, incentivized participation. Rather than attempt to dynamically adjust incentives to current needs, the cellular network will assign nodes to the P2P network with indeterminate future incentives. Implementing this model will require coordinating organizational, algorithmic, hardware, distributed accounting, and security elements.

12.5.1 Organization

We propose a voluntary participation regime. By default, customers will be candidates for participation in the P2P network. Although they may choose to never participate, they would then no longer be eligible to receive the participation incentives. When the cellular network actually becomes congested, and customers are assigned to the P2P network, they may also opt out through the user interface. But requiring users to explicitly opt out will make the default participation more prevalent.

The cellular network will need to determine the desired node density for the P2P network (to provide QoS comparable to that of the cellular network). This determination might be accomplished through a combination of initial context-based estimates and iterative convergence. In general terms, the trade-off is that increasing participation will initially improve performance (by providing greater end-to-end connectivity) but eventually cause congestion from the participants' own network traffic.

Smart selection of P2P network participants can maximize the benefit to the cellular network while minimizing the cost for the P2P network. For example, nodes that are both near the P2P infrastructure and are conducting data-intensive activities would be ideal candidates. This will improve the QoS of the cellular network for a while, because the routing path will be short, minimizing the incentives needed to deliver the requested content. More generally, network congestion within the area covered by a P2P network can be managed by a combination of shorter transmission ranges and smaller P2P networks. Both of those techniques will increase the network throughput per unit area.

12.5.2 Algorithms

Nodes participating in the P2P network will use our SBSD protocol [10] for information search and discovery. The SBSD protocol is built around demand-driven controlled flooding. It causes more popular information to be cached more widely, more persistently, and nearer to future query sources. Ultimately, this approach will reduce the costs of information search within the network; not all traffic will need to be routed all the way to the P2P network's infrastructure. Further, SBSD will automatically regulate the boundaries for competing traffic of nearby P2P infrastructure points without any centralized oversight.

12.5.3 Hardware

Ideally, the SBSD algorithms would be offered as chips or modules that can be easily swapped out of phones, although they might initially be offered as free or low cost software applications to encourage adoption. Each smartphone or other mobile computing device will contain a miniaturized module that provides the SBSD functionality (to allow participation in the P2P network) and tracks participation (for real-time conversion to incentives). This module will be a "black box" to prevent tampering with the dynamic pricing. As well, the mobile devices will require interoperable wireless technology components to enable interacting with the P2P network (various WiFi technologies are, of course, standard features on smartphones).

12.5.4 Distributed Accounting

The total of incentives to be distributed among the P2P network participants depends on the congestion level as well. That is, the P2P incentives are taken as a portion of the cellular network users' prorated subscription costs. The presence of heavier traffic in the cellular network indicates more intense demand for wireless services. If QoS can be maintained while adding additional capacity, greater revenues and profits can be realized. Further, this intense demand should translate into greater willingness to participate in the P2P network, which can then operate with a reduced incentivization plan.

The cellular network needs a mechanism for evaluating each node's participation as a fraction of P2P network activity. This has two applications. First, the P2P network must aggregate participation data as a baseline for prorating each node's incentives according the fraction of network traffic they forward. Second, in order to regulate activity within the P2P network, nodes will track their own generated traffic compared to that of their peers. Priority access will be granted to nodes that generate less traffic. Those nodes that generate more traffic will have their packets delayed the most if the P2P network becomes congested.

Because nodes may only participate in the P2P network for short periods, it may be desirable to offer an additional layer of incentives for more persistent participations. Alternatively, more redundant caching may be used to address intermittent node participation without sacrificing QoS (especially for continuous streaming applications). Yet, this redundancy also limits throughput, so a balance must be struck for more reliable delivery at reduced volume.

12.5.5 Network Security

Controls will be necessary to exclude noncustomers from the P2P network. When a node is assigned to the P2P network, it needs to have its registration confirmed. This will be achieved through direct interactions with nearby nodes; those with an appropriate encrypted ID (belonging to a participating cellular network) will have their traffic forwarded by other nodes and interact directly with the P2P network infrastructure. This functionality will be provided by the SBSD module.

12.6 INCENTIVE MODELING

Our proposed P2P incentive model was inspired by our earlier work in modeling congestion management and vehicular pollution in the transportation domain [15]. The similarities between transportation and wireless networking are clear: in both, individual users of a shared resource (whether a set of roads or the wireless communication medium) impose negative externalities on other users. The basis of our solution is a Pigouvian tax model in which users are charged according to their negative externalities in order to arrive at a socially optimal consumption of the shared resource.

In the transportation domain, traffic congestion is primarily relieved by charging tolls (to directly reduce the number of vehicles on the road, either by inducing them to take less congested routes, switch modes, or not drive during congested times). Related models have been developed to support searches and dynamic pricing for parking spaces, which can also suffer from congestion when demand exceeds supply [16]. Likewise, packet routing can be modeled as a set of data flows between sources and destinations and network congestion as a cost that rises with the volume of data the network handles.

For our P2P network, we wish to minimize network congestion primarily through autonomous node decisions. This is accomplished through a pair of models both of which offer users trade-offs between their individual QoS and their earned incentives. The first is a flow model that estimates the P2P network congestion generated by a given set of nodes and their generated network traffic. The second is a prioritization model, where some network traffic may be delayed during congested periods; this is analogous to charging tolls for faster travel routes or paying some vehicles to park during peak traffic. In both cases, users have the option to receive improved QoS by accepting a smaller quantity of incentives.

12.7 FLOW MODEL

Our flow model represents the P2P network as a graph of vertices and edges, in which data flows are managed to maintain congestion below an acceptable level determined by the cellular network. Each vertex represents a nonnull set of participating mobile devices and each edge represents a link between such groups of mobile devices. Depending on the granularity desired, nodes at a given vertex might all be in mutual communication range or be more dispersed. Edges represent direct communication links between at least one pair of nodes from adjacent vertices. Data flows over edges; when network traffic is heavier, the flows become slower.

Consider a graph network $G(V, A)$ with a set V of vertices and a set E of edges.

- For an edge $e \in E$, let x_e be its network traffic ($x \geq 0$) and $t(x_e)$ be the time required to traverse e, where we assume $t(x_e)$ is a nonnegative, strictly increasing, and convex function. Because our SBSD model uses controlled flooding, we adopt a first-order approximation of modeling network traffic as uniform across all edges in E.
- The planning horizon is M periods. For precise management, the planning horizon may be divided into an arbitrarily large number of periods but which will entail additional computational costs and more frequent information gathering about network topology and congestion. On the other hand, if peak network traffic is sufficiently uniform, it might be sufficient to simply set $M = 1$, that is, a single-period planning horizon.
- Let R and S, respectively, denote the sets of origin and destination nodes. The amount of network traffic to be routed from any $r \in R$ to any $s \in S$ is represented by q_{rs}.
- The set of paths connecting an origin–destination pair rs is denoted by K_{rs}, and during a period m, the flow on path $k \forall k \in K_{rs}$ is represented by f_{km}^{rs}.
- The P2P network participation incentives are equally shared among all participating nodes and are depleted as nodes contribute to network congestion. Nodes are permitted to purchase additional congestion rights from other nodes. Let the unit price of buying additional congestion rights for period m through the competitive bidding process be ρ_m multiplied by a transaction cost factor τ.
- Let $\varepsilon(m, x_e)$ denote the congestion (i.e., the delay on all other network traffic) inflicted by network traffic x on edge e in period $\in M$. Following our assumption

that $t(x_e)$ increases with x, $\varepsilon\left(m,x_e\right)$ is a nonnegative, strictly increasing, and convex function.

- Let C_m be the maximum network congestion permitted in a period m; the maximum congestion permitted over the entire planning horizon is $\sum_{m=1}^{M} C_m$.
- Each node in the P2P network is granted the right to inflict an average amount of congestion by injecting new network traffic. Let $A\left(\sum_{m=1}^{M} C_m\right)$ be the cost of assigning the $\sum_{m=1}^{M} C_m$ congestion rights over the planning horizon. As this communication will further increase network traffic, the function $A\left(\sum_{m=1}^{M} C_m\right)$ is a nonnegative, strictly increasing, and convex.

We formulate the equilibrium problem with network congestion costs as

Input Data. $C(\cdot), \varepsilon_n(\cdot), t(\cdot), \rho_m, \tau, q_{rs}, m,$ where $e \in E, r \in R, s \in S,$ and $M = 1, 2, \ldots, m.$

Decision Variables. $M_m, x_e, f_{km}^{rs},$ where $k \in K_{rs}, r \in R, s \in S, e \in E,$ and $M = 1, 2, \ldots, m.$

$$\min Z\left(x, C_m, f_{km}^{rs}\right) = \sum_{m=1}^{M} \sum_{e} \left(\int_0^{x_e} t\left(x_e\right) dt + x_e\left(\rho_m \tau\right) \right) + A\left(\sum_{m=1}^{M} C_m \right)$$

subject to:

$$\sum_{m=1}^{M} \sum_{k} f_{km}^{rs} = q_{rs} \quad r \in R,\ s \in S$$

$$\sum_{r} \sum_{s} \sum_{k} f_{km}^{rs} \delta_{e,k,m}^{rs} = x_e \quad e \in E,\ k \in K,\ r \in R,\ s \in S,\ m \in M$$

$$\delta \begin{array}{l} = 1 \text{ if } e \in K \\ = 0 \text{ otherwise} \end{array}$$

$$\varepsilon\left(m, x_e\right) \le C_m \quad e \in E,\ m \in M$$

$$f_{km}^{rs} \ge 0 \quad k \in K_{rs},\ r \in R,\ s \in S,\ x_e \ge 0,\ e \in E,\ m \in M$$

The above model extends [17], a single-period pollution credit trading model with a known total pollution cap M_1 and no transaction costs. Setting our model parameters as $m = 1$, $\varepsilon\left(x_1\right) = \sum_e \rho_m x_e$, and $A\left(\sum_{m=1}^{M} C_m\right) = \tau = 0$, and using a known constant C_m, our model reduces to the one from [17]. Our model is feasible, similar to the simpler classical traffic assignment problem, and contains features needed to consider the broader impacts of network congestion policies, which are needed to guard against unintended consequences of policy decisions. The following are few examples.

Variable Period and Horizon Caps. Letting $C_m(m = 1, 2, \ldots, M)$ change allows analyzing the consequences of dynamic congestion caps, such as for peak and off-peak loads. The overall cap $\sum_{m=1}^{M} C_m$ may also be changed. During many periods, no monitoring may be needed, for example, for light network traffic typical of late evening and early morning. Variable horizon caps allow considering transitions across temporal monitoring boundaries.

Realistic Congestion Levels. Congestion is not a simple linear function of node density but is affected by node behaviors, network topology, and sources of interference from outside the P2P network. The pollution constraint represented by $\varepsilon(m, x_e)$ is a nonnegative, strictly increasing, and convex function; it does not assume linearity.

Moving Beyond Simple Linear Relationships. Including congestion rights $A(C_1, C_2, \ldots, C_m)$ in the objective function lets us explicitly incorporate the effect of congestion rights on optimal flows in our analysis. This is particularly necessary for systems with many monitoring points, which increases system complexity.

Cellular Network Involvement. The model determines the optimal congestion cap but the cellular network might select a different congestion cap to achieve other goals, such as operating the P2P network to reach particular performance targets. The model also informs system planners what costs are required to impose a range of potential congestion cap distributions.

Incentive models can yield unintended consequences. As shown in Reference [18], price discrimination across different types of network traffic sometimes benefits the network provider at the expense of customers, sometimes the opposite, and in some cases benefits both. To better understand the likely consequences of network policy decisions, our flow model might be further developed to consider the following factors more deeply.

Balancing Congestion. To balance congestion across cellular and P2P components of the hybrid network, the congestion target for the P2P network must match that of the cellular network. However, the P2P network has the additional constraint of requiring a minimum node density to provide continuous routing paths.

Graph Modeling. The grouping of P2P network nodes into vertices is a necessary abstraction for our model; treating each node as a vertex would continuously require an enormous amount of communication to track network topology. However, the abstraction leads to difficult modeling questions. How many nodes should a vertex contain? What area should those nodes occupy? How many hops should traverse a vertex? One solution is for the cellular network may specifically select nodes that are densely located on or near major roads. This will simultaneously address P2P node density concerns while conforming to the link-based transportation model assumptions.

Density Modeling. In transportation models, vehicle density converts almost directly to traffic density. There might be some large trucks and buses that take up additional space but, typically, they compose a small fraction of total vehicular traffic. For wireless networking, however, it is important to distinguish between individual nodes. Individual node contributions to traffic can vary far greater than vehicle sizes (e.g., compare a customer watching streaming video vs one intermittently loading web pages).

Monitored Variables. Pollution in the transportation model corresponds to the network congestion produced by each node. The impact of congestion will be felt differently in the cellular and P2P networks. That is, in the cellular network, assuming a constant per-node network traffic quantity, congestion rises linearly with the node population governed by a given cell tower. In the P2P network, however, the benefits of node population initially outweigh the likely costs of congestion.

Predictive Congestion Cap. As congestion in the P2P network increases, some network traffic will simply expire before it reaches its destination. In terms of congestion management, it would be better for nodes to post their queries in subsequent periods. For such cases, a sensible policy would be to enforce delays on additional traffic for which congestion rights cannot be bought from other network participants.

Intermittent Node Participation. To improve the model's accuracy, the length of each period should be less than the typical duration of P2P participation. This will allow better assessments of incentive requirements and needs for new participations but will require additional tracking of nodes. Given those facts, the cellular network might opt instead to provide additional incentives for nodes willing to participate across multiple periods.

12.8 PRIORITIZATION MODEL

We have also formulated low overhead approaches to regulating P2P network traffic in terms of autonomous forwarding decisions by individual nodes. That is, how do nodes prioritize network traffic without exacerbating congestion by adding many new rounds of communication? When the P2P network is congested, the participation incentives can be adjusted according to the willingness of nodes to accept delays on their communications. While maintaining the same total incentives for the entire P2P network, nodes that accept such delays will obtain a share of participation incentives from other nodes.

We enable nodes to rank traffic through another utility model, this one reflecting the external costs of network congestion. Broadly, nodes accumulate incentives by forwarding network traffic for other nodes and may consume incentives by injecting their own traffic into the P2P network. Like our previous model for route assignments, this model is also inspired from the transportation domain with the goal of jointly capping vehicular pollution and congestion [16]. In that paper, models were presented for cases where tradable pollution credits were either divisible or indivisible, that is, a seller's supply of pollution credits would be sold either in whole or not at all. For example, a driver who elected to delay travel until congestion was relieved might prefer to sell his/her entire stock of credits. In this networking domain, it may likewise happen that transactions involve some of all of the node's participation incentives.

12.8.1 Divisible Incentives

In this case, each P2P network user will choose a priority level from a range of options offered by the network provider. According to the priority level chosen, a node's traffic will be more or less delayed, which will transfer credits from the high priority users to the low priority ones. As needed, nodes may purchase quantities of credits from other users or, in the absence of nearby sellers, from the P2P network provider.

Transactions are not actively conducted by individual nodes, nor are they necessarily chosen to maximize incentive accumulation for particular nodes. Instead, our model will rank potential transactions with a combination of individual node profit and awareness of a transaction's effect on network congestion, according to the utility function for a trade g between a seller i and a buyer j as follows:

$$g_{ij}(p,\alpha,c(\Delta \mathrm{E})) = \alpha_{ij}\left[\gamma_{ij}\left(c\left(\Delta E_i\right) - \overline{c\left(\Delta E_i\right)}\right) + \left(1-\gamma_{ij}\right)\left(\overline{p_j} - p_i\right)\theta\left(\overline{p_j} - p_i\right)\right.$$
$$\left.\theta\left(c\left(\Delta E_i\right) - \overline{c(\Delta E_i)}\right)\right],$$

where $\theta(z)$ is a step function (equal to 1 for $z > 0$ and 0 otherwise), ΔE is the reduction in network congestion from accepting a forwarding delay, and γ_{ij} is a parameter expressing buyer preferences (for a trade between a seller i and a buyer j), $0 < \gamma_{ij} < 1$. A buyer with $\gamma_{ij} = 0$ will prefer the lowest price seller from among those whose congestion costs exceed a floor $\overline{c\left(\Delta E_i\right)}$. In other words, buyer j prefers matches with high consumption of P2P network capacity. Without the loss of generality, the floor can be set to 0. A buyer with $\gamma_{ij} = 1$ will select the seller with the highest congestion contribution, so long as the seller i ceases initiating any new communications within the P2P network (which it no longer could, having sold all its incentives to buyer j).

The model makes several assumptions. Notably, it explicitly considers the possibility that a seller may sell all his/her credits. This may not actually happen in the above formulation; the sale may be partial, in which case, buyer j is simply overestimating the value of its matching with seller i. As a result, this may lead to a suboptimal solution or unfeasibility. A compensating parameter $\alpha_{ij} = \frac{d_j}{s_i}$ that encourages transactions that relieve network congestion is introduced to favor sellers with smaller supplies of pollution credits to sell.

Utility is highest when either (i) P2P network usage pricing maximizes the price differential between buyer and seller, that is, a lower selling price increases utility, or (ii) the incentive depletion rate is high, making network usage more expensive, then demand for incentives will rise and sellers will ask higher prices for theirs. These two cases suggest the existence of price equilibria between sellers and buyers. Setting arbitrarily high seller prices and low buyer prices will not promote trading. The utility function, therefore, encapsulates the market incentives for network users to either participate solely by assisting peers or use the network at shorter more expensive intervals by purchasing incentives at increasing prices.

12.8.2 Indivisible Incentives

When most transactions involve fully depleting a user's incentives, the assumption of indivisibility is a reasonable approximation. For example, many users in the P2P network might elect to participate only in order to earn the participation incentives. In that case, they might sell their credits immediately upon joining (or leaving) the P2P network. Alternatively, there might be a small number of very heavy network users who wish to buy large quantities of incentives from their peers. The case of indivisible incentives uses a slightly different utility function than the one given earlier for divisible incentives:

$$g_{ij}(p, \alpha, c(\Delta \mathrm{E})) = \alpha_{ij}\left[\gamma_{ij}\left(c\left(\Delta E_i\right) - \overline{c\left(\Delta E_i\right)}\right)\theta\left(\left(d_j + 1\right) - s_i\right)\right.$$
$$\left. + \left(\gamma_{ij}\right)\left(\overline{p_j} - p_i\right)\theta\left(\overline{p_j} - p_i\right)\right]\theta\left(c(\Delta E_i) - \overline{c\left(\Delta E_i\right)}\right),$$

where $g_{ij}(p, \alpha, c(\Delta \mathrm{E}))$ differs only in step function $\theta((d_j + 1) - s_i)$. Here the utility is higher if the transaction results in a seller surrendering all its incentives and thus ceasing to contribute to network congestion.

In both the divisible and indivisible models, the objective of transactions is to maximize the system utility. As the utility expresses the best matching for the whole system in terms of achieving network congestion relief, it must adjust prices in order to achieve system goals. That is, transactions must optimize congestion relief for all the partitions $\mathfrak{P}(S)$ of set S of sellers. For each partition $\mathfrak{P}(S)$, the best match between the elements of partition $\mathfrak{P}(S)$ (which are disjoint subsets of S covering S) and the buyers must be found so the total utility is optimal.

The aggregate utility of a set of matches can be computed in several ways and will determine whether the individual buyers and sellers are optimally matched. The complexity of the problem is NP-hard in both the divisible and indivisible formulations [19]. Note that not all partitions need to be considered. A partition is said to be feasible if at least one element can be matched to one buyer. Further, let E_i be the congestion contributed by seller i. Assume seller i has been matched to a buyer j. In other words, there exists a unique element of $\mathfrak{P}(S)$, that has been matched to buyer j and contains seller i. We say that a partition $\mathfrak{P}(S)$ is relevant if the set of transactions does not result in exceeding the system-wide congestion cap determined to provide adequate QoS.

12.9 CONCLUSION

Currently, both cellular and WiFi networks are ubiquitous in urban areas. The WiFi networks may be underutilized during peak cellular network usage, because of difficulties in joining or remaining in a WiFi network. This is especially true for highly mobile network users for which direct communication with infrastructure is infeasible. The problem is creating a framework for enabling the cellular network to shift some of its network load to a set of WiFi networks, thereby relieving congestion in the cellular network while offering adequate QoS to the nodes removed from it.

In this chapter, we have presented our model for a hybrid cellular–P2P network, which was inspired by our earlier work in modeling congestion management and vehicular pollution in the transportation domain. The similarities between transportation and wireless networking are clear: in both, individual users of a shared resource (whether a set of roads or the wireless communication medium) impose negative externalities on other users. The basis of our solution is a Pigouvian tax model, in which users are charged according to their negative externalities in order to arrive at a socially optimal consumption of the shared resource.

During times when the cellular network is congested, a set of P2P networks will be created to increase the entire system's capacity. Although the cellular network will select nodes to join the P2P network, they will have the option to decline (and remain in the congested cellular network). However, remaining nodes that agree to participate in the P2P network will be incentivized for their assistance in forwarding other nodes' network traffic. These incentives will be paid out from subscription fees, thus allowing P2P network participants to partially recoup their costs of cellular network usage in return for expanding its capacity.

REFERENCES

1. PricewaterhouseCoopers LLC, Real time: The growing demand for data—2012 North American wireless industry survey, 2013.
2. Federal Communications Commission, Mobile broadband: the benefits of additional spectrum, October 2010.
3. S. Sen, C. Joe-Wong, S. Ha, and M. Chiang, "A survey of smart data pricing: Past proposals, current plans, and future trends," *ACM Computing Surveys*, 46(2), 2014.
4. S. Ha, S. Sen, C. Joe-Wong, Y. Im, and M. Chiang, "TUBE: time-dependent pricing for mobile data," *Proceedings of SIGCOMM'12*, Helsinki, Finland, pp. 247–258, Aug. 2012.
5. P. Loiseau, G. Schwartz, J. Musacchio, S. Amin, and S. Sastry, "Incentive Mechanisms for Internet Congestion Management: Fixed-Budget Rebate Versus Time-of-Day Pricing," *IEEE/ACM Transactions on Networks*, 22(2), 2013, 647–661.
6. I. Akyildiz and X. Wang, "A survey on wireless mesh networks," *IEEE Radio Communications*, 2005, s23–s30.
7. Z. Wan, "Commercial issues in hybrid cellular ad-hoc networks," *IEEE 2011 Conference on Management of E-commerce and E-governance*, Shanghai, China, 2011.
8. J. Dyaberi, B. Parsons, K. Kannan, V. Pai, Y. Chen, R. Jana, D. Stern, A. Varshavsky, and B. Wei, "Managing cellular congestion using incentives," *IEEE Communications Magazine*, 50(11), 2012, 100–107.
9. Y. Nie, "Transaction costs and tradable mobility credits," *Transportation Research Part B: Methodological*, 46(1), 2012, 189–203.
10. A. Ouksel and D. Lundquist, "Demand-driven publish/subscribe in mobile environments," *Wireless Networks*,16(8), 2010, 2237–2261.
11. D. Lundquist, *A Context-Aware Paradigm for Information Discovery and Dissemination in Mobile Environments*, Advisor: Aris Ouksel, University of Illinois at Chicago, Chicago, IL, 2011.

12. A. Ouksel and D. Lundquist, "A context-aware cross-layer broadcast model for ad hoc networks," *Personal and Ubiquitous Computing*, 18(4), 2014, 851–864.
13. C. Silverstein, H. Marais, M. Henzinger, and M. Moricz, "Analysis of a very large web search engine query log," *ACM Special Interest Group on Information Retrieval*, 33(1), 1999, 6–12.
14. A. Ouksel and D. Lundquist, "DF100: CACL: A context-aware cross-layer broadcast scheduling scheme," Provisional UIC patent, 2012.
15. A. M. Ouksel and T. Lee, Public Policy Implications of Cap and Trade of Pollution Credits and Switching of Travel Modes Incentives, *University of Illinois: CISORS Technical Report #5*, 2011.
16. A. Ouksel, Pollution Measurements and Pollution Maps, *University of Illinois: CISORS Technical Report #12*, 2011.
17. H. Yang and X. Wang, "Managing network mobility with tradable credits," *Transportation Research Part B: Methodological*, 45(3), 2011, 580–594.
18. A. Lahiri, R. Dewan, and M. Freimer, "Pricing of Wireless Services: Service Pricing vs. Traffic Pricing," *Information Systems Research*, 24(2), 2013, 418–435.
19. C. Papadimitriou and K. Steiglitz, *Combinatorial Optimization: Algorithms and Complexity*, Prentice Hall, Inc., New Jersey, 1982.

13 To Tax or To Subsidize: The Economics of User-Generated Content Platforms

SHAOLEI REN and MIHAELA VAN DER SCHAAR

13.1 INTRODUCTION

Enabled by ubiquitous broadband connectivity and seamlessly accessible wireless connections, user-generated content platforms have witnessed in the past few years an explosive growth, allowing everyone to conveniently publish information online and share their knowledge, opinions, experiences, and so on with the rest of the world. Every day, millions of people "tweet" on Twitter, update their status on Facebook, ask and answer questions on Yahoo! Answers, and share their videos on YouTube. As user-generated content platforms are becoming an integral part of our lives, both Internet users and platform owners are eager to see the continuing growth of such platforms, which are nonetheless largely hindered by a number of obstacles, notably low quality content and lack of revenue sources. Typically, users can view the content for free while they post content on user-generated content platforms voluntarily. Thus, users often have little incentive to devote their efforts into improving content quality, resulting in low quality content published on the platforms. To tackle this, various mechanisms have been proposed to promote high quality content produced by users. For example, a common approach is to eliminate or hide low quality content as adopted by popular web sites such as Yahoo! Answers [1]. While incentivizing high quality content contributions on user-generated content platforms still remains an active research area (see References 1 and 2 and references therein), turning the user-generated content into profit has become increasingly important for platform owners, as advertising accounts for a major (and possibly the only) revenue source while the operational cost (e.g., bandwidth, energy, and marketing) for managing

Smart Data Pricing, First Edition. Edited by Soumya Sen, Carlee Joe-Wong, Sangtae Ha, and Mung Chiang.

user-generated content platforms is skyrocketing. Even though sophisticated algorithms may be employed to increase the advertising revenue (e.g., displaying more relevant advertisement or use auction mechanisms for selling advertisement space), the continuously increasing operational cost still catapults the exploration of alternative revenues as one of the key challenges for user-generated content platform owners, thereby motivating our study of the economics of user-generated content platforms from the perspective of profit maximization.

On a user-generated content platform as illustrated in Figure 13.1, the platform owner is referred to as an *intermediary*, while the users can be classified as content producers and content viewers, although a user can be both a producer and a viewer. While content producers post content on the platform voluntarily without any obligation, content producers still receive *benefits* by publishing their content: millions of users engaging daily in Internet activities such as blogs, for which they receive no monetary rewards, suggest that content producers may simply derive satisfaction from attracting content viewers' attention (see, e.g., References 1, 3, and 4). In addition to social satisfaction, the intermediary may also provide economic incentives by subsidizing (i.e., paying) content producers for their contribution. The logic behind "subsidizing" is that the intermediary gives away some of its advertising revenue to content producers in the hope that the total advertising revenue may increase because of the increased amount of content available on the platform. Opposite to subsidizing, the intermediary may tax (i.e., charge) content producers for using the platform's service (e.g., storage space and bandwidth): if content producers have intrinsic social incentives to produce content, taxing content producers will give the intermediary an additional revenue source while still attracting some content producers, although the intermediary may suffer from reduced advertising revenue because of decreased user traffic resulting from less content production. In general, subsidizing content producers can be considered as a reward for providing content, whereas taxing content producers can be considered as a usage fee for utilizing the intermediary's resources. While, in practice, "subsidizing content producers" can be observed more often (e.g., YouTube Partner and Squidoo) than "taxing content producers" (e.g., Google Picasa, which taxes its users for storage exceeding the free quota), there is no simple answer

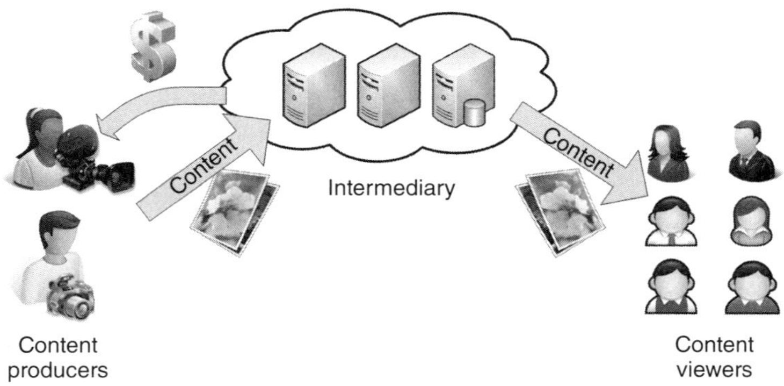

Figure 13.1 Illustration of user-generated content platform.

to the question of which strategy maximizes the intermediary's profit, which is crucial for the long-term growth of user-generated content platforms.

Concluding whether subsidizing or taxing content producers maximizes the intermediary's profit requires an appropriate model as the foundation. Naturally, with both content producers and content viewers as participants, user-generated content platform can be modeled as a two-sided market, where two user groups (i.e., content producers and content viewers) interact and provide each other with network benefits [5]. Besides user-generated content platforms, two-sided market can also model a variety of other markets such as a broadband communications market with content producers and end users [6] and payment card industry with merchants and customers [7]. Unlike other two-sided markets, however, user-generated content platforms have the following characteristics that have not been adequately addressed by the existing research. First, content producers compete for the content viewers' attention, that is, intragroup negative externalities, which have been neglected by most existing research on two-sided markets (see References 5 and 8 for a survey). Second, content viewers tend to have *love for variety*, that is, intergroup positive externalities, which have been incorporated by the existing literature [9] but not well suited to user-generated content platforms. Last but not least, user-generated content exhibits a diverse content quality and substitutability, that is, different content often has different qualities but may substitute each other to a certain extent from the content viewers' perspective. In what follows, in order to provide a formal analysis on when subsidizing or taxing content producers is profit maximizing for the intermediary, we provide a new model for user-generated content platform, which captures all the above three characteristics. As subsidizing content producers per content view is a common practice in the Internet industry (e.g., YouTube Partner) while not taxing users for viewing content is also a common practice (e.g., YouTube and Yahoo! Answers). we focus on a class of payment schemes in which the intermediary subsidizes or taxes content producers on a basis of per content view, while it provides the service for free to content viewers. Our analysis provides a formal guidance for the intermediary to decide its optimal payment to the content producers for profit maximization. In particular, we recommend that the intermediary should subsidize the content producers under the following circumstances: (i) there are few content viewers; (ii) content production cost is high; (iii) content producers do not receive strong social satisfaction; or (iv) there are more popular user-generated content platforms where the content viewers can go and view content.

The rest of this chapter is organized as follows. The model is described in Section 13.2. In Section 13.3, we analyze the problems of content viewers, content producers, and the intermediary. In Section 13.4, we outline how our analysis can be extended to the case of heterogeneous production costs. Finally, concluding remarks are offered in Section 13.5.

13.2 MODEL

We consider an online user-generated content platform owned by a profit-maximizing intermediary. Content on the platform is produced and viewed by individual users.

Our model is a three-stage game that is played by the intermediary, content producers, and content viewers in the following order.

- *Stage 1*. The intermediary sets a payment rate per content view, paid to content producers.
- *Stage 2*. Given the payment rate chosen by the intermediary, each content producer chooses whether or not to produce content on the platform.
- *Stage 3*. Given available content, content viewers, consolidated as a representative content viewer, allocate their total content views over available content.

While a single user can be a content producer and a content viewer at the same time, it is assumed in our analysis that the decision making of a user as a content producer is separable from that as a content viewer. In the following text, we describe the problems of the intermediary, content producers, and content viewers in detail.

13.2.1 Intermediary

It is well known that advertising is one of the most prevailing revenue sources in the Internet industry, especially for online content platforms such as YouTube and Yahoo! Answers. Hence, we consider a scenario in which the intermediary monetizes its content platform by displaying contextual advertisement to content viewers. Although a platform may use different bases (e.g., pay per click, pay per sale, and pay per impression) to charging advertisers, the advertising revenue is in general increases with the number of times that content with advertisement is viewed (or content views). For simplicity, we assume that the intermediary's advertising revenue is proportional to the total number of content views in the platform, as in Reference 10.

The intermediary chooses the amount of money, denoted by θ, which it subsidizes a content producer for each time the content producer's content is viewed. We refer to θ as the *payment rate* (to content producers per content view). We allow θ to be any real number. The intermediary subsidizes content producers if $\theta > 0$, while it taxes them if $\theta < 0$. Let b be the intermediary's profit per content view without accounting for payment to content producers and b can be interpreted as advertising revenue minus operation cost per content view. Note that b can be negative if the operation cost is larger than the advertising revenue. We assume that b is an exogenously given as a constant independent of the total number of content views, while the intermediary may increase its advertising revenue per content view by developing algorithms for displaying relevant advertisement or auction mechanisms for selling advertisement space (which is outside of the scope of this chapter). Let $\overline{x}(\theta)$ be the total number of content views, which is determined by the decisions of users, that the intermediary obtains when it chooses θ. Neglecting fixed operation cost, we can express the intermediary's profit as

$$\Pi(\theta) = (b - \theta)\overline{x}(\theta). \tag{13.1}$$

The intermediary's problem is to maximize its profit by choosing a payment rate.

13.2.2 Content Producers

As evidenced by the exploding number of YouTube users, a popular user-generated content platform can attract a huge number of content producers. To capture this fact, we consider a continuum of content producers of mass one as in prior work [11–13]. We index content producers by i (and sometimes j without ambiguity), where i is uniformly distributed over the interval [0, 1]. Content producer i can produce content of quality $q_i \geq 0$ while incurring production cost $c > 0$. The quality of content is represented by a nonnegative scalar, and we treat it as an internal feature of content (e.g., how fun or informative content is). Also, the production cost is the same for all content producers, and this assumption will be relaxed in Section 13.4. Content producers produce differentiated content, or in other words, no two content producers can produce identical content. We sometimes refer to content produced by content producer i as content i.

Given the payment rate chosen by the intermediary, content producers make decisions regarding whether to produce content or not. Once a content producer produces content, it posts the content on the intermediary's platform. In other words, the intermediary's platform is the only platform on which the content producers can provide their content. We assume that a user cannot provide content produced by others, for example, because of copyright restrictions. That is, content i can be posted on the platform only by content producer i. We represent the decision of content producer i by a binary variable y_i, which takes value 0 if content producer i chooses not to produce and 1 otherwise. The decisions of all the content producers are summarized in $y = (y_i)_{i \in [0,1]} \in \{0, 1\}^{[0,1]}$. Note that y determines content available on the platform.

There are two major types of benefits that a content producer obtains by posting its content on the platform. One is payment from the intermediary, which can also be interpreted as a cost when the payment rate is negative, while the other is social satisfaction. Millions of users engaging daily in Internet activities such as blogs, for which they receive no monetary rewards, suggest that content producers may simply derive satisfaction from attracting content viewers' attention (see, e.g., References 1, 3, and 4). We use content views to quantify the amount of received attention and assume that a content producer's satisfaction is proportional to the number of content views of its content. Let $s > 0$ be the social benefit, measured in a monetary unit, per content view that a content producer derives from content viewers' attention. Let $x_i(y)$ be the number of content views that content i attracts, given available content on the platform determined by y. Note that $(x_i(y))_{i \in [0,1]}$ is determined by content viewers. The payoff function of content producer i is given by

$$\pi_i(\theta, y) = \begin{cases} (\theta + s)x_i(y) - c & \text{if } y_i = 1, \\ 0 & \text{if } y_i = 0. \end{cases} \tag{13.2}$$

Content producers make production decisions simultaneously to maximize their own payoffs given the payment rate chosen by the intermediary.

Remark: In our model, each content producer can produce only one piece of content. However, if there is a content producer who can produce up to $m \geq 1$ pieces of content, it can be treated as m separate content producers in our model as long as the cost of producing any $\tilde{m} \leq m$ pieces is $\tilde{m}c$ (i.e., constant returns to scale, as assumed in Reference 9).

13.2.3 Content Viewers

We can expect that content viewers have diverse preferences toward content. In order to study their aggregate content viewing behavior conveniently, we adopt the widely used representative agent model [14]. That is, we analyze the optimal decision of a representative content viewer to determine the content views that each piece of available content receives. The representative content viewer has a fixed number of content views to be allocated to available content, which is denoted by $T > 0$. Note that T can be interpreted as the size of the representative content viewer or the market size on the viewer side. Besides the intermediary's platform, there are other platforms that offer (differentiated) content. For convenience, we assume that "outside" content has a fixed aggregate quality $q_a > 0$. In Section 13.3.2, we illustrate how the aggregate quality can be derived from a distribution of individual qualities. The assumption that q_a is independent of y can be justified by noting that there are many content platforms on the Internet and thus changes on a single content platform have negligible impacts on the other platforms.

Let x_i be the number of content views of content i for $i \in [0, 1]$ and x_a be the total number of content views of outside content (i.e., content on the other platforms). We use $x = ((x_i)_{i\in[0,1]}, x_a) \in \mathbb{R}_+^{[0,1]} \times \mathbb{R}_+$ to denote a content view allocation of the representative content viewer. Let $U(x)$ be the utility of the representative content viewer when its content view allocation is x. Then the representative content viewer's problem, given the decisions of the content producers, y, can be written as

$$\max_x U(x) \tag{13.3}$$

$$\text{s.t.,} \quad \int_0^1 x_i di + x_a = T, \tag{13.4}$$

$$x_i \geq 0 \text{ for all } i \in [0, 1], x_a \geq 0, \tag{13.5}$$

$$x_i = 0 \text{ for all } i \text{ such that } y_i = 0. \tag{13.6}$$

The constraint in Eq. (13.4) requires that the total number of content views be equal to T. The constraints in Eq. (13.5) are nonnegativity constraints for numbers of content views. The constraints in Eq. (13.6) impose that the content viewer cannot allocate a positive number of content views to unavailable content.

13.3 PROFIT MAXIMIZATION ON USER-GENERATED CONTENT PLATFORMS

13.3.1 Definition of Equilibrium

In this chapter, we analyze the optimal decision making of the intermediary, the content producers, and the representative content viewer. In Section 13.2, we have modeled the interaction among them as a three-stage game (with perfect information). Hence, we adopt the solution concept of subgame perfect equilibrium. The strategy of the intermediary is the payment rate θ, that of content producer i is the production decision $y_i(\theta)$ as a function of θ, and that of the content viewer is the content view allocation $x(\theta, y)$ as a function of (θ, y). As the payment rate θ does not affect the represent content viewer's problem Eqs. (13.3)–(13.6) directly, we restrict attention to the strategies of the content viewer that depend only on y, writing its strategy as $x(y)$. In the following, we define the equilibrium of the game played by the intermediary, the content producers, and the representative content viewer.

Definition 13.1 $(\theta^*, y^*(\theta), x^*(y))$ *is an* equilibrium *if*
(i) $x^*(y)$ *is an optimal solution to the content viewer's problem* (13.3)–(13.6), *given* y,
(ii) for each content producer $i \in [0, 1]$, $y_i^*(\theta)$ *is an optimal production decision, given* θ, $x_i^*(y)$, *and* $(y_j^*(\theta))_{j \neq i}$, *that is,*

$$(\theta + s)x_i^*(y^*(\theta)) - c \geq 0 \quad \textit{if } y_i^*(\theta) = 1 \tag{13.7}$$

and

$$(\theta + s)x_i^*(1, y_{-i}^*(\theta)) - c \leq 0 \quad \textit{if } y_i^*(\theta) = 0, \tag{13.8}$$

where $(1, y_{-i}^*(\theta))$ *is the production decision profile equal to* $y^*(\theta)$ *except that* $y_i = 1$, *and*
(iii) θ^* *is an optimal payment rate for the intermediary given* $y^*(\theta)$ *and* $x^*(y)$, *that is,*

$$(b - \theta^*)\overline{x}(\theta^*) \geq (b - \theta)\overline{x}(\theta) \quad \textit{for all } \theta \in \mathbb{R}, \tag{13.9}$$

where $\overline{x}(\theta) = \int_0^1 x_i^*(y^*(\theta))di$.

In the remainder of this chapter, we characterize the equilibrium by analyzing each agent's problem in detail.

13.3.2 Optimal Content Viewing

For analytical tractability, we impose the following assumptions on the utility function U of the representative content viewer.

Assumption 13.1 $U(x) = U(x')$ *for all* x *and* x' *such that* $x_i = x_i'$ *for almost all* $i \in [0, 1]$ *(with respect to the Lebesgue measure) and* $x_a = x_a'$.

Assumption 13.2 *U is continuous on its domain and twice continuously differentiable on the interior of its domain. It is additively separable in its arguments (i.e., all of its cross partial derivatives are zero).*

Assumption 13.3 *(i) For all $i \in [0, 1]$ such that $q_i = 0$, $\partial U/\partial x_i = 0$ for all $x_i > 0$.*
(ii) For all $i \in [0, 1]$ such that $q_i > 0$, $\partial U/\partial x_i > 0$ and $\partial^2 U/\partial x_i^2 < 0$ for all $x_i > 0$, while $\partial U/\partial x_i \to +\infty$ as $x_i \to 0$ and $\partial U/\partial x_i \to 0$ as $x_i \to +\infty$.
(iii) $\partial U/\partial x_a > 0$ and $\partial^2 U/\partial x_a^2 < 0$ for all $x_a > 0$, while $\partial U/\partial x_a \to +\infty$ as $x_a \to 0$ and $\partial U/\partial x_a \to 0$ as $x_a \to +\infty$.

Assumption 13.4 *$\partial U/\partial x_i$ is continuous and increasing in q_i for all $i \in [0, 1]$ and $\partial U/\partial x_a$ is continuous and increasing in q_a.*

We briefly discuss the above assumptions. Assumption 13.1 means that the content views of a single piece of content have an infinitesimal effect on the utility. Additive separability in Assumption 13.2 excludes complementarity between two pieces of content. Assumption 13.3(i) says that the marginal utility from content of quality 0 is 0. Assumption 13.3(ii) and (iii) have that the utility from content of a positive quality is increasing in its content views at a diminishing rate. Assumption 13.4 states that the higher quality of content leads to the larger marginal utility from content.

Assumptions 13.2 and 13.3 guarantee that for each $y \in \{0, 1\}^{[0,1]}$, there exists an optimal content view allocation $x^*(y)$ satisfying the following optimality conditions:

$$x_i^*(y) > 0 \text{ for all } i \text{ such that } q_i > 0 \text{ and } y_i = 1, x_a^*(y) > 0, \tag{13.10}$$

$$x_i^*(y) = 0 \text{ for all } i \text{ such that } q_i = 0 \text{ or } y_i = 0, \tag{13.11}$$

$$\int_0^1 x_i^*(y)di + x_a^*(y) = T, \tag{13.12}$$

$$\frac{\partial U}{\partial x_i}(x^*(y)) = \frac{\partial U}{\partial x_a}(x^*(y)) \text{ for all } i \text{ such that } q_i > 0 \text{ and } y_i = 1. \tag{13.13}$$

By Assumption 13.4, we have $x_i^*(y) > x_j^*(y)$ for all i and j such that $q_i > q_j$ and $y_i = y_j = 1$, that is, content of a higher quality attracts more content views. Consider a content view allocation $x'(y)$ satisfying $x_i'(y) = x_i^*(y)$ for *almost all* $i \in [0, 1]$ such that $y_i = 1$, $x_i'(y) = 0$ for all i such that $y_i = 0$ and $x_a'(y) = x_a^*(y)$. Then $x'(y)$ satisfies all the constraints of the content viewer's problem while achieving the same utility as $x_a^*(y)$ by Assumption 13.1. Thus, $x'(y)$ is also an optimal content view allocation. In this chapter, we focus on optimal solutions of the kind $x^*(y)$ in which the optimality conditions are satisfied for *all* $i \in [0, 1]$ such that $y_i = 1$. Note that such an optimal solution $x^*(y)$ is unique for each y under our assumptions.

A particular form of a utility function satisfying Assumptions 13.1–13.4, which we shall use extensively as an illustration, is the quality-adjusted version of the well-known Dixit–Stiglitz utility function [15], defined as

$$U(x) = \left[\int_0^1 q_i x_i^{\frac{\sigma-1}{\sigma}} di + q_a x_a^{\frac{\sigma-1}{\sigma}}\right]^{\frac{\sigma}{\sigma-1}}, \tag{13.14}$$

where $\sigma > 1$ measures the elasticity of substitution between different pieces of content. In the extreme case where $\sigma \to +\infty$, content becomes perfectly substitutable

[15], whereas when $\sigma \to 1$, content becomes unsubstitutable. Dixit–Stiglitz utility function provides an effective and tractable means of capturing constant elasticity of substitution in a market. More precisely, it provides a rigorous characterization of product diversity through the parameter $\sigma > 1$ and combines multiple types of product consumption into an aggregate utility, which is particularly suitable for user-generated content platforms where content producers produce diversified content. Next, by using the optimality conditions (13.10)–(13.13), we obtain the optimal solution

$$x_i^*(y) = \frac{Tq_i^\sigma}{q_a^\sigma + \int_0^1 y_j q_j^\sigma \, dj} y_i \quad \text{for all } i \in [0, 1], \text{ and} \tag{13.15}$$

$$x_a^*(y) = \frac{Tq_a^\sigma}{q_a^\sigma + \int_0^1 y_j q_j^\sigma \, dj} = \frac{T}{1 + \frac{\int_0^1 y_j q_j^\sigma \, dj}{q_a^\sigma}}, \tag{13.16}$$

assuming that the integral $\int_0^1 y_j q_j^\sigma \, dj$ exists. Note that $[\int_0^1 y_j q_j^\sigma \, dj]^{1/\sigma}$ can be interpreted as the aggregate quality of content on the intermediary's platform, which reflects not only the overall quality but also the quantity of available content.[1] Also, $[q_a^\sigma + \int_0^1 y_j q_j^\sigma \, dj]^{1/\sigma}$ can be interpreted as the aggregate quality of all available content. The total content views in the intermediary's platform at the optimal solution are given by

$$\bar{x}(y) = \frac{T \int_0^1 y_j q_j^\sigma \, dj}{q_a^\sigma + \int_0^1 y_j q_j^\sigma \, dj} = \frac{T}{1 + \frac{q_a^\sigma}{\int_0^1 y_j q_j^\sigma \, dj}}, \tag{13.17}$$

and the indirect utility of the content viewer is

$$U^*(y) = U(x^*(y)) = T \left(q_a^\sigma + \int_0^1 y_j q_j^\sigma \, dj \right)^{\frac{1}{\sigma - 1}}. \tag{13.18}$$

We can see that the optimal number of the content views of content i, if produced, is increasing in its quality, q_i, and the size of the content viewer, T, while it is decreasing in the aggregate quality of all available content. Let us treat the platforms other than the intermediary's platform as a single platform. Then the optimal total number of the content views in a platform (i.e., $x_a^*(y)$ and $\bar{x}(y)$) is increasing in the size of the

[1] Similarly, the aggregate quality of content on other platforms, q_a, can be derived from a distribution of individual qualities. We index content producers providing content on the other platforms by k, and suppose that the content producers on other platforms are uniformly distributed over [0, 1] with total mass n_a. Then the Dixit–Stiglitz utility function U can be modified as $U(x) = \left[\int_0^1 q_i x_i^{\frac{\sigma-1}{\sigma}} \, di + n_a \int_0^1 q_k x_k^{\frac{\sigma-1}{\sigma}} \, dk \right]^{\frac{\sigma}{\sigma-1}}$. If we set $q_a = (n_a \int_0^1 q_k^\sigma dk)^{1/\sigma}$, assuming that the integral exists, we obtain the same optimal number of content views in the intermediary's platform in both approaches, while x_a represents the total number of content views in the other platforms, that is, $x_a = \int_0^1 x_k dk$.

content viewer and the aggregate quality of content available on the platform, while it is decreasing in the aggregate quality of content available on the other platform. Finally, the per-capita indirect utility, $U^*(y)/T$, is increasing in the aggregate quality of all available content.

13.3.3 Equilibrium Content Production

At equilibrium, no content producer can gain by unilaterally changing its production decision. In other words, equilibrium content production decisions $y^*(\theta)$ can be considered as a Nash equilibrium of the game played by the content producers, given the payment rate θ and anticipating the optimal content view allocation $x^*(y)$ of the content viewer. In the following lemma whose proof is omitted for brevity, we provide some basic properties of equilibrium content production decisions.

Lemma 13.1 *Let $y^*(\theta)$ be an equilibrium strategy profile of the content producers.*
(i) If $\theta \leq -s$, then $y_i^*(\theta) = 0$ for all $i \in [0, 1]$.
(ii) If $\theta > -s$, $y_i^*(\theta) = 1$ and $q_j > q_i$, then $y_j^*(\theta) = 1$.

Lemma 13.1(i) shows that if the intermediary chooses $\theta \leq -s$, no content producer will produce content and thus the intermediary will obtain no content views, that is, $\bar{x}(\theta) = 0$. Hence, $\Pi(\theta) = 0$ for any $\theta \leq -s$. Now consider the case where $\theta > -s$. Let us define $\tilde{I}_q = \{i \in [0, 1] : q_i \leq q\}$ for all $q \in \mathbb{R}$. Assuming that $\tilde{I}_q$ is (Lebesgue) measurable for every q, we define $F(q)$ to be the measure of $\tilde{I}_q$. Then $F(\cdot)$ can be considered as the cumulative distribution function of the qualities of content that can be produced by the content producers. We assume that $F(\cdot)$ has a continuous probability distribution function $f(\cdot)$ with support $[0, 1]$.[2] Lemma 13.1(ii) implies that for any equilibrium strategy profile $y^*(\theta)$ of the content producers, there exists a threshold, or sometimes called the *marginal content quality*, $q_m^*(\theta) \in [0, 1]$, such that $y_i^*(\theta) = 1$ if $q_i > q_m^*(\theta)$ and $y_i^*(\theta) = 0$ if $q_i < q_m^*(\theta)$. If $q_m^*(\theta) \in (0, 1)$, a content producer who can produce content of quality $q_m^*(\theta)$ is indifferent between producing and not producing given payment rate θ. For convenience, we restrict attention to threshold strategies of the content producers in which all indifferent producers choose to produce, although given Assumption 13.1 and the assumption on $F(\cdot)$, the behavior of indifferent producers does not affect the utility of the content viewer as well as the profit of the intermediary.

At the optimal content view allocation, the content viewer allocates the same number of content views to two pieces of available content of equal quality. That is, if $q_i = q_j$ and $y_i = y_j = 1$, then $x_i^*(y) = x_j^*(y)$. For each $q \in [0, 1]$ and $q_m \in [0, 1]$, we define $z^*(q|q_m) = x_i^*(y)$ where $q = q_i$, $y_i = 1$, and $y_j = 1$ if and only if $q_j \geq q_m$ for all $j \neq i$. In other words, $z^*(q|q_m)$ is the number of content views that a content producer receives if it produces content of quality q while all the other content producers use threshold q_m. We can obtain the following properties of $z^*(q|q_m)$.

[2]More generally, we can have a finite support $[\underline{q}, \overline{q}]$ where $0 \leq \underline{q} < \overline{q} < +\infty$. We can think of using 1 instead of $\overline{q}$ as a normalization, while we use 0 instead of sufficiently small $\underline{q}$ to simplify our analysis.

Lemma 13.2 *For all $q_m \in [0, 1]$, $z^*(0|q_m) = 0$ and $z^*(q|q_m) > 0$ for all $q > 0$. $z^*(q|q_m)$ is continuous and increasing in each of q and q_m.*

With the Dixit–Stiglitz utility function in Eq. (13.14), we have

$$z^*(q|q_m) = \frac{T(\sigma + 1)q^\sigma}{(\sigma + 1)q_a^\sigma + (1 - q_m^{\sigma+1})}, \tag{13.19}$$

which obviously satisfies the properties in Lemma 13.2.

By Lemma 13.2, we have $(\theta + s)z^*(0|q_m) - c < 0$ for all $q_m \in [0, 1]$. If $(\theta + s)z^*(1|q_m) - c \geq 0$, then there exists a unique minimum element of the set $\{q \in [0, 1] : (\theta + s)z^*(q|q_m) - c \geq 0\}$ because $z^*(q|q_m)$ is continuous and increasing in q. We define a mapping $Q_\theta : [0, 1] \to [0, 1]$ by

$$Q_\theta(q_m) = \begin{cases} \arg\min\{q \in [0, 1] : & \text{if } (\theta + s)z^*(1|q_m) - c \geq 0, \\ (\theta + s)z^*(q|q_m) - c \geq 0\} & \\ 1 & \text{otherwise.} \end{cases} \tag{13.20}$$

$Q_\theta(q_m)$ is the threshold of the optimal production decisions, given payment rate θ when the content producers expect that only producers who can produce content of quality higher than q_m choose to produce. Thus, if q_m^* is a fixed point of the mapping Q_θ, that is, $q_m^* = Q_\theta(q_m^*)$, then no content producer can gain from unilateral deviation when the content producers use threshold q_m^*.

Definition 13.2 q_m^* *is an* equilibrium marginal content quality (or equilibrium production threshold), given payment rate θ, *if it satisfies* $q_m^* = Q_\theta(q_m^*)$.

By Lemma 13.2, $z^*(1|1)$ is the maximum number of content views that a content producer can attract in the platform. Hence, for $\theta > -s$, if $(\theta + s)z^*(1|1) - c \leq 0$, a content producer can never obtain a positive payoff from producing content. The condition $(\theta + s)z^*(1|1) - c \leq 0$ can be rewritten as $\theta \leq \underline{\theta}$, where

$$\underline{\theta} = \frac{c}{z^*(1|1)} - s. \tag{13.21}$$

We establish the existence and uniqueness of an equilibrium marginal content quality in Proposition 13.1, whose proof can be found in Reference 16 and references therein.

Proposition 13.1 *For any $\theta > -s$, there exists a unique equilibrium marginal content quality given payment rate θ, $q_m^*(\theta)$. $q_m^*(\theta) = 1$ for $\theta \leq \underline{\theta}$ and $q_m^*(\theta) \in (0, 1)$ for $\theta > \underline{\theta}$. Moreover, $q_m^*(\theta)$ is continuous and decreasing in θ on $[\underline{\theta}, +\infty)$ and approaches zero as $\theta \to +\infty$.*

Proposition 13.1 guarantees the existence of a unique equilibrium threshold and shows that if the the content producer who can produce content of the highest quality can never obtain a positive payoff (because of high production cost, low payment rate from the intermediary, or low satisfaction from attracting content views), then (almost) no content producers choose to produce content at equilibrium.

In practice, the content producers may not have complete information regarding each other (e.g., about the distribution of qualities), and hence, they may not be able to make decisions that strike the equilibrium at one shot. In such a scenario, the content producers may use an adjustment process to update their decisions based on the limited information. A natural and well-studied approach to modeling an adjustment process is the best-response dynamics, in which each decision maker chooses the best action in response to the decisions made by others. In this chapter, we consider the best-response dynamics based on the naive (or static) expectation. Specifically, time is discrete and labeled as $t = 1, 2, \ldots$ Let $q_{m,t}$ be the threshold used by the content producers in period $t = 1, 2, \ldots$ At the end of each period $t = 1, 2, \ldots$, the content producers observe $q_{m,t}$ (or the qualities of all content available on the platform in period t) and expect that $q_{m,t}$ is used when they make production decisions in period $t + 1$. That is, each content producer i expects to attract $z^*(q_i|q_{m,t})$ content views if it chooses to produce in period $t + 1$. Then content producer i chooses to produce in period $t + 1$ if $(\theta + s)z^*(q_i|q_{m,t}) - c \geq 0$, and given an initial belief $q_{m,0} \in [0, 1]$, the best-response dynamics induces a sequence of thresholds $\{q_{m,t}\}_{t=0}^{\infty}$ that evolve following the relationship

$$q_{m,t+1} = Q_\theta(q_{m,t}) \tag{13.22}$$

for $t = 0, 1, \ldots$ Similar decision processes have been adopted in the existing literature (e.g., [11–13] and references therein). Essentially, the dynamics defined by Eq. (13.22) is a fixed-point iteration for $Q_\theta(\cdot)$, and it converges regardless of the initial point if $Q_\theta(\cdot)$ is a contraction mapping [17]. Following the contraction mapping theorem, we can easily specify further a sufficient condition for convergence, and we omit the details here for brevity. It should be noted that, by considering the dynamics specified by Eq. (13.22), we implicitly assume that content produced in the previous periods has little value and will not significantly affect the content views in the current period (e.g., news content). Moreover, the dynamics specified by Eq. (13.22) requires that all the content producers update production decisions in every period. If only a randomly chosen fraction $\varepsilon \in (0, 1]$ of the content producers can update their decisions in each period, the sequence of thresholds is generated by $q_{m,t+1} = (1 - \varepsilon)q_{m,t} + \varepsilon Q_\theta(q_{m,t})$. This modification does not affect the equilibrium analysis although it slows down the convergence.

13.3.4 Optimal Payment Rate

From Proposition 13.1, we can see that $\overline{x}(\theta) > 0$ if and only if $\theta > \underline{\theta}$. Also, from Eq. (13.1), we can see that $\Pi(\theta) > 0$ only if $\theta < b$. Suppose that $b \leq \underline{\theta}$. Then $\Pi(\theta) = 0$ for $\theta \leq \underline{\theta}$ and $\Pi(\theta) < 0$ for $\theta > \underline{\theta}$. Hence, the maximum profit of the intermediary is zero, and any $\theta \leq \underline{\theta}$ is an optimal payment rate. In the remainder of this chapter, we shall assume $\underline{\theta} < b$. In this case, $\Pi(\theta) = 0$ for $\theta \leq \underline{\theta}$, $\Pi(\theta) > 0$ for $\theta \in (\underline{\theta}, b)$, and $\Pi(\theta) \leq 0$

for $\theta \geq b$. Hence, when searching for an optimal payment rate, we can restrict attention to the interval $(\underline{\theta}, b)$. Then the problem of the intermediary to find an optimal payment rate θ^* can be written as

$$\max_{\theta \in (\underline{\theta}, b)} (b - \theta)\overline{x}(\theta), \tag{13.23}$$

where $\overline{x}(\theta) = \int_{q_m^*(\theta)}^1 z^*(q|q_m^*(\theta))dF(q)$.

Using Lemma 13.2, we can show that $\int_{q_m}^1 z^*(q|q_m)dF(q)$ is continuous and decreasing in q_m on $[0, 1]$, and thus by Proposition 13.1, $\overline{x}(\theta)$ is continuous and increasing in θ on $[\underline{\theta}, +\infty)$. That is, an increase in θ on $[\underline{\theta}, +\infty)$ will encourage more content producers to produce content and make the intermediary to attract more content views from the content viewer. Then it follows that the intermediary's problem (13.23) has an optimal solution by extending the region of θ into a compact set $[\underline{\theta}, b]$. Moreover, if we assume that the objective function $\Pi(\theta)$ is strictly concave on $(\underline{\theta}, b)$, the intermediary's problem has a unique solution. In this case, the unique optimal payment rate θ^* satisfies the first-order optimality condition

$$\Pi'(\theta) = -\overline{x}(\theta) + (b - \theta)\overline{x}'(\theta) = 0, \tag{13.24}$$

assuming that $\overline{x}$ is differentiable. Moreover, if $0 \in (\underline{\theta}, b)$, we can find out whether the intermediary should subsidize or tax the content producers by examining the sign of $\Pi'(0)$. Specifically, θ^* has the same sign as that of $\Pi'(0)$. If $\Pi'(0) > 0$, or $(b - \theta)\overline{x}'(0) > \overline{x}(0)$, the first-order gain from increased content views dominates the first-order loss from payment to the content producers as the intermediary increases θ from 0. Thus, in this case, it is optimal for the intermediary to subsidize the content producers, that is, $\theta^* > 0$. On the other hand, if $\Pi'(0) < 0$, or $\overline{x}(0) > (b - \theta)\overline{x}'(0)$, it is optimal for the intermediary to tax the content producers, that is, $\theta^* < 0$.

In the remainder of this chapter, to gain insights on the optimal payment rate, we focus on the quality-adjusted Dixit–Stiglitz utility function and uniform distribution of content qualities. In this case, $z^*(1|1) = T/q_a^\sigma$ and thus $\underline{\theta} = cq_a^\sigma/T - s$. In Proposition 13.2, whose proof is given in Reference 16 and references therein, we study the optimal payment rate and its sign.

Proposition 13.2 *Suppose that the utility function $U(x)$ of the representative content viewer is given by the Dixit–Stiglitz utility function in Eq.* (13.14) *and that the qualities of content that can be produced by the content producers are uniformly distributed on* $[0, 1]$. *Then there exists a unique optimal payment rate $\theta^* \in (cq_a^\sigma/T - s, b)$ that maximizes the intermediary's profit. The equilibrium marginal content quality given θ^*, denoted by q_m^{**}, is the unique root of the following equation in the variable q_m on $(q_m^*(b), 1)$:*

$$-\frac{T(b+s)\,q_a^\sigma}{\left[(\sigma+1)\,q_a^\sigma + 1 - q_m^{\sigma+1}\right]^2} + \frac{c\left(\sigma + q_m^{\sigma+1}\right)}{(\sigma+1)^3\, q_m^{2\sigma+1}} = 0, \tag{13.25}$$

and θ^* *is given by*

$$\theta^* = \frac{c\left[(\sigma+1)\,q_a^\sigma + 1 - \left(q_m^{**}\right)^{\sigma+1}\right]}{T\,(\sigma+1)\,(q_m^{**})^\sigma} - s. \tag{13.26}$$

Moreover, if $0 \in (cq_a^\sigma/T - s, b)$*, then*

$$\begin{cases} \theta^* \in (0, b) & \text{if } \dfrac{\sigma}{q_m^*(0)} + [q_m^*(0)]^\sigma < \dfrac{c(b+s)(\sigma+1)q_a^\sigma}{Ts^2}, \\ \theta^* = 0 & \text{if } \dfrac{\sigma}{q_m^*(0)} + [q_m^*(0)]^\sigma = \dfrac{c(b+s)(\sigma+1)q_a^\sigma}{Ts^2}, \\ \theta^* \in \left(\dfrac{cq_a^\sigma}{T} - s, 0\right) & \text{if } \dfrac{\sigma}{q_m^*(0)} + [q_m^*(0)]^\sigma > \dfrac{c(b+s)(\sigma+1)q_a^\sigma}{Ts^2}. \end{cases} \tag{13.27}$$

Proposition 13.2 provides conditions under which the profit-maximizing intermediary subsidizes or taxes the content producers. We can see that the intermediary should subsidize the content producers in the following cases:

1. The total number of content views T is sufficiently small;
2. The production cost c is sufficiently large;
3. Social satisfaction per content view s is sufficiently small;
4. The aggregate quality of content on the other platforms q_a is sufficiently large;
5. Outside profit per content view b is sufficiently large.

The first four cases lead to a situation in which it is difficult for a content producer to make a profit from providing content on the intermediary's platform. Thus, in these cases, the intermediary tends to subsidize the content producers in order to encourage content production. The statements can also be mathematically verified. Let us take the first case as an example. When T is sufficiently small, $q_m^*(0)$ will approach 1, making the left-hand side of the first inequality in Eq. (13.27) reach $\sigma + 1$. Thus, the first inequality in Eq. (13.27) will hold when T is sufficiently small. In the last case, the intermediary derives high outside profit per content view, for example, from advertisement. Hence, it subsidizes off for the intermediary to subsidize the content producers in order to increase content production on its platform, while it has more revenue to share with them.

Numerical results illustrating the impacts of q_a, c, and T are plotted in Figure 13.2. It can be seen that the intermediary can improve its profit significantly by using the proposed payment scheme compared to the case in which it does not use a payment scheme at all (i.e., the case where $\theta = 0$). For example, we observe from Figure 13.2b that by optimally choosing the payment rate, the intermediary's profit increases from approximately 0.21–0.5 (nearly 150% increase) when $c = 1.5$.

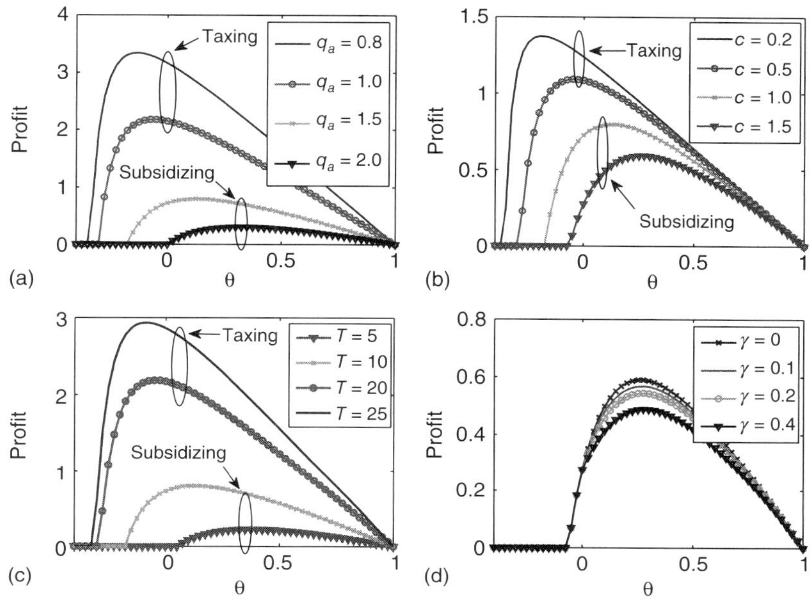

Figure 13.2 Profit versus price θ. $\sigma = 2$, $b = 1$, $s = 0.4$. (a) Impact of q_a. $T = 10$, $c = 1.0$. (b) Impact of c. $T = 10$, $q_a = 1.5$. (c) Impact of T. $c = 1.0$, $q_a = 1.5$. (d) Impact of γ. $T = 10$, $c = 1.5$, $q_a = 1.5$.

Finally, we conclude this chapter by discussing two extreme cases, $q_a \to 0$ and $\sigma \to +\infty$. First, consider the case where $q_a \to 0$. In this case, the aggregate quality of content on the other platforms is negligible (e.g., very low quality or little content available), and the intermediary becomes virtually a monopolist in the market. As the content producers who can produce content of the highest quality 1 have the strongest incentive to produce, almost all the T content views will be devoted to content on the intermediary's platform as long as there is some content on the platform. Note that $\overline{x}(\theta) > 0$ if and only if $\theta > \underline{\theta}$ and that $\underline{\theta} \to -s$ as $q_a \to 0$. Thus, as $q_a \to 0$, the optimal payment θ^* converges to $-s$ while the maximum profit converges to $(b + s)T$. Next, consider the case where $\sigma \to +\infty$. In this case, content becomes perfectly substitutable, and almost all the content views will be devoted to content of the highest quality among content on all the platforms. This can be verified by taking the limits of $x_i^*(y)$ and $x_a^*(y)$ given in Eqs. (13.15) and (13.16), respectively, as $\sigma \to +\infty$ with $\int_0^1 y_j q_j^\sigma dj = [1 - (q_m^*)^{\sigma+1}]/(\sigma + 1)$. When $q_a > 1$, we have $x_i^* \to 0$ for all i and $x_a^* \to T$. Thus, content produced on the intermediary's platform will attract no content views and the intermediary cannot obtain a positive profit regardless of the payment rate θ. On the contrary, when $q_a < 1$, we have $x_a^* \to 0$ and $\overline{x} \to T$ with $x_i^*/x_j^* \to +\infty$ for all i, j such that $q_i > q_j > q_a$. When $q_a < 1$, we again have $\underline{\theta} \to -s$ as $\sigma \to +\infty$. Hence, the optimal payment rate θ^* converges to $-s$, while the maximum profit converges to $(b + s)T$. To sum up, when $q_a \to 0$ or $\sigma \to +\infty$ with $q_a < 1$, the intermediary can attract almost all the T content views while extracting almost all the social satisfaction of the content producers.

Remark: Suppose that the aggregate quality of content on the other platforms q_a is derived from a distribution of qualities. Suppose further that the qualities are uniformly distributed on the interval $[q_l, q_h]$ with total mass n_a, where $0 \leq q_l < q_h$. Then by the relationship $q_a = (n_a \int_0^1 q_k^\sigma dk)^{1/\sigma}$, we obtain

$$q_a = \left[\frac{n_a\left(q_h^{\sigma+1} - q_l^{\sigma+1}\right)}{1+\sigma}\right]^{\frac{1}{\sigma}} = n_a^{\frac{1}{\sigma}} q_h^{\frac{\sigma+1}{\sigma}} \left[1 - \left(\frac{q_l}{q_h}\right)^{\sigma+1}\right]^{\frac{1}{\sigma}} \left(\frac{1}{1+\sigma}\right)^{\frac{1}{\sigma}}. \quad (13.28)$$

As $\lim_{\sigma\to+\infty} n_a^{1/\sigma} = \lim_{\sigma\to+\infty}\left[1 - (q_l/q_h)^{\sigma+1}\right]^{1/\sigma} = \lim_{\sigma\to+\infty}(1/(1+\sigma))^{1/\sigma} = 1$ and $\lim_{\sigma\to+\infty} q_h^{(\sigma+1)/\sigma} = q_h$, we have $q_a \to q_h$ as $\sigma \to +\infty$. Thus, when $\sigma \to +\infty$, the highest quality q_h determines the aggregate quality q_a, and we have $q_a > 1$ if $q_h > 1$ and $q_a < 1$ if $q_h < 1$.

13.3.5 Overjustification Effects

Now, we briefly discuss overjustification effects on the intermediary's equilibrium profit.

It has been argued that an external incentive such as money or prizes decreases a person's intrinsic motivation to perform a task, and such phenomenon is referred to as *overjustification effects* [18]. In the context of user-generated content platforms, content producers' internal incentive to produce content may decrease in the long term if the intermediary chooses to subsidize content producers. We model overjustification effects using the following formula

$$s' = \max\left[0, s - \gamma \cdot \max(0, \theta)\right], \quad (13.29)$$

where s' is the new social satisfaction per content view in the presence of overjustification effects and $\gamma \in [0, \infty)$ indicates the severity of overjustification effects. Note that if the intermediary taxes content producers, overjustification effects do not exist, that is, $s' = s$. We show the profits under various values of γ in Figure 13.2d and observe that even though overjustification effects exist, the intermediary can still increase its profit by applying the proposed payment scheme, as long as overjustification effects are not too strong (e.g., γ is not too large). In the presence of strong overjustification effects (e.g., $\gamma \geq 1$), a content producer's incentive to produce content will even decrease if the intermediary subsidizes content producers for content production (unless the intermediary subsidizes content producers sufficiently high).

13.4 EXTENSION TO HETEROGENEOUS PRODUCTION COSTS

In the analysis so far, it has been assumed that all the content producers incur the same production cost c when they choose to produce content. In this chapter, we extend our model and generalize the preceding analysis by relaxing this assumption and considering a scenario in which the content producers have different production costs. We assume that there are K groups of content producers and that the production cost of a content producer in group k is given by $c_k > 0$ for $k = 1, \ldots, K$. We assume that $c_1 < c_2 < \cdots < c_K$. We use I_k to denote the set of the indexes of the content producers in group k, for $k = 1, \ldots, K$. Then $\{I_1, \ldots, I_k\}$ forms a partition of $[0, 1]$. The mass of the content producers in group k is denoted by n_k so that $\sum_{k=1}^{K} n_k = 1$. We assume that for each k the qualities of content that can be produced by the content producers in group k are distributed according to a cumulative distribution function $F_k(\cdot)$, which has a continuous probability distribution function $f_k(\cdot)$ with support $[0, 1]$.

With heterogeneous production costs, the definition of an equilibrium is modified so that (ii) in Definition 13.1 becomes as follows: for each content producer $i \in I_k$, $k = 1, \ldots, K$, $y_i^*(\theta)$ is an optimal production decision, given θ, $x_i^*(y)$, and $(y_j^*(\theta))_{j \neq i}$, that is,

$$(\theta + s)x_i^*(y^*(\theta)) - c_k \geq 0 \quad \text{if } y_i^*(\theta) = 1 \tag{13.30}$$

and

$$(\theta + s)x_i^*(1, y_{-i}^*(\theta)) - c_k \leq 0 \quad \text{if } y_i^*(\theta) = 0, \tag{13.31}$$

while (i) and (iii) remain the same. When $\theta > -s$, equilibrium production decisions are characterized by thresholds, one for each group. We use $q_{m,k} \in [0, 1]$ to denote the threshold used by the content producers in group k, for $k = 1, \ldots, K$. We also use a vector notation $\mathbf{q}_m = (q_{m,1}, q_{m,2}, \ldots, q_{m,K}) \in [0, 1]^K$. As before, we use $z^*(q|\mathbf{q}_m)$ to denote the number of content views that a content producer receives if it produces content of quality q while all the other content producers use thresholds $\mathbf{q}_m$. Lemma 13.2 can be modified so that the following holds: (i) for all $\mathbf{q}_m \in [0, 1]^K$, $z^*(0|\mathbf{q}_m) = 0$ and $z^*(q|\mathbf{q}_m) > 0$ for all $q > 0$ and (ii) $z^*(q|\mathbf{q}_m)$ is continuous and increasing in q and each element of $\mathbf{q}_m$. We define a mapping $\mathbf{Q}_\theta : [0, 1]^K \to [0, 1]^K$ so that the kth element of $\mathbf{Q}_\theta(\mathbf{q}_m)$ is given by

$$Q_{\theta,k}(\mathbf{q}_m) = \begin{cases} \arg\min\{q \in [0, 1] : & \text{if } (\theta + s)z^*(1|\mathbf{q}_m) - c_k \geq 0, \\ (\theta + s)z^*(q|\mathbf{q}_m) - c_k \geq 0\} & \\ 1 & \text{otherwise,} \end{cases} \tag{13.32}$$

for $k = 1, \ldots, K$. Equilibrium content production is characterized by a fixed point of $\mathbf{Q}_\theta$, as defined in the following.

Definition 13.3 $\mathbf{q}_m^*$ *is an equilibrium marginal content quality vector (or equilibrium production threshold vector), given payment rate θ if it satisfies $\mathbf{q}_m^* = \mathbf{Q}_\theta(\mathbf{q}_m^*)$.*

The following properties of equilibrium thresholds can be readily established.

Lemma 13.3 *Let $\mathbf{q}_m^*$ be an equilibrium marginal content quality vector, given payment $\theta > -s$. Then $0 < q_{m,1}^* \leq q_{m,2}^* \leq \cdots \leq q_{m,K}^* \leq 1$ and*

$$\frac{z^*(q_{m,k}^*|\mathbf{q}_m^*)}{z^*(q_{m,l}^*|\mathbf{q}_m^*)} = \frac{c_k}{c_l}, \tag{13.33}$$

for all $k, l \leq k^*$, where $k^* = \max\{k \in \{1, \ldots, K\} : q_{m,k}^* < 1\}$.

In the previous analysis, we have defined $\underline{\theta}$ as the payment rate that induces the content producers to start producing content. That is, $\underline{\theta}$ was the payment under which a content producer who can produce content of the highest quality 1 is indifferent between producing and not producing. As there are K heterogeneous groups in the current analysis, we define K payment rates, $\underline{\theta}_1, \ldots, \underline{\theta}_K$, where $\underline{\theta}_k$ is the payment rate that induces the content producers in group k to start producing content. Following $c_1 < c_2 < \cdots < c_K$, it is obvious that $\underline{\theta}_1 < \underline{\theta}_2 < \cdots < \underline{\theta}_K$. First, the indifference condition satisfied at $\theta = \underline{\theta}_1$ is $(\theta + s)z^*(1|(1, \ldots, 1)) = c_1$. Hence, we obtain

$$\underline{\theta}_1 = \frac{c_1}{z^*(1|(1, \ldots, 1))} - s. \tag{13.34}$$

Now consider $k = 2, \ldots, K$. At $\theta = \underline{\theta}_k$, there are some content producers who choose to produce in each group $l < k$, and the following indifference conditions should be satisfied.

$$\left(\underline{\theta}_k + s\right) z^* \left(\overline{q}_{m,l}^k | \left(\overline{q}_{m,1}^k, \ldots, \overline{q}_{m,k-1}^k, 1, \ldots, 1\right)\right) = c_l, \text{ for } l = 1, \ldots, k-1, \text{ and} \tag{13.35}$$

$$\left(\underline{\theta}_k + s\right) z^* \left(1 | \left(\overline{q}_{m,1}^k, \ldots, \overline{q}_{m,k-1}^k, 1, \ldots, 1\right)\right) = c_k. \tag{13.36}$$

Note that Eqs. (13.35) and (13.36) together define a system of k equations in k unknowns, $\overline{q}_{m,1}^k, \ldots, \overline{q}_{m,k-1}^k$, and $\underline{\theta}_k$. To simplify our analysis, we impose the following assumption.

Assumption 13.5 *For any $q \in [0, 1]$, $q' \in (0, 1]$, and $\mathbf{q}_m \in [0, 1]^K$, the ratio $z^*(q|\mathbf{q}_m)/z^*(q'|\mathbf{q}_m^*)$ is independent of $\mathbf{q}_m$.*

We can verify that $z^*(q|\mathbf{q}_m)$ induced by the Dixit–Stiglitz utility function satisfies Assumption 13.5. By Assumption 13.5, we can write $z^*(q|\mathbf{q}_m)/z^*(q'|\mathbf{q}_m^*)$ as $\tilde{z}(q, q')$. By modified Lemma 13.2, given $q' \in (0, 1]$, $\tilde{z}(q, q')$ is continuous and increasing in q with $\tilde{z}(0, q') = 0$ and $\tilde{z}(q', q') = 1$. Hence, there exists unique $\overline{q}_{m,k-1}^k \in (0, 1)$ that satisfies $\tilde{z}(\overline{q}_{m,k-1}^k, 1) = c_{k-1}/c_k$. The remaining thresholds, $\overline{q}_{m,1}^k, \ldots, \overline{q}_{m,k-2}^k$, can be obtained by solving $\tilde{z}(\overline{q}_{m,l}^k, \overline{q}_{m,l+1}^k) = c_l/c_{l+1}$ from $l = k-2$ to $l = 1$ recursively. This yields the $k-1$ thresholds $\overline{q}_{m,1}^{k}, \dots, \overline{q}_{m,k-1}^{k}$ such that $0 < \overline{q}_{m,1}^{k} < \cdots < \overline{q}_{m,k-1}^{k} < 1$. Then using Eq. (13.36), we obtain

$$\underline{\theta}_k = \frac{c_k}{z^*\left(1 | \left(\overline{q}_{m,1}^{k}, \dots, \overline{q}_{m,k-1}^{k}, 1, \dots, 1\right)\right)} - s \tag{13.37}$$

for $k = 2, \dots, K$. On the basis of the discussion so far, we can obtain the following proposition, which is stated without a proof because of space limitation.

Proposition 13.3 *For any $\theta > -s$, there exists a unique equilibrium marginal content quality vector given payment rate θ, $\mathbf{q}_m^*(\theta)$. For each $k = 1, \dots, K$, the following properties hold. $q_{m,k}^*(\theta) = 1$ for $\theta \le \underline{\theta}_k$ and $q_{m,k}^*(\theta) \in (0,1)$ for $\theta > \underline{\theta}_k$. $q_{m,k}^*(\theta)$ is continuous and decreasing in θ on $[\underline{\theta}_k, +\infty)$ and approaches zero as $\theta \to +\infty$. For $\theta \in [\underline{\theta}_k, \underline{\theta}_{k+1})$ ($[\underline{\theta}_K, +\infty)$ when $k = K$), we have $0 < q_{m,1}^*(\theta) < \cdots < q_{m,k}^*(\theta) \le 1 = q_{m,k+1}^*(\theta) = \cdots = q_{m,K}^*(\theta)$.*

As in Section 13.3.3, we can consider a best-response dynamics for the content producers to adjust their production decisions based on the thresholds used in the previous period. Let $\mathbf{q}_{m,t} = \left(q_{m,1,t}, q_{m,2,t}, \dots, q_{m,K,t}\right)$ be the vector of the thresholds used in period $t = 1, 2, \dots$ Then starting from an initial belief $\mathbf{q}_{m,0}$, the dynamics yields a sequence $\{\mathbf{q}_{m,t}\}_{t=0}^{\infty}$ generated by $\mathbf{q}_{m,t+1} = \mathbf{Q}_\theta(\mathbf{q}_{m,t})$ for $t = 0, 1, \dots$ More specific results regarding the equilibrium marginal content quality vector and the convergence of the best-response dynamics can be obtained using the Dixit–Stiglitz utility function. The details are omitted for brevity.

Now we turn to the intermediary's problem of finding an optimal payment rate. Note that we have $\overline{x}(\theta) = \sum_{k=1}^{K} n_k \int_{q_{m,k}^*(\theta)}^{1} z^*(q|\mathbf{q}_m^*(\theta)) dF_k(q)$. We assume that $\underline{\theta}_1 < b$. Then $\Pi(\theta) = 0$ for $\theta \le \underline{\theta}_1$, $\Pi(\theta) > 0$ for $\theta \in (\underline{\theta}_1, b)$, and $\Pi(\theta) \le 0$ for $\theta \ge b$. Hence, we can restrict attention to the interval $(\underline{\theta}_1, b)$ when searching for an optimal payment rate. Also, if $\underline{\theta}_k \ge b$ for some k, then it is not profitable for the intermediary to induce content production by producers in groups $k, k+1, \dots, K$, and we can ignore these groups in the analysis. Hence, without the loss of generality, we assume that $\underline{\theta}_K < b$. For concreteness, we will focus on the Dixit–Stiglitz utility function and uniform distribution of qualities for each group. In this case, we have

$$z^*\left(q|\mathbf{q}_m\right) = \frac{T(\sigma+1)q^\sigma}{(\sigma+1)q_a^\sigma + \sum_{k=1}^{K} n_k \left(1 - q_{m,k}^{\sigma+1}\right)}. \tag{13.38}$$

Thus, we have $\tilde{z}(q, q') = (q/q')^\sigma$, from which we get $\overline{q}_{m,l}^{k} = (c_l/c_k)^{1/\sigma}$ for $l = 1, \dots, k-1$, for $k = 2, \dots, K$. Hence, we have $\underline{\theta}_1 = c_1 q_a^\sigma / T - s$ and

$$\underline{\theta}_k = \frac{c_k \left\{ q_a^\sigma + \frac{1}{\sigma+1} \sum_{l=1}^{k-1} n_l \left[1 - \left(\frac{c_l}{c_k}\right)^{\frac{\sigma+1}{\sigma}}\right] \right\}}{T} - s \tag{13.39}$$

for $k = 2, \ldots, K$.

Let $\underline{\theta}_{K+1} = b$. We consider the problems $\max_{\theta \in [\underline{\theta}_k, \underline{\theta}_{k+1}]} \Pi(\theta)$ for $k = 1, \ldots, K$. Once we obtain $\Pi_k^* = \max_{\theta \in [\underline{\theta}_k, \underline{\theta}_{k+1}]} \Pi(\theta)$, we have $\max_{\theta \in (\underline{\theta}_1, b)} \Pi(\theta) = \max\{\Pi_1^*, \ldots, \Pi_K^*\}$ and an optimal payment rate is the one that achieves the maximum profit. As in the previous analysis, when solving $\max_{\theta \in [\underline{\theta}_k, \underline{\theta}_{k+1}]} \Pi(\theta)$, we replace the choice variable θ with $q_{m,k}^*$, where the corresponding region of $q_{m,k}^*$ is $[q_{m,k}^*(\underline{\theta}_{k+1}), q_{m,k}^*(\underline{\theta}_k)]$. Note that $q_{m,k}^*(\underline{\theta}_k) = 1$ for $k = 1, \ldots, K$ and $q_{m,k}^*(\underline{\theta}_{k+1}) = \overline{q}_{m,k}^{k+1}$ for $k = 1, \ldots, K-1$. When the intermediary chooses $q_{m,k}^* \in [q_{m,k}^*(\underline{\theta}_{k+1}), q_{m,k}^*(\underline{\theta}_k)]$, we have $q_{m,l}^* = (c_l/c_k)^{1/\sigma} q_{m,k}^*$ for $l < k$ and $q_{m,l}^* = 1$ for $l > k$. The corresponding value of θ is given by

$$\theta = \frac{c_k \left\{ q_a^\sigma + \frac{1}{\sigma+1} \sum_{l=1}^k n_l \left[1 - \left(\frac{c_l}{c_k}\right)^{\frac{\sigma+1}{\sigma}} \left(q_{m,k}^*\right)^{\sigma+1} \right] \right\}}{T \left(q_{m,k}^*\right)^\sigma} - s, \tag{13.40}$$

while the total number of content views in the platform is

$$\overline{x} = \frac{T \sum_{l=1}^k n_l \left[1 - \left(\frac{c_l}{c_k}\right)^{\frac{\sigma+1}{\sigma}} \left(q_{m,k}^*\right)^{\sigma+1} \right]}{(\sigma+1) q_a^\sigma + \sum_{l=1}^k n_l \left[1 - \left(\frac{c_l}{c_k}\right)^{\frac{\sigma+1}{\sigma}} \left(q_{m,k}^*\right)^{\sigma+1} \right]}. \tag{13.41}$$

Hence, the problem $\max_{\theta \in [\underline{\theta}_k, \underline{\theta}_{k+1}]} \Pi(\theta)$ can be transformed into $\max_{q_{m,k}^* \in [q_{m,k}^*(\underline{\theta}_{k+1}), q_{m,k}^*(\underline{\theta}_k)]} \tilde{\Pi}_k(q_{m,k}^*)$, where

$$\tilde{\Pi}_k\left(q_{m,k}^*\right) = \frac{T(b+s) \sum_{l=1}^k n_l \left[1 - \left(\frac{c_l}{c_k}\right)^{\frac{\sigma+1}{\sigma}} \left(q_{m,k}^*\right)^{\sigma+1} \right]}{(\sigma+1) q_a^\sigma + \sum_{l=1}^k n_l \left[1 - \left(\frac{c_l}{c_k}\right)^{\frac{\sigma+1}{\sigma}} \left(q_{m,k}^*\right)^{\sigma+1} \right]} - \frac{c_k \sum_{l=1}^k n_l \left[1 - \left(\frac{c_l}{c_k}\right)^{\frac{\sigma+1}{\sigma}} \left(q_{m,k}^*\right)^{\sigma+1} \right]}{(\sigma+1)\left(q_{m,k}^*\right)^\sigma}. \tag{13.42}$$

By taking the second-order derivative of $\tilde{\Pi}_k$, we can show that $\tilde{\Pi}_k$ is strictly concave on $(q_{m,k}^*(\underline{\theta}_{k+1}), q_{m,k}^*(\underline{\theta}_k))$. Hence, there exists a unique optimal solution to $\max_{q_{m,k}^* \in [q_{m,k}^*(\underline{\theta}_{k+1}), q_{m,k}^*(\underline{\theta}_k)]} \tilde{\Pi}_k(q_{m,k}^*)$. As in the case of a homogeneous production cost, we can also analyze whether the intermediary should subsidize or tax the content producers. Nevertheless, we omit the result because of its similarity to Proposition 13.2.

13.5 CONCLUSION

In this chapter, we studied the economics of user-generated content platforms by focusing on the profit maximization problem of an intermediary who owns a user-generated content platform. We considered a class of payment schemes in which the intermediary subsidizes or taxes content producers per content view while it provides the service for free to content viewers. We analyzed a three-stage game using backward induction. First, we used the representative content viewer to determine how content viewers' attention is allocated across a variety of available content. Next, after establishing the threshold property of equilibrium production decisions, we showed the existence and uniqueness of an equilibrium production threshold used by the content producers and derived its properties. Lastly, we formalized the intermediary's profit maximization problem and studied the optimal payment rate by using the quality-adjusted Dixit–Stiglitz utility function and the uniform distribution of content qualities as a concrete example. We provided conditions under which the intermediary should subsidize or tax the content producers and discussed qualitatively the impacts of the aggregate quality of outside content and content substitutability on the intermediary's profit. We also discussed the effectiveness of the proposed payment scheme in the presence of overjustification effects. As an extension of our model, we considered heterogeneity in the content producers' production costs. There are various directions to extend our work, among which we mention only a few as follows. First, we can consider a more general class of payment schemes. In this chapter, we considered anonymous and linear payment schemes, but the intermediary may use personalized (e.g., quality dependent) and nonlinear payment schemes as well. Second, we can consider a scenario where a content producer can choose a production intensity, which determines the production cost and the content quality, and can choose a content platform to post its content among several competing platforms. Third, we can incorporate the intermediary's payment decision into its back-end resource management (e.g., data centers). As shown in our preliminary work [19], jointly optimizing the payment decision and resource management by exploring the interactions among them can significantly enhance the intermediary's profitability. Last but not least, the tree-stage game studied in this chapter can be extended to other application scenarios (e.g., customer-to-customer market [20]) or incorporate additional decisions to model the intermediary's investment strategy (e.g., infrastructure and technology investment on television content platforms [21]).

REFERENCES

1. A. Gosh and P. McAfee. "Incentivizing high-quality user-generated content," World Wide Web, 2011.
2. S. Jain, Y. Chen, and D. C. Parkes. Designing Incentives for Online Question and Answer Forums. In *ACM Conf. Electronic Commerce*, 2009.
3. V. K. Singh, R. Jain, and M. S. Kankanhalli. Motivating Contributors in Social Media Networks. In *ACM SIGMM Workshop on Social Media*, 2009.

4. Wikipedia, "User-generated content," http://en.wikipedia.org/wiki/User-generated_content
5. J. C. Rochet and J. Tirole. "Platform competition in two-sided markets," *Journal of the European Economic Association*, 1, 2003, 990–1029.
6. P. Hande, M. Chiang, A. R. Calderbank, and S. Rangan. Network pricing and rate allocation with content provider participation. In *IEEE Infocom*, April 2009.
7. J. C. Rochet and J. Tirole. "Cooperation among competitors: some economics of payment card associations," *Rand Journal of Economics*, 33, 2002, 549–570.
8. J. C. Rochet and J. Tirole. "Two-sided markets: a progress report," *RAND Journal of Economics*, 37, 2006, 645–667.
9. A. Hagiu. "Merchant or two-sided platform?" *Review of Network Economics*, 6(2), 2007, 115–133.
10. J. Musacchio, G. Schwartz, and J. Walrand. "A two-sided market analysis of provider investment incentives with an application to the net-neutrality issue," *Review of Network Economics*, 8(1), 2009, 22–39.
11. Y. Jin, S. Sen, R. Guerin, K. Hosanagar, and Z.-L. Zhang. Dynamics of competition between incumbent and emerging network technologies. In *NetEcon*, Aug. 2008.
12. S. Ren, J. Park, and M. van der Schaar. "Entry and spectrum sharing scheme selection in femtocell communications markets," *IEEE/ACM Transactions on Networking*, 21(2), 2013, 218--232.
13. S. Ren and M. van der Schaar. "Data demand dynamics in communications markets," *IEEE Transactions on Signal Processing*, 60(4), 2012, 1986–2000.
14. J. Nair, A. Wierman, and B. Zwart. "Exploiting network effects in the provisioning of large scale systems," *SIGMETRICS Performance Evaluation Review*, 39(2), 2011, 26–28.
15. A. K. Dixit and J. E. Stiglitz. "Monopolistic competition and optimum product diversity," *American Economic Review*, 67(3), 1977, 297–308.
16. S. Ren, J. Park, and M. van der Schaar. Maximizing profit on user-generated content platforms with participant heterogeneity. In *IEEE Infocom*, 2012.
17. J. R. Munkres. *Elements of Algebraic Topology*. Perseus Books Publishing, New York, 1993.
18. M. R. Lepper, D. Greene, and R. E. Nisbett. "Undermining children's intrinsic interest with extrinsic reward: a test of the 'overjustification' hypothesis," *Journal of Personality and Social Psychology* 28(1), 1973, 129–137.
19. S. Ren and M. van der Schaar. Joint design of dynamic scheduling and pricing in wireless cloud computing. In *IEEE Infocom*, 2013.
20. S. Ren and M. van der Schaar. Revenue maximization in customer-to-customer markets. In *GameNets*, 2012.
21. S. Ren and M. van der Schaar. "Pricing and investment for online TV content platforms," *IEEE Transaction on Multimedia*, 14(6), 2012, 1566–1578.

PART V
Managing Content Delivery

14 Spare Capacity Monetization by Opportunistic Content Scheduling

BELL LABS and ALCATEL-LUCENT

14.1 SUMMARY

In recent years, mobile data traffic has been growing at an exponential rate with traffic volume doubling every year and some studies predicting even more than 18-fold increase in data traffic in the next 5 years [1]. This rapid increase in data traffic is making it necessary to grow network capacity at a much faster rate than ever before. Unfortunately, the rate of growth of data revenue is much lower than the rate of growth of traffic, raising the specter that at some point capacity costs may even begin to outweigh revenues thus greatly lowering the incentives for capital investments. At the same time, the highly uneven usage by data users has resulted in significant network inefficiencies whereby even congested networks are heavily underutilized most of the time. In this chapter, we present the capacity monetization system PLUTUS that taps this stranded network capacity for additional revenue and enhanced network efficiency.

Many factors have contributed to the mobile data tsunami, including the introduction of smartphones along with a wide variety of multimedia-focused mobile applications and the increase in performance of the mobile technology as it moved from 2G on to 4G. The industry's bold but short experimentation with unlimited data plans allowed the subscribers armed with powerful devices to unleash an onslaught of uncontrolled traffic. The resulting data usage patterns that have emerged are chaotic at best creating random, intermittent bottlenecks and congested hotspots spread across time and space. This has created a situation where even at low average utilization, the peak network load can be significantly high. As CAPEX decisions are mostly driven by peak load, this has driven up the cost of carrying data over mobile network. Operators have responded by putting in place tiers, caps, and throttling

Smart Data Pricing, First Edition. Edited by Soumya Sen, Carlee Joe-Wong, Sangtae Ha, and Mung Chiang.

in order to regulate bandwidth demand. However, by over pricing data and over restricting usage, operators run the risk of stifling the demand for mobile data, thus loosing the opportunity to generate high revenue from increasing data usage.

There is growing recognition on the part of operators that they must tap into the abundant unused spare capacity in their network both to create new sources of revenue and to keep data delivery costs low. According to some studies [2, 3], peak load to average utilization ratio in mobile networks can be as high as four times with most of the traffic carried during a few busy hours while the network stays underutilized the rest of the time. As all the time periods where the network is underutilized is already sunk cost for the operator, additional traffic can be carried during such time periods at negligible marginal cost. As reported in [1,4] every other bit being carried on mobile networks is now video, with more than 25% coming from Youtube alone [1,4]. As audio and video traffic involves downloading large files or chunks thereof (e.g., using progressive download or adaptive streaming), it can be easily time shifted to periods with low network utilization. Likewise, content recommended on social networks (e.g., Facebook) and media sites (e.g., Netflix, Pandora) that has a high chance of being watched could be preloaded and cached onto users device during off-peak times. In addition, the abundant capacity of WiFi and Small Cell networks can also be tapped for carrying time-shifted traffic. Operators can, therefore, start generating additional revenues by selling their unused capacity at a relatively low price as well as increase network efficiency by encouraging subscribers to shift their usage to appropriate times on available networks. This controlled rearrangement of data usage patterns can lead to a reduction in the peak-to-average ratio ideally flattening the load curve, thus slowing down the rate of network expansion by accommodating more traffic without investing in capacity enhancements. Similar ideas have been tried in other industries such as smart electric grids [5, 6] for reducing peak load by shifting load toward off-peak hours when electricity is cheaper or for alleviating road congestion by the use of tolls or for increased efficiency by Yield Management [7] approaches practiced in airline, hotel, and other industries.

In this chapter, we present the design of PLUTUS, a capacity monetization and network efficiency enhancement system. The PLUTUS system lets the operator realize extra revenue by delivering additional data using unused capacity of diverse access networks. The PLUTUS system is best suited for the delivery of large files (multimedia content, app updates, etc.) that constitute the majority of the traffic on the mobile network and are also the biggest source of congestion. A data file marked for delivery using the PLUTUS system is opportunistically transferred at the right time using the best available network. This includes dynamic resuming or suspending of data transfers in reaction to the availability of unused capacity. It also includes seamless network switchovers while maintaining session continuity. In addition, the PLUTUS system data transfers are tightly controlled to make the best use of the spare capacity by evenly spreading out the network load. In PLUTUS data items transferred to the users handset are cached locally in the device storage. This enables local playback of multimedia content providing a high quality experience.

As 80% of the mobile network cost is in the radio access network (RAN), which is also the most capacity constrained portion of the network, monetizing the spare capacity in RAN can bring the most benefit. As reported in previous studies [2, 3], the cell loads fluctuate unpredictably and rapidly at the granularity of minutes and there may not be much correlation among cells in different locations. The PLUTUS system is designed for fast identification and reaction to changes in available capacity at the cell/sector/carrier level without introducing significant signaling or data overheads, which has been a challenging problem [8–10]. In addition, the PLUTUS system dynamically and quickly adapts to changes in the traffic patterns to avoid creating any new peaks or congestion in the network. All this is accomplished while factoring in the mobility of users.

In order to increase users' confidence and gain widespread acceptance, the PLUTUS system also provides reasonable estimates for the additional delays introduced by time shifting. These estimates are derived from data collected on the historic patterns of users' mobility, network access and usage. In order to stay within the estimated delays, a combination of offline- and online-controlled scheduling of data transfers that adapts dynamically to changes in network and user state is applied. This scheduling also aims to minimize battery drain and avoid tying up scarce RAN resources from data transfers that may get stretched out for a long time when there is limited capacity to share. By preloading and locally caching content in advance of consumption, PLUTUS can further reduce peak loads and also enhance the user experience. The PLUTUS system also addresses efficient ways of interfacing with existing applications and services that want to take advantage of the spare capacity in the network. The rest of the chapter is structured as follows. We start out by surveying the related work in this space. Next we describe the main ideas used in the PLUTUS system including how pricing plans can be designed to give the right incentive to the user while ensuring that the operators are able to profitably monetize the spare capacity in their network. Then we present the design and architecture of the PLUTUS system. Finally, we present implementation details and performance results from various trials with subscribers and operators around the world.

14.2 BACKGROUND

Pricing plans for wireless data have seen a number of changes over the past few years. First, to drive up user adoption, the expensive metered data plans were replaced by the popular unlimited all-you-can-eat plans. As initially the data traffic was not very high, these plans made good revenue for the Internet service providers (ISPs). But in recent years, these plans have contributed to the exponential growth in the data traffic making it unprofitable for the service providers to continue to offer them. The ISPs have responded by replacing these with tiered and sometimes usage-based plans [11, 12]. Some of these plans include surcharges for exceeding tier limits while others throttle the data transfers once the limit is crossed. ISPs are also trying other ways of charging for data, namely, based on time [13] and application [14].

Data pricing has been an active area of research with many ideas influenced by pricing work in other industries (e.g., utilities, road transportation, airlines). As early as in 1950s, Houthakker [15], Steiner, and others proposed a dynamic pricing scheme where pricing for peak periods is higher than the pricing for off-peak periods and traffic is charged based on when it traverses the network. Odlyzko [16] have proposed static partitioning of the network resources into separate logical traffic classes that are charged differently. Interested reader is referred to Sen et al. [17] for a comprehensive survey of the large body of work in this area. Unfortunately, so far there has been little adoption by ISPs of any of these proposed pricing schemes. However, this is likely to change as the ISPs must implement efficient schemes to manage their growing traffic costs and remain profitable. We believe that the key to adoption is a pricing scheme that cannot only be easily implemented by the ISP but is also predictable and simple for the user. This is the motivation behind the approach taken in this chapter.

The ability to quickly detect and react to congestion is key for efficient monetization of spare capacity. Many schemes have been proposed for wireline networks [18], but there is limited work for wide area broadband wireless networks. This is because the shared nature of the transmission channel, interference from external sources, and use of rate adaptation algorithms (e.g., proportional fair) makes it a challenging problem. It has been proposed that channel utilization may be a better measure of congestion in wireless network [19]. Solutions to determine channel utilization are based on both active probing [8, 9, 19, 20, 35] and passive probing [10]. We describe a client-based active probing approach that we have developed for PLUTUS.

ISPs are also implementing other solutions for managing traffic growth including video optimizations (e.g. transcoding, transrating, pacing, adaptive streaming), offloading (to WiFi, small cells), upgrading to more efficient IP and 4G technologies, and so on. These solutions are orthogonal to PLUTUS and their use does not eliminate the need for time shifting and, when used in conjunction with PLUTUS, can achieve even greater network utilization. Detailed discussion of these solutions, however, is out of scope of this chapter.

14.3 THE PLUTUS APPROACH

The PLUTUS system enables the operator to deliver additional data using capacity that is already built into the network and is not currently in use. The main idea is to classify data transfers into one of two classes, the "Any Time Data" (ATD) class for transfers that can happen at any time irrespective of the state of the network and the "Surplus Data" (SD) class for delivery using capacity that is not in use by the ATD class (however, some minimum capacity may be reserved for the SD class). Note that the ATD class corresponds to the present mode of operation without any time shifting, while the SD class represents data transfers at opportunistic times when the network has unused capacity or when other networks (e.g., WiFi, small cells) are available. The SD class data transfers may, therefore, incur additional delays but have negligible marginal cost for delivery and, therefore, can include price discounts to incentivize user adoption. The PLUTUS system manages all aspects of delivery for data items in the SD class including classification, opportunistic scheduling, temporary or long-term caching, differential accounting and charging, and so on. The ATD class data transfers, however, continue to operate as before.

The typical use case for PLUTUS is as follows. An application or user uses PLUTUS to issue a SD class data transfer request at any time irrespective of the loading state of the network. On submission, a delivery estimate is provided followed by status updates and notifications as the data is delivered. In that respect, PLUTUS has resemblance to the "Digital FedEx Service" [21] proposal for deeply discounted delivery of content using spare network capacity which in turn is inspired by courier-based delivery of physical goods scheduled to happen by a fixed set of deadlines. The PLUTUS system fully manages the data delivery including transfers at times when the network has unused capacity and scheduling to evenly spread out the network load. PLUTUS suspends or resumes the data delivery in reaction to the changes in the availability of unused capacity, and it caches the transferred data locally at the receiver. In PLUTUS, transferred data is consumed directly from local cache. This ensures high quality user experience that is not impacted by network impairments [22] with delivered data available for use even when not connected to the network.PLUTUS uses a progressive mode of delivery and can let the user start using the data before it is fully transferred (e.g., for multimedia content playback). This is unlike existing pricing approaches [16, 23] that only provide price incentives to time shift usage but have limited guarantees on the user experience during or after the time shift, PLUTUS can ensure, post-delivery, high quality user experience that is not impacted by network impairments with delivered data available for use even when not connected to the network. The PLUTUS system is, therefore, very well suited for the delivery of multimedia clips and files, e-mail attachments, application, software updates and so on, which constitute a significant portion of the traffic on the mobile network [4].

With PLUTUS, the network can be thought of being partitioned into separate logical channels that differ only in the price paid by the users to use them. The higher priced channel referred to as the *ATD channel* retains commonly used pricing structures that are currently in place for the mobile networks (flat, tiered, etc.). The discounted channel referred to as the *SD channel* offers discounts over the ATD channel. A typical SD channel pricing offering may include a tiered data plan at a 50% discount over an identical ATD plan or a zero-rated media application for a fixed monthly subscription such that the user is not charged for any data usage within the application all of which is carried over the SD channel. Thus PLUTUS maintains the simplicity of logical channels each with relatively constant and easily understood pricing. This is in contrast to much of the recent work [23] that relies heavily on incentives that are time and space dependent. We believe that staying with pricing plans that the users are already comfortable with as well as by not requiring significant enhancements to the network accounting and charging systems can make for an easier adoption. Section 14.3.1 describes a more formal mathematical model for setting the discounts on SD channel.

PLUTUS uses priority as one of the mechanisms for creating logical channels: data transfers over the SD channel are carried out using only the capacity which is not in use by the ATD channel. The SD channel capacity is, therefore, not fixed but can exhibit wide temporal and spatial variations depending on the usage on the ATD channel. This means that by itself the SD channel data sessions may not only experience wide throughput variations but also suffer from periods of interruptions when no capacity is available. In a modified form, some minimum capacity may be reserved for SD sessions, thereby partitioning network capacity between the SD and the ATD channels. In the rest of the chapter, we mostly focus on the priority-based approach and where needed point out the extensions for the capacity partitioning approach.

Even with time shifting, the network can quickly become unusable if the traffic load shifted from the peak gets concentrated to create new peaks. If this happens, then the pricing mechanisms will not be very effective in fully exploiting the available spare capacity in the network. PLUTUS solves this problem by load balancing the traffic on the SD channel and by caching the transferred data in the local cache thus avoiding loading the network at the time of consumption. Load balancing is managed using an efficient delay and interruption-tolerant protocol and via special proxies that tightly schedule the transfer of data over the SD channel. This scheduling also minimizes the amount of time the sessions are kept radio active. This reduces the battery drain on the users' device and network resources unnecessarily during the transfers. In addition, because data is routed via proxies, it can be closely tracked, identified, (de)prioritized, and differentially charged by the network.

The pricing mechanisms considered in this chapter bear close resemblance to the Paris Metro Pricing (PMP) proposal [16] that uses price differentiation. As in PLUTUS, the basic idea in PMP is to partition the network into separate logical channels such that user is charged different prices for using them. The rationale being that the channels with higher prices would stay less congested than those with lower prices, thus providing a better service. PMP's mechanisms for creating logical channels include fixed partitioning of network capacity among channels or giving higher priority to traffic carried over higher priced channels. The biggest advantage of PMP comes from its simplicity. However, PMP is not without its challenges [16]. This is especially so when PMP is used as a mechanism for flattening the peaks and for encouraging the use of spare capacity in mobile networks. First and foremost, the adoption of an approach such as PMP may be limited if the users are not satisfied with the even lower quality of service (QoS) for the lower priced channels especially in mobile networks where the capacity is already constrained. Second, such an approach may not be very effective in managing congestion because multimedia content, which is a significant portion of the mobile traffic, has high QoS requirement and is, therefore, not well suited for shifting to lower capacity channels. With PLUTUS, on the other hand, the quality of experience stays high especially with multimedia content, because the data is cached and served from local storage. Unlike PLUTUS, PMP does not include any additional provisions for evenly distributing traffic load. Thus with PMP, the network channels may quickly become unusable because of traffic getting concentrated around (new) peak periods. Other issues with PMP [16] are how to set prices and allocate capacities for the channels and how to ensure predictable performance of the channels. Finally, the interface modification needed for the applications and network to make use of the different channels is a challenge as well. These challenges are addressed by PLUTUS as we describe in the following text starting with how to set discounts for the SD channel.

14.3.1 Pricing Model

One of the biggest challenges faced by service providers in pricing data is how to set price-based incentives and discounts. There is the fear that highly discounted plans can end up cannibalizing the premium full price plans ending in lower overall revenue. On the other hand, if the discounts are not substantial, then they may not be effective in driving the shift of traffic over to surplus capacity. In this section, we present a pricing scheme for which the service provider revenue can only increase no matter how heavily the discounted plans are subscribed. This is possible with our approach as the traffic on the SD channel can be very evenly load balanced and, therefore, any revenue lost from discounts is gained by the stranded capacity that gets released when the traffic is moved over to the SD channel. In addition, with the tight control that can be exerted on the data transfer delays in our approach, the right amount of segregation can be created between the end-users experience on data transfer using the SD channel versus using the ATD channel. For example, content delivery could be artificially delayed even when there is spare capacity to make the SD channel less appealing to users who demand high quality. As pointed out by Odlyzko [16] and Deneckere and McAfee [24], such segregation mechanisms, which are widely practiced by many industries (e.g., Couriers, Hi-Tech), can result in stable performance of the different channels with even load distribution. Furthermore, it has been established [24] that such a mechanism when used for price discrimination can be a strict Pareto improvement: can promote social welfare with strict benefits for all parties (service providers and users).

As mentioned earlier, the network cost is driven by peak load as new capacity is added when peak load reaches the current capacity. However, even at peak loads, the overall network is heavily underutilized. Typically, the peak-to-average-load ratio P is in the range 3–4. This implies that the highest average network utilization N is mostly between 25% and 35% (assuming that capacity exhaust happens at peak loads corresponding to full 100% network usage). We show that with our pricing scheme, the spare capacity can be discounted by as much as N percent while maintaining or increasing the ISP's revenues. Therefore, under our scheme, the spare capacity can be sold at almost one-third the price of any time data in most networks. In reality, the actual discount offered by the ISP may be less because they would want to set the discount just enough to maximize the overall revenue. We also show how to compute this optimal discount under our scheme.

Theorem 14.1 *Let the maximum network utilization in the present mode of operation (without subsidies for off-peak usage) be N percent. Then as long as the price charged for data transfer on the SD channel is at least N percent of the price charged for an equivalent any time data plan, the service providers revenue can not go down.*

Proof:

Let us assume that in the present mode of operation, the average revenue per user is R units and there are U users. So the ISP makes a total revenue of UR units with the peak network load of PU units and average network load of U units. Here P is the peak-to-average-load ratio, and this is related to the highest average utilization N percent as $N = 100/P$.

Now let the ISP offer a surplus capacity plan at S percent discount. Then the fractional discount for using the SD channel is $s = S/100$. The ISP derives an average revenue of sR per "nonpremium" user that signs up to use the surplus capacity plan while they continue to get an average revenue of R from a "premium" user that stays on the ATD channel. Let us denote by $x(s)$ and $1 - x(s)$ the resulting fractional sizes of the partition of the users into premium and nonpremium classes, respectively. $x(s)$ is an indicator of the take rate of the discounted plan: the smaller the $x(s)$, the higher the popularity of the discount s. Note that it is possible that a user signs up for both the plans. We can think of such users as representing two logical users, one premium and the other nonpremium, thus ensuring that these two user classes are disjoint.

The difference between a premium and nonpremium users is that the former contributes P units to the peak load but only contributes 1 unit to the average load and generates R units of revenue for the ISP. On the other hand, the nonpremium user contributes $l < P$ units of peak load while contributing 1 unit to the average load and provides a revenue of sR units. Here $l = 0$ when using priority-based partitioning (i.e., the nonpremium users can use only capacity left over from premium users). We can assume $l = 1$ for capacity-based partitioning (i.e., minimum capacity is reserved for nonpremium users) because it represents even load distribution on the SD channel (as attained in our scheme) carried over reserved capacity proportional to the average load of nonpremium users. We start out by assuming $l = 1$ and show later how our results extend to $l = 0$.

Let us consider the point in the evolution of the system under the new pricing scheme where the total number of users (SD and ATD users) is equal to the number of users in the system before introducing the new pricing scheme. As can be seen in the following, it is not required that the system must go through such a state for the proof to hold but rather is only used to illustrate the impact of the pricing change on the system parameters (peak load, revenue). The peak load in the network in this state is, therefore,

$$\text{Peak load} = x(s)UP + (1 - x(s))U, \tag{14.1}$$

and the revenue is

$$\text{Revenue} = x(s)UR + (1 - x(s))sUR. \tag{14.2}$$

Note that this revenue can be lower than before (UR). However, because the peak has come down, the network now can sustain additional users without capacity expansion. The operator can keep adding new users (premium and nonpremium in the same ratio as before) till the peak load reaches PU the peak load we started with. Let $y(s)$ be the fractional increase in the number of users. That is, the network can now accommodate $(1 + y(s))U$ users. From Eq. (14.1), it, therefore, follows

$$[x(s)P + (1 - x(s))](1 + y(s))U = UP. \tag{14.3}$$

From Eq. (14.2), it, therefore, follows that for the combined revenue from all these users to be no less than the original revenue without subsidies, we must have

$$[x(s) + (1 - x(s))s](1 + y(s))UR >= UR. \tag{14.4}$$

From Eqs. (14.3) and (14.4), we can infer that as long as $s >= 1/P = N/100$, Eq. (14.4) holds for all $0 \leq x(s) \leq 1$. This means that as long as the SD channel price is at least N percentage of the ATD channel the ISP's revenue can only go up irrespective of the take rate of the discounted plan (i.e., the value $x(s)$).

In the $l = 0$ (priority-based partitioning), the nonpremium users do not contribute to the peak load, and hence, Eq. (14.3) for the maximum off-peak load (both premium and nonpremium users contribute equally to the average network utilization) is changed to

$$[x(s) + (1 - x(s))](1 + y(s))U \leq UP \quad \text{or} \quad 1 + y(s) \leq P \tag{14.5}$$

along with an additional constraint on the peak load owing to premium users alone:

$$[x(s)P](1 + y(s))U \leq UP \quad \text{or} \quad (1 + y(s))x(s) \leq 1, \tag{14.6}$$

with at least one of these two constraints holding tight at optimality. It can be seen that even for $l = 0$ at $s >= 1/P = N/100$, the inequality (14.4) holds for all possible $0 \leq x(s) \leq 1$ implying the ISP revenue can only increase. In particular, the ISP revenue is strictly higher unless there are no premium users (i.e., $x(s) = 0$) at the selected value of s. ■

Corollary 14.1 *For our scheme, the optimal ISP revenue is obtained by maximizing a function that depends on the popularity of the subsidized plan and the peak-to-average-load ratio P (before subsidized plan is offered).*

Proof: We only consider $l = 1$ because the other case is similarly handled. From the proof of Theorem 14.1, it follows that the ISP's revenue maximization problem is

$$\max \alpha[X_1(s) + X_2(s)s]$$

such that

$$\alpha \left[X_1(s) + \frac{X_2(s)}{P}\right] = 1$$

and

$$\frac{1}{P} \leq s \leq 1,$$

where at fractional discount s for SD channel, the ISP has $X_1(s)$ premium and $X_2(s)$ nonpremium users. As before, this is a logical partitioning of users into disjoint sets. The functions $X_1(s)$ and $X_2(s)$ depend very much on inherent value the users derive from a given discount s versus their experience on each of the two channels. As such user surveys, focus groups, and other means may be employed to estimate these functions. The parameter α represents the additional users that can be accommodated and can be determined based on the value of the functions $X_1(s)$ and $X_2(s)$. Once these functions are determined, the optimal fractional discount s can be derived by solving the optimization problem. ■

So far we assumed a single pricing tier for each of the premium and nonpremium class. In practice, pricing plans may offer multiple tiers (e.g., monthly data bundle of 5 GB for \$50 or 3 GB for \$30, etc.) and users may have many possible ways of adopting new pricing plans such as by moving from a higher to a lower tier of the premium class and making up for the reduced data bundle by supplementing with a nonpremium pricing plan tier. The analysis presented above can also be extended to this more general case; however, we omit the details for lack of space.

The pricing plan presented above can also be offered in a "bundled" mode where instead of offering new surplus capacity plans, the ISP may just increase the amount of data provided to the user at a slightly higher cost. The additional data would be allowed only at off-peak times (over the SD channel). The analysis presented above directly applies to this case by virtue of the logical partitioning of each user into the premium and the nonpremium classes.

14.3.1.1 Other Scenarios

14.3.1.1.1 Network Efficiency Enhancements Without Pricing Plan Changes With our SD-channel-based delivery approach, network peaks can also be flattened and the user experience enhanced by anticipating, delivering, and locally caching data in advance of usage. This can even be accomplished without price-based incentives as long as there are effective mechanisms for predicting user behavior. Although no prediction mechanism is perfect and, therefore, much of the data preloaded into the local cache may get wasted and not be consumed, this is not a significant problem because the marginal cost of delivering data using spare capacity is negligible. Note that in such a mode of operation, no user behavior change is required (adoption of new pricing schemes or delayed data consumption) and the user may even be oblivious to any such preloading and caching of data.

14.3.1.1.2 Sender-Pays or Two-Sided Pricing The approach described here can easily accommodate pricing schemes where the content provider is charged for all or portion of the content delivery. Such pricing models have been tried in the past (e-book delivery for Kindle) and are being considered as an equivalent of toll-free service for data traffic. For such pricing mechanisms to be appealing to the content providers, the delivery costs especially for large multimedia files (movies, shows) should be fairly low. This is possible when the content is delivered using surplus network capacity because it has negligible marginal cost. With this approach, the additional benefit is predictable delivery estimates and higher quality playback experience as the content is cached locally on the users device [potentially with digital rights management (DRM) and access controls for premium content]. Also with PLUTUS, the charging can be simple as the content provider can simply pay according to the size of the delivered data and PLUTUS can ensure that the users data plan is not charged for the bytes delivered on the SD channel. This addresses many of the issues with sender-pays plans [25].

14.4 ARCHITECTURE AND DESIGN

There can be wide variations in the SD channel capacity, and therefore, it may not be well suited for session-oriented protocols such as Transmission Control Protocol (TCP) that time out or exhibit poor performance under impairments. Moreover, keeping the sessions "alive" for long periods of time can drain the handset battery quickly and also unnecessarily tie up limited radio resources (e.g., traffic channels). Therefore, the implementation of SD channel requires more than just strictly prioritizing the ATD channel data packets over SD channel packets in the network routers and switches and RAN elements. The end applications that use the SD channel need to be resilient to session interruptions and must be capable of relaunching sessions after capacity becomes available again. Also the network must be able to convey capacity availability to the applications that must always stay ready in standby mode to be able to avail of this capacity. The system could, therefore, get complex if each application and the network were to be modified to work this way. A more scalable approach is to enhance the network to provide most of these features, thus minimizing the impact on individual applications. This motivates our design for PLUTUS system whose main components are shown in Figure 14.1.

The Control and Scheduling Components (CSC) track and control all the data transfers on the SD channel. These data transfers are routed via proxy components [Server Proxy Components (SPC) and Client Proxy Component (CPC)] that can

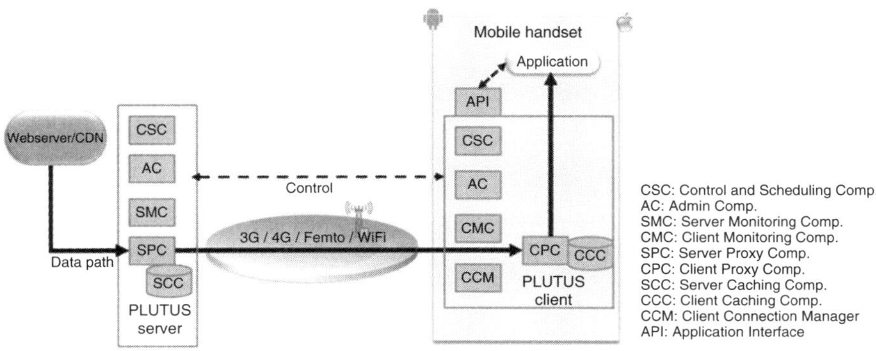

Figure 14.1 Architecture of the PLUTUS system.

suspend or resume data transfers as needed. The Server Monitoring Component (SMC) and Client Monitoring Component (CMC) together track the network state including the availability of capacity on the SD channel, the availability of other network accesses (e.g., WiFi, Femto), and also the user's radio link state and device state (battery, storage). The information on alternative network availability is used by the PLUTUS Client Connection Manager component (CCM) for automatically switching over data transfers to the best network. Owing to the possibility of long interruptions, the SD channel data sessions operate mostly in the background with data caching performed by the Server Caching Component (SCC) and Client Caching Component (CCC). PLUTUS exposes application programming interfaces (APIs) for applications to initiate data transfers over the SD channel as well as for getting notifications on the status of data transfer and for accessing their transferred data from the PLUTUS caches. The PLUTUS system has additional Admin Components (AC) for accounting and charging of data transferred over SD channels and for provisioning of operator policies and user preferences for controlling the data transfer process. We describe all these components in the following in detail.

14.4.1 Components

14.4.1.1 The Scheduling component The CSC is responsible for computing estimates on delays for data transfers on the SD channel and also for scheduling the data transfers within the estimated delays. Using historic data to predict anticipated user and network dynamics, the CSC constructs a globally efficient, low delay, data transfer schedule. Data transfers are then instrumented according to this schedule with changes applied based on the actual system dynamics.

First, the amount of capacity predicted to be available on the SD channel of each cell in each time period along with the predicted availability of the access points (e.g., cells, WiFi, small cells) to each device during each time period is computed. This is based on the past and current trends on network and device availability, user mobility, device usage, traffic load, etc. Each data transfer job is then characterized by the deadline by which it needs to finish (its delay estimate), the outstanding data to be transferred (size), and the set of other jobs that must be processed before it (e.g., based on a ordering specified by the user or their submission time). Next a global scheduling problem is solved for assigning the (pre-emptive) processing of the data transfer jobs on selected access points at times when the SD channel of the access points is predicted to be available to the respective jobs. It is acceptable to schedule a single data transfer job over several nonconsecutive times or different accesses. The objective is to complete jobs by their deadline such that all the job ordering constraints are enforced. The solution to the scheduling problem is the temporal assignments of jobs and portions thereof to the access points and the rates for the underlying data transfer. The solution to the scheduling problem also provides data transfer delay estimates in the case when a priori deadlines are not given.

We now show how to mathematically model and solve this scheduling problem. We assume that the scheduling is performed at the granularity of time slots each of which is τ seconds long. In other words, the deadlines for job completion have to be at the granularity of τ so either a job can be scheduled in this time slot or not. The choice for τ depends on whether we want the schedule to be more flexible in adapting to dynamic changes (τ should be small) or we want to reduce the overhead of recomputing the schedule more frequently (τ should be large). In practice, $\tau = 1$ second can provide a good trade-off for these two requirements.

Let A denote the set of access points (e.g., cells, WiFi, small cells). For an access point $a \in A$, let $I(a)$ denote the set of time slots when it is predicted to be available. Let $c_a(i)$ denote the available capacity in bytes of access point $a \in A$ in time slot $i \in I(a)$. Thus $c_a(i)$ is τ times the average available rate of access point a in time slot i. Let U be the set of users, and let $J(u)$ be the list of jobs $j_1(u), j_2(u), \ldots$ of user $u \in U$ arranged in the order in which they need to be processed. Each job $j_k(u)$ is characterized by its size $s_k(u)$ in bytes, its arrival time $a_k(u)$, and its deadline $d_k(u)$. Let $a_u(t)$ denote the preferred access point $a \in A$ of user u among the access points that are available to user $u \in U$ at time slot t. Note that $a_u(t)$ is not defined if user u is not in proximity of any available access point. In case there are multiple such access point, then we assume that there is a mechanism to pick the preferred one (e.g., prefer WiFi to 3G).

The schedule is determined by solving a max-flow problem in a graph constructed as follows. The graph has four types of nodes: user nodes (V_1), one per user $u \in U$; job nodes (V_2), which are one for each job for each user,

$$\cup_{u \in U} \{j_i(u) | j_i(u) \text{ is in list } J(u)\};$$

available time slot nodes (V_3) one per access point and time slot in which it is available

$$\cup_{a \in A} I(a);$$

and a pair of source and sink nodes s and t, respectively. In this graph, the edges and their capacities are as follows. From the source node s, there is a directed edge to each of the nodes in V_1 of infinite capacity. From a node $v_1 \in V_1$ that corresponds to user u, there is a directed edge to every node $v_2 \in V_2$ that corresponds to the job $j_k(u)$ for user u. This edge has capacity set to the size $s_k(u)$ of job $j_k(u)$. From every node $v_2 \in V_2$ that corresponds to the job $j_k(u)$ for user u, there is a directed edge to every time slot node $v_3 \in V_3$, which has the following properties. The time slot i corresponding to v_3 must be in the time interval $(a_k(u), d_k(u))$ in which the job is available to be processed. The node v_3 must correspond to the access point $a_u(i)$ that is available to user u in time slot i. The capacity of this edge is set to $c_a(i)$ corresponding to the available capacity of access point a in time slot i. Finally, there is a directed edge from every time slot node $v_3 \in V_3$, which corresponds to time slot i and access point a, to the sink node t of capacity $c_a(i)$. It can be seen that a feasible flow from s to t in this graph corresponds to a schedule that satisfies all constraints except possibly the constraint on the ordering among the jobs of a user. However, the latter can be easily attained by rearranging the schedule individually for each user so that the jobs are scheduled in the desired order on the set of access points and time slots determined using flow computation. The max-flow in this graph corresponds to a schedule that maximizes the total amount of processing of the jobs, and in particular, it can be used to test feasibility of processing all jobs. Other objective functions (besides maximizing the total flow) can also be used to enforce additional requirements (e.g., fairness).

The scheduling problem as defined can be solved using linear programming techniques but can quickly become intractable even for medium size networks with a few millions of users. However, we believe that fast heuristics with bounded performance should be possible under practical assumptions. Additional constraints may be introduced in the problem formulation for a more efficient schedule. These include constraints to minimize the delay for data transfers. It may also include constraints to avoid high signaling load on the network and high device battery drain because of frequent job interruptions (e.g., because of pre-emptions, lack of capacity, network switchovers) resulting in inefficient traversals between different radio states [26]. In addition, constraints may also be introduced to take advantage of the spectral efficient wireless broadcast and multicast capacity if available for data transfers. This involves identifying overlaps across different users jobs that can be scheduled together using multicast [27].

This “offline” solution to the scheduling problem serves as a guide for the “online” scheduling of the data transfers. It provides the necessary input to make the decision as to when to start, suspend, and resume each of the data transfer and at what rate. As the offline solution is based on predictions regarding the availability of capacity at the access points, mobility of the user, and the parameters of the currently submitted data transfer jobs (sizes, deadlines,) it can quickly become suboptimal unless it is dynamically updated to deal with the occasional deviations in user actions, mobility, network conditions, and unanticipated traffic loads. In addition, the schedule must be dynamically updated to accommodate new data transfer jobs. Most of the times, it should be possible to handle these changes efficiently locally with only incremental updates thus requiring very infrequent computationally intensive recomputations of the schedule.

Online scheduling optimizations in PLUTUS also include adjusting the schedule including the job priorities to give more opportunities to data transfers that have fallen behind or at times when they can make more efficient use of network resources (e.g., when radio link is of higher quality), thus increasing the network efficiency. An additional consideration is to keep the SD sessions radio active for the least amount of time in order to minimize battery drain and to avoid tying up scarce RAN resources. By scheduling data transfers, which share limited capacity (e.g., within one cell), one at a time rather than all at the same time, the CSC not only lowers the average transfer time but also reduces the average time spent in radio active state. The benefit of serial over parallel data transfer scheduling within a single cell is illustrated in Figure 14.2. When all three data transfers are scheduled together (as in Fig. 14.2a), they all take a long time to finish because each one gets only one-third of the cell’s capacity and,

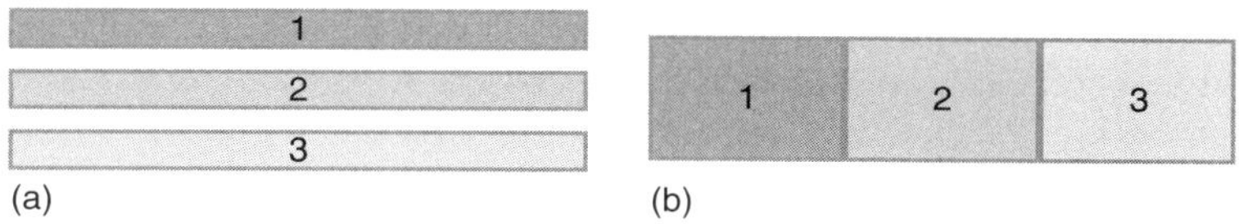

Figure 14.2 Serial scheduling is more efficient. (a) Parallel data transfer and (b) serial data transfer.

therefore, also stays active for the whole time. When scheduled one at a time (as in Fig. 14.2b), they all still finish within the same time period. However, some data transfers finish earlier (1 and 2), and more importantly, each of them stay active only one-third of the time. More generally, the CSC scheduler not only schedules a few data transfers at any given time but also pre-empts and switches data transfers in order to make fair progress across all of them. Such fairness is required to gain the trust of the end users possibly by also putting a neutral entity in charge of overseeing the scheduling operations.

14.4.1.2 The Proxy components and the Data Transfer Protocol In PLUTUS, the SD channel application's data session is proxied via the CPC that runs in the background on the client side and is optionally proxied via the network-resident SPC. The session data flow for a download involving both the CPC and the SPC is illustrated in Figure 14.1. Upload works in the same way and is, therefore, not described. The download can have two logical phases. In the first phase, data is downloaded by the SPC from the data source in the network and temporarily cached in the server cache SCC. In the second phase, the data cached in the SCC is downloaded by the CPC to the client cache CCC using the disruption-tolerant data transfer (DTDT) protocol running between the SPC and CPC. The second phase does not have to wait for the first phase to finish but rather the download to the client may start as soon as the data is available on the server cache. The DTDT takes care of suspending or resuming content transfer and stopping or restarting sessions in reaction to scheduling controls from the CSC. The DTDT can be implemented using any standard data transfer protocol (e.g., TCP) when enhanced with the additional capability of automatically stopping and starting data connections and rate limiting. All the state about how much data is transferred and where to resume data transfer from is maintained by the CPC because it has that information. The SPC component is not required if (for download) the data source server is capable of resuming the data transfer from any specified offset. Such capability is widely supported by http downloads from web servers on the World Wide Web. In such cases, the CPC makes the request to the data source server to resume the data transfer from the position where it terminated the last time (e.g., via range request for http downloads). This is, therefore, well suited for multimedia content because commonly used protocols such as progressive download and adaptive bitrate streaming are all based on http download. The preferred mode of operation for the PLUTUS system is to avoid proxying all the data transfers via a SPC as it results in a more distributed and scalable architecture.

14.4.1.3 The Monitoring and Data collection component The CMC and the SMC together monitor the surplus capacity of the network. Note that it is not just sufficient to use network-based QoS mechanisms to prioritize ATD traffic over SD traffic. Monitoring is needed to identify opportune times when SD channel data transfers can be started, otherwise data transfers must be kept alive all the time with serious impact on battery life and network performance. In PLUTUS the focus is mostly on RAN and in particular on cell level (per radio link) monitoring. Typical monitoring systems make the assumption that the available capacity can be determined as the difference of the capacity of the bottlenecked link to the offered load. However, in a mobile network, this is not always the case because the capacity of the radio link is very much dependent on the channel quality of the users and, hence, can be time varying. Thus even though the observed load on the cell may be low, the cell may still not have any capacity left over. Therefore, a monitoring solution has to measure the cell load in conjunction with the users channel quality. On the network side, such monitoring is best done from the cells base transceiver station (BTS) where both the users channel quality and the allocated BTS resource usage can be monitored. This results in more accurate monitoring but can be a challenge to deploy because of (i) the significant cost and challenge of upgrading all the BTSs many with proprietary interfaces, (ii) the additional cell (e.g., backhaul) overhead to send monitored data from the BTSs to the PLUTUS system, and (iii) the resulting high delays in informing the PLUTUS clients of the changes in network load thus slowing their reaction time.

On the client side, a combination of active and passive probing techniques can be used for monitoring local conditions such as the users channel quality as well for estimating global network state including cell congestion. Network availability is accurately estimated by passively monitoring the radio channel slot occupancy [28]. However, this can be challenging because it requires access to proprietary vendor-specific APIs for the radio modem of the device. Active probing, on the other hand, involves sending probe packets to estimate the available bandwidth and, therefore, is mostly independent of the device firmware, hardware, operating system, and so on. The challenge with active probing, however, is the additional signaling and data overhead it can introduce. We describe in Section 14.4.2 how active-probing-based client-side monitoring of available capacity is implemented in PLUTUS.

In addition to tracking the user and network state, the CMC monitors the device state including battery level, charging status, storage usage, processor occupancy, and so on. This is to ensure that device resource usage is according to user preferences and settings (e.g., SD channel data transfers only when battery level is high or the handset is getting charged). Data is also collected by the CMC on the availability of other networks (WiFi, small cells) so that data transfers can be moved to the least cost network. Data may also be collected by the CMC on the users network and content access and mobility patterns. This may include data on the locations much frequented by the user, the types of networks available there, their loading state, and their probability of being used by the user. Such historic data can be a sufficiently accurate predictor of the future [2, 29, 30] and, therefore, helps in deciding when and where to schedule users data transfers on the SD channel. Likewise data collected on the users content access patterns helps to predict contents of interest that can be preloaded in advance of users request. Although there can be privacy concerns related to the collection of such personal data, these may be ameliorated by opt-out and anonymization mechanisms [31].

14.4.1.4 The Admin component As transferring large files may not just incur significant cost to the user but may also result in substantial battery drain and quickly fill up the storage on their device, PLUTUS lets the user set preferences via the AC to control the transfer process. This may include allowing content transfer only above certain battery levels or only during charging periods or only when there is enough storage on the device. It may also include preferring the use of their home WiFi and requiring automatic switchover to make use of a preferred access network when available. Likewise the operator may configure policies to block data transfers during certain time periods or may allow the use of their managed WiFi hotspots. In order to enforce these preferences and policies, PLUTUS closely tracks the network and device state including the availability of network accesses.

The AC also interfaces with policy and charging functions (e.g., PCRF) in the network to convey session information (e.g., TCP connection's 5 tuple) for every data transfer over the SD channel. This helps the network gateways to identify, classify, differentially account, and charge all the packets belonging to the SD channel traffic.

14.4.1.5 The Application, User Interfacing component Applications interface via the PLUTUS APIs to request data transfer on the SD channel. The request may be triggered from a direct user action where the user explicitly selects a digital item to be delivered via their SD channel data plan. The request may also be triggered by an indirect user action where the user has allowed the application to automatically deliver data items that become available on their social networks, subscribed channels, playlists, video queues, camera folder and so on. The data transfer is then routed via the SD proxies and is scheduled for delivery by the CSC using the DTDT protocol. The PLUTUS APIs provide delivery state updates to the application including when the data item is available for use. Delivered item is cached locally and is accessed by the application using PLUTUS APIs.

14.4.1.6 The Prediction and Preloading Component The PLUTUS system is ideally suited for preloading content on users device in advance of consumption. An important challenge, however, is to be able to deliver content that has the greatest chance of being consumed by the user. However, more and more traffic on the mobile network is from recommendation-based content. This includes much of the usage on Pandora, Netflix, and even Youtube [32], three of the biggest source for mobile data traffic [4]. In addition, social-network-based recommendations can play a big role in this preloading as Facebook, Twitter, and other social networks are becoming a significant source of referrals [33] especially for video content [34]. This is in addition to subscription-based preloading of content from media channels, playlists, movie queues, watch it later applications, magazines, news web sites, and so on. The best delivery options for preloading popular content may be mobile multicast and broadcast. Thus a combination of unicast and multicast may be used for preloading content [27].

14.4.2 Client-Side Monitoring of Available Capacity

PLUTUS uses an active probing mechanism whereby traffic is occasionally downloaded by the client to allow it to sample the available bandwidth [18] on the SD channel. One of the parameters that the client measures is the download throughput. However, throughput by itself is not enough to estimate the available capacity. This is because in wireless networks, the throughput depends not only on the available capacity but also on the maximum cell capacity (which can vary from cell to cell based on its technology, frequency, number of carriers, etc.) and clients channel quality. In particular, even with only one client on the cell, the download throughput may be low if the client has poor channel quality. In addition, the throughput also depends on the number of users or level of sharing in the cell. This is because the base station or eNodeB uses proportionally fair scheduling to divide radio resources fairly and evenly between all active users attached to a cell.

The estimation of the available capacity by the client proceeds as follows. First, by correlating historic data about channel conditions, as measured by parameters such as signal-to-noise ratio (SNR) and received signal strength indication (RSSI) with user throughput, PLUTUS computes the maximum cell capacity for each channel condition. This is performed separately for each cell. Then in real time, the client measures the download throughput and simultaneously evaluates its channel conditions over a sequence of time windows. Each of these time windows is short enough that within each of these windows, either there is no (ATD channel) data transfer from any other user or the cell capacity is fully utilized from such data transfers. Thus in these time windows, a PLUTUS client gets 1/n-th of the maximum cell throughput, that is possible for the given channel condition. Here n is the number of cell users active during the time window, inclusive of this client. This is illustrated in Figure 14.3.

Let the client observe throughput $B_0, B_1, \dots, B_k$ over a time period defined by a sequence of time windows $t_0, t_1, \dots, t_k$. Let the max channel capacity in these time windows (as estimated by channel quality measurements) be $C_1, C_2, \dots, C_k$. Then the average available capacity over this time period spanning these k time windows is estimated as l/k, where $l \leq k$ is the number of time windows in which the throughput exceeds half the cells max capacity: those time windows t_i for which $B_i > C_i/2$. This is because in such time windows, there is no other active user, otherwise the throughput would be at most $C_i/2$. This fraction is then compared to a cell-specific capacity threshold τ. Only if the fraction is not less than τ, the cells available capacity can be utilized for SD channel data transfers in the subsequent time period.

Usually, the active probing as described earlier operates independent of client's data transfers. However, in PLUTUS, the two are combined. The PLUTUS proxy components monitor the capacity availability as it is transferring users data. In other words, no special data transfers are needed for active probing, thus eliminating the overheads. The only exception is when there is no available capacity on the cell and, hence, the users data transfers are kept suspended. In that case, there is some

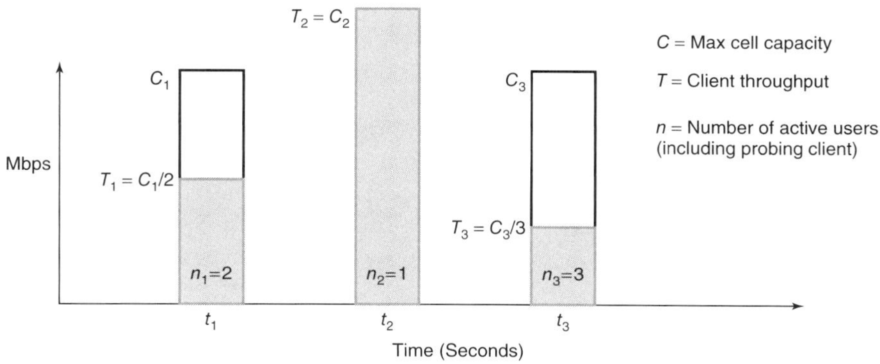

Figure 14.3 Active probing throughput in three time windows.

impact on the data transfers on the ATD channel by the active probing for SD channel capacity. However, this is minimized by appropriately spacing out the active probing activities while still ensuring that capacity availability can be detected in a timely manner. Further optimizations include scheduling probing activities across clients within a cell to ensure that only a small number of clients are actively probing in the cell at any given time. The selection of clients for active probing within each cell is dynamically adjusted to keep the probing load spread across the clients, thus minimizing their battery drain from probing.

14.5 PERFORMANCE EVALUATION

We implemented the PLUTUS system with client software running on Android and iOS handsets and other control and server software resident in the network. Here we present some results from analyzing the performance of PLUTUS on a commercial network for which data was collected using a tool developed internally. The data includes network load and other information for a day from more than 5000 3G cell sites. We also present results on the performance of the PLUTUS system in the controlled laboratory environment of mobile operators.

14.5.1 Network Utilization

The traditional understanding is that during peak hours the network is very heavily loaded and there is not much spare capacity during those times. As a result, most solutions for utilizing spare capacity are primarily designed to look for spare capacity during "off-peak" hours. However, we found that this view is not accurate and even busy cell sites can exhibit many short periods of network availability even during peak hours. This effect is particularly pronounced when the network load is analyzed at the smaller time granularity of minutes. As a result, we have designed PLUTUS to quickly identify and react to availability of spare capacity even at these short intervals thus maximizing the efficiency gains.

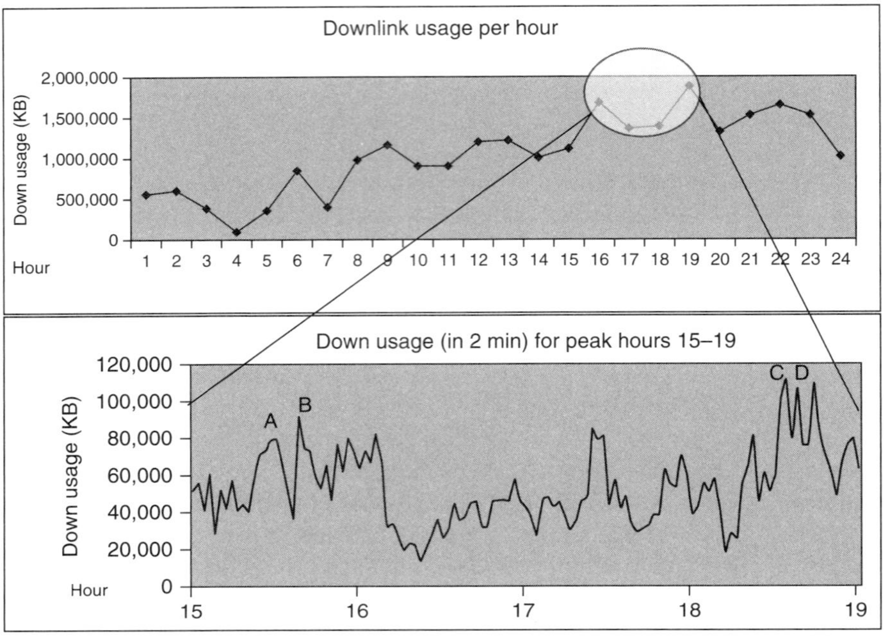

Figure 14.4 Network load.

Figure 14.4 shows the typical load of a congested cell. The top graph shows the hourly distribution of load where hours 15–19 are seen to be the most loaded and can be identified as peak hours. However, as we drill down to 2-min data intervals for the peak hours, it is clearly seen that there are many time intervals where the network load is quite low, for example, between peaks at A,B and C,D, and this spare capacity can be leveraged for data transfers on the SD channel even during peak hours. The PLUTUS system's capability to detect spare capacity even in peak periods not only enables it to maximize the utilization gains but also helps it to reduce the wait time for SD data session, making it more attractive to the user.

14.5.2 Delay

In PLUTUS, the SD channel data sessions have a lower priority than the ATD sessions and, hence, can incur additional delays as they are waiting for capacity to become available. We used the network load data from top 200 loaded cells of the aforementioned 3G network, to analyze the delay introduced by using the SD channel instead of the ATD channel. We used the current network load as baseline and introduced new data sessions of different sizes. The comparison was done for equal number of SD and ATD sessions over a 24-h period. However, the number of sessions varied in proportion to the existing load on the system at a given time to reflect higher user activity at busy times compared to nonbusy times. To compute delay, we compare the time difference between completing the new data session on an ATD versus SD channel.

The 200 cells used for this analysis have an average utilization of 35%. Although we do not have any data on how close these cells are to exhausting their capacity but given that these are the top loaded cells for a heavily utilized network, we can reasonably assume that their peak-to-load ratio is close to 3.

In our analysis, we allow the SD channel to use available capacity from the network only if the total load on the network at that time is 70% of the network capacity. If the network load is higher, the network is considered as congested and only ATD traffic is allowed. As seen in Figure 14.5, almost 87% of the smallest (10 MB) SD channel data sessions incur no additional delay compared to equivalent ATD channel data sessions. Also 65% of large data sessions (500 MB) incur no additional delays. In this network with maximum throughput of 60 MB/min, the available throughput at average network utilization of 35% is 39 MB/min. Therefore, a 500-MB ATD session can take approximately 13 min to finish while only 12% of equivalent SD sessions incur additional delays of 4 min or more. This means that 88% of 500-MB SD sessions incur no more than 30% additional delay. Also more than 80% of small and medium sized sessions incur no additional delay at all. The relatively short delay on the SD channel is due to the availability of enough intermittent network capacity even during peak periods as was shown in the previous section. In addition, during peak periods, because an ATD session gets additional service (compared to an equivalent *SD* session) only at times when the network is highly loaded (70% or more), it cannot finish much faster.

We analyzed the delay performance of PLUTUS in a laboratory environment under different congestion conditions. In particular, we considered two scenarios where network is congested either 20% or 50% of the time. In this laboratory, the maximum throughput of the clients ranged between 500 and 600 Kbps. As shown in Figure 14.6, when network is congested 20% of the time, there is an additional SD channel delay of less than a minute for a download of 63 MB data that takes 20 min on the ATD

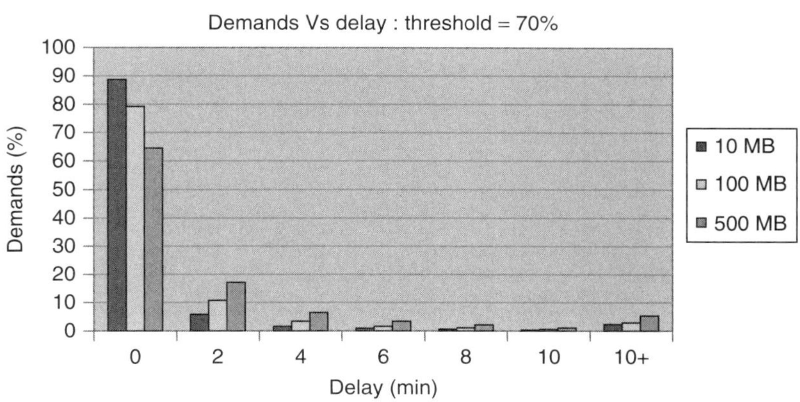

Figure 14.5 SD channel additional delay.

Delay Performance	Congestion = 20%	Congestion = 50%
Data downloaded on ATD channel in 20 min	62.9 MB	44.94 MB
Additional delay on SD channel (in minutes)	0.89	3.762

Figure 14.6 SD channel delay versus congestion.

channel. Even for networks with 50% congestion when the SD channel has spare capacity only half the time, a user download of 45 MB that takes over 20 min, sees an additional delay of just 3.7 min or 18%. As in most networks the peak period is less than 50%, it follows that the users will see very short additional delays between sessions on ATD and SD channel. Moreover, although an ATD session would stay active 100% of the time during the download, the SD session would be active only 68% of the time, 50% of the time when network is not congested and an additional 18% because of the delay. As SD sessions are active for a shorter duration during the download the PLUTUS system can not only improve device and battery performance but also reduce the impact on the network by not tying up precious resources for as long.

14.5.3 User Experience

In the laboratory environment, we also studied the impact on session quality of users moving some of their sessions to the subsidized SD channel. Figure 14.7 shows the network with four peaks created by ATD session activity. The grey curve depicts the

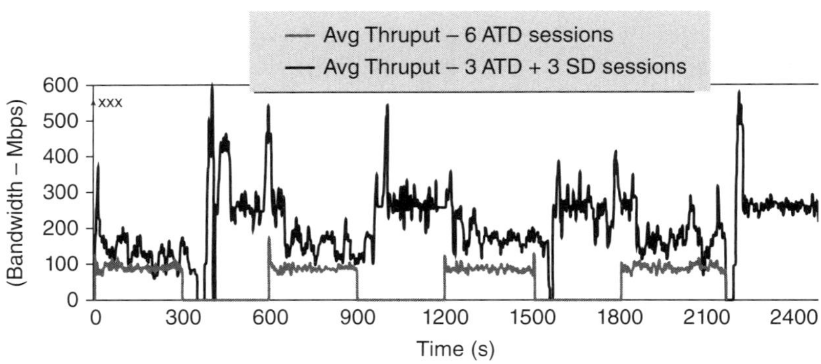

Figure 14.7 User performance.

average throughput of six ATD sessions that were active for 5 min and then inactive for the next 5 min thereby creating a pattern of 5-min long-peak and off-peak periods. The black curve depicts user throughput as three of the six ATD sessions were converted into SD sessions so that only three ATD sessions stayed active during peak times and the activity of the three SD sessions got shifted to off-peak times. As can be seen that this resulted in the throughput of the individual ATD sessions going up from 100 Kbps to more than 150 Kbps. This means that the premium ATD users see less buffering and better quality even during peak times. The ATD sessions that got time shifted as SD sessions to off-peak periods got even better average throughput of more than 250 Kbps. This is because the PLUTUS system scheduled at most two SD sessions simultaneously in order to avoid creating new peaks and to reduce the length of time that SD sessions are active. Thus with PLUTUS spreading out the load more evenly in the network, it results in improved device throughput across the board, thereby increasing user satisfaction.

14.6 CONCLUSIONS AND FUTURE WORK

In this chapter, we presented a capacity monetization and network efficiency enhancement PLUTUS system for efficiently and cost-effectively delivering multimedia content using unused capacity of diverse access networks. It works by scheduling the data transfers at the appropriate times using the appropriate access networks and by caching data on the large storage on end devices. We showed that with PLUTUS the operators revenue increase can be ensured using only simple and predictable pricing plans. We addressed the many challenges of designing the scheduling and monitoring components of the PLUTUS system. We also presented the initial results on the viability along with the efficiency gains from deploying the PLUTUS system both in commercial and in laboratory environments.

Future research directions include creating improved models for predictions of the network state and user behavior based on the data collected by the PLUTUS clients. Linear-programming-based approach to model and solve the basic scheduling problem can be quite powerful but can also be computationally intensive. Future research directions include the design of faster heuristics with bounded performance guarantees that can also flexibly accommodate the various constraints of the scheduling problem. Future work also includes personalized recommendation systems that can maximize the "hit" ratio while balancing the limits on the network capacity for delivering content along with the storage on users device for preloading content.

ACKNOWLEDGMENTS

We thank the entire Kaveri team at Alcatel-Lucent Ventures for their contributions to the PLUTUS system.

REFERENCES

1. Cisco. Cisco visual networking index: global mobile data traffic forecast, 2011-2016.
2. U. Paul, A. P. Subramanian, M. M. Buddhikot, and S. R. Das. Understanding traffic dynamics in cellular data networks. In *Proceedings IEEE INFOCOM*, 2011
3. M. El-Sayed, A. Mukhopadhyay, C. Urrutia-Valdés, and Z. J. Zhao. Mobile data explosion: monetizing the opportunity through dynamic policies and QoS pipes. *Bell Labs Technical Journal*, 16(2), 2011, 79–100.
4. Sandvine. Global Internet Phenonomena Report 1H 2012. Available at: http://www.sandvine.com/downloads/documents/Phenomena_1H_2012/Sandvine_Global_Internet_Phenomena_Report_1H_2012.pdf, 2012.
5. P. Du and N. Lu. Appliance commitment for household load scheduling. *IEEE Transactions on SmartGrid*, 2(2), 2011, 411–419.
6. A. H. Mohsenian-Rad, V. W. S. Wong, J. Jatskevich, and R. Schober. *Optimal and Autonomous Incentive-Based Energy Consumption Scheduling Algorithm for Smart Grid*. Innovative Smart Grid Technologies 2010. IEEE, 2010.
7. A. Ingold, I. Yeoman, and U. McMahon. Yield Management: Strategies for the Service Industries, 2001.
8. K. Lakshminarayanan, V. Padmanabhan, and J. Padhye. Bandwidth estimation in broadband access networks. In *Proceedings of ACM Internet Measurements Conference, Taormina*, Oct. 2004.
9. D. Koutsonikolas and Y. Charlie Hu. On the feasibility of bandwidth estimation in 1x evdo networks. In *MICNET*, 2009.
10. A. Gerber, J. Pang, O. Spatscheck, and S. Venkataraman. Speed testing without speed tests: estimating achievable download speed from passive measurements. In *IMC*, 2010.
11. Kevin C. Tofel, Gigaom. ATT ends Flat-rate mobile plans. Available at: http://gigaom.com /mobile/att-shuts-down-the-mobile-broadband-buffet/2010
12. arstechnica. Verizon confirms the future of 3G data is tiered. Available at: http://arstechn ica.com/gadgets/2010/09/verizon-confirms-the-future-of-3g-data-is-tiered/2010
13. Stacey Higginbotham, Gigaom. Differential data rates based on Time and Apps. Available at: http://gigaom.com/2010/12/14/mobile-operators-want-to-charge-based-on-time-and -apps/2010
14. MTN. MTN provides free facebook. Available at: http://www.mtn.co.ug/MTN-Services/Communication/Facebook.aspx, 2011.
15. H. S. Houthakker. Can Speculators Forcast Prices. The Review of Economics and Statistics, 1957.
16. A. Odlyzko. Paris metro pricing for the Internet. *Proceedings of the 1st ACM Conference on Electronic Commerce*, pp. 140–147, ACM, 1999.
17. S. Sen, S. Ha, C. Joe-Wong, and M. Chiang. Pricing Data: A Look at Past Proposals, Current Plans, and Future Trends. Available at: http://arxiv.org/abs/1201.4197
18. R. S. Prasad, M. Murray, C. Dovrolis, and K. Claffy. "Bandwidth estimation: metrics, measurement techniques, and tools," *IEEE Network*, 17, 2003, 27–35.
19. A. Jardosh, K. Ramachandran, K. Almeroth, and E. Belding-Royer. Understanding congestion in IEEE 802.11b wireless networks. In *Proceedings of the Internet Measurement Conference*, Oct. 2005.

20. P. Acharya, A. Sharma, E. M. Belding, K. C. Almeroth, and K. Papagiannaki. "Rate adaptation in congested wireless networks through real-time measurements," *IEEE Transactions on Mobile Computing*, 9(11), 2010, 1535–1550.
21. S. Humair. Yield management for telecommunication networks: defining a new landscape. PhD Thesis, MIT, 2001.
22. F. Qian, K. S. Quah, J. Huang, J. Erman, A. Gerber, Z. M. Mao, S. Sen, and O. Spatscheck. Web caching on smartphones: Ideal vs. Reality. In *Mobisys*, 2012.
23. S. Ha, S. Sen, C. Joe-Wong, Y. Im, and M. Chiang. TUBE: time dependent pricing for mobile data. In *SIGCOMM*, 2012.
24. R. J. Deneckere, R. P. McAfee. "Damaged goods," *Journal of Economics and Management Strategy*, 5(2), 1996, 149–174.
25. D. Bubley. New Report: 10 Reasons Why the "toll-free" 1-800 Apps Concept Won't Work. Available at: http://disruptivewireless.blogspot.com/2012/07/new-report-10-reasons-why-toll-free-1.html, 2012.
26. J. Huang, F. Qian, A. Gerber, Z. M. Mao, S. Sen, and O. Spatscheck. A close examination of performance and power characteristics of 4G LTE networks. In *Mobisys*, 2012.
27. R. Bhatia, G. Narlikar, I. Rimac, and A. Beck. UNAP: user-centric network-aware push for mobile content delivery. In *IEEE INFOCOM 2009*, April 2009.
28. N. Shah, T. Kamakaris, U. Tureli, and M. Buddhikot. Wideband spectrum sensing probe for distributed measurements in cellular band. In *International Workshop on Technology and Policy for Accessing Spectrum, ACM, vol. 222*, Aug. 2006.
29. P. Deshpande, A. Kashyap, C. Sung, and S. R. Das. Predictive methods for improved vehicular WiFi access. In *Proceedings MobiSys '09*, June 2009.
30. A. J. Nicholson and B. D. Noble. BreadCrumbs: forecasting mobile connectivity. In *Proceedings of the Annual International Conference Mobile Computing and Networking (MobiCom)*, pp. 46–57, 2008.
31. C. Shepard, A. Rahmati, C. Tossell, L. Zhong, and P. Kortum. "LiveLab: measuring wireless networks and smartphone users in the field," ACM SIGMETRICS Performance Evaluation Review, 38(3), 2010, 15–20.
32. R. Zhou, S. Khemmarat, and L. Gao. The impact of YouTube recommendation system on video views. In *Internet Measurement Conference*, 2010.
33. Zoe Fox, Mashable Impact of Social Media Referrals. Available at: http://mashable.com /2012/02/01/pinterest-traffic-study/
34. REELSEO. Facebook is the 2nd Largest Referral Source for Online Video. Available at: http://www.reelseo.com/facebook-2nd-video/2010
35. S. Sen, J. Yoon, J. Hare, J. Ormont, and S. Banerjee. Can they hear me now?: A case for a client-assisted approach to monitoring wide-area wireless networks. In *Internet Measurement Conference*, 2011.

15 Asynchronous Content Delivery and Pricing in Cellular Data Networks

VIJAY GABALE, UMAMAHESWARI DEVI, RAVI KOKKU, and
SHIVKUMAR KALYANRAMAN

15.1 INTRODUCTION

Cellular and wireless data traffic have been growing at unprecedented rates over the past few years. According to the Cisco Visual Networking Index [1], in 2012, mobile data traffic grew by 70% and was nearly 12 times greater than the total global Internet traffic seen during the year 2000. This growth trend is expected to continue for at least another 5 years, with predictions of 66% compound annual growth rate (CAGR) for mobile data traffic during 2012–2017. In particular, video constituted half of the total mobile data during 2012 and is projected to rise to two-thirds level by 2017.

15.1.1 Surging Mobile Data Traffic and Declining Operator Profits

The exponential growth in mobile data traffic, especially in the form of video, is severely stressing the cellular backhaul and core networks. On one hand, emerging applications are bandwidth heavy and/or pose stringent requirements for end-user quality of experience (QoE); on the other hand, mobile network operators (MNOs) are unable to charge in proportion to the costs incurred by the higher demands and, hence, are faced with declining profits. MNOs are, therefore, on the lookout for solutions that can help to manage the traffic growth without degradation to end-user QoE and also boost their revenues. Some solutions in this direction that are in active exploration are (i) capacity expansion via infrastructure upgrades, for example, increased number of base stations, deployment of Femto cells, and enabling WiFi offload [2]; (ii) design of sophisticated resource allocation, multimedia content management, and delivery techniques such as caching, including at the cell edge [3–5], transcoding [6],

Smart Data Pricing, First Edition. Edited by Soumya Sen, Carlee Joe-Wong, Sangtae Ha, and Mung Chiang.

and just-in-time delivery [7]; and (ii) moving away from flat pricing toward pricing schemes that are *smarter*—adopting variable, dynamic, and adaptive pricing methods to manage growth such as usage-based pricing, tiered wireless data pricing [8], and time-of-the-day pricing and providing economic incentives to promote access patterns and behaviors that can alleviate congestion.

While capacity expansion and upgradation can become inevitable at a certain stage, rolling out upgrades is often expensive, and hence, any alternate techniques that delay expansion without impacting user QoE can result in significant cost savings for MNOs.

15.1.2 Traffic Variations and Peak-Time Congestion

One aspect of the traffic demand observed in operational networks that can be leveraged for better traffic management (and, hence, delaying capacity expansion) is the significant variation in utilization levels over time. Such variations are primarily due to cyclic and regular patterns in everyday human activity; for example, networks see more utilization during days than nights [9, 10], cellular base stations in residential areas are more loaded during mornings and late evenings, whereas base stations near commercial areas are busy during office hours [11], etc. In fact, measurements at a commercially deployed base station (reported in Section 15.3.2) point to large variations in achievable throughput even at short timescales of the order of minutes. This aspect has been reported and corroborated in certain other works [9, 12]. Consequently, while networks are overloaded and congested during certain peak times of a day, they remain sparingly used during certain other times. In other words, *while the peak demand is higher than the base station capacity, the average demand is much lower.*

Flows in session during peak periods compete for bottleneck resources (such as the spectrum bandwidth), which reduces the bandwidth achieved by each flow and increases the packet delay. Such ineffective use leads to a lower *yield* of the network resources. The yield of a network deployment is a function of the number of "useful" bytes carried by the network during any interval of time, that is, bytes that do not get dropped, get delivered in time, and do not lead to degradation of end-user QoE, and, hence, are of value.

15.1.3 Yield Management through Smart Pricing

When the average demand is observed to be much lower than the capacity, and congestion occurs only at certain points in time, such peak-time congestion can be eased without expanding network capacity if part of the peak-time traffic is *time shifted* to periods of lighter use. For this purpose, data traffic can be considered to be of two broad types: (i) delay-inelastic traffic—which users cannot tolerate any delay for any reason—and (ii) delay-elastic traffic—which can be deferred or for which users may be willing to wait, in exchange for incentives. Large multimedia and other file downloads, cloud synchronization, video uploads, and software updates are some examples of traffic that is delay elastic. Note that delay-elastic traffic is in turn of two types: in the first type, the delivery is awaited by users (even if time shifted) for direct consumption, e.g., as in the cae of multimedia files, and in the second one, users are agnostic about transfers and consume content only passively, e.g., software updates.

As video traffic dominates the data carried on mobile networks and content consumption penetrates into larger sections of the society, many of whom react positively to incentives, there is much scope for lowering peaks by including mechanisms to time-shift traffic. In particular, designing incentive and delivery mechanisms for time-shifting traffic can help both delay-elastic and delay-inelastic traffic. For delay-elastic traffic, consumers can also benefit in terms of QoE if their expectations are negotiated and appropriately set in exchange for incentives, such as price discounts, and the altered expectations are adequately met. User-agnostic traffic such as software updates can be shifted by setting the right policies, such as performing updates during nighttime, without affecting the QoE of a user in any way. When some delay-elastic traffic is moved away from peak period, delay-inelastic traffic during that time also benefits because of less contention for bottleneck resources, thereby improving the overall network yield.

Time-shifting traffic, however, introduces a radically different delivery model than what consumers are normally used to and may also require changes to the applications to adhere to the new delivery semantics. As a first step to understand the implications of this changed delivery model and to gauge the sensitivity of consumers to the various aspects of such a delivery model, we conduct a detailed user survey with a representative set of user population. We discuss the setup and results of our study in Section 15.2 and derive interesting insights that are useful for designing traffic time-shifting mechanisms.

Details of our user survey is followed by a description of approaches to time-shifting traffic and their high level comparison in Section 15.3. Section 15.4 describes a pricing scheme for one of the approaches and its integration with the MNO's infrastructure. Section 15.5 evaluates the time-shifting approaches through simulations, while Section 15.6 concludes.

15.2 USER SURVEY

This section describes the survey that we conducted to specifically understand how users would react to a model that offers delay versus price trade-off and summarizes the results obtained.

15.2.1 Setup and Goals

The survey page, available at [13], includes several relevant questions to collect inputs from more than 100 users. The set of people taking the survey includes engineers, Internet-savvy nonengineers, people active on social networks such as Facebook, and other tech-savvy users. About 20% respondents are from North America, about 10% from Europe, and the remaining are from India.

The intention of our survey is to find answers to the following questions.

- *Current State of User QoE.* Do users experience degraded QoE during peak hours?
- *Delay Tolerance.* What fraction of the users is willing to accept delays to content delivery if offered discounts?
- *Delay Elasticity by Traffic Type.* What kind of traffic are users willing to delay if offered incentives?
- *Price Sensitivity.* For a given discount, how long can users wait for the content to be available for consumption?
- *Adoption.* Given discounts on content delivery, would users increase content consumption (i.e., network usage) further?
- *Pricing Interface.* Given discounts, how frequently would users like to make choices of pricing?

While a larger user base will make the results even more representative and authoritative, our current observations shed light on several interesting aspects that motivate and guide the design of solutions for time-shifting traffic.

15.2.2 State of User QoE

In our survey, we asked users to report the top three connectivity problems that they typically face when using cellular data network. A few representative responses reported by many users are as follows.

> *I get very low download speed on my wireless broadband connection in the morning during office hours. The 4G (LTE) network connection on my phone often slows down and sometimes breaks when I am commuting, especially, close to the office. Video streaming on my phone over 3G is usually faster during early mornings or late evenings (off peak hours); however, because the data plans are relatively more expensive than wired broadband, I restrict my usage.*

While some of the above problems may have several other possible root causes than network congestion, most of the responses suggest the impression the users carry about network connectivity: *poor QoE during peak hours*. These responses indicate the viability of a pricing/delivery model that can make the users aware of network congestion *explicitly* and then provide them with the option to delay content delivery for a discount.

15.2.3 Delay Tolerance

Next, to understand the delay-tolerance thresholds of users, we asked the following question: If provided with a delivery promise in terms of an expected delivery time (EDT), how long can you wait for the content to be delivered and available?

Users' responses are summarized in Figure 15.1. The figure shows that roughly 50 users when choosing delays are sensitive to discounts offered (indicated by f(discount)), while more than 40 are willing to delay the download (by 1–6 h) for an

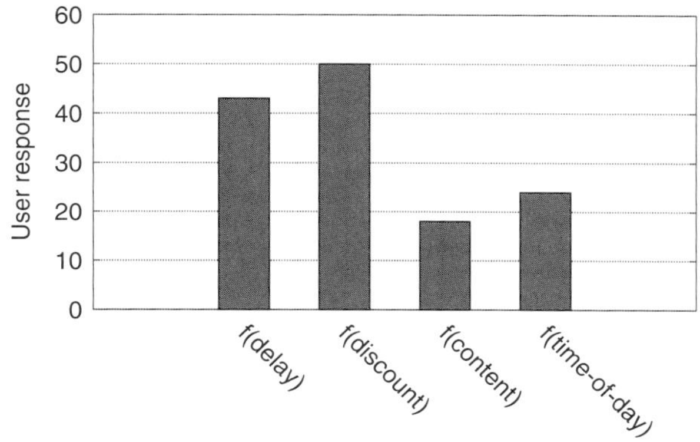

Figure 15.1 Tolerance to delays.

EDT promise for better QoE even *without* any discount (indicated by f(delay)). This is possibly because of the existing low bandwidth experience in the network and lack of good QoE [14]. About 20% of the users are also sensitive to the specific content type (indicated by f(content)) and time of the day (indicated by f(time-of-day)). Note that a user could select multiple options for this question. For example, delay that is acceptable to a user can depend on both discount and time of the day.

15.2.4 Delay Elasticity by Traffic Type

To get a sense of the content types that are delay elastic, we asked users to select the traffic classes for which they would accept delays to the delivery for discounted rates. Users' responses are summarized in Figure 15.2. The figure shows that only

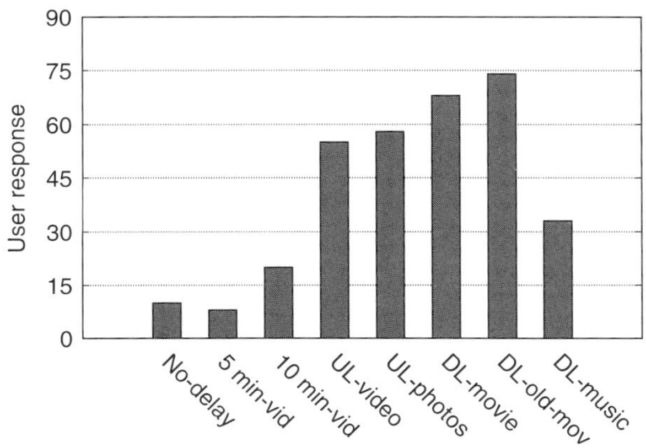

Figure 15.2 Tolerance for content types.

about 10% of the users are particular about receiving all the content with no extra delay. Others show increased tolerance to different content and transfer types. As expected, people are more willing to tolerate longer delays for larger transfers. Also, more than 50% of the respondents are willing to delay video and photo uploads, possibly because they themselves do not consume the content.

15.2.5 Price Sensitivity

We then asked the users to indicate the discount levels acceptable to them. Response summary in Figure 15.3 shows that most users are willing to wait for up to 2 h if given 50% discount, but not as many users are willing to tolerate higher delays even for 75% discount. However, it shows that many users are price sensitive, and they can trade a delay of 1–2 h, for a sizeable discount.

15.2.6 Adoption

Next, to get a sense of the additional network usage and revenue that the new pricing scheme can lead to, we asked the users as to whether they would be accessing more content, if given a 50% discount on the original price and a promise of delivery within an EDT. Figure 15.4 shows that more than 50% of the respondents would increase usage by 2× if given 50% discount. As this increased adoption is barely sufficient to balance the discount offered, it does not lead to an increase in the operator revenue. However, the same figure also shows that more than 20% of the respondents would increase usage by 3× for the same 50% discount, which has the potential to increase operator revenues. Thus, by providing appropriate discounts and managing expectations on delivery times, the operators can not only manage congestion but also expect their revenues to grow.

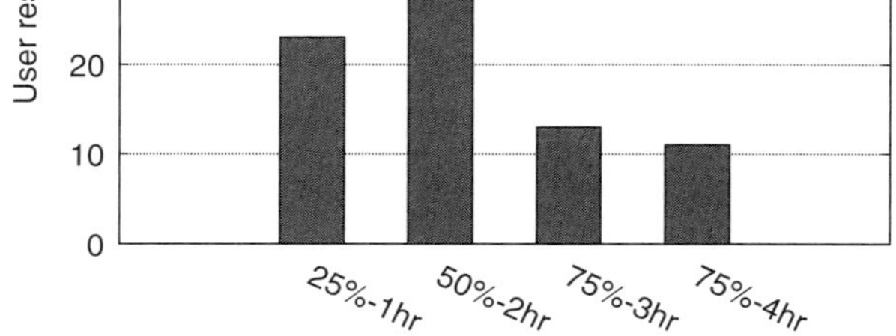

Figure 15.3 Response to discounts.

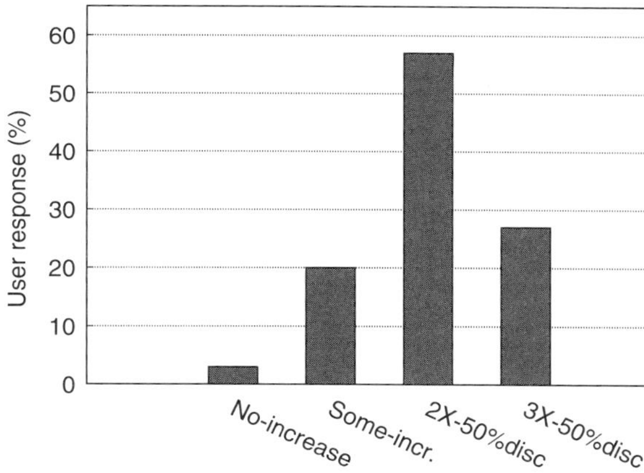

Figure 15.4 Adoption of discounts.

15.2.7 Pricing Interface

We finally asked users for their choice of interaction to avail discounts: per-request discounts, day-ahead notification of discounted time periods, or monthly discount plans. Figure 15.5 shows that most respondents are interested in getting discounts: About 45% preferred the maximum flexibility provided by per-object pricing and delivery-time decisions (i.e., choosing the price of an object just before downloading), while about 30% chose to keep it simple with monthly discounts. Only about 10% chose to get day-ahead discounts. The main reason could be that day-ahead discounts require users to keep track of daily price changes, which could get cumbersome, while month-ahead discounts obviates such day-to-day tracking.

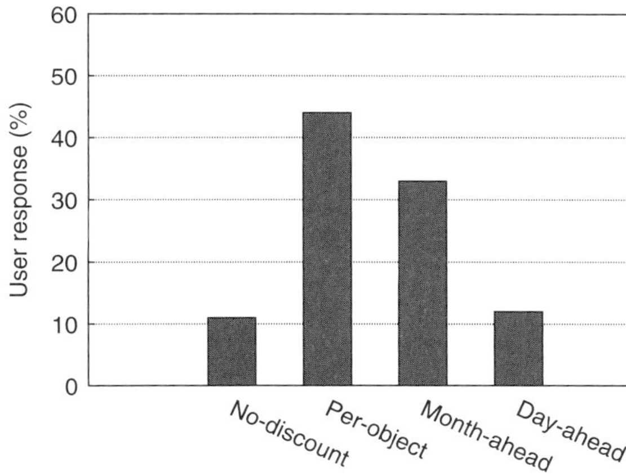

Figure 15.5 Interface to pricing.

15.3 TIME-SHIFTING TRAFFIC

The results of the user survey in the previous section suggest that a majority of the users is amenable to traffic deferrals by bounded time durations, in exchange for proportionate discounts. In this section, we discuss approaches for realizing such deferrals.

Internet service providers and MNOs have traditionally offered flat-rate data pricing schemes that charge a fixed amount of money for data service and place no caps on data usage. The explosion in data traffic in recent years, however, has led network operators to move away from a flat pricing model to tiered and usage-based pricing [8]. Usage-based and tiered pricing schemes can limit per-user demands but do not explicitly attempt to reduce traffic congestion periods. This is because such schemes do not include any measures that can promote requests to be spread out over time.

To help to reduce peak-time congestion and delay the requirement of network capacity upgrades, an MNO requires more sophisticated differential pricing schemes that will aid in intelligent time shifting of traffic. A survey of such pricing techniques can be found in Reference 15. For example, Murphy and Murphy [16] and Paul and Subramanian [17] propose to smoothen traffic by providing incentives in the form of higher access rates during off-peak hours. In a similar vein, the TUBE system [12] has helped in renewing this direction of research for cellular networks by developing a time-of-the-day pricing scheme. This scheme provides price discounts for off-peak accesses and incentivizes increased usage.

15.3.1 Time-Shifting Taxonomy

Request-Shift Versus Delivery-Shift. Consider the typical browsing and content consumption pattern of any user: (i) selection, the user selects an object to view (and clicks on a hyperlink); (ii) request, the user device issues a request to the content server for the object on a transport connection; (iii) delivery, the content server delivers the object on the transport connection; and (iv) consumption, the user uses (e.g., views) the object.

In such a setting, time-shifting traffic can be realized using two possible approaches (see Fig. 15.6):

- *Request-Shift.* In this approach, if there is congestion at a certain point in time, users are provided an incentive to stop making requests and return back at a later point in time during a low price period and reissue their object requests. With Request-shift, the three steps of request, delivery, and consumption happen close together in time during a low price period of user's (or their agent's) choosing. Note that with this approach, delivery is expected to be confined to the chosen low price period, which can be away from the time the user makes a selection (or intends to select) by a large duration.
- *Delivery-Shift.* In this approach, an object request is collected immediately when the user intends to view the object, but the user is provided an EDT of

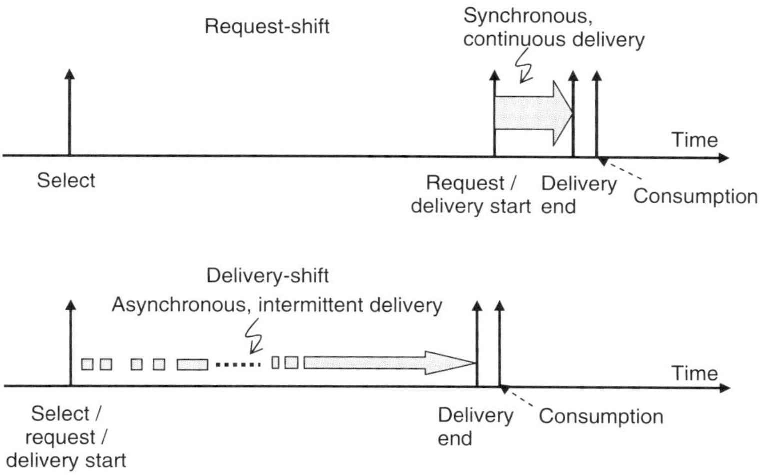

Figure 15.6 Request-shift and Deliver-shift approaches of time-shifting traffic.

when the object will be ready to consume. The object's delivery is spread over time, and the request–response transaction is asynchronous to be able to opportunistically transmit during noncongested periods at short time scales.

An instance of the Request-shift approach is the TUBE system [12]. TUBE computes time-dependent prices a priori, typically a day in advance, using past history of network traffic by time and a user-behavior model that predicts user responses to prices offered. TUBE uses a feedback loop between the price-computation and user-behavior models to adapt and optimize price computation based on user's reaction to the prices. To simplify the task of choosing a time for issuing a request so that the overall usage is optimized, TUBE also includes an autopilot mode that can schedule requests on behalf of the users. The prices are computed over time durations of the order of 30 min to 1 h.

An instance of the Delivery-shifting approach is the Async system [18]. Async introduces a "sachet" model in which users (or automated user agents) are allowed to choose the delivery price and *expected delivery time* (EDT) of different objects at fine granularity. For realizing this model, as shown in Figure 15.7, for each object request, Async presents users with a set of (price, EDT) options to choose from. Thus, the Async system facilitates an MNO to negotiate a delay in content delivery with users in exchange for incentives. The negotiated delay is used by the MNO to actively manage content delivery in such a way that QoE improves for both deferred flows and regular flows (which are not deferred). The price for a request for a given EDT is computed by determining the total traffic levels during the times at which the request is likely to be scheduled. By charging a price that is in proportion to the expected traffic levels, the MNO can offer higher discounts for longer delays. Note that deferred deliveries under Async are not just best-effort attempts but are associated with EDTs. By letting

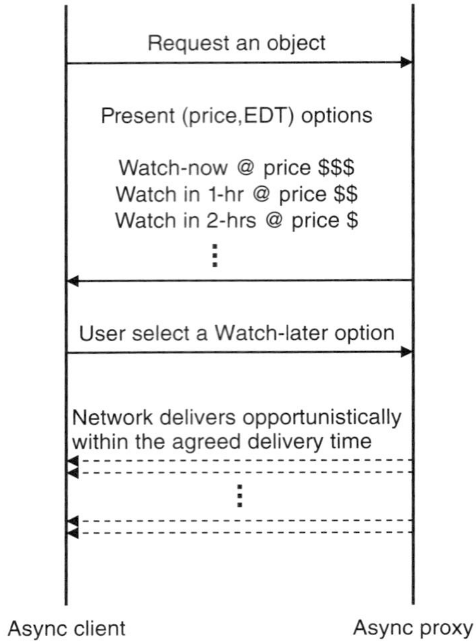

Figure 15.7 Async user interface.

the users to choose the (price, EDT) combination, users get flexibility in choosing the level of service and price they want. At the same time, the MNO also gets to control *when* and at *what* price content is delivered.

15.3.2 Comparison of the Time-Shifting Alternatives

The Request-shift approach leads to a simple system implementation, requiring no changes to the network elements, except for setting different charging rules dynamically. The onus is on the user (or a user agent) to issue requests at appropriate times. On the other hand, Request-shift approaches can have the following drawbacks. (i) As the network has no control over the amount of traffic that would arrive in the low price period, time-shifted traffic may again encounter congestion, thereby disappointing the user who chose to defer the traffic. (ii) Transmission opportunities that become available over short time scales in the time intervening between the object selection and final consumption cannot be utilized, lowering the overall network utilization. (iii) Owing to lack of active network control, the system cannot adapt satisfactorily if actual conditions deviate from the predictions. In other words, reasonably accurate estimates of future traffic conditions are required a priori for the system to compute time-of-the-day prices and be effective in practice.

To understand the variations over time in the usage of cellular data links and the capacity that is available, we conducted a set of experiments over a week. In these experiments, we measure and characterize the achievable throughput across different

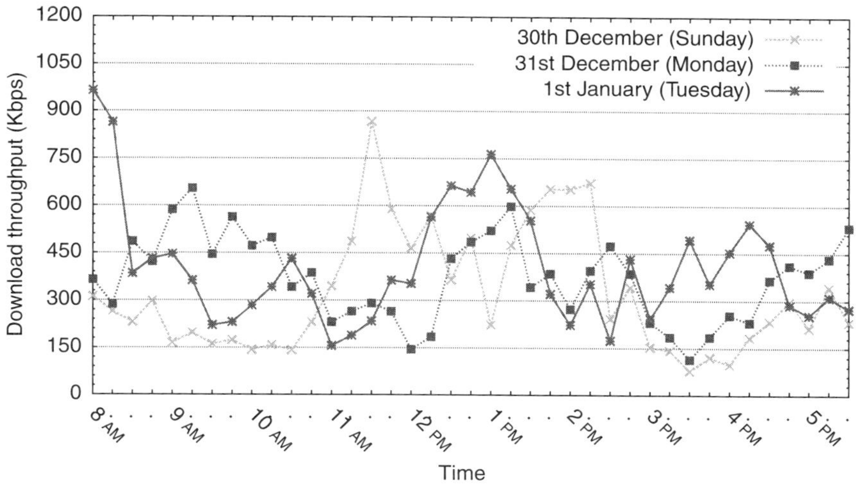

Figure 15.8 Achievable throughput using downlink probes.

base stations in Bangalore, India. Figure 15.8 shows measurements of throughput with backlogged flows (for a duration of 50 s) using the cellular network of a leading wireless service provider in India from two different locations (in Bangalore). Figure 15.8 shows that there is a lack of any temporal predictability in the achievable throughput (and, hence, load) at cellular base stations; as can be seen, the load can vary significantly across two consecutive half-hour time periods and can be very different at a given time of the day on different days. We observed similar variations at other commercially deployed base stations also. These observations indicate that meeting user expectations with a Delivery-shift approach may be easier than that with a Request-shift approach.

The Async system, which is an instance of Delivery-shifting, strives to overcome the shortcomings of Request-shifting approaches but has to address the following issues. First, it requires active management of content delivery using an in-network element (essentially a content proxy) in order to opportunistically transmit deferred flows during off-peak times. Second, because object delivery can be spread over longer durations, in comparison to Request-shift, and can be intermittent, the number of outstanding requests in the system can be higher. Further, the system has to tolerate disconnections because of mobility and user activity such as device switch off, necessitating that state be managed. Finally, Async requires a reliable sensing method to estimate periods of low utilization to transmit deferred traffic to improve spectrum utilization in comparison to Request-shift. Async successfully addresses these issues by designing an application layer protocol that can manage content delivery, with client-side maintenance of content transfer state [18]. Additionally, such active management allows the system to be more dynamic and adaptive in that delivery times and associated prices can be computed using current traffic conditions in association with coarse estimates of nondeferrable traffic.

15.4 PRICING TO ENABLE DELIVERY-SHIFTING

In order to enable Delivery-shifting, we now turn to describing a method for computing a price for a specified delay for flows arriving at a base station. This method can be used for putting together a set of (price, EDT) options that Async uses for negotiation with end users (refer Section 15.3.1 and Figure 15.7). While such options can be computed in several different ways, we present a simple scheme to illustrate the feasibility of the Delivery-shift approach and quantify possible gains.

Pricing schemes that can facilitate time shifting fall under the broad category of congestion pricing. Congestion pricing methods have been proposed in the past for wired networks [19, 20] and in other domains such as electricity supply [21, 22]. The prior methods, however, are not based on an asynchronous opportunistic delivery and do not consider determining EDT-dependent prices.

The pricing method that we present in the following computes prices so that congestion cost for the operator is minimized (assuming convex cost functions). One can in addition consider a feedback loop that accounts for the effect of the prices on users' demand, which in turn can influence the price computation. It is also possible to consider maximizing benefits to the user or profits to the operator. (In our evaluation, we do consider a user-behavior model in determining a user's response to the options they are presented with and the (price,EDT) option that they choose.) Nevertheless, we show that the following approach can effectively alleviate congestion for a given discount budget and thereby demonstrate the potential of the overall method. (Consideration of the aspects of benefits to user, profits to operator, and user-behavior model can only improve the effectiveness of the pricing scheme.)

15.4.1 Computing (Price, EDT) Options

For every incoming request, Async presents a set of (price,EDT) options to the users. EDTs can be a fixed set of values, such as 1, 2, and 4 h, or chosen based on the context and length of the request. Price for a given EDT is computed based on an estimation of the total traffic level or congestion during the times that the incoming request is likely be scheduled. To estimate congestion, we determine an allocation for the incoming request with respect to the EDT under consideration in the manner of water-filling algorithm, so that the maximum traffic at any point in time is minimized. This gives a reference schedule for the incoming request that can potentially be used by a network-side entity for actively managing its delivery.

Note that computing a reference schedule requires some knowledge of nondeferrable traffic, that is, non-Async traffic, arriving in the future. We assume that historical loads are available for estimating such future traffic. Such historical information can be obtained using the total cell-level load information (available with the network operator) and details of Async traffic available with the Async system (because all Async requests pass through the Async system). To account for deviations in the actual traffic from the estimated one and be adaptive, we recompute reference schedules for outstanding Async requests periodically or when a new request arrives, using their pending transmission sizes. (We, however, do not make changes to negotiated prices or EDTs of outstanding requests.) Such recomputation aids in reducing errors while determining prices for future flows. Further, because prices are computed at runtime, updates to the actual traffic can be made use of while computing prices. Hence, unlike approaches such as TUBE [23], prices in our scheme can evolve based on the load observed at runtime.

In what follows, we describe our pricing methodology in greater detail.

15.4.1.1 Assumptions and Definitions Time is assumed to be divided into discrete *periods*. The duration of a period can be set appropriately based on the desired granularity, for example, we set each period to be 30-min long. Let C denote the base station capacity in terms of aggregate bytes that can be transmitted in a period. We denote the total aggregate load owing to regular flows (i.e., non-Async flows) in time period t as D_t. The total number of Async flows active at time t is denoted by N_t and the number of bits allocated to Async flow i in period t is denoted by $x_{i,t}$. Thus, the total traffic at the base station at time t is then given by $\alpha_t = D_t + \sum_{i=1}^{N_t} x_{i,t}$.

To quantify the degree of congestion, we classify the congestion experienced at a base station into one of a set of L distinct *levels*. To enable such a classification, the base station's transmission capacity $[0, C)$ is partitioned into L discrete subranges, given by $[T_0, T_1), \ldots, [T_{L-1}, T_L)$, where $T_{\ell-1} < T_\ell$ for $1 \le \ell \le L$. Thus, the congestion level at a base station in period t is given by the subrange in which the total traffic it sees in that period, α_t, lies and is said to be ℓ if $\alpha_t \in [T_{\ell-1}, T_\ell)$. For each level ℓ, we assign a constant but monotonically increasing congestion weight K_ℓ. We then define a convex congestion-cost function (which is piecewise linear) for time period t as follows.

$$B_t = \sum_{\ell=1}^{L} K_\ell \cdot \max\left(\min(\alpha_t, T_{\ell+1}) - T_\ell, 0\right). \tag{15.1}$$

It is easy to see that this function is nondecreasing with α_t. This cost function allows the MNOs to choose different level thresholds (T_ℓ) and appropriate cost (K_ℓ) for the level based on their experience, for example, based on the probability of facing overload and related operational expenditure.

15.4.1.2 Allocating Async Flows to Minimize Congestion Cost To compute a price for an EDT, we first formulate and solve the problem of allocating N Async flows over P time periods so that the aggregate congestion cost over the time periods is minimized. This optimization problem is formulated as follows.

PACKING

$$\begin{aligned}
\text{Minimize} \quad & \textstyle\sum_{t=1}^{P} B_t, \\
\text{such that} \quad & \textstyle\sum_{i=1}^{N} x_{i,t} \le C - D_t, \ \forall t = 1, \ldots, P \\
& F_i = \textstyle\sum_{t=1}^{P} x_{i,t}, \quad \forall i = 1, \ldots, N \\
& x_{i,t} \ge 0, \qquad \forall t = 1, \ldots, P, \forall i = 1, \ldots, N,
\end{aligned} \tag{15.2}$$

where F_i is the size (in bits) of Async flow i. The constraints in Eq. (15.2) ensure that the base station capacity is not exceeded and the flow is served completely. The capacity of a base station depends on the channel quality seen by the users and, hence, in general, is dynamic for cellular systems. For the above optimization problem, we assume an average capacity given by the average signal quality experienced.

Note that even though the objective cost function is piecewise linear, the problem can be solved optimally because the slope of the objective is bounded and all the constraints are linear. We assume that at least one of the congestion weights and congestion-level widths is strictly increasing and the other is monotonically increasing.

15.4.1.3 Computing Per-Flow Options When a new flow arrives, we solve the optimization problem PACKING once for each EDT to be presented to the user (with P set to that EDT E). In solving the problem, we use the allocations in the future periods to previously accepted Async flows (obtained as solutions to instances of PACKING solved when those flows arrived) and estimates of D (traffic because of regular flows). If the problem is infeasible for the largest acceptable EDT, then we declare that the incoming request cannot be scheduled and send a notification back to the client. We compute a price for EDT E, p_E, as follows:

$$p_E = \frac{\sum_{t=1}^{E} B_t}{\sum_{t=1}^{E} \sum_{\ell=1}^{L} K_\ell \cdot (T_\ell - T_{\ell-1})}. \tag{15.3}$$

We assume that $B_t = 0$ if $x_{i,t} = 0$ to avoid a large value of D_t (regular traffic) for some t (with no Async traffic allocated) affecting the price computation; i is the new request under consideration. Also, p_E lies between 0.0 and 1.0 and can be used along with a base price. For instance, a fraction p_E of the base price may be charged for the request.

It can be shown that the prices computed in the above manner are monotonically decreasing with increase in EDTs, ensuring higher incentives for waiting for a longer time. This, unlike models in prior work such as in TUBE [23], gives users an easy-to-understand interface: *wait more, gain more* [17]. Moreover, Async offers higher discounts to the requests arriving in low congestion periods, thus incentivizing users to increase their usage.

15.4.2 Integration with an MNO's Infrastructure

In the 3GPP specification for cellular networks, differential pricing per flow is generally achieved by adding *rules* to a system referred to as *Policy and Charging Rules Function* (PCRF) [24]. A rule specifies the price to be paid per byte by a flow identified by the standard 5-tuple:

```
<src-ip,src-port,dst-ip,dst-port,proto>
```

In the above, ip addresses and ports can be provided as wildcards. When a flow passes through the Gateway GPRS Support Node (GGSN), where it needs to be billed, a

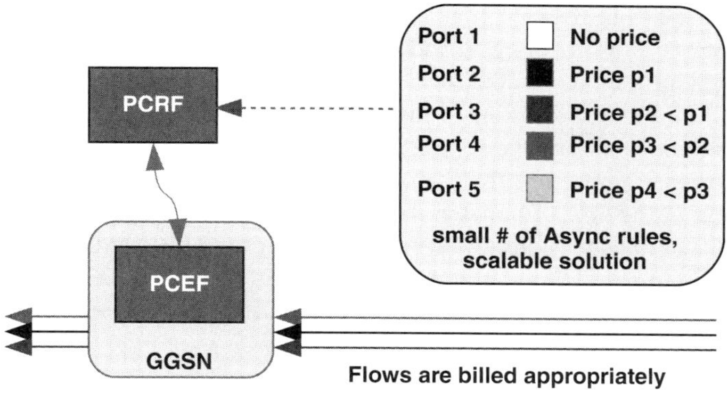

Figure 15.9 A simple mechanism for billing flows in Async.

module called *Policy and Charging Enforcement Function* (PCEF) [24] finds a rule matching the flow and uses it for determining the charges.

To realize our pricing scheme, a small set of ports, each of which maps to a price per byte, can be assigned for the Async flows at a network element, which we shall refer to as the Async *proxy*. Async flows are made to pass through the proxy so that they can be managed. One rule per port *i* of the form (`src-ip`=proxy-IP, `src-port`=*i*, `price`=$$) can then be added to the PCRF. By assigning ports to flows based on the download price chosen by the user, ports at the proxy can be used for price differentiation. This approach is minimally intrusive to the cellular network deployment and also scalable, because it only adds a small set of rules (as many as the number of discrete pricing levels) to the PCRF. Note that this change to the set of PCRF rules does not affect PCEF (its scalability, in particular), because the charging enforcement functionality of PCEF is independent of how the rule for a flow is specified in PCRF or retrieved from it. Our approach for billing Async flows is illustrated in Figure 15.9.

Recall that for each incoming Async flow, a set of (price, EDT) options is computed as described in Section 15.4.1.3. To facilitate billing as described earlier, only those options with prices close to the prices specified in PCRF rules (based on a difference threshold) are considered. For each option included in the final set using the above rule, the final price is rounded to the highest discrete price in PCRF not exceeding the computed price. On the basis of the price that the user agrees to, the port on which the user can connect to for downloading the content is communicated. Thus, in a simple end-to-end manner, we can enable deadline and price aware transfer of content.

To enable a Request-shift approach such as TUBE, time-dependent rules can be added to PCRF; for instance, a price-per-byte rule can be added for each period of the day. In such a case, every byte of traffic arriving in a particular time period will be billed according to the pricing rule set for that time period. Further, because the number of periods in a day is small, of the order of a few tens, it is sufficient to add and delete a small set of rules per day, thus incurring only a small overhead. Hence, it is quite simple for a mobile network operator to deploy a time-dependent pricing scheme. On the flip side, such time-dependent pricing is flow agnostic. For instance, if a flow starts in a low price period because of the discounted price but does not finish within the period and continues in the next period, whose price is high, it needs to either be torn down, resulting in incomplete downloads or poor streaming, or pay a higher price to continue the service. In contrast, because Async's pricing method provides a price agreement for each object request, which maps to a flow, and is independent of the time of transfer, it is immune to such time-based issues.

Note that retransmissions from the proxy may be billed multiple times by the MNO's infrastructure. Peng et al. [25] studied and discussed the problem and provided a few solutions for MNOs to avoid overcharging for the retransmitted content.

15.5 SIMULATION RESULTS

We now present a simulation study to evaluate the efficacy of the Delivery-shift and Request-shift approaches. We consider recent work in TUBE [23] as a representative pricing scheme for the Request-shift approach. In this section, we first describe the performance measures and simulation setup used in evaluating the two approaches. Subsequently, we describe the results of the evaluation.

15.5.1 Performance Measures

The main idea in both the Request-shift and Delivery-shift approaches is to alleviate congestion by deferring delay-elastic flows. Hence, to compare the two approaches, we define the following two performance measures: (i) *aggregate base station load (over time)* and (ii) *total (end-to-end) delivery time for delayed flows*.

To compute the first performance measure, we divide time into multiple periods and compute the aggregate traffic transmitted from the base station during each period. We then plot a cumulative distribution function (or CDF) of the load values. The second performance measure, total delivery time, is defined as the difference between the time a user selects an object and the time at which the corresponding content is completely delivered (i.e., all the bytes of the requested object are delivered at the client). Thus, the total delivery time measures the extent to which a scheme time shifts traffic to ease congestion. Both the schemes use discount as a means to achieve traffic shifts, and the higher the discount, the higher the chances of traffic shifts. To ensure that the traffic shifts are not just due to discounts but also due to effective delivery mechanisms, we also compare the total discount offered by the two schemes.

15.5.2 Simulation Setup

For the evaluation, we first choose an aggregate traffic pattern depicting traffic variations over a day. For this, we observe the traces from a leading service provider in India and consider different base traffic instances. One such instance is shown in Figure 15.10. The pricing schemes that we consider for both Delivery-shift and

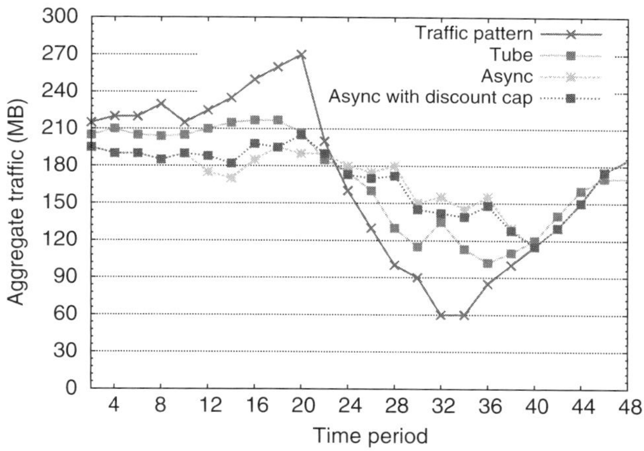

Figure 15.10 Aggregate traffic (load per time period) with a Delivery-shift and a Request-shift mechanisms. The base station capacity is 270 MB/period.

Request-shift divide a day's time into multiple time periods. We consider 30-min time *periods* and choose an aggregate traffic load for 48 continuous periods for 24 h. We assume base station capacity to be 270 MB/period. We then generate flow requests period by period to match the aggregate load. We distribute the flow arrival times uniformly randomly within each period. Each generated flow is one of the four content types: web request, short video, long video, and file download. The mean sizes for the four types are chosen from different sets, for example, 50 KB, 5 MB, 25 MB, and 50 MB.

We implement the Async pricing scheme for Delivery-shifting described in Section 15.4 using CVXOPT library [26]. In the Async framework, flows are of two types. (i) Flows that are amenable for deferrals, referred to as Async *flows*. Upon arrival, such flows are presented with a set of possible (price, EDT) options. (2) Flows that are not amenable for deferrals and do not wish to pass through the Async system, referred to as *regular* or *non-*Async *flows*. The Async pricing program takes the following as input: (i) details of Async flows (time of arrival, object size, type of flow) and (ii) an estimate of the cumulative non-Async traffic (based on historic trends). It should be pointed out that non-Async flows are not presented to the in-network control element, and hence, unlike Async traffic, the total amount of actual non-Async traffic is not available at runtime to the pricing component. The pricing scheme only uses coarse estimates based on historic data (as described in Section 15.4.1). The parameters of the pricing scheme (described in Section 15.4.1) are set as follows: we partition the base station capacity into $L = 10$ equal congestion levels. For each level, we set $K_l = K_{l-1} + 1$ for $l \geq 2$ with $K_1 = 1$.

As output, the pricing program determines the delivery method (NOW or deferred) and, if deferred, the EDT, for each Async flow. Async flows for which deliver NOW option is chosen are delivered without delay, just like a regular flow. We assume a proportional fair scheduling scheme (which is commonly used by cellular base station schedulers) for making allocations to the regular flows while computing (price, EDT) for the deferred flows.

For use with Request-shifting, we also implement the pricing scheme described in Reference 23 to determine per-period discounts based on the aggregate loads specified. We employ the user-response model described therein to determine the aggregate amount of traffic shifts across periods. For fair comparison, we use the same user-behavior model for the Async pricing scheme mentioned earlier to determine the final (price,EDT) options selected by end users. Specifically, for each of the content types described earlier, we choose a patience index (which models the amount of time by which users can shift their requests) as described in Reference 23. For the four content types mentioned earlier, we use 0.5, 1, 1.5, and 2 as the respective patience indices. These patience indices are then used in choosing the end-user's (price, EDT) option from among the set computed for a request. For both the schemes, we compute the discount as a fraction between 0.0 and 1.0, which essentially provides the discount rate per byte served. It should be pointed out that the TUBE pricing approach of Ha et al. [23] computes byte-level shifts, that is, the fraction of bytes that shift from one period to a later period. When extended to flows, a flow that does not complete within a period carries over into the next period. Owing to this, TUBE time periods are constrained to be reasonably granular, for example, at least 15-min long; else, the actual price charged for a flow may not match the price that a user expects to pay when deferring a flow. On the other hand, because the price computed for Delivery-shifting is on a per-object basis, it is not impacted by the times at which the object is actually delivered, and hence, no constraints are placed on period granularity.

In this chapter, our focus is on the scope of the Delivery-shift and Request-shift approaches in depeaking congestion while improving the network yield. Hence, as mentioned earlier, for Delivery-shift, we assume that an entity in the network (such as a proxy) can schedule the deferred flows such that the reference schedule computed by the pricing module is followed. The Delivery-shift approach is better able to utilize short-term transmission opportunities and improve yield (as described in Section 15.1) because of the following: (i) not restricting a flow to a single or few time periods, as described earlier—hence, traffic spread can be wider—and (ii) active network control, which can sense such opportunities. One scheme that is capable of adapting to and making use of capacity variations is described in Reference 18.

We performed our evaluation for two cases: predictable and unpredictable traffic patterns. As TUBE computes per-period prices in an offline manner assuming a predictable traffic pattern, the first case allows us to compare Async with TUBE in a fair manner. For the second case, we scale the traffic pattern (e.g., increase the aggregate traffic in each period by up to 30%) to introduce unpredictability. The second case evaluates the adaptive nature of Async to runtime deviations to the traffic pattern assumed.

15.5.3 Results

15.5.3.1 Predictable Traffic Conditions Figure 15.10 shows the input traffic assumed and the aggregate traffic as observed with the two approaches for time-shifting traffic. Given that the base station capacity is 270 MB per time period, periods 0–22 are congested in comparison to the remaining periods. Instead of shifting the flows entirely, the schedule computed by Async pricing can be used to serve the flows over multiple time periods and utilize the transfer opportunities whenever they become available. Further, such a schedule is computed and adapted periodically at runtime by considering the state (i.e., residual content and deadline) of active Async flows. Hence, as can be observed from the figure, the traffic curve for Async is smoother than that for TUBE. Thus, because of runtime determination of congestion and computation of appropriate price and EDT, the Delivery-shift mechanism in Async is more effective in alleviating congestion.

15.5.3.1.1 Flow Deferrals and Discount To evaluate Async pricing, we generated a total of 750 flows for the duration of 24 h to conform to the aggregate input traffic chosen. (Recall that TUBE pricing operates at the aggregate traffic level and does not make use of flow-level details.) Table 15.1 shows the mean percentage of bytes deferred over several runs under Async and TUBE. Note that, although the input trace is the same, the number of bytes deferred by the two mechanisms are different because Async computes the deferrals in response to the runtime traffic schedule (considering both the deferred and nondeferred flows), while TUBE precomputes the deferrals. In addition to the deferrals, Table 15.1 also shows the mean values of total discounts. In the case of Async, for every flow deferred, we compute the discount as the product of the rate of discount per byte and the total size in bytes of the flow. For TUBE, we compute the total discount using per-period discounts and aggregate per-period traffic. Note that the discounts offered by Async are comparable to that of TUBE. Further, for a fair comparison with TUBE, we also compute an Async schedule wherein we restrain Async from offering the discounts once the cumulative discount offered becomes equal to the total TUBE discount. Figure 15.10 shows that the traffic shift with above-mentioned discount cap is almost similar to that without any discount cap. This shows that, in comparison to the Request-shift mechanism in TUBE, the Delivery-shift mechanism in Async can better depeak the traffic for a given discount budget. For fair comparison, we let the discount cap remain active in presenting the results for the two performance measures.

15.5.3.1.2 Aggregate Load In Figure 15.11, we plot the CDF of aggregate traffic per time period. In our setup, the maximum aggregate load a base station can handle in one period, that is, the base station capacity, is 270 MB. From the figures, we can see that Delivery-shift is more effective in reducing the peaks with a load less than 65% of the capacity in 90% of the periods. We observe a similar behavior when we remove

TABLE 15.1 Comparison of Deferrals and Discounts

Mechanism	Deferrals	Total Discount, Units
TUBE	7.4	313
Async	11.2	389
Async with discount cap	9.8	313

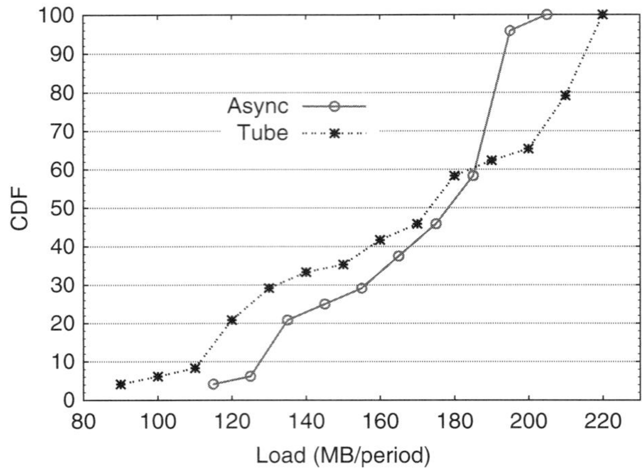

Figure 15.11 Better congestion management with Delivery-shift because of runtime price computation.

the discount cap. However, with TUBE's Request-shifting, more than 30% of time periods are faced with load exceeding 65% of capacity, reaching up to 80%. This shows that the Delivery-shift mechanism of Async is more effective in alleviating congestion.

15.5.3.1.3 Total Delivery Time Figure 15.12 shows the CDF of the total delivery times of flows under the two delivery mechanisms. As the Delivery-shift mechanism of Async does not wait for low price time periods, but computes schedules for flows in a feasible manner whenever there is capacity available, the EDTs are much lower than the flow deferrals under the Request-shift mechanisms. The median delivery time of flow deferrals in Async is about 18 periods, which is 50% lower than the

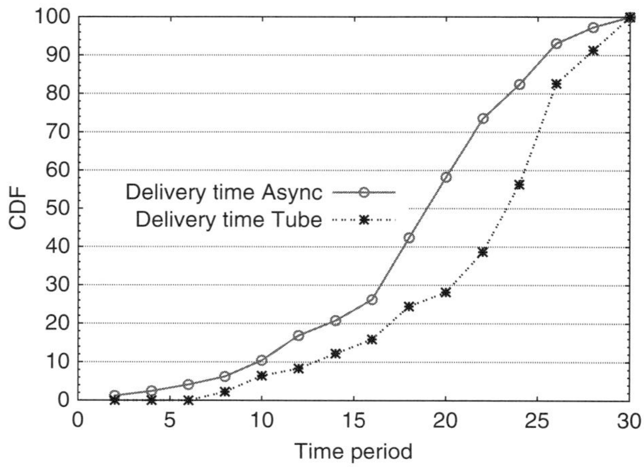

Figure 15.12 Smaller total delivery time with Delivery-shift than Request-shift.

median delivery time of flows deferral in TUBE. Thus, for a given discount budget, the Delivery-shift mechanism of Async reduces the delivery time of flows while also depeaking the traffic in an effective manner.

15.5.3.2 Handling Deviations to Estimated Traffic In another set of experiments, we consider deviations to actual traffic from that estimated by scaling (increasing by 30%) the input traffic pattern in Figure 15.10. Note that, to enable request shifts, TUBE computes prices for different time periods by assuming the original (non-scaled) traffic pattern. In comparison, by design, Async reacts to the changes in traffic at runtime. As a result, as we can see in Figure 15.13, even when there is unpredictable increase in traffic, Delivery-shift outperforms the Request-shift mechanism by better smoothing the traffic and managing the congestion. It can be observed that with the scaled input traffic, there is a peak with TUBE in time period 32, whereas the traffic is smoother with Async. Moreover, even with a 30% increase to the input traffic, the median delivery time in Async is lower than that of TUBE by 45%. Thus, because of the lack of network-side control, the Request-shift approach does not react effectively to the deviation in input traffic patterns. Further, flows shifting to the presumed low traffic periods may suffer from increased contention.

A comparison summary of the salient features of the Request-shift and Delivery-shift approaches discussed in this chapter is presented in Table 15.2

15.6 CONCLUSION

In this chapter, we have considered and evaluated time shifting of traffic to ease the peak-time congestion and increase the overall yield of cellular data networks. We have discussed two approaches for time shifting, namely, Request-shifting and Delivery-shifting, and, using appropriate instantiations, evaluated them qualitatively

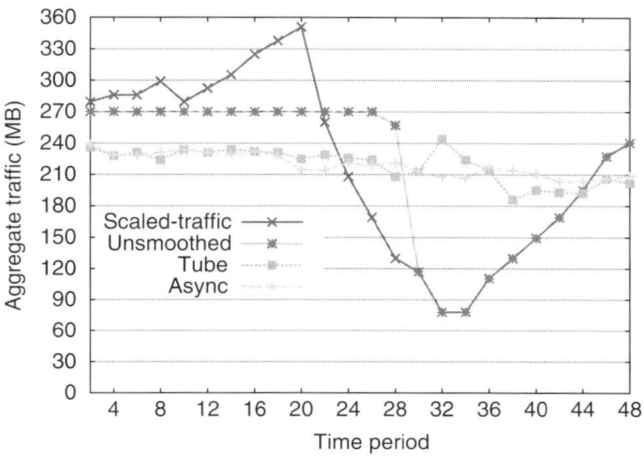

Figure 15.13 Better adaptation to deviations from assumptions to input traffic with Delivery-shift.

TABLE 15.2 Overall Comparison of the Request-Shift and Delivery-Shift Approaches

Feature	Request-Shift (e.g., TUBE)	Delivery-Shift (e.g., Async)
Ease of implementation	Simpler. Involves offline computation and dissemination of time-of-day prices	More complex, requiring runtime computation of prices at fine granularity, down to the level of individual requests
Runtime network control	None	Requires active intervention to control when deferred flows get transmitted
Yield management	Cannot make use of short time-scale transmission opportunities	Can put short time-scale transmission opportunities to good use
Depeaking traffic	Less effective	More effective
Robustness	Requires reasonable estimates of future traffic to be effective	Can adapt at runtime to deviations from expected traffic
Ease of integration with PCRF	Easy to provide time-of-day prices but flow-level expectations may be violated	Easy to provide a few discrete prices. Pricing agreements can be made and honored at flow level

and quantitatively. Our evaluations suggest that while the Request-shift approach is simple to instantiate, it may not effectively alleviate congestion, whereas the Delivery-shift approach exercises greater network control over how flows are deferred and, hence, achieves better congestion and yield management, leading to higher QoE, albeit with extra system complexity. Our user survey sheds light on several design considerations that are important for implementing either of the approaches in real networks.

REFERENCES

1. CISCO. Cisco visual networking index: global mobile data traffic forecast update, 2012–2017. Available at: http://www.cisco.com/en/US/solutions/collateral/ns341/ns525/ns537/ns705/ns827/white_paper_c11-520862.html.
2. K. Lee, J. Lee, Y. Yi, I. Rhee, and S. Chong. Mobile data offloading: how much can WiFi deliver? In *Proceedings of the 6th International COnference, Co-NEXT '10*, 2010.
3. J. Erman, A. Gerber, M. Hajiaghayi, D. Pei, S. Sen, and O. Spatscheck. "To cache or not to cache: the 3G case," *IEEE Internet Computing*, 15(2), 2011, 27–34.
4. S. Woo, E. Jeong, S. Park, J. Lee, S. Ihm, and K. S. Park. Comparison of caching strategies in modern cellular backhaul networks. In *ACM MobiSys*, 2013.
5. Altobridge. Wireless network caching. Available at: http://www.altobridge.com/data-at-the-edge/wireless-network-caching/. May 2014.

6. J. Xin, C.-W. Lin, and M.-T. Sun. Digital video transcoding. *Proceedings of the IEEE*, 93(1), 2005, 84–97.
7. ByteMobile. Advantages of bytemobile's video optimization solution. Available at: http://www.bytemobile.com/docs/WP_VideoOptimizationSolution.pdf. May 2013.
8. G. Kesidis, A. Das, and G. de Veciana. flat-rate and usage-based pricing for tiered commodity Internet services. In *42nd Annual Conference on Information Sciences and Systems, 2008*, pp. 304–308, 2008.
9. U. Paul, A. P. Subramanian, M. M. Buddhikot, and S. R. Das. Understanding traffic dynamics in cellular data networks. In *INFOCOM*, 2011.
10. Y. Zhang and A. Arvidsson. Understanding the characteristics of cellular data traffic. In *CellNet*, 2012.
11. C-Ran. The road towards green radio access network. Available at: http://labs.chinamobile.com/cran/. May 2014.
12. S. Ha, S. Sen, C. Wong, Y. Im, and M. Chiang. TUBE: time-dependent pricing for mobile data. In *ACM SIGCOMM*, 2012.
13. V. Gabale. Affordable pricing for cellular data networks. Available at: https://docs.google.com/forms/d/12Ac6LUzJ7qJI-I8r-xSGlpquBBcKUrG8edHPaDGQOyE/viewform. May 2014.
14. X. Liu, F. Dobrian, H. Milner, J. Jiang, V. Sekar, I. Stoica, and H. Zhang. "A case for a coordinated internet video control plane,". *SIGCOMM Computer Communication Review*, 24 (4), 2012, 359–370.
15. S. Sen, C. Joe-Wong, S. Ha, and M. Chiang. Pricing Data: Past Proposals, Current Plans, and Future Trends. arXiv, July 2012. Available at http://arxiv.org/abs/1201.4197.
16. P. Key, L. Massoulie, and M. Vojnovic. Farsighted users harness network time-diversity. In *Proceedings of IEEE INFOCOM, vol. 4*, pp. 2383–2394, 2005.
17. N. Laoutaris and P. Rodriguez. Good things come to those who (can) wait or how to handle delay tolerant traffic and make peace on the internet. In *ACM HotNets-VII*, 2008.
18. V. Gabale, U. Devi, R. Kokkku, V. Kolar, M. Madhavan, and S. Kalyanaraman. Async: de-congestion and yield management in cellular data networks. In submission, May 2013.
19. J. Murphy and L. Murphy. Bandwidth allocation by pricing in atm networks. In *IFIP TC6 Second International Conference on Broadband Communications II*, pp. 333–351, 1994.
20. A. Ganesh, K. Laevens, and R. Steinberg. Congestion pricing and adaptation. In *IEEE INFOCOM*, pp. 959–965. IEEE, 2001.
21. M. Roozbehani, M. Dahleh, and S. Mitter. Dynamic pricing and stabilization of supply and demand in modern electric power grids. In *International Conference on Smart Grid Communications*, pp. 543–548. IEEE, 2010.
22. P. Samadi, A. Mohsenian-Rad, R. Schober, V. W. S. Wong, and J. Jatskevich. Optimal real-time pricing algorithm based on utility maximization for smart grid. In *International Conference on Smart Grid Communications*, pp. 415–420. IEEE, 2010.
23. S. Ha, S. Sen, C. Joe-Wong, Y. Im, and M. Chiang. TUBE: time-dependent pricing for mobile data. In *ACM SIGCOMM*, 2012.
24. 3GPP. TS 23.203 Policy and charging control architecture. http://www.3gpp.org/ftp/Specs/html-info/23203.htm.
25. C. Peng, G.-H. Tu, C.-Y. Li, and S. Lu. Can we pay for what we get in 3g data access? In *Mobicom*, 2012.
26. CVXOPT. Python Software for Convex Optimization. http://cvxopt.org. May 2014.

16 Mechanisms for Quota Aware Video Adaptation

JIASI CHEN, AMITABHA GHOSH, and MUNG CHIANG*

16.1 INTRODUCTION

16.1.1 Two Conflicting Trends

Two recent and conflicting trends in Internet applications motivate the discussion in this work: video traffic becoming dominant and usage-based pricing becoming prevalent.

Video Traffic Becoming Dominant. Video consumption is on the rise. For example, Cisco predicts that 70% of all mobile traffic will be from video alone by 2016 [1]. Likewise, wireline traffic is also dominated by video, with NetFlix accounting for almost 30% of wireline Internet traffic [2]. Together with YouTube, Netflix, Hulu, HBO Go, iPad personalized video magazine apps, and news webpages with embedded videos, video traffic is surging on both wireline and wireless Internet.

Usage-Based Pricing Becoming Prevalent. Tiered pricing, or usage-based pricing, is becoming increasingly commonplace in the United States as well as in other countries for both wireless and wireline broadband. Table 16.1 compares the usage fees from various international Internet Service Providers (ISPs).

These two trends, *video traffic becoming dominant* and *usage-based pricing becoming prevalent*, are at odds with each other. On the one hand, videos, especially on high resolution devices (e.g., iPhone 5, iPad 3, and Android tablets), consume much more data than other types of traffic. For instance, 15 min of low bitrate YouTube video per day uses up 1 GB a month; likewise, one single standard-definition movie can take up to 2 GB. On the other hand, usage-based pricing threatens the business model of delivering entertainment via the high-speed Long-Term Evolution (LTE). These factors can result in high overage charges by the service provider, subscription to more expensive data plans, or even discontinuation

*Work was done while the author was a postdoctoral research associate at the Princeton University.

Smart Data Pricing, First Edition. Edited by Soumya Sen, Carlee Joe-Wong, Sangtae Ha, and Mung Chiang.

TABLE 16.1 Usage Fees for Wireline and Wireless ISPs Around the World

Carrier	Country	Wireline/ Wireless	Baseline Quota, GB	Overage Charge
AT&T	USA	Wireless	2	10 USD per GB
Verizon	USA	Wireless	2	10 USD per GB
Reliance [3]	India	Wireless	2	0.01 Rupee per 10 kB
Rogers [4]	Canada	Wireline	80	2 CAD per GB
AT&T	USA	Wireline	250	10 USD per 50 GB

of data service by disgruntled users. Given this conflict, a natural question to ask is: *Can a consumer stay within his/her monthly data quota without suffering a noticeable drop in video quality (distortion)?*

16.1.2 Current Approaches in Practice

In today's practice, there are two main approaches to balancing the competing goals of delivering high quality video while consuming less data.

16.1.2.1 Decrease Video Consumption Through Self-Imposed Warnings Consumers may be warned by (i) service providers or (ii) self-imposed warnings to stop watching more videos once their usage-based charges become too high. In the first case, consumer concern over "bill shock" has resulted in legislation against service providers. For example, the European Union passed regulations in 2010 requiring mobile ISPs to provide warnings when a consumer has reached 80% of his/her quota and to block data access once the usage fee reaches €50 [5]. Similarly, the Federal Communications Commission (FCC) worked with commercial ISPs on a program to provide warnings to consumers when their data quota limits are reached [6].

Mobile applications that can report real-time data usage are also becoming increasingly prevalent. External applications, such as Onavo and DataWiz, provide consumers with attractive graphical user interfaces (GUIs) for monitoring data usage [7, 8]. Google also introduced a built-in data monitoring application in a recent version of the Android operating system [9]. Overall, however, these straightforward warning systems can be undesirable, as they could result in dissatisfied users.

16.1.2.2 Decrease Data Consumption by Lowering Video Quality Content providers (CPs) can take a *one size fits all* approach of cutting back bitrates across all video requests, for all users, and at all times. For example, the YouTube mobile app automatically chooses a low quality video for requests made over a cellular data network, versus high quality video when a WiFi connection is used. Netflix implemented a similar approach in Canada in 2011 for both wireless and wireline customers in the light of expensive usage-based charges by Canadian ISPs. Netflix allows consumers to choose between three different qualities: low (0.3 GB/h), medium (0.7 GB/h), and high (1 GB/h); however, this approach cuts down the quality across all videos, without taking user preferences and remaining quota into account.

16.2 RELATED WORK

In this section, we give a brief background on existing video adaptation techniques and streaming protocols.

16.2.1 Video Adaptation

Video adaptation is the process of modifying a video to suit the user preferences, usage environments, resource constraints, content characteristics, or digital rights among other factors. In the context of this work, we consider video adaptation as a possible solution to the conflict between the growing demand for video traffic and usage-based data pricing. Adapting video quality with respect to resource constraints and user utility has been extensively studied since the 1990s, as surveyed, for example, by Chang and Vetra [10]. Liu et al. also considered video adaptation to maximize users' quality of experience (QoE), using the control knobs of video bitrate and content delivery network (CDN) selection [11].

16.2.2 Video Streaming Protocols

In order to perform video adaptation, the appropriate technologies must be enabled in practice. The algorithms and systems to perform such video adaptation have long been studied in the research community [11–14].

Traditionally, in practice, videos were streamed using the User Datagram Protocol (UDP) because of their time-sensitive nature. More recently, video streaming over the Hyper Text Transfer Protocol (HTTP) has become common, because of HTTP's widespread acceptance, its ability to traverse Network Address Translations (NATs), and its support for caching at CDNs [15]. Various protocols, such as Scalable Video Coding (SVC), have also been proposed [16]. With SVC, each video is encoded into a base and multiple enhancement quality layers. Within the video, the video quality may be modulated.

However, SVC never gained widespread adoption. Increasingly, but increasingly interest from industry has focused on *dynamic adaptive HTTP video streaming over HTTP*. Apple, Microsoft, and Adobe have developed proprietary HTTP video streaming protocols that perform intra-video bitrate switching to adapt to varying channel conditions. These protocols are supported by various web technologies:

- *Apple HTTP Live Streaming*. Quicktime, iOS (iPhone, iPod Touch, iPad), Android;
- *Microsoft Smooth Streaming*. Microsoft Silverlight, also adopted by Netflix;
- *Adobe OSMF*. Adobe Flash.

Recent progress by an industrial consortium has resulted in the Moving Picture Experts Group–Dynamic Adaptive Streaming over HTTP (MPEG-DASH) standard [17], which aims to address the lack of interoperability between current video streaming protocols by providing an open interface to access the quality levels of a video. These lines of work still focus on modifying the video quality based on *channel* conditions, but underscore the increasing interest, in both academia and industry, in video adaptation technologies. Specifically, encoding and storing multiple bitrate versions of each video enables video quality adaptation with respect to each user's *data quota* in the context of usage-based pricing.

16.2.3 Quota Aware Video Adaptation

Jan et al. [18] developed a system for compressing images on webpages under a data quota. Although the motivation of addressing quota concerns is similar, its application is only to images and webpages. Recently, systems such as Onavo [7] enable users to save on data plans by forwarding all images and text data through a proxy server, where they are compressed. However, they do not exploit consumer usage patterns or deal with video traffic, which has a much larger range of compressibility (than images and text) and comprises the bulk of mobile data consumption. These can provide substantial additional data saving opportunity.

16.3 A POTENTIAL SOLUTION: QAVA

16.3.1 Trading off Quality Versus Cost Versus Volume

As a potential solution to the problem of video consumption under capped data plans, we propose the QAVA (Quota Aware Video Adaptation) system. Our premise is the following: *Not every video bit is needed for every consumer* and the bitrates can be adjusted not only based on screen resolution and channel conditions but also based on usage patterns. QAVA can be customized to each user's demand and monthly data quota by adaptively choosing an appropriate bitrate for each video request, thereby *shaping the supply* for the user. We will show that by leveraging *video compressibility* and profiling *usage behavior*, QAVA can significantly mitigate the conflict between the growing demand for video traffic and usage-based pricing.

At the heart of QAVA is a *Stream Selector* (SS), which takes inputs from a *User Profiler* (UP) and a *Video Profiler* (VP), to select a particular bitrate and pre-emptively compress the more compressible videos early in the billing cycle. The VP provides information related to a video, such as its compressibility, which measures the extent to which the size of a video can be reduced without a significant distortion in quality. The UP predicts consumer usage patterns from past history and customizes the system to every user's flavor for watching certain types of videos. The SS then uses the information provided by both VP and UP to optimize QAVA for each user based on his/her monthly data quota.

The benefits to a QAVA-enabled user include the ability to watch more videos under a given monthly data plan without suffering a noticeable distortion in video

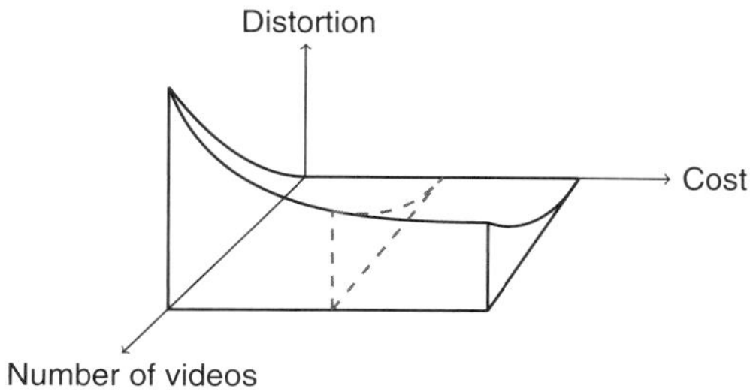

Figure 16.1 A three-way trade-off between distortion, cost, and the number of videos watched. For a fixed cost, QAVA enables a user to watch the desired videos while minimizing distortion.

quality, as compared to a non-QAVA user. Or, phrased differently, if a user's demand for video traffic remains the same or goes down, QAVA tries to save money for the user with a minimum impact on video quality. This three-way trade-off is illustrated in Figure 16.1. Across the three competing goals of minimizing cost, maximizing the number of videos watched, and minimizing distortion, QAVA strikes a graceful, tunable trade-off.

16.3.2 Incentives for Players in QAVA Ecosystem

A natural question to our proposed approach is: *What are the incentives for different players in the ecosystem to use QAVA?* We address this from the perspective of three major players: users, ISPs, and CPs.

Users. A user has the most obvious incentive, because QAVA enables him/her to stop worrying about his/her monthly data plan and watch all the videos he/she wants with minimal distortion.

ISPs. An ISP has two options: it may wish to (i) reduce data traffic to lower network congestion and thereby cut down on operational and capital expenditure costs, or (ii) preserve traffic to continue receiving overage charges from customers and/or usage fees from CPs. In the first case, QAVA ensures that all customers remain below their quota, which indirectly lowers the traffic rate. In the second case, the ISP can set quota parameter to ensure that its users still consume the same amount of data as non-QAVA users but receive better video quality.

Content Providers. The advantages for a CP are threefold. Firstly, as QAVA allows a user to access more content under the same data plan, the CP achieves a greater profit from advertising revenue through increased content consumption. Secondly, the CP improves customer satisfaction by removing his/her worries about exceeding the quota, which can be marketed as a competitive advantage over other CPs. Thirdly, QAVA reduces the potential need for the CP to pay the ISP for customer data charges.

QAVA is thus mutually advantageous from the perspectives of all three players. In the remainder of this work, we discuss QAVA's contribution along three different dimensions:

- *Architecture.* We design a modular system architecture for QAVA comprising three functional modules: stream selector, user profiler, and video profiler. Our design makes QAVA a deployable system that can be used by real-world consumers.
- *Algorithms.* We design an online bitrate selection algorithm for the SS module based on the concepts of finite-horizon Markov Decision Process (MDP) [19]. This algorithm runs at the heart of QAVA, enabling it to provide a graceful, tunable control of the quality, cost, and volume trade-off.
- *Experiments.* We evaluate the performance of QAVA through simulations using traces of real video requests.

16.3.3 Design Considerations

There are several alternatives and variants for designing the QAVA architecture. We briefly describe these alternatives, their advantages and disadvantages, and our design decisions.

16.3.3.1 Availability of Video Versions We assume that all the videos are pre-encoded and stored on the CP's server. An alternative to this is on-the-fly transcoding and adaptation, which requires compressing each video dynamically at the particular bitrate determined by the SS module. This can be a time-consuming operation and has significant implementation challenges; however, it would be capable of adapting video feeds for live events. In contrast, pre-encoded streams can be selected with minimal computation but cannot handle video streams of live events.

We also assume that the different versions of a video chosen by the SS module are supported by the channel in terms of bandwidth requirement. These sustainable video versions may be preselected by the content provider based on typical wireless or wireline bandwidth, or chosen on-the-fly based on bandwidth estimation techniques currently proposed for use in adaptive HTTP video streaming algorithms [14].

16.3.3.2 Time Scale of Video Adaptation There are two choices for the time scale of video adaptation: (i) intervideo adaptation and (ii) intravideo adaptation. Intervideo adaptation is choosing a single bitrate stream for the entire duration of the requested video, whereas intravideo adaptation involves dividing each video into smaller clips and choosing the correct adaptation operation for each clip.

Intervideo adaptation is suitable for video clips of short duration (e.g., Youtube videos of less than 5 min), because the spatial and temporal activities tend to be similar throughout the duration. However, for longer videos such as movies, it is more appropriate to stream different bitrate versions for different parts of the video depending on the spatial and temporal activities. The algorithms presented in this work apply equally to inter- or intravideo rate switching. QAVA can be used for intravideo adaptation by considering each smaller segment as a separate video request. Such intravideo switching requires synchronous bit stream switching, which can be achieved with the advent of new video streaming protocols such as MPEG-DASH [17]. QAVA can also work with existing channel-based switching algorithms by optimizing and restricting the rates available as input to those algorithms.

16.3.3.3 Heterogeneous Data Quota. Data usage under a single data plan can be decomposed into three usage layers: (i) multiple users, (ii) multiple devices per user, and (iii) multiple traffic types per user per device. QAVA's control knob is on video traffic per user per device; thus the "video quota" per device must be set. To accommodate non-video data traffic, the video quota should be set to a percentage of the total data quota based on historical video data usage. Running QAVA per user is also possible by aggregating video request logs across devices. This results in coarser granularity user profiling but may improve the performance by decreasing sensitivity to noise. For the remainder of this work, we focus on the case of a single fixed video quota and a single user with a single device, but QAVA can easily be extended to encompass the other cases just outlined.

16.4 QAVA SYSTEM DESIGN

16.4.1 A Modular Architecture Design

The architecture of QAVA comprises three different modules, each one is responsible for a specific function. The modules work together to enable QAVA to optimize across the three performance goals shown in Figure 16.1. We first describe the motivation for each of the modules.

16.4.1.1 Selecting Right Bitrates The basic operation of QAVA is to choose an appropriate bitrate for every video request made by a user. This bitrate selection is based on two factors: (i) the user's data consumption pattern and (ii) the particular video requested. This job is performed by a *Stream Selector* (SS) module running at the heart of QAVA on the CP's server, as shown in Figure 16.2. We focus on the pre-encoded bit stream scenario, where each video has multiple copies; each copy is pre-encoded in a different bitrate and stored on the CP's server. The number of copies of different bitrates of a video is predetermined by the CP.

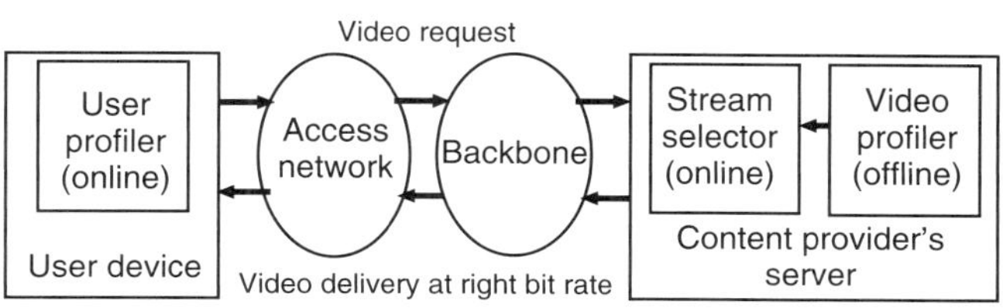

Figure 16.2 QAVA's modular system architecture: The UP module sits on a user device, whereas the SS and VP modules are located on a content provider's server. A video request originating from a user device travels through the access network and the backbone to the server, which then runs a stream selection algorithm to choose an appropriate bitrate to deliver to the user.

16.4.1.2 Profiling Usage Behavior A user's past data consumption history gives an indication of his/her future usage pattern. As the video requests from a user can arrive at different times without any prior knowledge, a prediction of these arrival times is helpful to QAVA so it can choose the appropriate bitrates to serve the requests. A simple usage pattern could be watching on an average x_d number of videos (or, equivalently, y_d bytes) on day d of a week. When the number x_d (or y_d) remains approximately the same for the same day d across different weeks, we may notice a weekly pattern. More complex usage behavior can lead to small-scale seasonal variations (daily or hourly) as well as *trends*, which are long-term variations due to habitual changes that lead to a steady increase or decrease in data usage over months or years.

QAVA employs a *user profiler* (UP) module to find patterns in usage behavior and to predict future video requests. In particular, the UP module estimates the probability of a video request arriving in a given time interval. The length of this interval can be uniform or variable depending on the past data consumption history and is configured by a system parameter. We design the UP module as an application that can be installed on a user device (client), as shown in Figure 16.2.

16.4.1.3 Estimating Video Compressibility In addition to staying within a monthly data quota, the SS algorithm also aims to minimize video distortion. For this, the algorithm needs to know to what extent the requested video can be compressed and how much distortion it would cause in doing so. Different videos have different levels of compressibility depending on their spatial and temporal activities as well as on the choice of encoder. For example, a talk show that has very little motion in it can be greatly compressed using an H.264/AVC encoder, whereas a motion-rich sports video may not be compressible to the same extent.

The SS algorithm should be careful in choosing the right bitrate for every video request to avoid the following undesirable situation. Suppose the algorithm chooses a high bitrate for an easily compressible video when the user has a lot of quota left, possibly in the beginning of a month. Then it might be forced to choose low bitrates for some not-so-easily compressible videos near the end of the billing cycle in order to stay within the monthly budget, thus causing significant distortion. A possible remedy is to choose low bitrates for easily compressible videos *even when* there is sufficient quota left. However, such intelligent online decisions can be made only if the system knows about the distortion versus bitrate trade-off for every video and can learn the quota consumption pattern over a billing cycle for each user. QAVA employs an offline *video profiler* (VP) module to compute this distortion for every bitrate and store it on the CP's server, as shown in Figure 16.2.

We now summarize the three modules of QAVA.

- *Stream Selector (SS).* The SS module is at the heart of QAVA and is located on the CP's server. It is an online module that decides the right bitrate for every video request.
- *User Profiler (UP).* The UP module predicts the probability of future video requests from past usage history and also computes a user-specific distribution

of video types reflecting a user's taste of watching different types of videos. It is an online module and is located on the user device.

- *Video Profiler (VP).* The VP module is located on the CP's server and is an offline entity. It computes the distortion for every bitrate version of all the stored videos. We loosely call this the "*compressibility*" of the videos.

Other types of module separation and interconnections are also possible. For example, the VP and SS modules can be combined into a single module performing joint optimization. The SS requests videos of a certain compressibility, and the VP optimizes over various codecs to generate videos with desired characteristics. This would provide a finer decision granularity to the SS, but is computationally complex as the VP also runs video encoding operations.

16.4.2 Module Placement

In this section, we discuss the location of the various modules in QAVA. Although we refer to the specific modules of QAVA, our discussion of SS module placement is applicable to any general video adaptation system that makes online decisions.

The placement of the UP and VP modules is fairly intuitive: by necessity, the VP module is located on the CP's server, as only the CP knows about the video characteristics. Profiling the video on the user device is not feasible because of the CPU (central processing unit) and battery limitations. The UP module logs user data, so it should be placed on the user device to alleviate user concerns over data collection and privacy. The placement of VP and UP modules is shown in Figure 16.2.

16.4.2.1 ContentProvider-Based or ISP-Based Architectures The location of the SS module that runs the bitrate selection algorithm is, however, not so intuitive. For every video request, the SS module requires inputs from both UP and VP. One possibility is to place the SS module in the ISP's access network. Then, in order to satisfy a video request, the SS module first needs the video compressibility information to be sent from the CP, as well as user information from the UP module. After receiving these inputs, the SS module runs the SS algorithm to choose the right bitrate. It then sends another request to the server to transmit the actual video in the selected bitrate. Overall, this results in unnecessary messages and potential delay, which is undesirable for delay-sensitive traffic such as video.

Alternatively, the SS module could be placed on the CP's server. In order to satisfy a video request, the SS module still needs user information from the UP module but already has access to information from the colocated VP. The selected bitrate can thus be transmitted immediately after the SS algorithm is run. As this reduces the amount of message passing, placing the SS module on the CP is more desirable. This also makes QAVA complementary to other video adaptation approaches [11].

Placing the SS and VP modules on the CP's server incurs some monetary cost to the CP, which must be overcome by the advantages discussed in Section 16.3.2. We argue that the cost to the CP is small: it must install the SS module on its server (one time) and compute the video profiles of all videos (a small amount of text data compared to video data size). And regardless of where the SS module is placed, our algorithms are equally applicable.

16.4.2.2 Client-Based Architectures In case QAVA operates without the support from a CP or an ISP, *client-based* architectures are possible. We discuss two potential architectures: (i) a transcoding-based system and (ii) a throttling-based system.

In a transcoding-based solution, all traffic from the client is routed through a web proxy. By inspecting the packets, video traffic can be distinguished and transcoded on the fly by the proxy server to the correct bitrate. This approach has the advantage of using standard web proxy technology and can handle all types of videos as long as the transcoder has the appropriate codecs. However, there are significant implementation challenges in terms of latency minimization: the proxy and transcoder must perform quickly in order to satisfy the demands of delay-sensitive video traffic. Moreover, implementing a transcoder for HTTP video streaming is also difficult.

A throttling-based solution leverages the emerging popularity of adaptive HTTP video streaming [17]. In this technology, the videos automatically adjust their bitrates based on their estimates of TCP throughput. By limiting the bandwidth observed by the rate switching algorithm, QAVA can force the video bitrates to automatically settle to the rates determined by the SS module. This throttling can be performed on the client device or on a web proxy. In terms of implementation, this approach is simpler than the transcoding-based solution but suffers from the limitation that only adaptive HTTP video streams can be modified in this way.

16.4.3 QAVA Operational Example

In this work, we focus on the case where the SS module is located on the CP's server. The following description does not change significantly for other placement options. The relationship between the different modules describing their inputs, outputs, and update frequencies is shown in Table 16.2. The input to the SS module is the compressibility of the requested video, the user's remaining monthly budget, and the user's profile as determined by the UP module. For every request, it runs the stream selection algorithm and outputs the selected bitrate version. The input to the VP module is the set of all videos stored on the CP's server, and its output is the compressibility of each video. The input to the UP module is the time stamps for the user's past video requests, as well as the compressibility of those videos. Its output is the predicted video request probability and the video type distribution (i.e., compressibility distribution) specific to that user. These quantities characterize the user's behavior and are fed to the SS module. These predictions can be made in the beginning of the billing cycle, or updated more periodically throughout the billing cycle.

To be concrete, we give an operational example.

1. The CP computes and stores the compressibility of all the videos on the server.
2. At the beginning of a billing cycle, the UP makes predictions (to be used in the current cycle) of the video request probability and the compressibility distribution, based on its log of the past requests of a user.

TABLE 16.2 Three Key Functional Modules of QAVA with Their Inputs, Outputs, and Update Frequency

Module	Input	Output	Frequency
SS	Compressibility of video request, remaining monthly budget, and the user profile	Selected bitrate version to deliver for the video request	Every request
UP	Time stamps and compressibility of all past video requests	Probability of a video request arriving at a time interval, and the video compressibility distribution for all past requests. Together these comprise the user profile	Every billing cycle
VP	All videos stored in the content provider's server	Compressibility of all videos	Offline

Abbreviations: SS, stream selector; UP, user profiler; VP, video profiler.

3. When a user requests a video in his/her current billing cycle, the request is sent to the SS module, which selects the bitrate to be delivered. The CP also sends the compressibility of the requested video to the UP. The UP logs this as well as the timestamp of the request.
4. Once the current billing cycle is over, the UP updates its predictions based on the recent request logs. Steps 2–4 repeat for the next billing cycle.

16.5 STREAM SELECTION

In this section, we first describe the video request, utility, and cost model and then formulate the bitrate selection problem. We also introduce the key notation, as summarized in Table 16.3.

16.5.1 Video Request, Utility, and Cost Model

We divide the length of a billing cycle (e.g., month) into T time intervals, indexed by $t = 1, \ldots, T$, and assume that a user has a total budget B (measured in bytes) in one billing cycle. In each time interval t, a video request arrives with a certain probability, denoted by p_t. The remaining budget of the user at time t is b_t. The request probability p_t and budget b_t are provided by the UP module for each user.

Each video is encoded into M different bitrates, indexed by $j = 1, \ldots, M$. Associated with each video request arriving at time t is a vector of utilities $\mathbf{u}_t = (u_{t1}, \ldots, u_{tM})$

TABLE 16.3 Table of Key Notation

Symbol	Meaning
u_{tj}	Utility of bitrate version j for a video request arriving at t
c_{tj}	Cost of bitrate version j for a video request arriving at t
x_{tj}	Indicator variable (1 if bitrate j is chosen for video request at t; 0 otherwise)
M	Number of bitrates for each video
T	Number of time intervals in a billing cycle
$P(\mathbf{u}, \mathbf{c})$	User-specific joint probability distribution of video types based on past history.
p_t	Probability of a video request arrival at t
b_t	Remaining user budget at t
B	Total user budget in one billing cycle

and a vector of costs $\mathbf{c}_t = (c_{t1}, \ldots, c_{tM})$ for different bitrate versions of the video.[1] When there is no video request arrival at time t, the vectors $\mathbf{u}_t$ and $\mathbf{c}_t$ are null vectors with all components being zero, because then no bitrate is selected. The VP module provides the utility and cost of each video, $\mathbf{u}_t$ and $\mathbf{c}_t$, respectively.

Each user might prefer different types of videos with different compressibilities. For example, he/she might want to watch news clips that have different compressibilities than sports videos. To capture this effect, we introduce a joint probability distribution $P(\mathbf{u}, \mathbf{c})$, which is user specific and represents the probability that a user requests videos with certain utility–cost characteristics. This distribution is provided by the UP module for each user.

In Figure 16.3, we show a typical utility versus cost function for a video encoded using the H.264/AVC codec with a resolution of 720×480 pixels. Such utility–cost

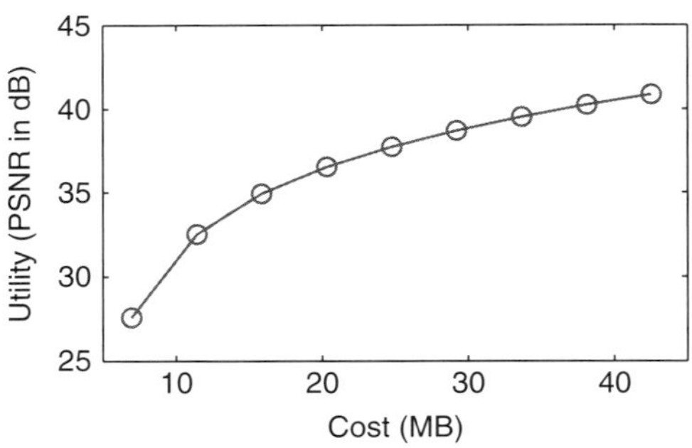

Figure 16.3 Video utility versus cost showing diminishing returns for increasing cost. The x-axis represents the size of the video when encoded from 100 to 900 Kbps in 100 Kbps increments.

[1]We note that these utility and cost vectors are fixed and not time dependent. The use of the time index t in u_{tj} and c_{tj} is purely for the ease of exposition.

curves are usually concave with diminishing returns for utility at higher cost (or equivalently, higher bitrate, as bitrate is proportional to data size for a fixed-length video). A video with a flat utility–cost curve is "easily compressible," because lowering the bitrate decreases the utility only slightly. In contrast, a "hard-to-compress" video has a steep curve. We measure the utility u_{tj} of bitrate version j as its *peak signal-to-noise ratio* (PSNR) × the duration of the video [20]. The cost c_{ij} is the size in bytes. Discussion of the utility and cost metrics is reserved for Section 16.6.2.

16.5.2 Stream Selection as Knapsack Problems

We now formulate the problem of choosing the right bitrate by the SS module as different versions of the well-known knapsack problem [21] studied in combinatorial optimization. We first present an offline formulation, which is easy to understand, and then motivate the need for an online stochastic formulation.

16.5.2.1 Offline Multiple-Choice Knapsack Problem The goal of the stream selector is to choose the right bitrate for every video request made by a user in a single billing cycle. We aim to maximize the sum of the utilities across all the video requests without exceeding the user's quota. In other words, we maximize the *average* video utility. An alternative formulation is to maximize the minimum utility across all videos requested during the billing cycle. However, as the high level goal of QAVA is to maximize the overall user satisfaction, we optimize the average utility over time instead of the worst-case experience, as in the alternative formulation.

For a video request arriving at time t, we define a decision vector $\mathbf{x}_t = (x_{t1}, \ldots, x_{tM})$, where each x_{tj} takes the value 1 if bitrate version j is chosen and 0 otherwise. Then our problem is

$$\begin{aligned} \text{maximize} \quad & \sum_{t=1}^{T} \sum_{j=1}^{M} u_{tj} x_{tj} \\ \text{subject to} \quad & \sum_{t=1}^{T} \sum_{j=1}^{M} c_{tj} x_{tj} \leq B \\ & \sum_{j=1}^{M} x_{tj} = 1, \ \forall\, t \\ \text{variables} \quad & x_{tj} \in \{0, 1\}, \ \forall\, t, j, \end{aligned} \tag{16.1}$$

where the first constraint says that the cost of the selected bitrates for all the videos requested in a billing cycle must not exceed the total budget B, and the second constraint says that only one bitrate version may be selected for each video.

This optimization problem is known as the *multiple-choice knapsack problem* (MCKP) [21]. In the regular single-choice knapsack problem, we are given a set of items, each with an associated value and weight. The objective is to pick a subset of the items such that the total value is maximized while not exceeding the knapsack capacity. In our stream selection problem, the items are the individual bitrate versions of the videos, and the multiple choices arise because exactly one version of each video must be selected.

The traditional offline version of MCKP, where all the input items are known in advance, is well studied. The problem is NP-hard, but pseudo-polynomial time dynamic programming (DP) solutions exist [21]. Contrary to this offline version, the SS module does not know the video requests in advance, and so needs to make decisions in an online manner. This requires a modification to the above formulation to handle online requests.

16.5.2.2 Online Stochastic Knapsack Problem Unlike the traditional offline MCKP, in our scenario, the video requests are revealed one by one in an *online* manner. Thus, existing DP solutions to the offline knapsack problem cannot be used. Online algorithms handle this situation by making a decision on the fly when a new video request arrives, without any prior knowledge of future requests. Once a decision is made, it cannot be revoked or changed in the future.

As the data quota is reset after a billing cycle is over, there is a time deadline before which all the actions must be made. We also note that the bitrate decisions for future intervals should not depend on the decisions taken at previous intervals, given the current remaining budget. This implies the Markov property. The problem can naturally be modeled as a finite-horizon MDP. A key assumption is that the video requests are independent of time and, therefore, the transition probabilities are stationary.

The MDP formulation allows the SS module to make foresighted bitrate selection decisions by taking into account the future impact of its current decisions on the long-term utility. This is better than just an online algorithm that makes myopic decisions at every time step. For example, a greedy solution might choose a bitrate that maximizes only the utility of the current request without overshooting the quota.

Figure 16.4 shows a simple example of choosing between three different bitrates over one time step. The state of the system is defined as the four-tuple $s_t = (t, b_t, \mathbf{u}_t, \mathbf{c}_t)$, comprising the current time interval t, the remaining quota b_t, and the utility and cost vectors $\mathbf{u}_t$ and $\mathbf{c}_t$, respectively. There are three possible actions: (i) choose the lowest bitrate, that is, $\mathbf{x}_t = (1, 0, 0)$; (ii) choose the second bitrate, that is, $\mathbf{x}_t = (0, 1, 0)$; or (iii) choose the third bitrate, that is, $\mathbf{x}_t = (0, 0, 1)$. If the lowest bitrate is chosen, the system moves to time $t + 1$ with remaining budget $b_t - c_{t1}$. The algorithm collects utility (reward) u_{t1} and receives the new video request with utility and cost vectors $\mathbf{u}_{t+1}$ and $\mathbf{c}_{t+1}$. If the second bitrate is chosen, the system moves to time $t + 1$ but now subtracts the cost c_{t2} from its remaining budget, leaving it with $b_t - c_{t2}$. It also collects utility u_{t2}. A similar state transition results from choosing the third bitrate.

The set of actions $\{\mathbf{x}_1, \dots, \mathbf{x}_T\}$ taken by the algorithm at every time step is called a *policy*. A policy that solves the MCKP of Eq. (16.1) is called an *optimal policy*. If the arriving video requests are known, an optimal policy can be determined using the traditional offline techniques previously mentioned. However, because the

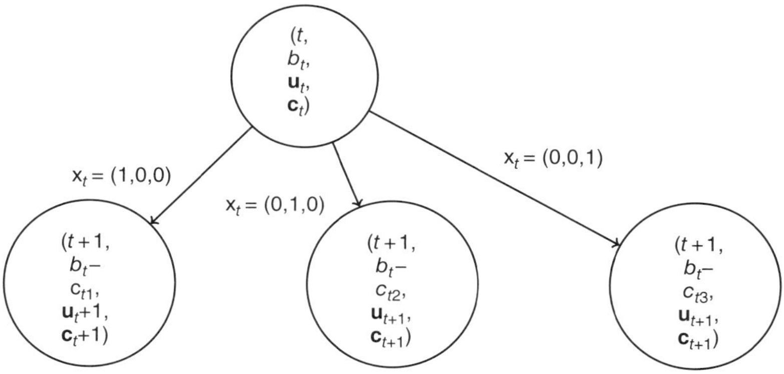

Figure 16.4 Stream selection modeled as a finite-horizon Markov decision process. A one step state transition is shown with three bitrate choices.

video requests are not known a priori, the MDP finds a policy that instead maximizes the *expected* sum of the utilities. We develop a solution using DP and online optimization.

16.5.3 Solving Finite-Horizon Markov Decision Process

The optimal policy can be found using the well-known backward induction techniques for finite-horizon MDPs [19]. We first define $U_t(b_t)$ as the expected utility accumulated from time t until the end of the billing cycle at time T, when the remaining quota is b_t. This expected utility assumes that the optimal action is applied in each state. Then, the optimal action at each time step t is found by solving

$$\begin{aligned} \text{maximize} \quad & u_{tj}x_{tj} + U_{t+1}(b_t - c_{tj}x_{tj}) \\ \text{subject to} \quad & c_{tj}x_{tj} \le b_t \\ & \sum_{j=1}^{M} x_{tj} = 1 \\ \text{variables} \quad & x_{tj} \in \{0, 1\}, \ \forall j, \end{aligned} \tag{16.2}$$

where the first constraint ensures that the cost of the selected bitrate is less than the remaining quota. The objective function has an intuitive meaning: it maximizes the current utility plus the sum of the expected utilities, subject to the remaining quota. The problem can be solved in $O(M)$ time by discarding the bitrates that violate the constraints and then picking the bitrate j^* that maximizes the objective function. It is solved every time a video is requested.

Solving Eq. (16.2) requires the computation of $U_t(b_t)$. It can be shown that the solution can be found by dynamic programming using

$$U_t(b_t) = p_t \left(u_{tj^*} + E_{(\mathbf{u},\mathbf{c})}\left[U_{t+1}\left(b_t - c_{tj^*}\right)\right]\right) + \left(1 - p_t\right) U_{t+1}(b_t), \tag{16.3}$$

with boundary condition is that the expected accumulated utility is 0 when the billing cycle is over, or the remaining budget is less than 0 [22]. In this work, we choose the budget granularity to be 1, so b_t takes on possible values $1, \dots, B$. The running time of computing the $U(\cdot)$ matrix is $O(TBM\Gamma)$, where Γ is the cardinality of the set $\{(\mathbf{u}, \mathbf{c})\}$ in the video type distribution.

The online and offline components of the MDP stream selection algorithm are summarized in Algorithm 16.1. In the special case of two bitrates ($M = 2$), our algorithm reduces to that of Papastavrou et al. [23]. With accurate user and video profiling, Algorithm 16.1 maximizes the sum utility while staying under the quota. In the case of inaccurate inputs, however, the algorithm may exceed the quota. In that case, the algorithm should simply choose the lowest bitrate, although this case never occurs in our numerical simulations.

Algorithm 16.1

MDP Stream Selection Algorithm

DP Computation of Utility Matrix

Input: Video type distribution $P(\mathbf{u}, \mathbf{c})$, quota B, and billing cycle length T.

Output: A matrix $\mathbf{U}$ of size $T \times B$.

1. Initialize $U_t(b_t) = 0, \forall\, t > T, b_t < 0$.
2. Compute each entry $U_t(b_t)$ of $\mathbf{U}$, using (16.3).

Online Bitrate Selection

Input: Utility–cost vectors $\mathbf{u}_t$ and $\mathbf{c}_t$, remaining budget b_t, quota B, billing cycle length T, and matrix $\mathbf{U}$.

Output: Optimal bitrate j^*.

if $b_t < 0$

  Choose $j^* = 1$.

else

  Discard the bitrates with cost greater than b_t.

  Compute j^* that maximizes the objective function of (16.2).

end

Update $b_{t+1} = b_t - c_{tj^*}$.

16.6 USER AND VIDEO PROFILERS

In this section, we detail the functionality of the UP and VP modules. We first describe different patterns in user behavior and tastes and then propose the UP algorithm. The VP module is also briefly explained.

16.6.1 Profiling User Behavior

The UP runs on the client device and characterizes each user through (i) the video request probability at each time interval and (ii) the distribution of video types preferred by the user.

16.6.1.1 User Viewing Pattern and Taste Depending on their lifestyles, different users have different time preferences for watching videos. For example, some users prefer watching videos on weekends rather than on weekdays, whereas others watch more in the evening after working hours than in the mornings. The taste in content of the users can also be different. For example, some users watch sports videos more often than movie trailers, whereas some others watch more news clips than music videos. Such preferences in user behavior can lead to well-defined patterns, in terms of both the viewing times and the types of the videos being watched.

The job of the UP is to estimate these temporal viewing patterns and video type preferences for each user. The UP module does this based on the user's past video request records, spanning either the previous billing cycle or the entire history. In this work, we consider requests from the last billing cycle.

16.6.1.2 Computing Video Request Probability In each time interval t, there is a certain probability p_t that the user requests a video. This request probability can either vary with each interval or be constant. As a first attempt, we compute the *average* request probability per interval, and set p_t for each interval equal to this average. The average request probability is computed by summing the number of requests in the previous billing cycle and dividing by the number of periods T. The time interval should be set small enough so that the average request probability is less than 1 but not so small as to inhibit the computation of Eq. (16.3).

There are several alternative approaches, including fitting distributions and prediction-based techniques. The arrival rate of videos might follow a particular known distribution (e.g., Poisson), in which case the probability of an arrival can be computed directly from the distribution itself. Alternatively, one can use more sophisticated time series analysis techniques. For example, at the beginning of the billing cycle, one can predict the sequence of future viewing times in the upcoming billing cycle and then compute the average request probability by adding up the predicted number of requests and finally dividing that by the number of intervals. One can also design online algorithms, such as predicting the sequence of viewing times for intervals $t+1, t+2, \dots, T$, while at interval t, and updating the predictions when a new video request arrives. Such alternatives trade off accuracy versus computation need.

We have developed one such online algorithm based on "triple exponential smoothing" [24]. However, here we will employ the simple averaging technique previously mentioned. The resulting computation requires less memory and power and can be performed easily on a resource-constrained (in terms of battery and memory) client device. Our goal is not necessarily to develop the best UP but to find a method that works well in the system as a whole. To establish this, we run trace-driven simulations in Section 16.7 to compare the performance of QAVA when the UP module uses the average request probability, to a scenario when the UP module has perfect knowledge of future arrivals. We find that our technique, while simple, achieves close to optimal performance (more than 95% on an average).

16.6.1.3 Computing Video Type Distribution The joint probability distribution $P(\mathbf{u}, \mathbf{c})$ reflects a user's preference for watching different types of videos. For example, a user who watches a lot of sports videos (which are not so compressible) will have a different distribution from a user who watches a lot of talk shows (more compressible). This video type distribution can remain the same over the length of a billing cycle, or can be time dependent, reflecting, for instance, the fact that a user watches more sports videos at night and more news clips in the morning. As a first-order approximation, we assume that the distribution does not change with time. The distribution is computed once at the beginning of a billing cycle based on the video requests in the last billing cycle.

Our method is as follows: Each video request arriving at time interval t in the previous billing cycle has a $(\mathbf{u}_t, \mathbf{c}_t)$ pair associated with it. The probability distribution is calculated by counting the frequency of each $(\mathbf{u}_t, \mathbf{c}_t)$ pair from the last billing cycle and then normalizing appropriately to form a probability distribution. As the utility and cost are continuous variables, they can be binned for greater computational efficiency; however, in our dataset, we find this optimization unnecessary. Through simulation, we show that this estimate performs very well, compared to the ideal scenario when the distribution of the requested videos is perfectly known ahead of time. Our user profiling algorithms are summarized in Algorithm 16.2.

Algorithm 16.2

User Profiling Algorithm

Input: Time stamps and utility–cost vectors $(\mathbf{u}_t, \mathbf{c}_t)$ of each video request in the previous billing cycle.
Output: Video request probability p_t, $\forall t$, and the video type joint probability distribution $P(\mathbf{u}, \mathbf{c})$.

1. Count the number of requests n_r, and the number of time intervals T in the last billing cycle.
2. Compute average request probability as $\overline{p} = n_r/T$, and set $p_t = \overline{p}$, $\forall\ t$.
3. Count the number of times each $(\mathbf{u}_t, \mathbf{c}_t)$ pair appears in the past; denote this count by $n_{(\mathbf{u}_t,\mathbf{c}_t)}$.
4. Construct the joint probability distribution by computing the individual probabilities as $p(\mathbf{u}_t, \mathbf{c}_t) = n_{(\mathbf{u}_t,\mathbf{c}_t)} / \sum_{(\mathbf{u}'_t,\mathbf{c}'_t)} n_{(\mathbf{u}'_t,\mathbf{c}'_t)}$

16.6.2 Profiling Video Cost and Utility

The purpose of the VP running on the VP module is to estimate the utility and cost for all the bitrate versions of all videos stored on the CP's server. There are many estimation techniques for computing the quality of a video. One standard objective metric is the PSNR, which we employ. The PSNR is a well-known objective metric for measuring video quality. A video typically comprises a sequence of images, and the PSNR for each image is defined as a function of the mean square error (MSE) between the original and compressed image. Mathematically, it is expressed in the logarithmic unit of decibel (dB) as PSNR $= 10\log_{10}(Q^2/D)$, where D is the pixel-wise MSE between the original and reconstructed image and Q is the maximum pixel value (usually 255).

We compute the video PSNR as the PSNR averaged over all images in the sequence. Typical PSNR values for lossy video compression are between 20 and 30 dB, where higher is better. To account for the duration of the video, we set the utility equal to the PSNR × the video duration. An example utility–cost curve is shown in Figure 16.3.

There exist other, potentially more accurate, metrics of video utility (e.g., mean opinion scores or MOS [25]), as well as means of calculating subjective metrics from objective measurements [26–28]. However, we choose PSNR as it can be easily computed using a mathematical formula (in contrast to MOS that requires time-consuming human experiments) and is very well known to the multimedia community.

Measuring the cost of a video in bytes naturally follows from the fact that the data quota is measured in bytes. The VP calculates the utility and the cost in megabytes for all the videos only once. These utility and cost vectors are stored alongside the videos on the CP's server.

16.7 PERFORMANCE EVALUATION

We evaluate the performance of QAVA stream selection by comparing it with three alternatives: (i) a hindsight-based offline, optimal algorithm that knows all the video requests in a billing cycle ahead of time; (ii) a worst-case online algorithm; and (iii) a naïve method (used by, e.g., Netflix). We also explore the sensitivity of QAVA to UP prediction errors.

16.7.1 Experimental Setup

Our simulations are based on the public-domain traces of 2 weeks of YouTube video requests from a wireless campus network [29]. The data comprises 16,337 users making a total of 611,968 requests over 14 days. YouTube is the largest source of video traffic, so the dataset captures a major portion of video viewing behavior [2]. The first week of trace data is used to train both VP and UP. The second week emulates a billing cycle where the stream selector is run for each user's requests.

Each video is encoded in H.264/AVC at 100, 200, 300, 400, and 500 Kbps. The stream selector chooses one bitrate from the first four choices. The 500 Kbps version is treated as a reference for computing the PSNR of the other bitrates. The cost of each video is its size in megabytes. We set the user's quota at the halfway point between the minimum data usage (always selecting 100 Kbps) and the maximum data usage (always selecting 400 Kbps) and also sweep across quotas when appropriate. The period length is set to 30 min, as we experimentally find that varying the period length does not greatly change system performance.

One limitation of this evaluation is that not all videos were available from YouTube at the time of this study. In the training phase, missing videos are not included while generating the video type probability distribution. In the test phase, the utility–cost curves of missing videos are sampled from the video type distribution of the training phase. This gives an advantage to our algorithm, because the probability distribution of the training phase is similar to that of the test phase. We also examine the effect of mis-estimation of the distribution.

16.7.2 Comparing Stream Selection Algorithms

We first evaluate the offline algorithm that solves problem (16.1) with the knowledge of all future video requests. It achieves the best possible performance and is treated as the benchmark against which we compare the performance of the online algorithms. We call this the *hindsight offline optimal* algorithm.

Zhou et al. [30] present an online algorithm to solve Eq. (16.1) with a worst-case performance guarantee regardless of the sequence of video requests. We call this the online *MCKP* algorithm and chose this to compare with our MDP algorithm because it optimizes for the *worst-case* performance, while our algorithm optimizes for *expected* performance. The MCKP algorithm, however, uses less information than our MDP algorithm, needing only the maximum and minimum utility-to-cost ratio across all requested videos, and an estimate of the sum data of the smallest bitrates. The MCKP algorithm does not use prediction or time deadlines but requires only the quota.

The second online algorithm we compare with is the naïve solution currently used by Netflix. Netflix allows subscribers to select a default streaming bitrate. We assume that a Netflix user chooses one bitrate for the entire billing cycle and is also intelligent enough to presciently choose the maximum bitrate that fits all videos in the quota. *Clearly, this algorithm is an ideal algorithm and not suitable for practical use, because it assumes advance knowledge of the number of videos to be watched.* Comparison with this Netflix method allows us to evaluate how our MDP algorithm performs against an existing practical solution.

16.7.3 Single-User Examples

To understand the operation of the stream selector, we first run the algorithm for a single user with a target quota of 1426 MB (see Fig. 16.5). The video requests arrive in bursts, and each algorithm selects the bitrate of each video. The Netflix method always chooses 200 Kbps. The MDP algorithm has foresight and thus chooses lower bitrates in the beginning of the billing cycle, knowing to save for later. The MCKP algorithm does not use time deadlines, only the remaining quota, and so it chooses high bitrates in the beginning before cutting back as it starts depleting the quota. The offline optimal algorithm chooses a variety of bitrates over time.

We also sweep across different monthly quotas and measure the utility obtained by each algorithm (see Fig. 16.6). The offline algorithm performs the best, with MDP and MCKP close behind. The Netflix method exhibits a staircase-like shape, because as the quota increases, the default user-selected bitrate can increase. In all cases, the algorithms use less data than the target quota.

16.7.4 Multiuser Stream Selection

16.7.4.1 Average Performance We now present the evaluation results of the MDP algorithm for multiple users. Each trial takes as input a fixed set of video requests and a quota and computes the utility obtained by each algorithm. Some input combinations achieve higher utility than others. In order to fairly compare across multiple

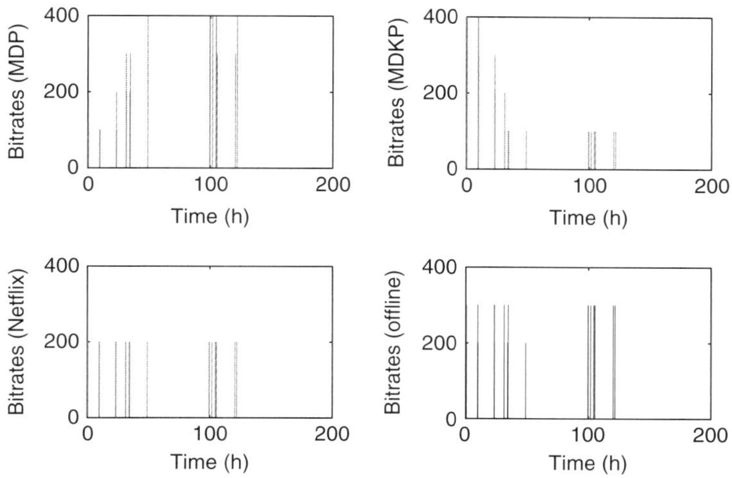

Figure 16.5 Bit rate selection by different algorithms for a single user. The MDP and MCKP algorithms choose different bitrates over time, while Netflix chooses a constant bitrate and the offline algorithm chooses the optimal bitrates.

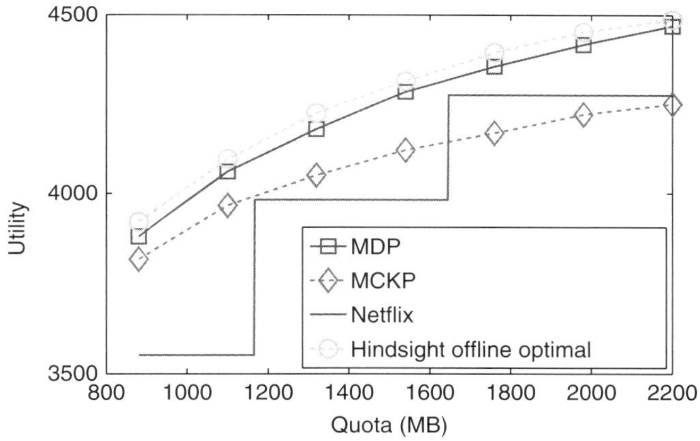

Figure 16.6 Quality–cost trade-off for a single user with different quota and fixed video requests. MDP obtains close to optimal utility, while MCKP and Netflix perform suboptimally.

trials, we normalize the utility across different users by measuring the utility of the online algorithm as a fraction of the offline optimal utility. To normalize the quota, we measure data as a fraction of the total data if the 400 Kbps bitrate were always selected.

Figure 16.7 shows the average utility across 10 different users. We observe that, on an average, the MDP algorithm performs better than the MCKP algorithm. The Netflix method resembles a staircase function as in the single-user case and obtains especially low utility for low quotas. This is arguably the most important scenario:

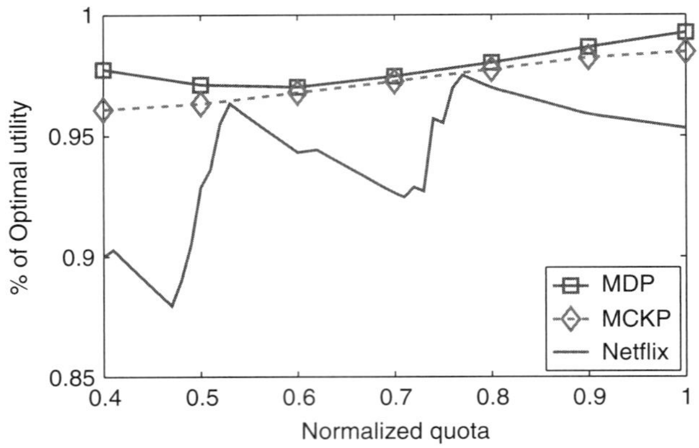

Figure 16.7 Quality–cost trade-off averaged and normalized over multiple users. The MDP algorithm achieves nearly optimal performance, with the MCKP algorithm close behind. Both algorithms outperform the naïve Netflix method.

When the user's quota is small compared to the number of videos he/she wishes to watch. For these low quotas, the MDP has a definite advantage over MCKP, which in turn outperforms the Netflix method.

16.7.4.2 Performance Variability To examine the utility distribution, and not just the average, we plot their cumulative distribution functions (CDFs) in Figure 16.8a across multiple quotas and users. The ideal result is a step function at 1, indicating 100% of the trials result in optimal utility. We see the Netflix method performs the worst, obtaining, for example, less than 95% utility in 50% of the time. The MCKP curve is steeper, indicating it has less performance variation, which makes intuitive sense as the algorithm optimizes for the worst case. The MDP algorithm optimizes for the average-case performance but not the spread and thus exhibits greater variation but is closer to the ideal step function.

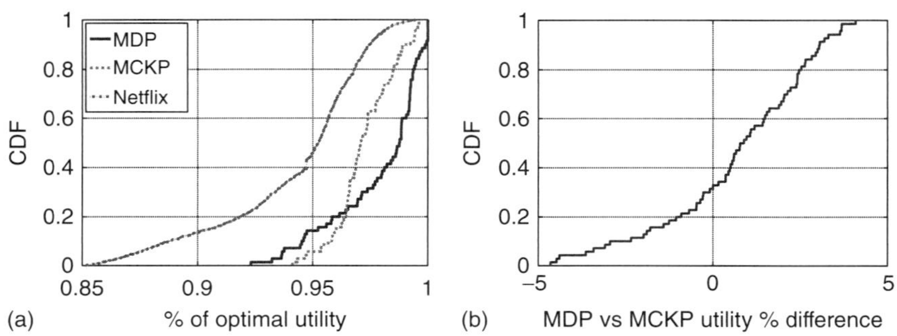

Figure 16.8 (a) CDF of the utility achieved by MDP, MCKP, and Netflix algorithms. (b) Distribution of MDP performance improvement over MCKP.

The MDP and MCKP algorithms are further compared in Figure 16.8b, which shows the CDF of their percentage utility difference. If the MDP were always better than the MCKP, we would see a step function at 0. However, we observe that the MCKP sometimes outperforms the MDP algorithm. This may occur because the MCKP optimizes for worst-case performance, so when a user's behavior exhibits high variability, the MCKP is better able to handle the user's requests.

On the surface, these simulations seem to suggest the Netflix method performs reasonably well in general. It achieves above 85% of the optimal utility, suggesting that this naïve solution is acceptable for most users. However, the caveat is that our simulated Netflix method assumes perfect knowledge of the number of video requests in the billing cycle, so that the user knows how to correctly set the default bitrate given the quota. This is represented by the sharp jumps with increasing quota in Figures 16.6 and 16.7. If the user sets the default bitrate too high, he/she will overshoot his/her quota. If the user sets the default bitrate too low, he/she will obtain suboptimal utility. A main advantage of QAVA is that it automatically adjusts the bitrate, so the Netflix user does not need to estimate his/her usage to set his/her default bitrate, which might result in over- or undershooting the quota.

16.7.5 Sensitivity to Prediction Error

It is important to examine how errors from the UP module affect the performance of the SS algorithm. There are two possible sources of error: video request probability and video type distribution. To test sensitivity to request probability error, we measure the utility obtained by the MDP algorithm when it uses (i) the estimated average arrival rate trained on historical data, (ii) the true average arrival rate of the test data, and (iii) the true request times. These results are averaged across multiple users and shown in Figure 16.9a, with the error bars indicating standard deviation. We observe that the greater the information accuracy, the greater the average utility. The performance difference is quite small, which suggests that the MDP algorithm

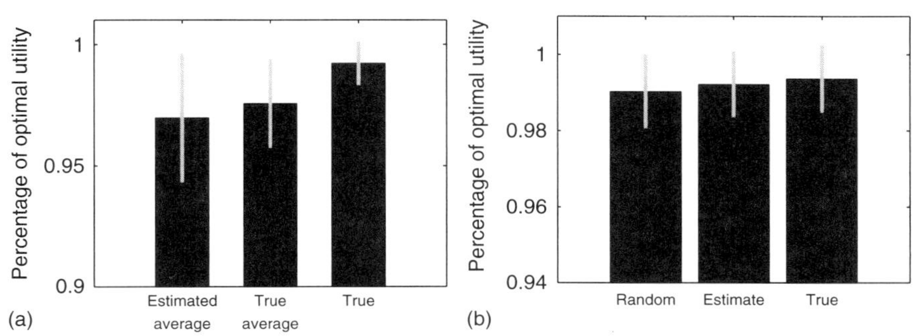

Figure 16.9 Effect of user profiler prediction error. Our video and user profiler use imperfect estimates of user behavior, but achieve good performance compared to the case with perfect information. (a) Arrival rate. (b) Video type.

performs well independent of arrival probability accuracy. In addition to the performance results shown in Figure 16.9a, the average difference between the true and estimated arrival rate (the input to the algorithm) is 8%. This suggests our user profiling technique is sufficient to capture variability of user behavioral patterns.

To analyze the sensitivity of the SS to video type prediction errors, we perform the following experiment. We calculate the true video type distribution of the test data and also randomly generate a video type distribution. This random distribution has both random videos (drawn from the pool of requested videos from all users) and random probabilities. The utility obtained by the MDP algorithm using the random, estimated, and true distributions, averaged across all users, is shown in Figure 16.9b with the error bars indicating standard deviation. We find that the average utility increases only slightly as more accurate information is known, suggesting robustness of the MDP algorithm to video type distribution errors.

16.8 CONCLUSIONS

Two emerging trends of Internet applications, *video traffic becoming dominant* and *usage-based pricing becoming prevalent*, are at odds with each other. Current approaches in practice center on either decreasing video consumption through self-imposed warnings or decreasing data consumption by lowering video quality, both of which degrade user quality of experience. Given this conflict, is there a way for users to stay within their monthly data plans (data quotas) without suffering a noticeable degradation in video quality? We proposed an online video adaptation mechanism, QAVA, that manages this trade-off by leveraging the compressibility of videos and by predicting consumer usage behavior throughout a billing cycle. QAVA automatically selects the best bit rate to enable the consumer to stay under his/her data quota, while suffering minimal distortion. We developed the QAVA architecture and its main modules, including SS, user profiling, and video profiling. Online algorithms were designed through dynamic programming and evaluated using real video request traces. Empirical results based on real video traces showed that QAVA performed better than existing approaches in literature and practical solutions. This suggests that QAVA can provide an effective solution to the dilemma of usage-based pricing of heavy video traffic.

REFERENCES

1. CISCO. "Cisco Visual Networking Index: Global Mobile Data Traffic Forecast Update," 2011–2016.
2. Sandvine. "Global Interet Phenomena Report," Sandvine, 2012.
3. Reliance. Reliance 3G Plans & Pricing. Available at: http://www.rcom.co.in/Rcom/personal/3G/HTML/PostpaidDataPlans.html, 2012.
4. ROGERS hi-speed internet. FAQ. Available at: http://www.keepingpace.ca/faq.html, 2012.

5. Matthew Lasar arstechnica. “EU cracks down on “bill shock” roaming horror stories,” *arstechnica*. Available at: http://arstechnica.com/tech-policy/2010/03/eu-cracks-down-on-bill-shock-roaming-horror-stories/, 2010.
6. Cecilia Kang and Hayley Tsukayama. “FCC holds off on “bill shock” rule as industry plans alerts for reaching monthly limits,” *Washington Post*. Available at: http://www.washingtonpost.com/blogs/post-tech/post/fcc-holds-off-on-bill-shock-rule-as-industry-plans-alerts-for-reaching-monthly-limits/2011/10/17/gIQAF0CbrL_blog.html, 2011.
7. Onavo. Available at: http://www.onavo.com/.
8. DataMi. DataWiz. Available at: http://www.datami.com/.
9. “Ice Cream Sandwich—Android Developer. Available at: http://developer.android.com/about/versions/android-4.0-highlights.html, 2012.
10. S. F. Chang and A. Vetra. “Video adaptation: concepts, technologies, and open issues,” *Proceedings of the IEEE*, 93(1), 2005, 148–58.
11. X. Liu, F. Dobrian, H. Milner, J. Jiang, V. Sekar, I. Stoica, and H. Zhang. A case fo a coordinated internet video control plane. In *ACM SIGCOMM*, 2012.
12. R. Rejaie, M. Handley, and D. Estrin. Quality adaptation for congestion controlled video playback over the Internet. In *ACM SIGCOMM*, 1999.
13. J. Liu, B. Li, and Y. Zhang. “An end-to-end adaptation protocol for layered video multicast using optimal rate allocation," *IEEE Transactions on Multimedia*, 6(1), 2004, 87–102.
14. C. Liu, I. Bouazizi, and M. Gabbouj. “Rate adaptation for adaptive HTTP streaming,” In *ACM MMSys*, 2011.
15. S. Akhshabi, A. C. Began, and C. Dovrolis. “An experimental evaluation of rate-adaptation algorithms,” In *MMSys*, 2011.
16. H. Schwarz, D. Marpe, and T. Wiegand. “Overview of the scalable video coding extension of the H.264/AVC standard,” *IEEE Transactions on Circuits and Systems for Video Technology*, 17(9), 2007, 1103–20.
17. MPEG-DASH. http://dashpg.com/.
18. R. H. Jan, C. P. Lin, and M. S. Chern. “An optimization model for Web content adaptation,” *Computer Networks*, 50(7), 2006, 953–65.
19. M. L. Puterman. *Markov Decision Processes: Discrete Stochastic Dynamic Programming*. John Wiley & Sons, Inc., New York, 2005.
20. A. Bovik. *The Essential Guide to Video Processing*. Elsevier, Burlington, MA, 2009.
21. H. Kellerer, U. Pferschy, and D. Pisinger. *Knapsack Problems*. Springer, Heidelberg, Germany 2004.
22. J. Chen, A. Ghosh, J. Magutt, and M. Chiang. QAVA: quota-aware video adaptation. In *ACM CoNEXT*, 2012.
23. J. D. Papastavrou, S. Rajagopalan, and A. J. Kleywegt. “The dynamic and stochastic knapsack problem with deadlines,” *Management Science*, 42(12), 1996, 1706.
24. P. R. Winters. “Forecasting sales by exponentially weighted moving averages,” *Management Science*, 6(3), 1960, 324–42.
25. ITU-R. Recommendation BT.500: Methodology for the subjective assessment of the quality of television pictures. In *International Telecommunication Union*, 2012.

26. Y. Wang, M. Schaar, S. Chang, and A. C. Loui. "Classification-based multidimensional adaptation prediction for scalable video coding using subjective quality evaluation," *IEEE Transactions on Circuits and Systems for Video Technology*, 15(10), 2005, 1270–8.
27. F. Dobrian, V. Sekar, A. Awan, I. Stoica, D. A. Joseph, A. Ganjam, J. Zhan, and H. Zhang. Understanding the impact of video quality on user engagement. In *ACM SIGCOMM*, 2011.
28. K. T. Chen, C. Y. Huang, P. Huang, and C. L. Lei. Quantifying skype user satisfaction. In *ACM SIGCOMM*, 2006.
29. M. Zink, K. Suh, Y. Gu, and J. Kurose. Watch global cache local: YouTube network traces at a campus network - measurements and implications. In *IEEE MMCN*, 2008.
30. Y. Zhou, D. Chakrabarty, and R. Lukose. "Budget Constrained Bidding in Keyword Auctions and Online Knapsack Problems," *WWW*, 2007.

17 The Role of Multicast in Congestion Alleviation

ALAN D. YOUNG

17.1 CONGESTION IN CELLULAR NETWORKS

At the AT&T's analyst conference in New York on November 7th, 2012, Ralph de la Vega, the President and CEO of their Mobility Business said that AT&T's mobile network traffic has grown by 25,000% over the past 5 years. According to Cisco's Visual Networking Index (VNI) [1], mobile network traffic in North America will grow 17-fold between 2011 and 2016. These staggering increases in data consumption are of course mirrored in the wireline world too but increasing capacity in the wireline world can be achieved by upgrading infrastructure with more and faster equipment, to do the same in the wireless world requires more radio frequency spectrum. This is a resource, which is in increasingly short supply, and it is not at all clear that enough spectrum to satisfy the expected growth in demand can be obtained (at any price).

Indeed, the growth in data usage has been so dramatic that the unlimited data plans of a few years ago, used to drive the rapid growth of smartphones, are gone for good, as the operators try to ensure that they can support future growth and get a good return on their huge investments in wireless spectrum and infrastructure. However, merely trying to ration the remaining bandwidth and/or increasing the price might not be enough to deliver that return. Increase the price too much and growth will decline, limit the bandwidth user can use too much and the same depression in growth will result. Failure to manage the use of bandwidth properly will result in the equally undesirable result of congestion.

In some places, we have already started to see issues. For instance, in New York City, because of the dense population and concentration of smartphone users, the data rates available at certain times of the day are very restricted. So much so in fact that it is almost impossible to exceed one's monthly data cap, leading to situations where people are paying for data they cannot use.

Cisco's VNI predicts that video will grow from 58% of all mobile data traffic in North America in 2011 to 69% in 2016. In light of this, the current chapter will begin by looking at video as an application and its impact on network resources,

Smart Data Pricing, First Edition. Edited by Soumya Sen, Carlee Joe-Wong, Sangtae Ha, and Mung Chiang.

before investigating how multicast architectures can be integrated into the overall mobile network architecture in an attempt to alleviate the congestion problem. It will inevitably get worse if nothing is done.

17.2 VIDEO, THE APPLICATION

Video applications are the main culprit for the massive increase in data usage on mobile networks over the past few years and are likely to be the main driver of bandwidth going forward. It is, therefore, important to study video as an application in a little depth.

Video applications are the most bandwidth-intensive applications contemplated for use over mobile networks. There are, of course, many types of video applications, which will be explored in more detail in the following text, but in general, video has the following characteristics and requirements that should be taken into account when designing a network to transport it.

- The video decoder in a device converts the incoming compressed video into a frame size and rate that can be displayed on its screen. It needs a sustained stream of data with a very low bit error rate (i.e., typically better than 10^{-10} or 1 bit error in 10 billion) in order to operate properly. Failure to achieve this will result in screen freezes and/or loss of synchronization of any accompanying audio, which will obviously be objectionable to the viewer even at a relatively low rate of occurrence. Buffering the video in the device to smoothen temporary losses in transmission from the cell tower to the smartphone (or elsewhere in the network) helps but only to a certain extent. If there are too many gaps in transmission or the incoming rate on an average is not sufficient to sustain the rate the video decoder needs, there will still be breaks in the video.
- The network must be able to deliver the entire video (a relatively large amount of data) to the device within the period of time the video content is to be watched. The degree of flexibility the network operator/content provider has in delivering the content to the viewer depends on several things, such as the quality of the video required, the length of the clip, and probably, most importantly, whether or not the content is live. If the content is live, there is little flexibility—the bandwidth available from server to smartphone must be at least equal to the rate of compressed video data for the duration of the live content. Buffering is of limited use because the content is live. One can hardly transmit more data when the bandwidth is available to cover times when it is not, because the data has not yet been produced! If too much buffering is used the event cannot be considered live anymore. If the content can be delayed by the content provider or network operator, or buffered by the device, or both, there are many more options and the requirements on the network can be relaxed a little. In such circumstances, the content (or parts thereof) can be compressed in time (as well as spatially) and transmitted in burst form to the device in blocks, for storage in the device's buffer. Alternatively, if there is limited bandwidth and the device has sufficient

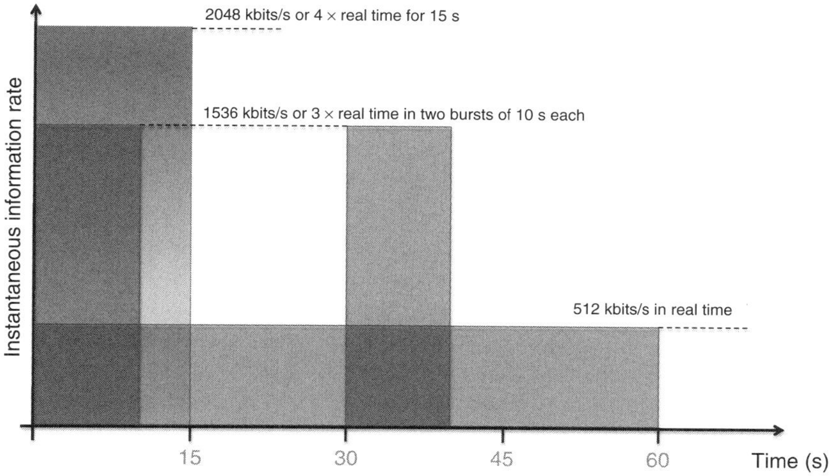

Figure 17.1 Different data transfer schemes for the same block of data.

storage, the data can be sent at a slower rate (thereby taking longer to transmit than the actual time the video content would take to play). Clearly, the quality and length of the video are also important factors, because they dictate how much total data needs to be transferred (see Fig. 17.1).

- The business model for video is perhaps not considered when designing a mobile network, but it should be! Video is, bit for bit, perhaps the most expensive application to transmit over certain networks, and this is especially true for network architectures that were not designed with video in mind in the first place. Not only is the amount of bandwidth high compared practically to any other application for video but also it is, in most applications (save, perhaps, for video conferencing), highly asymmetrical. In other words, the bandwidth is all one way with very little (if anything) going in the opposite direction. Most mobile phone networks were designed for much more symmetric traffic flows. This means that there is a lot of inefficiency built in which only compounds the expense of transporting video over such networks. More subtly, however, from the content provider's perspective, the cost of acquiring or producing content can only be paid back if sufficient people watch the content in a given time. Revenue comes from either direct or indirect payment for viewing the content, or, more likely, through advertising. Either way, the amount of revenue is directly linked to the number of people who watch it. In order to turn a profit, the cost of making and delivering the content must be less than the revenue obtained. If the economics of video were based on the bandwidth consumed, there is no way it could profit as currently configured. An SMS text message is typically charged on a basis, which is literally thousands of times more per bit than what is charged for video. Yet it is video and not SMS that determines the cost of provisioning and operating a network, and it is video that has the growing proportion. If this fundamental problem is not resolved in the long term, significant breakage will occur, as the costs of transporting video in the manner it is currently transported on mobile networks cannot be sustained.

17.3 WHY IS UNICAST NOT IDEAL FOR ALL VIDEO?

In the past, the underlying infrastructure used for any telecommunication network was designed for the application, or the primary use. Because different applications have different needs and characteristics, the network infrastructures built for them looked very different and were largely incompatible with other applications. Now, of course, as everything converges toward an all-IP future, all applications will also need to converge to using a common unified IP network for communications.

This is generally considered to be a beneficial trend, but it does create a mismatch, particularly in the case of video. This mismatch for video is primarily because of the general architecture of the Internet, rather than with the IP protocol suite on which the Internet is based. The Internet relies on unicast transmission, where each packet of data originates from one unique network address and is sent to a unique destination address. This is fine for most applications, but for video, it starts to break down when large numbers of network addresses all request the same video at the same time. If the number of requests for a piece of content exceeds the capability of the server to deliver those requests, the server will at best cap the number of simultaneous connections and deny addition requests; at worst, it will crash and deny access to everyone. President Obama's first inauguration speech on January 20, 2009, will be remembered not only for its obvious historic significance but also because the live streaming feed on the Internet was accessed by so many people that it caused web servers to crash and backbones to be stretched to the limit. As the amount of video on the Internet increases, together with the number of devices that can access this content, the likelihood of server request and/or bandwidth overload situations increases as well.

To date, the approach to solving this particular issue has been to build ever-larger data centers so that more and more requests for video can be serviced. However, this is becoming unsustainable, as the cost of building, operating, and maintaining such facilities (even with the economies of scale) rises faster than the revenues that support the services because facilities have to be sized in order to cope with peaks of demand rather than the average demand. As a result, the operating efficiency of large data centers is actually decreasing because most of the time the load is much lower than the peak.

As already hinted, a similar situation exists on IP networks. As requests for video are serviced, the amount of bandwidth coming out over each server rapidly multiplies. In fact, the amount of bandwidth required is generally the product of the number of viewers and the bandwidth of the stream. Using the example of President Obama's inaugural address, which was streamed live over the Internet and watched by more than 7 million viewers simultaneously, the required bandwidth was more than a terabit per second (150 kbits/s $\times$ 7,000,000). Content delivery networks (CDNs) have helped alleviate this fundamental problem to a certain extent, but this only helps the backbone. CDNs do nothing to resolve the problem on the last mile. As such, the delivery of video over unicast networks is very inefficient for situations where the same content is watched by a large number of people on a local network (e.g., a cell tower) within a short period of time as the unicast protocol requires the transmission of the same data multiple times.

It can be argued that as video consumption is moving from linear broadcast, in which everyone receives the same stream of data, to on demand, a unicast environment is necessary. While this is true for a significant portion of video, it is also true that even in a totally on-demand environment, unicast fails when the number of users demanding a piece of content (whether it is live or not) causes the bandwidth to exceed the capacity of the system. Imagine if the Super Bowl was only available as a unicast stream—it would require colossal resources in both the data center and network environments. It should also be remembered that linear television (TV) still accounts for the vast majority of revenue for the video industry and, as such, could provide a valuable source of revenue for the network operators who "tune" their networks to be able to carry such content. Consumers have shown little inclination so far to pay for content delivered over the Internet; however, they still expect to be able to access content over the Internet from any device. Many content providers (i.e., HBO® with HBO GO®) have responded by allowing limited on-demand access to their libraries on a variety of devices (including smartphones and other devices using the cellular networks) conditioned on the consumer paying for a subscription to their linear services. This has led to increased demand for bandwidth on cellular networks (and indeed all networks) but with little-to-no additional revenue for the network operators. Figure 17.2 illustrates the issue.

There is little doubt that the amount of data transmission capacity required to support the growth in users, devices, and content is not going to be easy to deliver. Spectrum is not limitless. What does exist is in high demand and, therefore, expensive. Modulation technology is approaching the Shannon limit (a fundamental telecommunication theory that limits the information transfer rate across a given amount of spectrum with a given amount of power), and so realistically, the only way to increase data capacity on cellular networks is to reduce the size of each cell. This would be very costly. A better approach may, therefore, be to address the inefficiencies of using only unicast delivery for all applications, and especially video, thereby creating more headroom for growth. This, if achieved, could at least put off more costly alternatives or be combined with alternatives to bring down the overall cost (provided, of course, that there are ways to address the inefficiencies at a lower cost).

17.4 WHY IS MULTICAST BETTER FOR VIDEO IN SOME CIRCUMSTANCES?

Multicast transmission allows more than one destination address on a network to receive data packets originating from a single source address on the network. It differs slightly from broadcast: whereas broadcast packets are received by all network addresses regardless of whether or not they request them (similar to

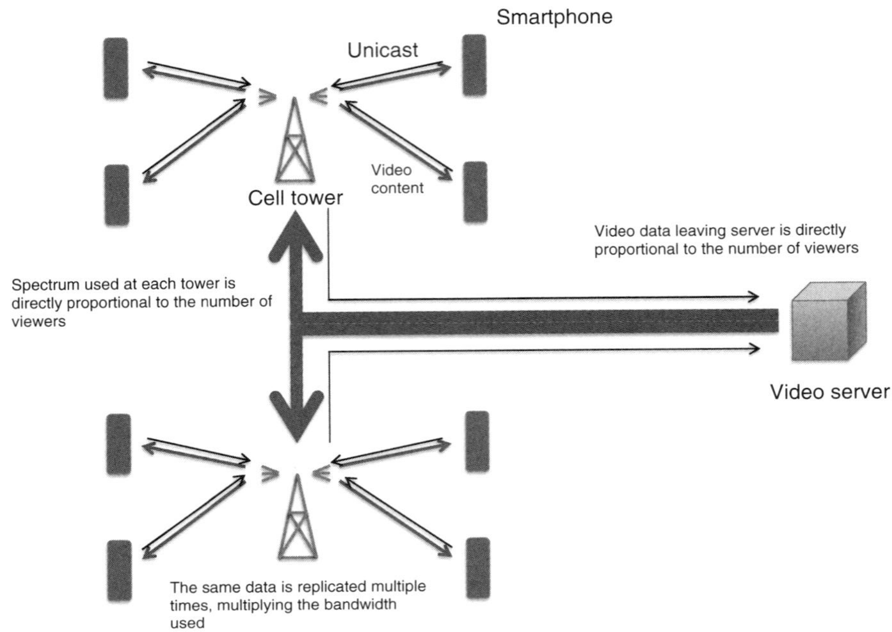

Figure 17.2 Unicast delivery.

broadcast television on a cable TV network), multicast packets are only received by those addresses that request (or "join") the multicast stream. Unlike unicast, which is, as already discussed, a point-to-point mechanism, multicast has no means of acknowledging receipt of a data packet. If this is required, separate provisions need to be made. Fortunately, multicast and unicast packets can coexist on IP networks, so this is relatively straightforward to implement.

The requirement to join a multicast stream provides a very efficient mechanism for delivery of video over a network with bandwidth constraints. When a receiving network node wishes to receive multicast packets from a particular source address it sends a join request into the network, and the first router it encounters with packets from the requested source address simply replicates those packets and sends them to the requesting node. In this manner, the problem of replicating traffic unnecessarily (i.e., in situations when large numbers of network nodes are all requesting the same content at the same time) is avoided. This is illustrated in Figure 17.3.

Of course, not every scenario is suited to multicast. But by intelligently and perhaps dynamically switching between multicast and unicast, it should be possible to make much more efficient use of server and network resources in any CDN.

The remainder of this chapter will cover various potential content delivery options and configurations. It is not intended to be an exhaustive list, but rather intended to stimulate thinking in how to better match the network and server resources to the applications that utilize those resources.

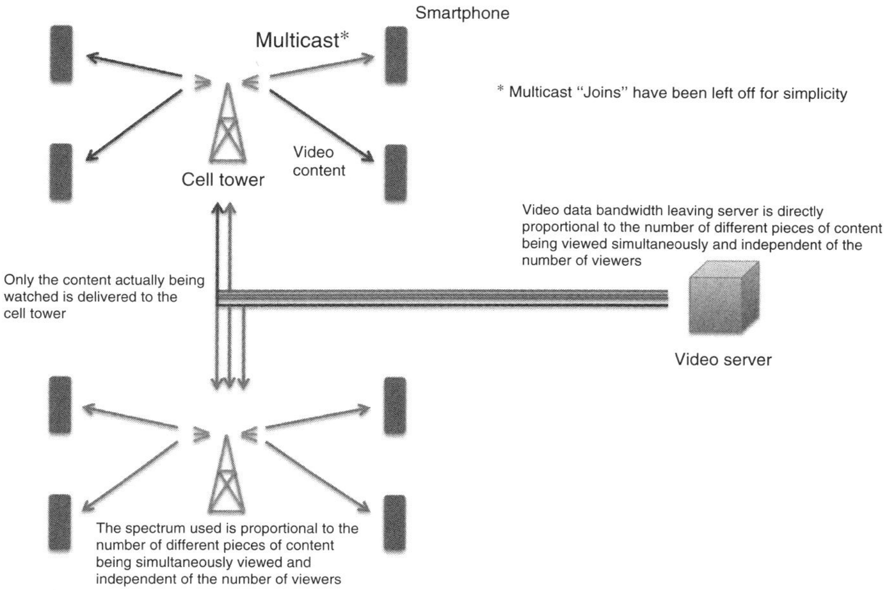

Figure 17.3 Multicast delivery.

17.5 BROADCAST, MULTICAST, AND UNICAST ARCHITECTURES FOR THE DELIVERY OF VIDEO

The most common delivery architecture used for video today is, by far, broadcast. However, over time, many delivery mechanisms have evolved to include a mixture of broadcast, multicast, and unicast. Those that rely on only one architecture tend to struggle, as the nature of the video application has changed from linear only to a mixture of on-demand, linear, and interactive.

Cable TV systems, for instance, were built based solely on an end-to-end broadcast architecture until relatively recently. Programmers for the most part delivered their content via satellite, often uplinked directly from their playout facilities to thousands of cable headends (point-to-multipoint broadcast architecture). Each cable headend would also receive local broadcasts transmitted terrestrially in the very high frequency (VHF)/ultra high frequency (UHF) spectrum (a lot of which has now been reassigned to the cellular network operators). The headend would then redistribute the video content through a tree and branch broadcast infrastructure to end users (i.e., consumers). Every single TV channel the cable operator offered was delivered to every single customer's home, whether they watched the channel or not. As the number of channels grew, cable operators expanded the bandwidth delivered to each home in steps from 330 MHz to 450 MHz, then to 550 MHz, on to 750 MHz, then 860 MHz, and finally 1000 MHz, which is the practical limit.

As the Internet grew and cable modem services were developed to deliver always-on broadband Internet to homes as well, operators found that they had to reuse their bandwidth more and more. Originally, one headend could serve thousands of homes with all of the bandwidth shared by each. In a broadcast architecture, this is fine. However, with broadband Internet, which is of course based on unicast architecture, the available bandwidth per home soon became an issue. To counteract this, cable operators began to migrate to an architecture where the bandwidth of the system was shared between a smaller and smaller number of homes, so as the demand of the system rose, the number of homes that bandwidth was shared between would fall. Now some operators share 1000 MHz of bandwidth between as few as 50 homes.

As these changes were happening, there was a period of time between about 2005 and 2010 when operators were challenged because there was pressure to deliver increasing numbers of high definition channels (in order to compete with satellite broadcasters) as well as to deliver faster and faster broadband speeds to satisfy demand from consumers. To resolve the problem, many cable operators started to incorporate multicast to alleviate the pressure on their networks. The idea was to assign an amount of bandwidth to the linear channels that was significantly less than the bandwidth required for all of the channels. Only if at least one person on a particular node was watching a channel would it be transmitted on that node segment. Clearly, the amount of bandwidth that was necessary depended on how many different channels would be watched by a particular group of people at the same time. This is a statistical problem that can, fortunately, be modeled using the Pareto's distribution. The reduction in bandwidth needed to deliver essentially the same selection of channels as before allowed operators to grow their broadband businesses much faster than would have been possible otherwise.

As cable operators migrate their networks to an all-IP architecture, it is likely that they will continue to utilize a combination of unicast and multicast, to help optimize bandwidth resources while delivering the full suite of services demanded of them.

The telephone operators became interested in video only after it became clear that the cable operators were going to take a substantial portion for their traditional telephony and data business. Unfortunately, the telephone operators found that it was substantially more challenging for them to deliver video using twisted pair than it was for cable operators to deliver telephony over coaxial cables! The fact that the last mile bandwidth to the home was limited forced the telephone operators to use a multicast architecture from day 1, and today, the only multichannel video operators that use a completely IP architecture from end-to-end are telephone operators. Linear video is broadcast to headends, as with cable operators, and is then delivered in an IP multicast format to homes using the existing twisted pair cables via DSLAMs (digital subscriber loop access multiplexers). Because the resulting networks are all IP, telephone operators are able to deliver other video services such as video on demand, and, of course, broadband Internet, using unicast IP over the same infrastructure.

Cellular networks, in contrast to both the cable and wireline telephone operators, have maintained their use of a unicast-only architecture throughout. Their approach has been to acquire more spectrum and utilize more spectrally efficient transmission techniques. It is ironic that most of the spectrum acquired by the cellular providers to meet the increasing demand for transmission capacity came from the TV broadcasters. While this spectrum was previously used to deliver video in broadcast format to tens of millions of homes, it is now used to deliver essentially the same application (video) but in unicast format. This is the main reason why AT&T saw a 25,000% increase in traffic on its mobile networks in just 5 years—had they used a mixture of multicast and unicast, this figure would likely have been a lot lower. Logic would seem to suggest that if we are to migrate to an all-IP infrastructure, it should not be an all-unicast IP infrastructure.

17.6 FUTURE POTENTIAL ARCHITECTURES MIXING BROADCAST, MULTICAST AND UNICAST

Some of the possible architectures that could be considered for cellular networks will be investigated in the following text.

The falling cost per gigabyte of storage has made it feasible to consider adding storage at cell sites. This would allow popular content to be stored (cached) at the cell site, therefore, reducing the demand for backhaul capacity. In a similar manner, broadcast channels could be delivered to the cell towers. There would still be a need to determine what content to cache at each tower, and how long to keep it there. But initial algorithms could be as simple as triggering a caching operation after a certain number of users at a cell site request the same content within a defined period of time. The question of how long content could be stored would be dependent on how much storage was available and how long it takes before the rate of requests to view drops below a predetermined point. Many content providers would also restrict what content could be stored and for how long, and although this would clearly need to be taken into account, it will not be considered further here. Another issue that would need to be addressed in order to implement this solution would be how to manage handover between cells. For instance, suppose a user is viewing a piece of content that is being drawn from the cache at the cell tower serving that user. If this user moves out of range of that tower, and into range of another tower that does not have that content cached, there would be a need to communicate to the new tower what content it needs to acquire and what bit rate it needs to be able to deliver the content at.

Caching content at the cell site does not do anything to alleviate congestion on the last hop, between the cell tower and the mobile handset. If, however, a combination of multicast and local storage on mobile handsets was used, it should be possible to make better use of the available spectrum. For content, which is consumed live by a sufficient number of people within a single cell site, or more likely within a common locality, an operator could leverage the multicast capability of all towers in that locality. This would clearly be more bandwidth efficient than if the same number of users (assuming a sufficiently large number of users) were each to make unicast requests at the same time for the same streaming feed—which is how it is done today.

For on-demand content that is not live, storage on the device can be leveraged. Most smartphones today come with multiple gigabytes of storage, and the amount of storage on devices is increasing steadily. In a similar way to the above scenario where popular content is cached at the cell tower, popular content could be cached directly on smartphones. There are several differences, however. First, the algorithm to determine what to cache and what not to cache would need to be more sophisticated. It would need to be able to predict what content a particular user was likely to request in the near future. Assuming that such an algorithm were to be developed, the content could be downloaded to the user's smartphone during an off-peak period. If the content was unique to the user, then unicast would be the best method to use. However, if a large enough number of smartphone users were predicted to request the same content, it would be more efficient to multicast the content to all users (again during off-peak periods).

In this scenario, and assuming the algorithm was able to correctly predict which content a user was likely to want to watch, there would be no need for the content to be downloaded again. In addition to reducing the load on the cellular network, there are a number of other advantages. For one, the quality of the content could be enhanced, which would facilitate the user projecting the video to another device, such as a big screen TV or a computer screen. Another advantage from a business perspective is that providers could push content to users' phones and then advertise its availability. This could even become a revenue stream for the operators—popular TV shows could be pushed at night to users' smartphones in advance of broadcast, and the keys to decrypt the programming could be distributed with the content and activated at a particular time (e.g., by a text message). There are all sorts of potential business models where users would get access to the keys at different times depending on what subscription they have, or as part of a promotion or competition.

Another potential option for content distribution is the use of ad hoc networks for direct device-to-device communication and redistribution of content. This would not use any spectrum from the tower. However, it would require much greater intelligence in the network to manage as well as agreement on new protocols and spectrum that could be used locally. It would be best utilized in areas where the cell phone density is relatively high and where content that a user desires is available on one or more nearby smartphones. A peering network could be formed between the devices, and the content transferred, either all from one device or in parts from multiple different devices. Clearly, this is only plausible for certain types of content, and controls would need to be carefully managed to prevent the redistribution of private or personal material. But, despite the difficulties, it could nonetheless be another tool for the network operator to efficiently manage the limited spectrum available.

17.7 CONCLUSIONS

It is clear that the recent growth in data consumption on mobile networks has been tremendous and that video accounts for the lion's share of that growth.

We have seen that an all-unicast network architecture cannot efficiently meet the needs of all video applications. In particular, the video applications that generate the most income for the content providers (linear TV) are also the most inefficient to deliver via unicast. Although on-demand video is a growing portion of the content viewed, it is by no means the entirety. Even if it was, the use of unicast end to end to service every request is highly inefficient.

Usage patterns change over time, and it is important that the network architectures of the future are flexible enough to adapt to these changes. As we continue to migrate to an all-IP infrastructure, we are going to move from a situation where the network infrastructure was designed specifically for the application [e.g., a cable TV network infrastructure for TV or a public switched telephone network (PSTN) for telephony], so that it was optimized for that one application and incompatible with pretty much everything else, to a situation where the network infrastructure is based on IP and is generic. Along with the benefits of such a migration come the challenges of optimizing the network for all applications and implementing business models, which are sustainable.

From a pricing point of view, the idea of charging users a flat rate for an unlimited amount of data (or at least as much data as they can use) has failed because users had no disincentive to streaming as much video as they wanted leaving the operators struggling to keep up. Going forward, as unpopular as it may be, the reality is that the more data/bandwidth/spectrum resources that an application and/or user uses, the more money needs to be charged. This will force everyone in the ecosystem (operators, providers, and users) to consider (whether directly or indirectly) the resources required to deliver/receive each particular service. From a pricing point of view, users could be charged a per gigabyte fee for unicast data and a much lower flat fee for the ability to receive multicast data and content providers would be charged on the same basis except that the multicast fee would be for the right to deliver a certain amount of data using multicast. This pricing structure (or something similar) would incentivize everyone to be more conservation minded and would fuel innovation and development in this area.

REFERENCE

1. Cisco Visual Networking Index. Available at http://www.cisco.com/web/solutions/sp/vni/vni_mobile_forecast_highlights/index.html. Accessed April 4, 2014.

PART VI
Pricing in the Cloud

18 Smart Pricing of Cloud Resources

YU XIANG and TIAN LAN

Large investments have been made in recent years in data centers and cloud computing. A survey by KPMG International [1] in February 2012 shows that more than 50% of senior executives find the most important impact of cloud computing to their business models to be its cost benefits. Accordingly, much research has been focused on demystifying various economic issues in data center and cloud computing, for example, economic and pricing models, cost-optimization techniques, tariff and charging structures, and business objectives.

As in today's clouds the demand for resources in clouds are dynamically fluctuating because of many factors, thus offering fixed resources to all the users regardless of their demands may either not be able to meets some users' demand or leave some resources wasted. Therefore, one thing attracts most attention in today's Cloud computing is the idea of optimizing the utilization of computing resources among all cloud users, so that the computing resources can be distributed in the most effective manner and restrict the underutilization to a certain level. The intuitive idea to deal with this is that the resources should be regulated by the law of supply and demand, that is, when demand rises but supply remains the same, the price of types of virtual machines (VMs) should go up, and when demand drops, then prices should fall accordingly.

To implement this idea, cloud providers such as Amazon EC2 [2, 3] employs this sort of market-driven resource allocation for unused capacity. In Amazon EC2's spot markets, instead of renting a server for fixed price per hour for each hour that the server runs, users can make a bid for a certain number of hours of a certain type of server. The spot price is set by the cloud provider, which fluctuates in real time according to spot instances supply and demand. When the bid exceed the spot price, the instance will run until the spot price exceed the bid, in which case the instance will be terminated without notice. To use spot instances, users shall place a request that specifies the instance type, the availability zone desired, the number of spot instances desired, and the bidding price per instance hour. Amazon EC2 API provides the spot

Smart Data Pricing, First Edition. Edited by Soumya Sen, Carlee Joe-Wong, Sangtae Ha, and Mung Chiang.

Figure 18.1 Amazon EC2 VM spot instance price History: price of a m1.large linux spot instance in US Southeast-1a from May 17 to May 23, 2012 [4].

price history for the past 90 days to help customers decide their bids [4], as shown in Figure18.1.

Cloud resources can be dynamically allocated not only based on just the demand and supply such as in the case of spot instances but also based on service level agreement (SLA), that is, the providers has to ensure users can get certain service level when dynamically allocating resources based on demand and supply to maximize revenue. SLA is a critical bridge between cloud providers and customers, as in clouds, the providers aim to obtain the revenues by leasing the resources and users rent these resources to meet their application requirements. SLA provides a platform for users to indicate their required quality of service (QoS) [5], which specifies responsibilities, guarantees, warranties, performance levels in terms of availability, response time, and so on. Owing to the fast-growing power consumption and application heterogeneity of data centers, cost optimization and smart service pricing are also becoming ever more urgent and important. In contrast to the adoption of low power devices and energy saving solutions in References 6 and 7, a cost-optimization approach in References 8 and 9 exploits the diversities of electricity prices over time and geographic regions. In particular, a load-balancing algorithm is proposed in Reference 9 to coordinate cloud computing workload with electricity prices at distributed data centers, in order to achieve the goal of minimizing the total electricity cost while guaranteeing the average service delay experienced by all jobs.

These queuing-based models for price data center services based on an average service delay may be unsatisfactory to cloud users, not only because they may have heterogeneous job requirements and spending budgets but also because of a fundamental limitation of the average delay approach—individual job completion times are still randomly scattered over a wide range of values. Delayed response may frustrate users and, consequently, result in revenue loss. Therefore, the ability to deliver according to predefined job deadlines increasingly becomes a competitive requirement [10, 11]. Cloud providers and users can negotiate individual job deadlines to determine costs and penalties based on the desired performance and budget. This presents an opportunity to offer deadline-dependent pricing, which generates an additional source of revenue for cloud providers. It immediately raises the following question: How to jointly optimize both electricity cost of distributed data centers and total revenue from deadline-dependent pricing, in order to maximize the net profit cloud providers receive? Wang et al. [12] studied the problem of data center net profit optimization (NPO), which aims to maximize the difference between the total revenue from deadline-dependent pricing and the electricity cost of distributed data centers.

This chapter studies cloud resources allocation based on three types of pricing models, that is, to find out the optimal resource allocation that maximizes the total revenue, where the revenue function depends on various pricing models we employed; for instances, as we discussed earlier, price of cloud resource can be determined by three major factors such as demand and supply in current markets, the SLA level cloud servers provided, and the predefined job deadline users required. We proposed the detailed pricing model with respect to these three factors separately in the following sections and then obtain the optimal cloud resource allocation to maximize the revenue function formulated by these pricing models, an comparison of overall revenue with heuristic static pricing scheme is shown for each pricing model.

18.1 DATA CENTER VM INSTANCE PRICING

Amazon's spot instances mechanism can be described as a continuous seal-bid uniform price auction, where VMs of the same type are sold at identical price and the providers assigns resources to bidders in decreasing order of their bids until all available resources have been allocated, and the spot price may be adjusted when there are no instances running at the current spot price [13]. In this section, we formulate the problem of dynamic resource allocation for simultaneous spot markets, and here the goal is to allocate unused data center resources to each spot market in a timely manner that maximize the total revenue, subject to the capacity constraints of individual machines. In the rest of this section, we shall present our solution approach to achieve this goal for both fixed pricing scheme, where price of a VM type is fixed and thus is independent from market situation, and the uniform pricing scheme, where the price of a VM type is adjustable according to market demand and supply.

18.1.1 Dynamic Scheduling and Server Consolidation for Fixed Pricing Scheme

In the fixed pricing scheme, each VM type has a fixed price that does not fluctuate with the current supply and demand situation. Hence, the virtual machine revenue maximization problem (VRMP) can be modeled as a multiple knapsack problem (MKP) as follows: given a set of machines M and D resource types (differed in CPU, memory, and disk), where each machine $m \in M$ has a capacity c_{dm} for each type of VM $d \in D$. The set of VMs to be scheduled is V and each VM i has a size a_{id} for each $d \in D$ and a value v_i. To formulate the optimization problem, we would first choose the optimization variable (which is the VM scheduling parameter x_{im}), then define the objective and constraints. As we mentioned earlier, the objective is to maximize the total value, which is the summation of all of the chosen VMs(for a chosen VM, $x_{im} = 1$; for an abandoned VM, $x_{im} = 0$), which is $v_i x_{im}$. While the constraints for this goal is the machine capacity, that is, the summation of the size of all chosen VMs should be smaller than or equal to c_{dm}. On the basis of the intuition above, the problem can be formulated as

$$\begin{aligned} &\text{maximize} \sum_{i \in V} \sum_{m \in M} v_i x_i m \\ &\text{subject to} \sum_{i \in V} a_{id} x_{im} \leq c_{dm}, \quad \forall m \in M, d \in D, \\ &\qquad\qquad x_{im} \in \{0, 1\}, \qquad \forall i \in V, m \in M. \end{aligned} \tag{18.1}$$

It can be noticed that this MKP formulation is an NP-hard combinatorial optimization problem, we propose a $\frac{1}{2} - \varepsilon$ local search algorithm [14] that can approximate the optimal solutions, which is specified by Algorithm 18.1. In this algorithm, S is the current set of VMs on machine m, S' is the set of VMs chosen to maximize the total value, and v_j is the value of VM j. As it is depicted, the algorithm proceeds in rounds. For each round, first maximize the total revenue among pending requests and current running VMs on machine m; then if there is a potential new configuration for m by scheduling, first try to schedule all VMs using available free capacity and then preempt and migrate VMs when the total demand is reaching capacity. To achieve fast convergence rate, we require each local search operation to improve solution quality for each machine m by at least $(1 + \varepsilon)$.

Algorithm 18.1 Local Search Approximation Algorithm *Local(P)* for VRMP

1: **for** $\forall m \in M$ **do**
2: Find a set of VMs S' that among pending requests and the current running VMs on the machine m, maximize the total value $R(S') = \sum_{j \in S'} v_j$.
3: **end for**
4: **while** $\exists$ a machine m such that $R(S') \geq R(S)$ **do**
5: Schedule the requests in $S' \backslash S$, preempt and migrate VMs in $S \backslash S'$ if necessary.
6: **end while**

18.1.2 Price Estimation for the Uniform Pricing Scheme

We now consider the case in which the price of each instance vary with the demand–supply curve and the resource availability. In this scenario, above all, we need to periodically analyzes the market situation and predict the demand and supply curves, on which the price of instances depend. The prediction of the demand curves can be constructed by capturing the relationship between the quantity of acceptable requests and the bidding price(as shown in Figure 18.2) over time period

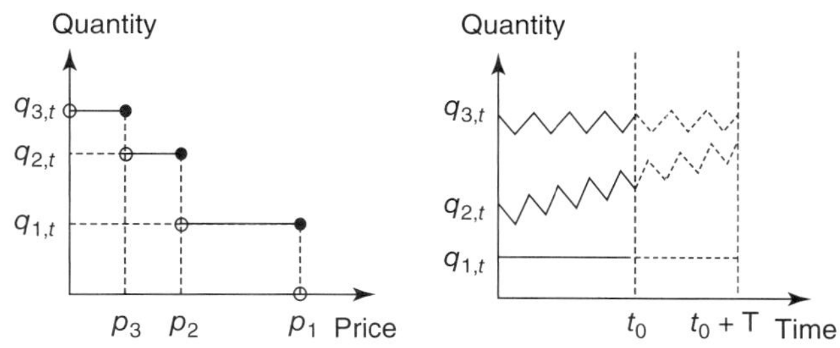

Figure 18.2 Example demand curve at time t and over time.

$[t_0, t_0 + T]$ at sampling interval Δt, where t_0 denotes the current time and T denotes the prediction period. Then we can decide the expected price for each type of VM in each market based on the prediction. Finally, we will be able to make online scheduling decisions for revenue maximization.

Let p_i denote the ith bidding price in decreasing order, $d_{i,t}$ denote the demand that bids at price p_i, and $q_{i,t}$ denote the demand that bids at price at least p_i at time t. As the spot instances mechanism allows requests with bidding prices higher than or equal to the spot price p_s to be accepted, then when p_s equals p_i, we can schedule $q_{i,t}$ VM requests. (Demand curve as shown in Fig. 18.2.) As the bidding price of each individual VM request is independent of the current market situation, we can model the demand quantity to $d_{i,t}$ independently for each p_i to predict the expected demand curve. Then we are able to predict the future demand $d_{i,t}$ for each p_i from t_0 to $t_0 + T$. Our approach to do this is to adopt an autoregressive (AR) model [15] that estimates $d_{i,t}$ from the historical values $d_{i,t-1}, ..., d_{i,t-k}$ as

$$d_{i,t} = \sum_{j=1}^{k} \phi_j d_{i,t-j} + \varepsilon_t,$$

where ϕ_j with $j \in 1, 2, .., k$ is the set of parameters of the historical values and ε_t is uncorrelated white noise with mean 0 and standard deviation σ_ε, and all these parameters can be obtained from historical demand data. This is an appropriate model because it is not only lightweight and easy to implement but also capable of capturing short-term trends to let us compute the expected demand curve.

Now we have the demand and supply curve, our objective is now to schedule requests of each spot market to maximize the expected revenue over the next prediction period. This dynamic revenue maximization with variable price (DRMVP) problem is identical to the former VRMP except that price of individual VMs is determined by the estimated demand curve $R_\tau(q_\tau)$. Again to formulate the optimization problem, first, we choose the same optimization variable as in Section 18.1.1, which is the scheduling vector of VMs. The difference is this time the variable varies with time because the price varies with time based on the demand curve, so it is defined as $x_{\tau m}$. And the objective is the sum of the revenue over demand prediction period $\sum_{\tau \in T} R_\tau(q_\tau)$, subject to the constraints that the total VMs scheduled during time τ

should be smaller than or equal to the total number of VM requests in τ, and the total size of all chosen VMs should not exceed the machine capacity c_{md}, noticed that each VM size $a_{\tau d}$ here also varies with time because of demand curve prediction. Now we can model this problem as follows:

$$\begin{aligned}
\text{maximize} \ & \sum_{\tau \in T} R_\tau(q_\tau) \\
\text{subject to} \ & \sum_{m \in M} x_{\tau m} = q_\tau, && \forall \tau \in T, \\
& \sum_{\tau \in T} a_{\tau d} x_{\tau m} \le c_{md}, && \forall m \in M, d \in D, \\
& x_{\tau m}, q_\tau \in \mathbb{N} \cup \{0\}, && \forall m \in M, \tau \in T. \qquad (18.2)
\end{aligned}$$

As the objective function of this optimization problem is nonlinear, the problem is even more difficult to solve that of the fixed priced case; however, the revenue function $R_\tau(q_\tau)$ for a single VM type is a piecewise linear function, and it can be observed from the $R_\tau(q_\tau)$ using examples in Figure 18.2. In some situations, scheduling a VM can cause the current market price to be lowered, resulting in a sharp drop in total revenue. Thus motivated by similar work on market clearing algorithms for piecewise linear objective functions, our approach to deal with this issue is to approximate $R'_\tau(q_\tau) = \max_{q \le q_\tau}\{R_\tau(q)\}$ using a concave envelope function $R_{e_\tau}(q_\tau)$, which is computed by constructing a upper convex hull using the extreme points in $R'_\tau(q_\tau)$. $R_{e_\tau}(q_\tau)$ has the following property.

Lemma 18.1 *$R_{e_\tau}(q_\tau) \le 2 \cdot R'_\tau(q_\tau)$ for any q_τ.*

Proof: For any q_τ, assume $R_{e_\tau}(q_\tau)$ is on a linear segment with extreme points (q_i, R_{e_i}) and (q_j, R_{e_j}) when $i \le j$. By definition of convex hull, R_{e_i} and R_{e_j} are also points on R'_τ with slope $p_{i\tau}$ and $p_{j\tau}$, respectively. Thus we have $p_{i\tau} \ge p_{j\tau}, R_{e_\tau}(q_i) = q_i \cdot p_{i\tau}$, and $R_{e_\tau}(q_j) = q_j \cdot p_{j\tau}$. The slope p_s of the segment of $R_{e_\tau}(q_\tau)$ can be bounded by $p_s = \frac{q_j \cdot p_{j\tau} - q_i \cdot p_{i\tau}}{q_j - q_i} \le p_{j\tau}$. The total overestimate of $R_{e_\tau}(q_\tau)$ at $(q_\tau, R_{e_\tau}(q_\tau))$ can be bounded by $\Delta = R_{e_\tau}(q_\tau) - R'_\tau(q_\tau) \le (q_\tau - q_i) \cdot p_s \le q_\tau \cdot p_{j\tau}$. Then we get $R_{e_\tau}(q_\tau) - R'_\tau(q_\tau) \le R'_\tau(q_\tau)$. The lemma follows.

Define $v_{q\tau} = R_{e_\tau}(q_\tau) - R_{e_\tau}(q_\tau - 1)$ as value for scheduling the qth request for type τ. As $R_{e_\tau}(q_\tau)$ is concave, $v_{q\tau}$ is a nonincreasing function of q. Now we have the value of each request, we can construct an instance VRMP' where each VM has value according to differentials of $R_{e_\tau}(q_\tau)$. ■

Theorem 18.1 *Running Algorithm 18.1 using $v_{q\tau}$ defined above is a $\frac{1}{4} - \varepsilon$ approximation algorithm DRMVP.*

Proof: As the values of VMs are nonincreasing, any solution of VRMP' that schedules q_τ VMs of type τ has a corresponding solution in DRMVP that schedules the

TABLE 18.1 Average Revenue Achieved by Different Policies

Policy	Metric	Income	Revenue Loss	Net Income
Static	Mean	67,030.44	399.01	66,631.43
	Standard deviation	13,573.32	172.45	13,400.87
Dynamic	Mean	78,026.33	3398.36	74,627.97
	Standard deviation	15,173.28	1083.63	14,089.65

q_τ most valuable VMs with at least $\frac{1}{2}$ the revenue. As Algorithm 18.1 is a $\frac{1}{2}$ approximation algorithm of VRMP', we have shown that every solution of VRMP' has a corresponding solution in DRMVP with at least half the total revenue.

The implementation results using eight type of VMs provided by Amazon EC2 simulated over a 1000 machine data center is shown in Table 18.1. From the collected data, we can see that the dynamic resource allocation has contributed to a great revenue gain. ■

18.2 DATA CENTER SLA-BASED PRICING

Cloud resources can not only be dynamically allocated based on just the demand and supply such as in the case of spot instances but also based on SLA, that is, the providers has to ensure users can get certain service level when dynamically allocating resources based on demand and supply to maximize revenue. SLA is a critical bridge between cloud providers and customers, as in clouds, the providers aim to obtain the revenues by leasing the resources and users rent these resources to meet their application requirements. SLA provides a platform for users to indicate their required QoS [5], which specifies responsibilities, guarantees, warranties, performance levels in terms of availability, response time, and so on. Usually, cloud providers charge users according to their requirement level for their tasks. For example, Amazon EC2 offers spot instances at a much lower price than that of reserved resources. Service instances (including the user and the VMs he/she rented for service) may have different attributes such as arrival rate, execution time, and pricing mechanism. The challenge here is how much resources must be distributed to a VM to maintain the promised performance in SLAs when trying to maximize the total revenue. Allocate more resources to those who have high arrival rate and high price will certainly ensure revenue, but in practice, there are cases when users have low arrival rate but high price and vice versa.

Our goal in this section is to maximize the SLA-based revenues [16] by proper resource allocation, that is, to schedule resources among different service instances in an adaptive manner based on the dynamic situation. For the rest of the section, first, we formulate the resource allocation problem based on the queuing theory to model user's requirement using parameters such as resource quantity, request arrival, service time, and pricing models. Then we propose our optimal SLA-based resource allocation algorithms, by which providers can maximize their revenues.

In order to find how many servers should be assigned for each service instance in order to achieve maximum revenue for a pricing model, we first formulate our mathematical model for this as follows: A data center consists of N homogeneous servers that are grouped into clusters dynamically; each cluster is virtualized as a single machine. The provider signs SLAs with m users, the number of servers allocated to each service instance is $n_1, n_2, ..., n_m$, we assume that the more servers assigned, the more powerful the virtual machine is. We also assume that the requests from users are Poisson distributed with arrival rate λ and service rate μ, thus the average service time is exponentially distributed with mean $\frac{1}{\mu}$, and the service rate of a VM with n servers is $n\mu$. Thus based on the above description, each service instance can be modeled as a FIFO (first in, first out) M/M/1 queue. Here we define service intensity ρ as the ratio of arrival rate to service throughput of one server:

$$\rho = \frac{\lambda}{\mu}.$$

The pricing mechanism specifies how service requests are charged, mostly provider oriented. Here we propose two user-oriented pricing mechanisms MRT (mean response time) and IRT (instant response time), in which users are charged according to achieved performance in terms of MRT, which is the mean of the time interval between a request arrives at the system and the instant at which the service is completed. Usually providers divide time into time slots, and we calculate the mean response time of every time slot. For the MRT pricing model, let F be an offset factor of the actual response time to benchmark, which is defined as

$$F = \frac{r}{R},$$

where r is the measured average response time during a time slot and R is a benchmark of response time defined in SLA, which determines by users' requirement. Then the pricing mechanism can be formulated as

$$B = b(1 - F),$$

where B is the price of each service provision and b is the price constant. As shown in Figure 18.3a, B is the linear function of r and when $r \geq R$, the provider will be penalized, also b/R is the slope of the price function,

$$q = \frac{b}{R}.$$

As MRT is of less practical value as a performance metric when response time varies quite a lot, we propose the IRT pricing model, where request is charged according to measured response time. That is,

$$B = \begin{cases} b & \text{if } r \leq R, \\ 0 & \text{otherwise.} \end{cases} \tag{18.3}$$

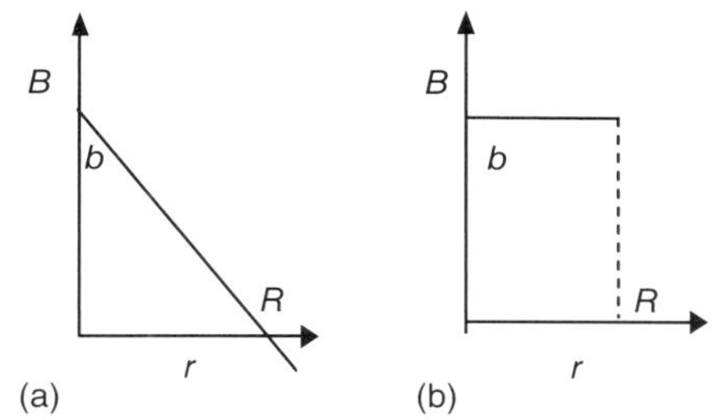

Figure 18.3 Price models. (a) Price model in terms of MRT and (b) price model in terms of IRT.

The pricing model can be illustrated as in Figure 18.3b, in which the charge under this model is determined by the number of service provisions with response time within required R.

Now we are ready to study the optimal allocation based on MRT and IRT. On the basis of some mathematical background of the most popular M/M/1 model in queuing theory, the average response time r_i of service instance i at the steady system state is

$$r_i = \frac{1}{n_i\mu_i} - \lambda_i$$

and the service performance level is

$$F_i = \frac{1}{(n_i\mu_i - \lambda_i)R_i}.$$

The mean revenue g_i is calculated as

$$g_i = b_i\left(1 - \frac{1}{(n_i\mu_i - \lambda_i)R_i}\right).$$

The overall revenues during a time slot from service instance i is

$$G_i = \lambda_i g_i = \lambda_i b_i\left(1 - \frac{1}{(n_i\mu_i - \lambda_i)R_i}\right).$$

Then the optimization problem can be formulated as

$$\begin{aligned} &\text{maximize} \sum_{i=1}^{m} \lambda_i b_i\left(1 - \frac{1}{(n_i\mu_i - \lambda_i)R_i}\right) \\ &\text{subject to} \sum_{i=1}^{m} n_i = N. \end{aligned} \tag{18.4}$$

This optimization problem can be solved using Lagrange method. The Lagrange composite function can be constructed as

$$L(n_i) = \sum_{i=1}^{m} \lambda_i b_i \left(1 - \frac{1}{(n_i\mu_i - \lambda_i)R_i}\right) + \overline{\lambda}\left(N - \sum_{i=1}^{m} n_i\right).$$

where $\overline{\lambda}$ is the Lagrange multiplier constant. Letting $dL/dn_i = 0$, with $i = 0, 1, 2, \ldots, m$,

$$\frac{\lambda_i b_i}{R_i} \frac{\mu_i}{(n_i\mu_i - \lambda_i)^2} - \overline{\lambda} = 0$$

$$n_i = \sqrt[2]{\frac{1}{\lambda}} \sqrt[2]{q_i\rho_i} + \rho_i;$$

then substituting this into the constraint of the optimization problem, we have

$$N = \sqrt[2]{\frac{1}{\lambda}} \sum_{j=1}^{m} \sqrt[2]{q_j\rho_j} + \sum_{j=1}^{m} \rho_j$$

$$\sqrt[2]{\frac{1}{\lambda}} = \frac{N - \sum_{j=1}^{m} \rho_j}{\sum_{j=1}^{m} \sqrt[2]{q_j\rho_j}}.$$

Then we yield the final answer, which is the number of servers used for each service instance,

$$n_i = \frac{N - \sum_{j=1}^{m} \rho_j}{\sum_{j=1}^{m} \sqrt[2]{q_j\rho_j}} \sqrt[2]{q_i\rho_j} + \rho_j.$$

This holds when the request arrival rate of each service instance is less than the service processing rate according to M/M/1 queuing model, otherwise the length of request queue will not converge. Thus our result only holds when

$$\lambda_i \leq n_i\mu_i.$$

Thus we have

$$n_i \geq \rho_i.$$

What is more, Figure 18.3 shows that the providers would be penalized once mean response time cannot meet the requirement in SLA, which is R_i; hence, our results ensures that

$$r_i = \frac{1}{n_i\mu_i - \lambda_i} \leq R_i.$$

Finally, we have the lower bound of resources that can be assigned for each service instance:

$$n_i \geq \frac{1}{\mu_i R_i - \lambda_i} + \rho_i.$$

Next we will consider the optimal solution for IRT pricing model, the sojourn time probability distribution is

$$\omega(t) = (\mu - \lambda)e^{(\lambda-\mu)t}.$$

Assuming that service instance i is allocated to n_i servers, then the mean revenue brought by a service provision is

$$g_i = \int_0^{R_i} b_i \omega(t)dt = b_i(1 - e^{(\lambda_i - n_i\mu_i)R_i}).$$

Then the overall mean revenue from service instance i during a time slot is

$$G_i = \lambda_i g_i = \lambda_i b_i(1 - e^{(\lambda_i - n_i\mu_i)R_i}).$$

Thus our optimization problem can be formulated as

$$\text{maximize} \sum_{i=1}^{m} \lambda_i b_i(1 - e^{(\lambda_i - n_i\mu_i)R_i})$$

$$\text{subject to} \sum_{i=1}^{m} n_i = N. \tag{18.5}$$

Constructing Lagrange composite function

$$L(n_i) = \sum_{i=1}^{m} \lambda_i b_i \left(1 - e^{(\lambda_i - n_i\mu_i)R_i}\right) + \overline{\lambda}\left(N - \sum_{i=1}^{m} n_i\right).$$

Letting $dL/dn_i = 0$, where $i = 0, 1, 2, ..., m$,

$$\lambda_i \mu_i b_i R_i e^{(\lambda_i - n_i\mu_i)R_i} - \overline{\lambda} = 0$$

$$n_i = \frac{\ln(\lambda_i \mu_i b_i R_i)}{\mu_i R_i} = \frac{\ln\overline{\lambda}}{\mu_i R_i} + \rho_i.$$

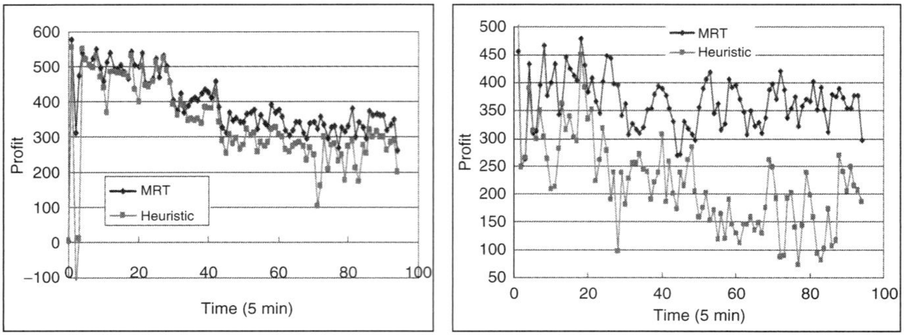

Figure 18.4 Evolution of revenue during 5 min over time with traces.

Then we have

$$N = \sum_{j=1}^{m} \frac{\ln(\lambda_j \mu_j b_j R_j)}{\mu_j R_j} - \ln\bar{\lambda} \sum_{j=1}^{m} \frac{1}{\mu_j R_j} + \sum_{j=1}^{m} \rho_j$$

$$\ln\bar{\lambda} = \frac{\sum_{j=1}^{m} \frac{\ln(\lambda_j \mu_j b_j R_j)}{\mu_j R_j} + \sum_{j=1}^{m} \rho_j - N}{\sum_{j=1}^{m} \frac{1}{\mu_j R_j}}.$$

Finally, we get

$$n_i = \rho_i + \frac{\ln(\lambda_i \mu_i b_i R_i)}{\mu_i R_i} - \frac{\sum_{j=1}^{m} \frac{\ln(\lambda_j \mu_j b_j R_j)}{\mu_j R_j} + \sum_{j=1}^{m} \rho_j - N}{\mu_i R_i \sum_{j=1}^{m} \frac{1}{\mu_j R_j}}.$$

And bounds should be obtained similarly as for the MRT case.

Experiments have been run on a C-based simulator we developed, and events been simulated include arrival, departure, resource allocation for both the MRT and IRT pricing models with the synthetic dataset, and the traced data set. Experimental results for both MRT and IRT shown in Figure 18.4 proved that our algorithms outperform heuristic method.

18.3 DATA CENTER TIME-DEPENDENT PRICING

In this section, we consider the problem of data center NPO, which aims to maximize the difference between the total revenue from deadline-dependent pricing and the electricity cost of distributed data centers [12]. Time-dependent pricing, which charges users based on not only *how much* resources are consumed but also *when* they are consumed, has been widely studied in the electricity industry [17–20] and the Internet service provider (ISP) industry [21–23] as an effective solution to even out resource consumption peaks and reduce operating costs. However, the problem for data center pricing is largely different because the prices are determined by job deadlines (i.e., completion times), whereas job duration or progress are irrelevant.

The goal here is to maximize the data center net profit through job scheduling. It requires (i) to maximize the total revenue as a function of individual job completion deadlines, (ii) to minimize the electricity cost by scheduling jobs to different time and locations, and (iii) to satisfy a capacity constraint at each distributed data center. We formulate this problem as a constrained optimization [12], whose solutions characterizes an interesting trade-off between delay and cost—while completing a job earlier generates a higher revenue, it restricts the set of feasible scheduling decisions, causing higher electricity cost. While being a complex, mixed-integer optimization due to the scheduling formulation, the NPO problem remains to be hard because of the coupling of scheduling decisions and job completion deadlines. There is no closed-form, differentiable function that computes job completion deadlines by scheduling decisions, which is often necessary if standard optimization techniques are to be applied.

As we mentioned earlier, deadline-dependent pricing charges users based on both the amount and the instance time at which the resources are consumed. Suppose that a cloud computing service consists of a set of N distributed data centers and that its billing cycle is divided into T periods. Let K be the total number of jobs submitted for the next cycle. To enable deadline-dependent SLA and pricing, each user request i not only contains the number of demanded VM instances (i.e., r_i) and the total subscription time (i.e., η_i) but is also associated with a bid function $U_i(t)$, which measures the payment user i is willing to make if his/her job is accepted and scheduled to complete by period t. For example, a bid function is flat when a user is indifferent to job deadlines, whereas it becomes strictly decreasing when a user is willing to pay more to get his/her job completed early.

18.3.1 Electricity Cost

In the electricity market of North America, Regional Transmission Organization (RTO) is responsible for transmit electricity over large interstate areas and electricity prices are regionally different. Electricity prices remain the same for a relatively long period in some regions, whereas they may change every hour, even 15 min in the regions who have wholesale electricity markets [9, 17, 18]. In this paper, we consider cloud computing services consisting of distributed regional data centers, which are subject to different electricity prices. When servers at each data center are homogeneous, the total electricity consumption at each data center j over period t can be calculated directly by multiplying the number of active servers $m_j(t)$ and the electricity consumption per server c_j. Let $P_j(t)$ be the electricity price of data center j at time t. We also assume that $P_j(t)$ for $t = 1, \dots, T$ is known noncausally at the beginning of each billing cycle. In practice, this can be achieved by stochastic modeling of electricity prices [24, 25] or by purchasing forward electricity contracts in the whole sale market [26, 27]. Thus the total electricity cost for N data centers over the next cycle is given as

$$\mathbf{D}C_{\text{total}} = \sum_{j=1}^{N} \sum_{t=1}^{T} m_j(t) c_j P_j(t). \tag{18.6}$$

18.3.2 Workload Constraints

We denote the number of VM instances received by job i from data center j at period t as $x_{i,j}(t)$. We consider two types of jobs: divisible jobs and indivisible jobs. An indivisible job that requires r_i VM instances and a subscription time η_i cannot be interrupted and must have r_i VM running continuously for η_i periods, whereas a divisible job can be partitioned arbitrarily into any number of portions, as long as the total VM instance hour is equal to the demand $r_i \eta_i$. Each portion of a divisible job can run independently from other portions. Examples of applications that satisfy this divisibility property include image processing, database search, Monte Carlo simulations, computational fluid dynamics, and matrix computations [28]. Given the scheduling decisions $x_{i,j}(t)$ of all jobs, the aggregate workload for each regional data center must satisfy a service rate constraint

$$\sum_{i=1}^{K} x_{i,j}(t) \leq \mu_j m_j(t), \ \forall j, t, \tag{18.7}$$

where μ_j is the service rate per server at regional data center j and $m_j(t)$ is the number of active server at time t. There is also a limitation on the number of servers at each location. Therefore, we have

$$m_j(t) \leq M_j, \ \forall t. \tag{18.8}$$

All user requests for the next cycle are collectively handled by a job scheduler, which decides the location and time for each job to be processed in a judiciary manner. This scheme can be viewed as a generalization of existing on-spot VM instances in Amazon EC2, which allows users to bid for multiple periods and different job deadlines (or job completion times). The system model is illustrated in Figure 18.5. Let d_i denote the scheduled completion time of job i, the total revenue received by the cloud provider over the next cycle is given as

$$\mathbf{D}U_{\text{total}} = \sum_{i=1}^{K} U_i(d_i). \tag{18.9}$$

The goal of this paper is to design a job scheduler that maximizes the net profit $U_{\text{total}} - C_{\text{total}}$ over feasible scheduling decisions $x_{i,j}(t)$, subject to the workload constraints. For a given job completion time d_i, we denote the set of all feasible job

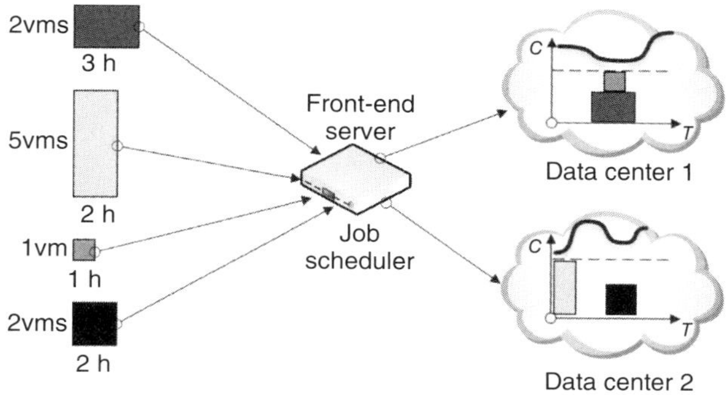

Figure 18.5 An illustration of our system model. Four jobs with different parameters are submitted to a front-end server, where a job scheduler decides the location and time for each job to be processed, by maximizing the net profit that a data center operator receives.

scheduling decisions by a set $\mathscr{X}(d_i)$. In particular, for a divisible job, a user is only concerned with the total VM instance hour received before time d_i. This implies

$$\mathscr{X}(d_i) = \left\{ x_{i,j}(t) : \sum_{j=1}^{N} \sum_{t=1}^{d_i} x_{i,j}(t) = r_i \eta_i, \right\}, \tag{18.10}$$

where $r_i \eta_i$ is the total demand of job i. On the other hand, an indivisible job must be assigned to a single regional data center and run continuously before completion. It will have a feasible scheduling set as

$$\mathscr{X}(d_i) = \left\{ x_{i,j}(t) : x_{i,j}(t) = 0 \text{ or } r_i \cdot \mathbf{1}_{(d_i - \eta_i \le t \le d_i)} \right\}, \tag{18.11}$$

where $\mathbf{1}_{(d_i - \eta_i \le t \le d_i)}$ is an indicator function and equals to 1 if t belongs to the scheduled execution interval $[d_i - \eta_i, d_i]$ and 0 otherwise. We suppress the positivity constraints of $x_{i,j}(t)$ for simplicity of expressions.

We formulate the data center NPO problem as follows:

Problem NPO :

$$\text{maximize} \quad \sum_{i=1}^{K} U_i(d_i) - \sum_{j=1}^{N} \sum_{t=1}^{T} m_j(t) c_j P_j(t) \tag{18.12}$$

$$\text{subject to} \quad \sum_{i=1}^{K} x_{i,j}(t) \le \mu_j m_j(t), \forall j, t \tag{18.13}$$

$$m_j(t) \le M_j, \ \forall j \tag{18.14}$$

$$\{x_{i,j}(t)\} \in \mathscr{X}(d_i), \forall i \tag{18.15}$$

$$\text{variables} \quad d_i, x_{i,j}(t), m_j(t). \tag{18.16}$$

The above problem can be looked at graphically as illustrated in Figure 18.5. It requires a joint optimization of revenue and electricity cost, which is a mixed-integer optimization because the scheduling decisions $x_{i,j}(t)$ are discrete for indivisible jobs. Further, because of our deadline-dependent pricing mechanism, the maximization of revenue over completion time d_i is coupled with the minimization of electricity cost over feasible scheduling decisions $\mathcal{X}(d_i)$. However, there is no closed-form, differentiable function that represents d_i by $\mathcal{X}(d_i)$. Therefore, off-the-shelf optimization algorithms cannot be directly applied.

It is worth noticing that our formulation of Problem NPO in Eq. (18.12) also incorporates sharp job deadlines. For instance, if job i must be completed before time t^*, we can impose a utility function with $U_i(t) = -\infty$, for all $t > t^*$, so that scheduling decisions with a completion time later than t^* becomes infeasible in Problem NPO.

Problem NPO is a complex joint optimization over both revenue and electricity cost, whose optimization variables; scheduling decisions $\{x_{i,j}(t)\} \in \mathcal{X}(d_i)$ and job completion time d_i are not independent of each other. There is no closed-form, differentiable function that relates the two optimization variables. Let $x_i(t) = \sum_j x_{i,j}(t)$ be the aggregate service rate received by job i at time t. For given scheduling decisions $x_{i,j}(t)$, we could have replaced constraint (18.15) by a supremum

$$d_i = \sup_{\tau} \left\{ \tau : \sum_{t=1}^{\tau} x_i(t) < r_i \eta_i \right\}, \tag{18.17}$$

where $r_i \eta_i$ is the total demand of job i. However, it still requires evaluating supremum and inequalities and, therefore, does not yield a differentiable function to represent d_i by scheduling decisions $\{x_{i,j}(t)\}$.

To overcome this difficulty, an approximation of the job completion time function in Eq. (18.17) by a differentiable function is proposed in Reference 12:

$$\hat{d}_i = \frac{1}{\beta} \cdot \log \left(\frac{1}{r_i \eta_i} \sum_{t=1}^{T} e^{\beta t} \cdot x_i(t) \right), \tag{18.18}$$

where β is a positive constant and $1/(r_i \eta_i)$ normalizes the service rate $x_i(t)$. The accuracy of this approximation is given as follows.

Theorem 18.2 *For a positive constant β and a bounded service rate $0 \le x_i(t) \le x_{\max}$, we have*

$$r_i \eta_i \cdot \left[1 - \frac{\log(1 + \beta / x_{\max})}{\beta / x_{\max}} \right] \le \sum_{t=1}^{\hat{d}_i} x_i(t) \le r_i \eta_i. \tag{18.19}$$

Hence, $\lim_{\beta \to \infty} \hat{d}_i = d_i$, that is, the approximation in Eq. (18.18) becomes exact.

By the approximation in Eq. (18.18), we obtain a closed-form, differentiable expression for $\hat{d}_i$, which guarantees an approximated completion of job i off by a logarithmic term $x_{\max} \log(1 + \beta/x_{\max})/\beta$ in the worst case. The approximation becomes exact as β approaches infinity. However, often there are practical constraints or overhead concerns on using large β. We choose an appropriate β such that the resulting optimality gap is sufficiently small.

An algorithmic solution for Problem NPO with divisible jobs is derived in Reference 12. The problem has a feasible set:

$$\mathcal{X}(d_i) = \left\{ x_{i,j}(t) : \sum_{j=1}^{N} \sum_{t=1}^{d_i} x_{i,j}(t) = r_i \eta_i \right\}. \tag{18.20}$$

In order to solve the problem, we leverage the approximation of job completion time in Eq. ([18.18]) to obtain an approximated version of Problem NPO. The resulting problem has an easy-to-handle analytical structure and is convex for certain choices of U_i functions. Further, when the problem is nonconvex, we then convert it into a sequence of linear programming and solve it using an efficient algorithm. It provides useful insights for solving Problem NPO with indivisible jobs.

Rewriting Eq. (18.15) in Problem NPO using the approximation in Eq. (18.18) and the feasible set in Eq. (18.20), we obtain a net profit optimization problem for divisible jobs (named Problem NPOD) as follows:

Problem NPOD :

$$\text{maximize} \sum_{i=1}^{K} U_i(\hat{d}_i) - \sum_{j=1}^{N} \sum_{t=1}^{T} m_j(t) c_j P_j(t) \tag{18.21}$$

$$\text{subject to} \sum_{i=1}^{K} x_{i,j}(t) \le \mu_j m_j(t), \forall j, t \tag{18.22}$$

$$m_j(t) \le M_j, \ \forall j \tag{18.23}$$

$$\hat{d}_i = \frac{1}{\beta} \log \left(\frac{1}{r_i \eta_i} \sum_{t=1}^{T} \sum_{j=1}^{N} e^{\beta t} x_{i,j}(t) \right) \tag{18.24}$$

$$\sum_{j=1}^{N} \sum_{t=1}^{T} x_{i,j}(t) = r_i \eta_i \tag{18.25}$$

$$\text{variables } \hat{d}_i, x_{i,j}(t), m_j(t), \tag{18.26}$$

where completion time becomes a differentiable function of $x_{i,j}(t)$. The optimization variables in Problem NPOD may still be integer variables, that is, the number of active servers $m_j(t)$ and the number of demanded VM instances $x_{i,j}(t)$. We can leverage rounding techniques (e.g., [29]) to relax Problem NPOD so that its optimization variables become continuous.

As inequality constraints in Eqs. (18.22), (18.23), and (18.25) are linear, we investigate the convexity of the optimization objective in Eq. (18.21), which is a function of $x_{i,j}(t)$ and $m_j(t)$, by plugging in the equality constraint (18.24). It is shown [12] that for $\hat{d}_i$ in Eq. (18.24) and a positive differentiable $U_i(\cdot)$, the objective function of Problem NPOD is convex if $U''(y) \geq \beta U'(y)$ for all $y > 0$ and concave if $U''(y) \leq \beta U'(y)$ for all $y > 0$. The two conditions of $U_i(\cdot)$ capture a wide range of functions in practice. Let $a > 0$ and b be arbitrary constants. Examples of $U_i(\cdot)$ that result in a concave objective function include certain exponential and logarithmic functions. Examples that result in a convex objective function include linear $U_i(y) = b - ax$ and logarithm $U_i(y) = a - b\log(y)$. We remark that when $U''(y) \leq \beta U'(y)$ for all $y > 0$, Problem NPOD is a concave maximization. It can be solved efficiently by off-the-shelf convex optimization algorithms, for example, primal-dual algorithm and interior point algorithm [30].

On the other hand, when Problem NPOD maximizes a convex objective, an iterative solution based on *difference of convex programming* is proposed [12]. Relying on a linearization of the objective, the following iterative algorithm is used to solve Problem NPOD via a sequence of linear programming. It starts at some random initial point $x_{i,j}^{(0)}(t)$, solves the linear programming with $x_{i,j}^{(k-1)}(t)$, and, therefore, generates a sequence $\{x_{i,j}^{(k)}(t)\}_{k=0}^{\infty}$. This procedure is indeed a special case of the *difference of convex programming* and has been extensively used in solving many nonconvex programs of similar forms in machine learning [31]. It is shown that the sequence $\{x_{i,j}^{(k)}(t)\}_{k=0}^{\infty}$ converges to a local minimum or a saddle of Problem NPOD, as in Theorem 18.2. We further apply a primal-dual algorithm [30] to obtain a distributed solution to each linear programming. The algorithm is summarized in Fig. 18.6.

Finally, we evaluate our algorithmic solution for Problem NPOD over a 24-h cycle, divided into $T = 144$ 10-min periods. The goal is to obtain empirically

Initialize a feasible point $\{x_{i,j}^{(0)}(t)\}$, $k = 0$, and step size $\delta > 0$.

for each k, iteratively solve the linearization:

Find effective profit for each data center j and period t:

$$\Gamma_j(t) = \frac{U_i'(\hat{d}_i^{(k-1)})}{r_i \eta_i \beta e^{\beta \hat{d}_i^{(k-1)}}} e^{\beta t} - \frac{c_j P_j(t) + \lambda_j^{(k)}(t)}{\mu_j}$$

for each job i

Schedule job i to maximize $\sum_{j,t} \Gamma_j(t) \cdot x_{i,j}^{(k)}(t)$, satisfying

NPOD: $\sum_{i,j,t} x_{i,j}^{(k)}(t) = r_i \eta_i$

NPOI: $x_{i,j}^{(k)}(t) = 0$ or $r_i \cdot \mathbf{1}_{(y-\eta_i \leq t \leq y)}$ for some y

end for

Obtain $m_j^{(k)}(t) = \sum_i x_{i,j}^{(k)}(t)$

Update price $\lambda_j^{(k)}(t) = \left[\lambda_j^{(k-1)}(t) + \delta(m_j^{(k)}(t) - M_j)\right]^+$

$k \leftarrow k + 1$

end for

Figure 18.6 Algorithmic solution for Problem NPOD.

validated insights about the feasibility/efficiency of our solution using synthesized data from a recent study [9]. We construct $N = 3$ regional data centers, which reside in different electricity markets and have electricity prices $P_1(t) =\$56/MWh$, $P_2(t) =\$(51 + 15\cos 0.05t)/MWh$, and $P_3(t) =\$(45 + 20\sin 0.06t)/MWh$. The data centers host $M_1 = 2000$, $M_2 = 1000$, and $M_3 = 1000$ servers. Each server is operating at $c_j = 1200$ W with $\mu_j = 4$ VMs per server for $j = 1, 2, 3$.

We construct two types of jobs: elephant jobs that subscribes $r_i \in [50, 100]$ VMs for $\eta_i = [10 - 20]$ periods and mice jobs that subscribes $r_i \in [5, 20]$ VMs for $\eta_i = [1 - 10]$ periods. In our simulations, both r_i and η_i are uniformly randomly generated from their ranges. We fix the total workload to be $K = 1200$ jobs, each being an elephant job with probability 20% and a mice job with probability 80%. Job i is associated with a nonlinear bid function, given by

$$U_i(d_i) = r_i\eta_i \cdot (a - b\log t) \text{ (dollars)}, \tag{18.27}$$

where $a \in [0.01, 0.02]$ and $b \in [0.001, 0.002]$ are uniformly distributed. Using this nonlinear bid function, the approximation of job completion time will result in a convex objective function in Eqs. (18.12) and (18.21). All numerical results shown in this section are averaged over five realizations of random parameters.

To provide benchmarks for our evaluations, we also implement a greedy algorithm that sequentially schedules all jobs with a *largest job first (LJF)* policy and a heuristic algorithm NPOI (net profit optimization problem for indivisible jobs) by assuming indivisible jobs in problem NPOI. Figure 18.7 compares the optimized net profit of NPOD, NPOI, and LJF algorithms. When jobs are indivisible, our NPOI algorithm improves the net profit by 12% over the baseline LJF algorithm (from \$1868

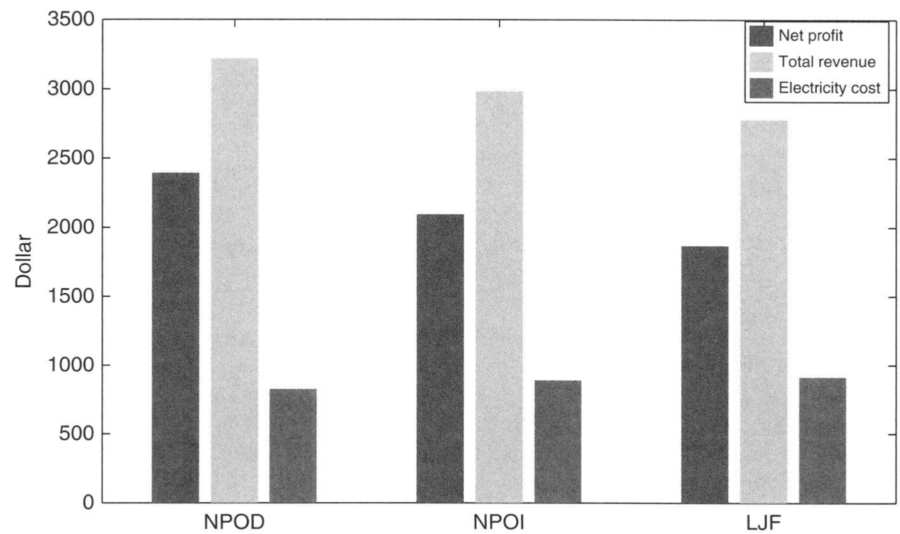

Figure 18.7 A comparison of optimized net profit of NPOD, NPOI, and LJF.

to \$2095), while an additional 16% increment (to \$2394) can be achieved by NPOD if all jobs are divisible. We also notice that our NPOI algorithm is able to simultaneously improve total revenue and cut down electricity cost, compared to the baseline LJF algorithm. This is achieved by the joint optimization over job completion time d_i and scheduling decisions $x_{i,j}(t)$.

18.4 CONCLUSION AND FUTURE WORK

This chapter discussed three different approaches on smart pricing of cloud services and resources: VM-based pricing, SLA-based pricing, and time-dependent pricing. There is a large number of open problems on topics discussed in this chapter. Many of them are due to the delicate trade-off between operators' goal of profit maximization and users' demand for individual performance optimization. In a multiuser environment, while maximizing their total revenue, cloud operators may need to consider establishing a social scheduling policy that would prevent "poor" users from being entirely deprived of resources. Algorithmic solutions to the nonconvex problem of job scheduling and NPO has not been found. Further, to analyze different pricing policies, user payment models must be constructed to reflect not only users' satisfaction and willingness to pay but also their budget and financial means. Further challenges will arise as stochastic arrivals/departures of users' demand and cloud resources are incorporated.

REFERENCES

1. KPMG International. "Clarity in the Cloud", *Online technical report at http : //www.kpmg.com/*, Nov. 2011.
2. N. Chohan. See spot run: using spot instances for mapreduce workflows. In *USENIX HotCloud Workshop*, 2010.
3. S. Yi, D. Kondo, and A. Andrzejak. Reducing costs of spot instances via checkpointing in the amazon elastic compute cloud. *IEEE International Conference on Cloud Computing*, 2010.
4. Amazon Elastic Compute Cloud User Guide. Getting Started with Spot Instances. *docs.amazonwebservices.com/AWSEC2*, Oct. 2008.
5. R. Buyya, Cs. Yeo, J. Broberg, and I. Brandic. "Cloud computing and emerging IT platforms: vision, hype, and reality for delivering computing as the 5th utility," *Future Generation Computer Systems*, 25, 2009, 599–616.
6. X. Fan, W. Weber, and L. A. Barroso. Power provisioning for a warehouse sized computer. In *Proceedings of the 34th Annual International Symposium on Computer Architecture*, 2007.
7. S. Nedevschi, L. Popal, G. Iannaccone, S. Ratnasamy, and D. Wetherall. Reducing network energy consumption via sleeping and rate-adaptation. In *Proceedings of the 5th USENIX Symposium on Networked Systems Design & Implementations (NSDI)*, 2008.
8. L. Parolini, B. Sinopoli, and B. Krogh. Reducing data center energy consumption via coordinated cooling and load management. In *Proceedings of ACM Workshop on Power Aware Computing and Systems*, 2008.

9. L. Rao, X. Liu, L. Xie, and W. Liu. Minimizing electricity cost: optimization of distributed internet data centers in a multi-electricity market environment. In *Procedings of IEEE Infocom*, 2010.
10. P. Patel, A. Ranabahu, and A. Sheth. Service level agreement in cloud computing. In *Proceedings of the Workshop on Best Practices in Cloud Computing*, 2009.
11. C. Wilson, H. Ballani, T. Karagiannis, and A. Rowstron. Better never than late, meeting deadlines in data center networks. In *Proceedings of SIGCOMM*, 2011.
12. W. Wang, P. Zhang, T. Lan, and V. Aggarwal. Data center net profit optimization with individual job deadlines. In *Proceedings of CISS 2012 (Invited paper)*, March 2012.
13. A. Danak and S. Mannor. Resource allocation with supply adjust- ment in distributed computing systems. In *International Conference on Distributed Computing Systems (ICDCS)*, 2010.
14. L. Fleischer. Tight approximation algorithms for maximum general assignment problems. In *ACM Symposium on Discrete Algorithms*, 2006.
15. W. W. S. Wei. *Time Series Analysis: Univariate and Multivariate Methods*. Addison Wesley, Reading, MA, 1990.
16. C. Gong, J. Liu, Q. Zhang, H. Chen, and Z. Gong. The characteristics of cloud computing. *39th International Conference on Parallel Processing Workshops*, pp.275–279, 2010.
17. S. Littlechild. Wholesale spot price pass-through. *Journal of Regulatory Economics*, 23(1), 2003, 61–91.
18. S. Borenstein. "The long-run efficiency of real-time electricity pricing," *The Energy Journal*, 26(3), 2005, 93–116.
19. K. Herter. "Residential implementation of critical-peak pricing of electricity," *Energy Policy*, 35(4), 2007, 2121–2130.
20. A. Faruqui, R. Hledik, and S. Sergici, "Piloting the smart grid," *Electricity Journal*, 22(7), 2009, 55–69.
21. H. Chao. "Peak load pricing and capacity planning with demand and supply uncertainty," *Bell Journal of Economics*, 14(1), 1983, 179–190.
22. G. Brunekreeft. *Price Capping and Peak-Load-Pricing in Network Industries*. Diskussionsbeitrage des Instituts fur Verkehrswissenschaft und Regionalpolitik, Universitat Freiburg, vol. 73, Freiburg, Germany, 2000.
23. H. Sangtae, S. Soumya, J. Carlee, I. Youngbin, and C. Mung. "TUBE: time-dependent pricing for mobile data," In *Procedings of ACM SIGCOMM 2012 Conference on Applications, Technologies, Architectures, and Protocols for Computer Communication, SIGCOMM '12*, pp. 247–258, 2012.
24. P. Skantze, M. D. Ilic, and J. Chapman. Stochastic modeling of electric power prices in a mult-market environment. *IEEE Power Engineering Society Winter Meeting*, 2000.
25. S. Fleten and J. Lemming, "Constructing forward price curves in electricity markets," *Energy Economics*, 25, 2007, 409–424.
26. C. K. Woo, I. Horowitz, and K. Hoang. "Cross hedging and value at risk: wholesale electricity forward contracts," *Advances in Investment Analysis and Portfolio Management*, 8, 2001, 283–301.
27. R. Weber. Cutting the electric bill for Internet-scale systems. In *Proceedings of SIGCOMM*, pp. 123–134, 2009.
28. B. Veeravalli and W. H. Min. "Scheduling divisible loads on heterogeneous linear daisy chain networks with arbitrary processor release times," *IEEE Transactions on Parallel and Distributed Systems*, 15, 2010, 273–288.

29. A. Neumaier and O. Shcherbina. "Safe bounds in linear and mixed-integer linear programming," *Mathematical Programming*, 99, 2004, 283–296.
30. S. Boyd and L. Vandenberghe. *Convex Optimization. Cambridge University Press*, Cambridge, United Kingdom, 2004.
31. B. K. Sriperumbudur and G. R. G. Lanckriet. "On the Convergence of the Concave-Convex Procedure", *Online technical report at http : //cosmal.ucsd.edu/ ~ gert/papers/nips_cccp.pdf*, 2009.

19 Allocating and Pricing Data Center Resources with Power-Aware Combinatorial Auctions

BENJAMIN LUBIN and DAVID C. PARKES

19.1 INTRODUCTION

As data centers become ever more essential to not only our economy but also our daily lives, it is increasingly important for them to be well managed. Three specific aspects are of paramount importance: ensuring that the right resources be allocated to the right use, ensuring that power is used only when it is providing real benefit, and making sure that the prices charged to users for service are providing incentives that support such allocative and power efficiency. In this chapter, we look at a method for employing a combinatorial market mechanism to achieve these ends.

In 2006, US data centers used about 61 billion kWh; that is, 1.5% of the 4 trillion kWh consumed in total. This is the amount of energy used by 5.8 million average US households (5% of all households) [1]. Producing this power resulted in 37 million metric tons of CO_2, or 0.6% of the 5.9 billion metric tons released from all sources. That is roughly 16% of that produced by the burning of jet fuel and more than that used to power TVs. This electricity cost $4.5 billion and required a peak load capacity of about 7 GW, more than double the level of consumption in 2000 [2, 3]. This pace of rising data center energy consumption has slowed, with global data center power usage rising 56% over its 2005 level by 2010 [4], but this modestly slower growth rate is still alarmingly fast and a serious cause for concern. Peak usage has reached approximately 30 GW, and of this, an average of only 6–12% of electricity goes to actual computation, the rest powering idle computers and support systems such as cooling [5].

Given rising power consumption and its associated financial and environmental costs, data-center operators realize that the established practice of running large

Smart Data Pricing, First Edition. Edited by Soumya Sen, Carlee Joe-Wong, Sangtae Ha, and Mung Chiang.

numbers of significantly underutilized servers is no longer acceptable and are eager for energy-saving solutions.

Market paradigms have often been proposed as useful for allocating limited computational resources and satisfying multicriteria objectives. The earliest work on such markets was for time sharing on the PDP-11 in the 1960s by Sutherland [6]. In the intervening years, there have been proposals to use such methods in high performance computing and grid computing, as well as in data centers.

However, existing proposals have deficiencies that can render them impractical for modern data centers. Here, we describe a general method for overcoming these concerns and illustrate its applicability to one specific environment. Our method provides the following:

A Realistic Model of Resources. We support a fine granularity of computational entity (e.g., core vs server, which is especially important as multicore machines become the norm) and fine control over the power state of machines (e.g., Dynamic Voltage and Frequency Scaling (DVFS), not just on/off). We also handle heterogeneous applications running on heterogeneous classes of servers.

A Realistic Representation of Goals. We use a general form of utility function expressed in terms of outcome percentile and derived from standard long-term service-level agreements (SLAs) that are programatically interpreted as short-term utility functions in a dynamic environment.

Principled Optimization. We adopt mixed-integer programming (MIP) for optimization of data center resources, providing a carefully formulated MIP model that can scale to large problem sizes.[1]

We show that a market-based approach provides a natural, feasible, and advantageous framework for representing the milieu of physical and computational resources, and the applications that consume these resources, in modern day data centers.

Experimental results indicate that such a system can robustly and scalably improve net profits of our data-center prototype by up to 137%. We envision reallocating resources in the data center in roughly 10-min periods. So it is important that our mechanism should be able to compute an allocation within one time period, permitting the allocation to be realized on time. For large instances of 1000 machines and 10 applications (all associated with different customers), each with a realistic demand of 2700 transactions a second, when limiting MIP solve time to 10 min, we achieve an average solve time of 5.16 min, with a negligible approximation imposed on the instances that timeout. Thus a dedicated machine can optimize the usage on 1000 others, providing an acceptable 0.1% overhead factor.

19.1.1 Related Work

In many ways, the modern problem of allocating data center resources is a recapitulation of the problem of allocating slices of the early time-sharing systems of

[1] We leverage recent advances in MIP technology that enable combinatorial markets to be cleared quickly in practice by exploiting problem structure and despite NP-hard winner determination problems [7].

the 1960s. The modern need for expressive ways for users to describe their requirements has parallels in early methods such as IBM's Job Control Language (JCL). Consequently, there is a long history of proposed methods for describing user's computational requirements and for allocating resources based on those requirements. Most of these systems have not taken a market approach. Rather, they have assumed that resource owners require effective ways to allocate their hardware and that they would thus gather enough information to approximately solve an appropriate bin packing problem. Pricing has typically been a separate consideration—either because the system owner and users are in the same firm or because it has been handled by a separate human-mediated contracting process.

There are a number of limitations to such an approach. First, just because the hardware may be owned by the same firm as the users does not guarantee that appropriate prioritization of jobs will be manifest in the system—competition for resources occurs even within cooperative organizations. Thus hardware ownership may have less importance than is often assumed as, inevitably, there are two sides to the allocation problem: those who need resources and those who have them. If the allocator can make an accurate estimate of user needs, then treating the system as monolithic is without loss. However, this is difficult in practice, and thus it is often essential for users to provide this information. As soon as they do so, they may be incentivized to misrepresent their needs, in order to obtain more resources or a better price. If, as is ideal, we wish to confront this potential gaming of the system, the full allocation problem falls into the field of *mechanism design*, which considers the problem of designing systems for gathering information necessary for optimal allocation when agents may attempt to game the system. Viewed from this lens, pricing creates powerful incentives for participants and is thus also essential for moderating manipulation of the system and thereby enabling high quality outcomes. In this chapter, we adopt this view and construct a market mechanism that attempts to balance the "buyers" and "sellers" view of the data center. Our method is directly scalable to the allocation of a thousand machines and could be replicated in a straightforward way to handle arbitrary numbers of such pools. We view this approach as better capturing the full scope of the allocation problem, although clearly the simpler problem of allocation on behalf of a single monolithic owner is still important and challenging in its own right.

Chase et al. [8] and Chen et al. [9] present a market-based system for data center resource allocation and are able to experimentally demonstrate significant energy savings over static allocations. However, their greedy clearing mechanism imposes restrictions on the form of utility that can be modeled, SLAs are not directly represented, and demand/utility computations occur with respect to mean statistics instead of distributional information. Their model does not handle the heterogeneity of data-center machines or modern power-throttling architectures (instead simply turning machines on and off) and allocates servers rather than cores. The nonlinear cost model that we use is related to the one provided by Chen et al. [9]. But rather than identifying total-value maximizing allocations with respect to SLAs, they treat SLAs as constraints and attempt to find the cheapest allocation subject to meeting implied quality constraints.

One branch of work on resource allocation in data centers has focused on *utility computing*, which seeks to provide access to data centers in a way that is analogous to that of a public utility (e.g., gas, water, power) [10, 11]. In general, utility computing views computational resources as easily substituted for one another, whereas we allow only particular machines to be configured for, and capable of running, certain applications. Rather than argue for a more radical shift in how computation is bought, sold, and deployed, we propose a more gradual evolutionary step by creating a market that handles the heterogeneity in present data centers and encompasses and generalizes the present contract format (SLAs).

There is an extensive literature on using market-based methods in related contexts, including *computational grids* and *high performance computing*. Yeo and Buyya [12] provide an abstract model of a market-based resource manager as comprised by its various parts, including a market model, an economic model, pricing, tendering, bidding, buyer/seller focus, a trading environment, QoS attributes, resource composition, an execution service/support, an accounting mechanism, management control, and a job model. The authors offer an extensive survey of existing systems according to these properties. In another survey, Broberg et al. [13] contrast several different market-based allocations systems and evaluate their pros and cons. These authors argue that fixed pricing can be problematic, that optimal combinatorial bidding and pricing can be expensive, and propose a particular ad hoc pricing mechanism that seeks a compromise between expressivity and scalability. In general, all allocation methods must grapple with this trade-off; we handle it by appeal to modern mixed-integer optimization algorithms that are "anytime" and can thus be stopped early at high quality approximate solutions.

A market in our setting is combinatorial in nature; participants must be able to bid (indirectly, via SLAs) on "packages" of items, where the value for a package need not be a linear combination of the value for its constituent items. There is a long literature on combinatorial auctions; an excellent overview is provided in the book by Crampton et al. [14].

19.2 A MARKET MODEL OF DATA CENTER ALLOCATION

In this section, we describe how to represent the energy-aware data-center allocation problem as a market. Using a market for the internal allocation both ensures that resources will be used in an economically optimal way and opens a path to exposing customers to the mechanism as well. In turn, this can enable better information revelation and thus higher efficiency allocations and/or higher revenue. To begin, we describe the type of operation we see such a system operating within and define some needed terminology.

Typical data centers have thousands of servers, many of which will share the same hardware and software configuration. We call such equivalence classes *"machine groups"* and assume that this partitioning is performed by a separate offline process. The owner of a data center typically contracts (either internally or externally) to provide these resources to a set of applications (each associated with a customer), each with time-varying load and utility and a range of resource requirements and importance.

In present use, each application is associated with an SLA that is negotiated between the customer and the data-center provider. Such an agreement specifies a price, a performance objective (e.g., a cap on the 95th percentile of response time), and a penalty for failing to meet the objective. The SLA is useful for assigning a relative importance to the applications, but despite its quantitative nature, it is generally used at present in only a qualitative way, as a guideline for personnel when manually configuring data-center operations. Yet, SLAs suggest a direction toward application utility functions that are highly relevant to obtaining reasonable performance in a power-constrained environment [15, 16]. Here, we introduce a system that adopts SLAs for the purpose of utility-based optimization of resource configurations.

When allocating resources in the data center, we seek to optimize the operator's business value for the data center: that is, the revenue net of costs. This means assigning (portions of) the machines from discrete machine groups to the various applications as well as specifying the power for each machine and thus restraining overall consumption. For this, we use a detailed model of the power saving modes available to modern servers and assume access to monitors of both power consumption and application usage.

Our market allocates goods (cores of machines from the various machine groups) to applications, as illustrated in Figure 19.1. The market is repeatedly cleared over brief periods, allocating resources to maximize the value reflected in short-term bids.

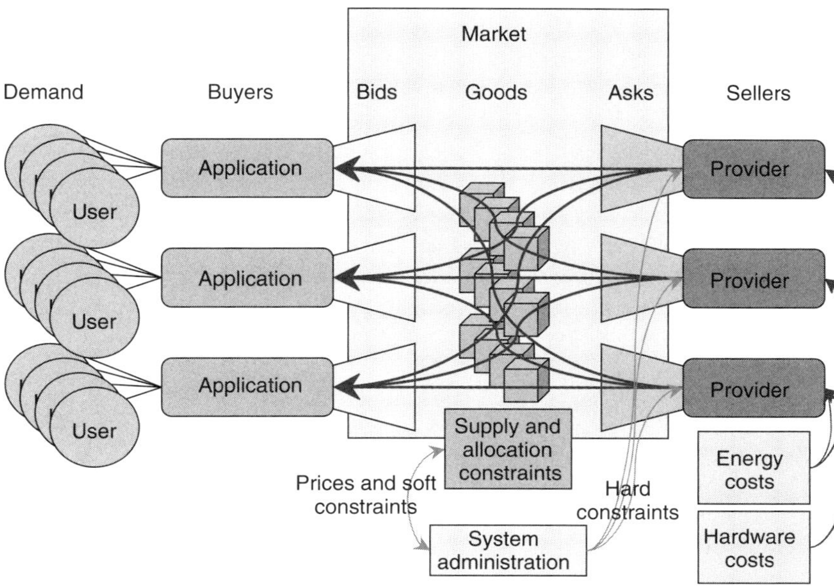

Figure 19.1 The data center market model.

These short-term bids are generated automatically from predictions about future supply and demand as well as applications' long-term SLAs. In our experiments, we have used 10-min intervals, chosen to match the demand volatility we see in the applications we are modeling. The winner-determination problem requires optimization and could be approximately solved via heuristic search. We choose to formulate it as a mixed-integer program and solve it via a commercial solver, specifically IBM ILOG CPLEX. The form of the buyer (or customer's) value model and seller cost model have been chosen to ease formulation of the problem as a MIP, as described in the following.[2]

The formulation presented here is myopic in that it predicts demand in the next period and seeks the optimal allocation for that period. Instead, we could formulate a stochastic optimization problem that predicts multiple time steps forward and seeks an optimal allocation over that longer time period. However, such a formulation will be complex and difficult to solve with much fidelity over the short time frame available to the solver, and thus we view a single time-step solution as a reasonable simplification. Moreover, if reasoning about future time periods, we would ideally permit participant's bids to capture changes in value over time, leading to a scheduled market instead of the *spot* market presented here. We leave this important extension for future work.

For each period, we use a myopic net revenue maximization objective:

$$\max \quad \sum_{a \in A} V_a - \kappa E_a^{\text{TOTAL}} - H_a^{\text{TOTAL}}, \tag{19.1}$$

where V_a is the value of the chosen allocation of machines to application (associated with a particular buyer) $a \in A$, κ is the dollar cost of a kiloWatt-hour of energy, E_a^{TOTAL} is the total energy used to establish and maintain the chosen allocation for the current period, and H_a^{TOTAL} is the dollar cost for the hardware.

The objective is thus quite straightforward, and the complexity comes from the constraints that define and restrict these variables. We begin by defining the buyer value, V_a, that is, the value associated with application a of some buyer. We note that throughout this section we treat V_a as the buyer's true value for the service he/she receives. In practice, this value is not known a priori to the mechanism and must instead rely on the buyer's reported values. The existence of strategic opportunities to customers for misreporting values will depend on the interaction with the pricing rule used to charge for the service. We defer discussion of pricing rules and their incentive properties to Section 19.5.

19.2.1 Buyer Valuation Model

The contracts signed for data center provisioning are typically in terms of SLAs. We model a per-application SLA contract as a piecewise linear function for the value of

[2]The size of the formulation will grow linearly with the number of applications and machine groups. However, it will grow with the square of the number of power modes (because it encodes a transition matrix from one mode to the next). Fortunately, polynomial growth in the size of the MIP need not imply exponential growth in practical computation time, and we examine scalability in Section 19.3.1.

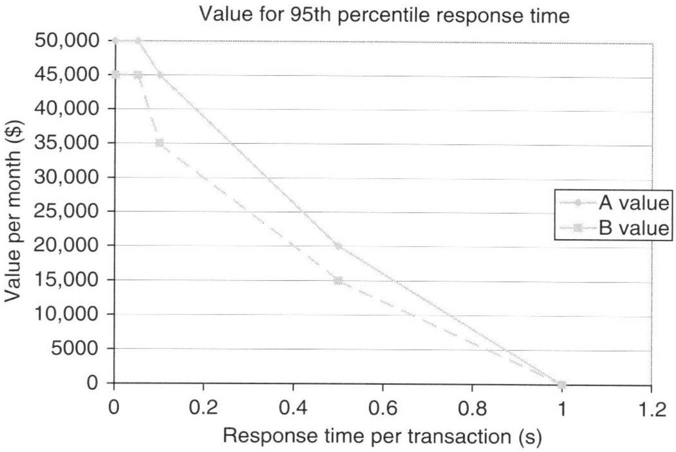

Figure 19.2 SLAs as provided by two different applications.

receiving a given response time at a given demand percentile. This form has a number of advantages. As in contracts commonly used in practice, it is expressed in terms of an outcome metric percentile (e.g., 95th percentile). This enables the most extreme cases to be ignored as outliers but captures "bad-case" behavior in a way that the mean does not. While typical industry SLAs specify a single value for service exceeding a threshold quality at this percentile, we generalize this notion to allow a range of possible service levels. We are thus able to capture the value of degraded-but-still-useful service in a way that traditional SLAs cannot. Figure 19.2 shows an example of an SLA value curve of this form for two different applications A and B.

Rather than having bidders interact directly with the market, we use a proxy that operates on their behalf. This shields the human buyers from having to deal with low level details, instead of letting them to concentrate on the high level value function with which they are familiar. The bidding proxy takes the SLA thus provided and combines it with supply and demand prediction and an application performance model, to represent the SLA as a short-term bid; that is, a bid that is appropriate for the period of time for which the market is cleared.

In order to construct this short-term bid, the bidding proxy needs a model of how a given supply of machines (and thus transaction capacity) and application demand for the next planning episode will translate to the long-term response-time distribution (and in turn to, e.g., its 95th percentile) and thus to the value curve associated with an SLA.

As a very simple example, let us assume that transaction processing is described as an M/M/1 queue (exponential interarrival and service times). In this case, the response-time distribution is exponential with mean response time $1/(\mu - \lambda)$, where μ is the supply and λ is the demand, both in transactions per unit time. The fraction of response times above a percentile P is given by the exponential quartile function: $-\frac{\ln(1-P)}{(\mu-\lambda)}$. The proxy composes the customer's SLA (Fig. 19.2) with this response-time

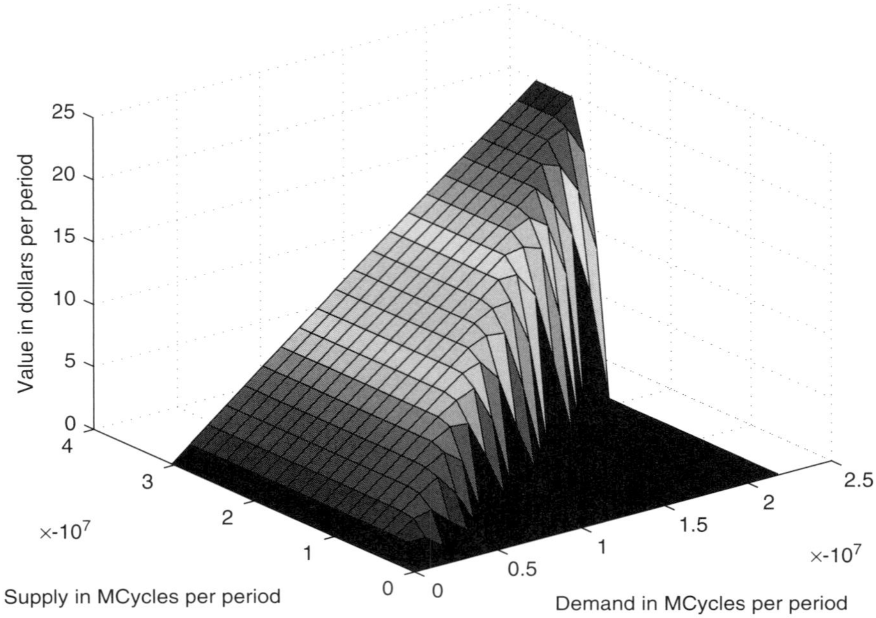

Figure 19.3 Application value by supply and demand.

model, resulting in a value function over both supply and demand at, for example, the 95th percentile (Fig. 19.3). As depicted in the figure, the value is scaled by predicted demand relative to the mean statistic. This is necessary to translate from the long-term value specified in the SLA into a short-term bid. Thus, we, on behalf of the bidder, estimate future demand and figure out an appropriate bid for the transactions expected in the current period. With this conversion, the market tracks the long-term statistics that are a customer's true concern (as reflected in the SLA). We note that this bid conversion is not done for strategic reasons, but rather to generate value-based information that can be used to drive good short-term allocation decisions.[3]

In addition, the bidding proxy needs a predictive model of application demand over the next period to construct its short-term bid. We have found it sufficient to simply use statistics gathered over a small window of previous periods to provide a Gaussian model of the distribution of possible demand in the next period via a maximum likelihood estimation (MLE) of mean and standard deviation. The bidding proxy draws equal weighted samples from this Gaussian demand prediction model and takes a slice from the value model (Fig. 19.3) for each sample, in a process of inverse

[3]Depending on the pricing structure ultimately used by the market, incentives for strategic bidding may or may not be present. Where such incentives remain, the consequent bidder strategizing will be outside the scope of the bid conversion process described.

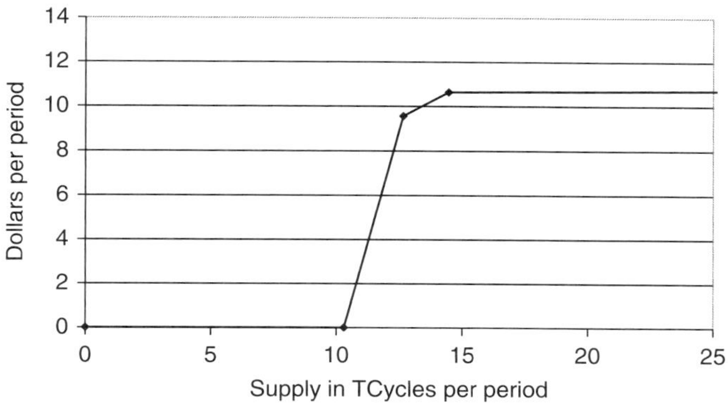

Figure 19.4 Expected short-term value for a single application.

transform sampling. These slices are averaged to produce a single consumption-value curve that manifests our demand model. By taking a piecewise linear approximation of this curve (obtained by chaining the control points of the originally supplied response-time/value curve through these transformations), we arrive at a utility curve, which is provided to the market in a given period as a bid. An example of such a bid is shown in Figure 19.4.

Through these transformations, we arrive at a value function, which is a mapping from potential resources bundles to agent value and used by the winner-determination algorithm to clear the market. In particular, the bidder proxy creates a function

$$V_a = F_a(Q_a), \tag{19.2}$$

for application a, where F_a is the piecewise linear function we have established through the process just described and Q_a is the number of cycles provided to application a by the chosen allocation. To formulate this function, any standard MIP representation for a piecewise linear function can be used, which will induce auxiliary constraints and variables in order to account for the various line segments.

Importantly, the value function is specified in terms of cycles, not in terms of the underlying machine groups and power modes that are supplying them. This results in a significant dimensionality reduction and simplifies the winner-determination problem. However, in cases where differences in the underlying hardware architecture are important to application performance (e.g., all cycles are *not* equal and xeon cycles perform differently from atom cycles), an extension to a more complex higher dimensional model is available; see Reference 17 for details.

Next, we turn our attention to modeling the argument to our function F_a, the total number of cycles provided to application a in a period, defined as

$$Q_a = \sum_{g \in G_a} \sum_{f \in M_g} \sum_{t \in M_g} \gamma_{g,t}(\tau - \delta_{g,f,t}) C^{\text{SOLD}}_{g,f,t,a} \ \forall\, a \in A, \tag{19.3}$$

where G_a is the set of machine groups that can support application a, M_g is the set of power modes available to machines in group g, $\gamma_{g,t}(\cdot)$ is the number of cycles provided by a machine from group g in mode t over a period of time given by its argument, τ is the amount of time in the current period, $\delta_{g,f,t}$ is the amount of time it takes to transition from mode f to mode t, and each $C^{\text{SOLD}}_{g,f,t,a}$ variable defines a quantity of cores (i.e., goods) allocated from group g that were in mode f and are now in mode t (described in more detail in the following text). With these definitions, the expression then captures the total cycles provided to application a, accounting for reductions due to time spent in transitioning power level and considering all machine groups and power levels.

19.2.2 Defining The Goods in the Market

Within each machine group, we track only the number of cores in each power state. An allocation of some quantity of such cores is ultimately mapped into an assignment of cores on physical machines in postprocessing. This avoids the creation of immaterial distinctions that would only complicate winner determination and is similar to complexity-reduction methods that have been used to good effect in recent high profile government bandwidth auctions [18, 19]. We currently use a fast but potentially only approximate greedy assignment for this postprocessing, but more sophisticated methods could be used if the identities of machines in a group are important.

To properly encode the data-center cost model, described in the next section, we need a representation of goods that captures not just static power states but power state *transitions*, enabling us to account for resultant changes in energy usage, cycle loss, and increases in failure rate. Consequently, our goods are manifest in $C^{\text{SOLD}}_{g,f,t,a}$ variables that capture the number of cores in a given machine group starting in mode f in the previous period, transitioning to (the possibly identical) mode t in the current period and assigned to a given application a.

Constraints are defined to ensure that an allocation of these goods will be physically implementable; for example, on present day platforms it is required that all cores on the same physical machine be at the same power level:

$$|\text{CORES}_g| \sum_{f \in M_g} M^{\text{SOLD}}_{g,f,t} = \sum_{f \in M_g} \sum_{a \in A} C^{\text{SOLD}}_{g,f,t,a} + C^{\text{PARTUNSOLD}}_{g,t}, \tag{19.4}$$

$$|\text{CORES}_g| \sum_{f \in M_g} M^{\text{UNSOLD}}_{g,f,t} = \sum_{f \in M_g} C^{\text{UNSOLD}}_{g,f,t} - C^{\text{PARTUNSOLD}}_{g,t}, \tag{19.5}$$

$$\forall\, t \in M_g\ \forall\, g \in G,$$

where $|\text{CORES}_g|$ is the number of cores per machine in group g, $C^{\text{UNSOLD}}_{g,f,t}$ are variables counting the unassigned cores, $M^{\text{SOLD}}_{g,f,t}$ and $M^{\text{UNSOLD}}_{g,f,t}$ count sold and unsold machines, respectively, and $C^{\text{PARTUNSOLD}}_{g,t}$ count the unsold cores on partially sold machines.

Additionally, we need to restrain available supply, through the machine counts:

$$|\text{MACHINES}_g| = \sum_{f \in M_g} \sum_{t \in M_g} M_{g,f,t}^{\text{SOLD}} + M_{g,f,t}^{\text{UNSOLD}} \;\forall\; g \in G, \tag{19.6}$$

where $|\text{MACHINES}_g|$ is the number of machines in group g.

19.2.3 Seller Cost Model

On the supply side of the market, we explicitly model both the hardware and energy costs of running the data center's machines in their various power states. Our model captures the power consumed and performance attained by each machine as a function of the number of active and inactive cores, as measured empirically on an IBM BladeCenter HS21 Server (Figs. 19.5 and 19.6). Modern *dynamic voltage and frequency scaling* (DVFS) enabled machines can have their most efficient state at less than full power: for example, a maximum of 64 versus 50 MCycles/W with four cores active (taking the ratio of the curves in each figure).

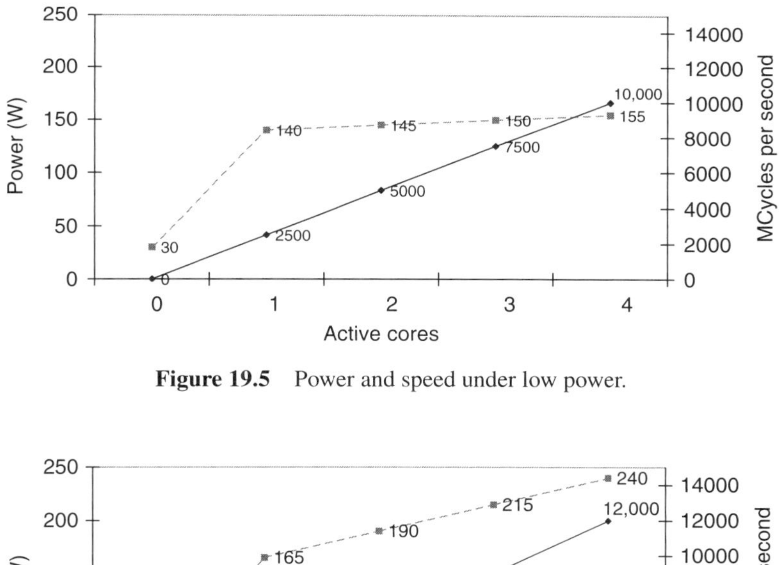

Figure 19.5 Power and speed under low power.

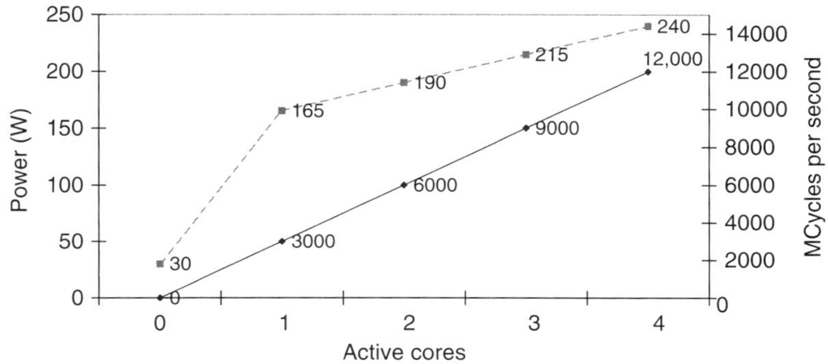

Figure 19.6 Power and speed under high power.

We define the energy requirements (i.e., power over the time period) of the active machines and cores as follows:[4]

$$E^{\mathrm{SOLD}} = \sum_{g\in G}\sum_{f\in M_g}\sum_{t\in M_g} E^{\mathrm{MULT}}(E^{\mathrm{TRANS}}_{g,f,t} + E^{\mathrm{BASE,ACTIVE}}_{g,\tau,t})M^{\mathrm{SOLD}}_{g,f,t}$$
$$+ \sum_{g\in G}\sum_{f\in M_g}\sum_{t\in M_g}\sum_{a\in A} E^{\mathrm{MULT}} E^{\mathrm{CORE,ACTIVE}}_{g,\tau,t} C^{\mathrm{SOLD}}_{g,f,t,a}, \tag{19.7}$$

where $E^{\mathrm{TRANS}}_{g,f,t}$ is the energy required to go from power state f to t, $E^{\mathrm{BASE,ACTIVE}}_{g,\tau,t}$ is the base power for an active machine, and $E^{\mathrm{CORE,ACTIVE}}_{g,\tau,t}$ is the incremental energy needed to run a fully loaded core in this power state. Both of there are parameterized by τ so as to only include energy for that period of time when the machine has reached its destination power level, excluding that used in the transition period (energy for the transition is accounted for by $E^{\mathrm{TRANS}}_{g,f,t}$ and on some hardware may occur at full power regardless of the proximal power levels). Here E^{MULT} accounts for the typically two- to threefold increase in energy needed to run power supply units, uninterruptible power supplies, network switches and storage, and, most importantly, cooling equipment. All elements of this expression are ultimately constants, except for the $M^{\mathrm{SOLD}}_{g,f,t}$ and $C^{\mathrm{SOLD}}_{g,g,f,t}a$ variables.

We stipulate the hardware costs for active cores across the full data center as follows:

$$H^{\mathrm{SOLD}} = \sum_{g\in G}\sum_{f\in M_g}\sum_{t\in M_g} (H^{\mathrm{BASE}}_{g,\tau,g} + H^{\mathrm{TRANSITION}}_{g,f,t})M^{\mathrm{SOLD}}_{g,f,t}, \tag{19.8}$$

where $H^{\mathrm{BASE}}_{g,\tau,g}$ is the prorated cost for each machine (again accounting for only that period of time cycles are actually being provided, as opposed to power level transitions) and includes not only the amortized server cost but also supporting equipment, buildings, and personnel; and $H^{\mathrm{TRANSITION}}_{g,f,t}$ accounts for the cost associated with an increased failure rate on a state transition due to, for example, spinning up/down hard drives. We expect each of these numbers to be easily obtainable through a principled evaluation of existing business practices and capital investments.

Episodic formulations have a common problem in that they may not bear large transition costs when they create a temporary loss in utility, despite a long-term gain. Consequently, we also include sell-side tracking of the power state of machines over previous periods (similarly to the buyer-demand prediction), which can be used to predict the expected time that a machine transitioned to a new state will stay within it. This can be used to amortize transition costs over a good estimate of their appropriate time frame. This adjustment can be calculated exogenously to the rest of the system and is used to adjust the $E^{\mathrm{TRANS}}_{g,f,t}$ and $H^{\mathrm{TRANSITION}}_{g,f,t}$ constants.

A system administrator might, in addition, wish to specify additional restrictions on the allocation based on technical requirements that are not externally visible or

[4]Note this expression captures the entire seller side of the market, so its E^{SOLD} is unparameterized.

those that are unstated by buyers but can reasonably be inferred as being part of their intent. Considerable flexibility is possible; some examples include min/max cores/machines for a given application, min/max energy used in a given machine group or for a given application, and max cores in a given machine group that can be allocated to a given application if a certain number is already allocated to specific alternative application (anti-colocation).

19.3 EXPERIMENTAL RESULTS

We have evaluated our market-based system in a set of simulation experiments to establish computational tractability and effective allocative behavior over a wide range of environments. Each experiment has been performed on a 3.2 GHz dual-processor dual-core workstation with 8 GB of memory and IBM ILOG CPLEX 11.1. Each data point is the mean of 10 randomly generated time-dependent demand traces.

Our synthetic traces are the sum of two sinusoidal curves (e.g., 1-day period with 9000 peak transactions/min plus 1-week period with 36,000 peak transactions/min) and a noise term drawn from a Gaussian with a standard deviation equal to 25% of the signal. These match well with real customer traces, where request density is time dependent and oscillates over both days and weeks [20]. Unlike captured data, synthetic traces enable us to test not only robustness to wide variation in absolute load level but also different amounts of correlation among applications. Each transaction is assumed to use 300 MCycles, which is representative of the processing needed to produce a custom database report. Lastly, each allocation period is 10 min, which is fast enough to react to dynamic changes in the load, but without being so short as to require hysteresis in the model beyond that implicit in the transition costs.

Because no allocator in the literature has comparable capabilities, we adopt as a benchmark a sophisticated *greedy allocator*, which operates as follows:

1. Put all the machines in their highest efficiency state.
2. Determine the target supply for each application by calculating what is required to produce its ideal response time at its 95th percentile of demand.
3. Allocate cores (from feasible machine groups) to the applications, weighted by the marginal value of supply to each application. If an application's supply of high efficiency cores is exhausted, then bump one of the machines supporting it into a higher power state. Stop when either all the targets have been met or all the cores/states have been allocated.
4. Consider each application in turn and trim the allocation until the expected value at the 95th percentile of demand is greater than or equal to the expected cost.
5. Place remaining machines in their lowest power state.

For exposition purposes we consider a simple scenario with two applications (i.e., two customers) and three machine groups (each capable of supporting the first,

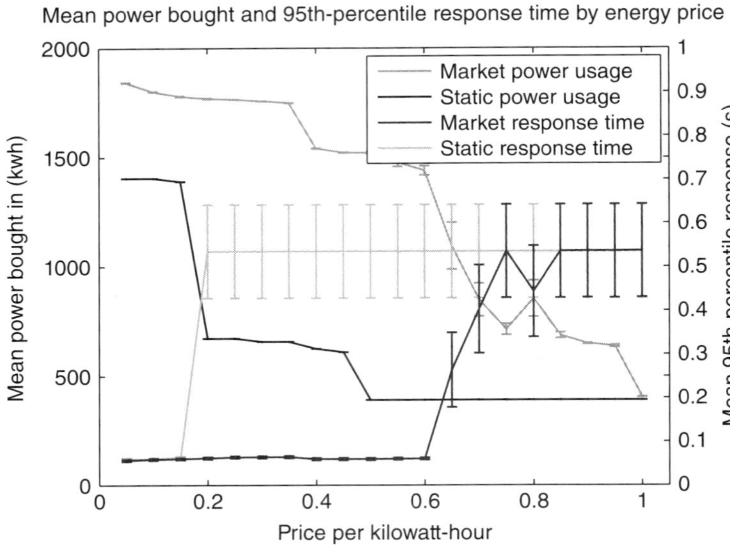

Figure 19.7 Energy used and response time as the cost of energy is varied under market and heuristic algorithms. Bars indicate one standard error over 10 random traces at each price point.

second, and both applications, respectively), for a simulated week of time-varying demand. We also provide results where we vary the price of energy, to demonstrate the flexibility that a market-based allocation scheme can bring to bear.

Figure 19.7 shows the effect of varying the price of energy under both the market and the static-allocation algorithm. We can see that, as expected, under both algorithms, the energy used falls and consequently the mean response time rises as the price of energy is increased. However, bidding proxies in the market (representing customers) find it profitable to purchase enough energy to maintain a near-optimal response time until the price finally reaches such a point that such high energy usage can no longer be sustained and more energy-frugal allocations are chosen.

In Figure 19.8, we see the impact of the choice of these allocations on buyer (i.e., customer) and seller value, as judged by SLAs and revenue net of cost, respectively.

The greedy allocation is cheaper to provide because of the static power levels and also results in significantly lower buyer value over a wide range of prices. The maximum revenue net cost improvement is 137% higher in the market model, although margins become slim when energy is expensive.

It is also important to consider distributive effects to customers in the data-center setting. In this scenario, the A application induces a larger load than B, but with a smaller marginal value for cycles. Consequently, as energy prices rise, the static allocator quickly devotes the limited resources that can be afforded to the B allocation, thereby starving the A application, as seen in Figure 19.9. The market allocation not only maintains the allocation for the B application but also recognizes that some resources can profitably be given to A. This is made possible by switching machines

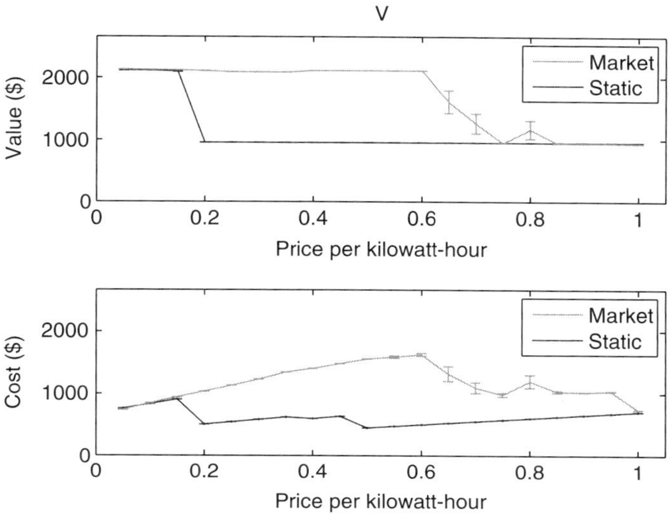

Figure 19.8 Buyer value and seller revenue net cost as the cost of energy is varied under market and heuristic algorithms. (a) Buyer value by energy price and (b) seller cost by energy price.

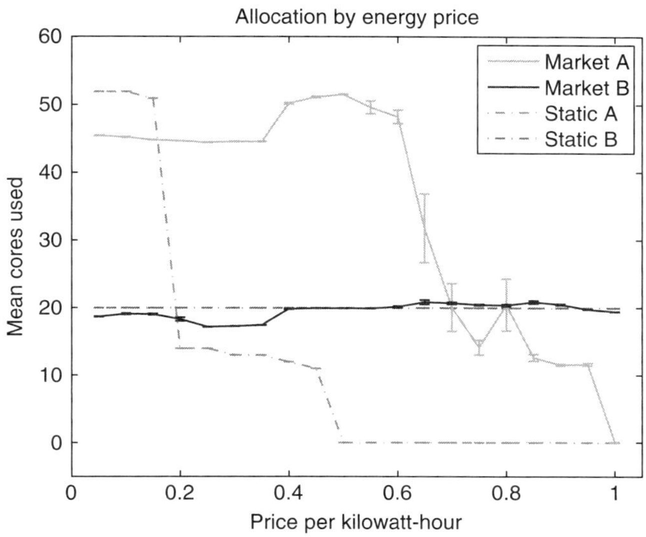

Figure 19.9 Application allocation by energy cost under market and static algorithms.

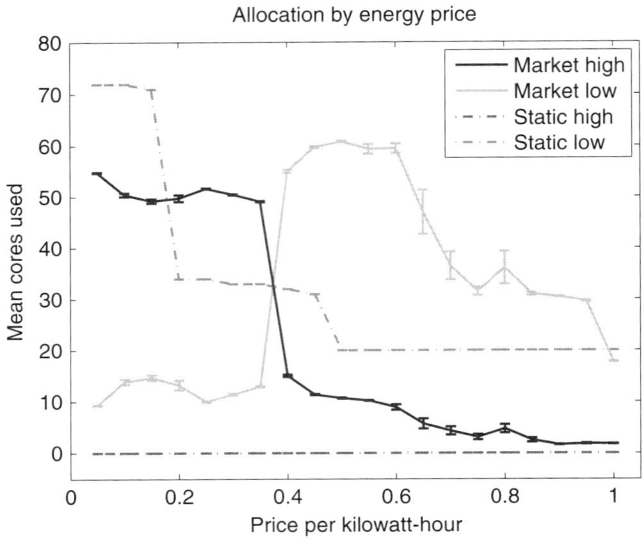

Figure 19.10 Power mode allocation by energy cost under market and static algorithms.

between their most efficient modes to conserve energy and their high power modes to track spikes in demand. Figure 19.10 shows that in this setting the static allocator has placed all the machines in the high efficiency "low power" mode, whereas the market has made use of both modes. When the price for power is low, the most efficient allocation is to maintain quite a few machines in the high power state. However, as the price crosses ¢40/kWh, there is a phase change and it becomes much more efficient to run mostly in the low power mode. Beyond about ¢60/kWh, it becomes impossible to afford the energy needed to maintain a supply sufficient to keep a low response time and the optimal allocation shrinks.

19.3.1 Scalability and Complexity

To evaluate the scalability of our MIP formulation, we evaluated 10 instances of a scenario with 200 quad-core machines in each of five groups for a total of 1000 machines. We configured 10 applications, each with a demand for some 2700 transactions per second, to draw upon these resources with each group supporting three of the applications in a ring topology. We restricted the optimizer to no more than 10 min of computation per instance, taking advantage of the anytime nature of modern MIP solvers. Four of the instances were capped to 10 min, and the average solve time was 5.16 min over all 10 instances, well within the time of a single period. Further, the approximation resulted in only a 0.08% revenue loss when compared to the optimal solution, which would have taken an additional 29 min on an average for these difficult cases. Thus a single machine is capable of optimizing the usage on 1000 others, providing an acceptable 0.1% overhead factor. For a data center with many more machines, one could decompose them into multiple machine pools, each of a size around 1000.

We have also investigated the effect on runtime of the structure of the bipartite graph that defines which application can be supplied by each machine group. For this, we use a scenario with five applications and five machine groups, where supply is set so as to be just sufficient to meet demand. The average solve time of the winner-determination problem increases nonlinearly as we vary the number of edges in the bipartite graph. A graph with 30% of the edges (already highly connected for current data centers) takes only 3.8% of the time needed to clear the market with a complete graph. With 50% connectivity, the computation time has risen to 58.8%, and with 60% connectivity, the timing has already risen to 86.6%. Interestingly, eliminating even a relatively substantial number of edges does not produce a correspondingly large increase in the application response time, as the system can still find an allocation that supports the demand. With 60% of the edges, we are only 8% above the response time of the complete graph. Consequently, moderately restricting the number of machine groups on which an application can run (e.g., for administrative reasons) will only modestly decrease solution quality.

We have shown that a market mechanism can allocate resources under realistic load conditions and more efficiently than the static allocators consistent with current practice. Further, we have shown that such a market can be cleared using a modest amount of computing power within an allocation window that is short enough to capture the essential dynamics of typical loads. So far we have described and evaluated methods for participants to describe their value to the market (the "bidding language" problem) and a method for clearing the market (the "winner-determination" problem). In the rest of the chapter, we first discuss an important class of generalizations to our model and then turn to the essential task of devising prices for the market that have the necessary computational, economic, and game theoretic properties.

19.4 GOING BEYOND PROCESSING AND POWER

So far we have considered only two properties of the data center in our resource allocation: CPU cycles and the energy needed to produce them. In general, such a system can handle any single aspect of the computational resources: CPU, memory, disk space, or bandwidth. In many cases, loads are bound by only a single such attribute, and the system described will be effective; it need not be CPU as we have illustrated but could be another attribute with suitable changes to the response-time model. However, more generally, there will be situations where application performance is heavily dependent on multiple attributes or where the data center is being used by a heterogeneous mixture of applications each with its own attribute performance profile.

To model the cost structure for more attributes is straightforward. We introduce additional variables that track how much of each attribute from a given machine group has been allocated to an application a. As these attributes can be reasonably modeled as having linear costs, these variables can be included in the objective with cost constants that vary by machine group (e.g., some groups have expensive memory or high speed bandwidth; others do not).

In the formulation, we have described up until now, the aggregate goods transferred from one side of the market to the other are total CPU cycles, and these are then assumed by the buyer proxy to be evenly divided across its application's load. There is a simplifying assumption here, in that we have modeled all cycles as fungible (see Reference 17 for a relaxation of this assumption). If we maintain this assumption in our multiple attribute model, we have a variable for the total amount of each attribute assigned to each application. Consequently, the buyer proxy must submit a more complex buyer value function:

$$V_a = F_a(Q_a^{\text{cpu}}, Q_a^{\text{memory}}, \ldots), \tag{19.9}$$

where each of the Q_a^{X} variables is the aggregate amount of attribute X given to application a. As before, we decompose F into a response-time model and a value model over the resulting time:

$$V_a = V_a^{\text{Response/Time}}(R_a(Q_a^{\text{cpu}}, Q_a^{\text{memory}}, \ldots)). \tag{19.10}$$

Thus, moving to multiple attributes does not change the piecewise linear SLA value curve $V_a^{\text{Response Time}}$, as may, for example, be seen in Figure 19.2, that the buyer enters into the system. We still assume value to be a direct function of response time.

However, we previously assumed the response time, $R_a(\cdot)$, was a one-dimensional function of CPU cycles, and now it must be a multidimensional function of multiple attributes. The most straightforward model for this is multiplicative, where the previous queuing model over CPU is multiplicatively altered by a factor for each of the attributes. Adopting a softmax function, to encode a threshold of quantity needed for good performance, is a reasonable approach for modeling the factors.

The bidder proxy would then compose this response-time model with the user-specified value model, yielding a nonlinear multivariate value function. This is sampled to produce a multivariate piecewise linear value function, which is submitted into the market as the buyer's bid.

So there is a straightforward way to extend both the buyer and seller side of the market. The catch comes from the complexity of clearing this market. Methods for encoding a single-dimensional, piecewise linear function into a MIP are very effective if there are a modest number of inflection points. Such methods extend to multiple dimensions but scale in an exponential manner. So working with a small number of attributes can be coupled with exact (or almost exact) MIP solutions. But this approach will be too slow for more than a few attributes, and it will make more sense to skip the linearization step and approximately solve the resulting nonlinear programming problem directly. Such an approach will not guarantee the fully efficient solution, but in the data-center context, local optima may well be sufficient.

19.5 PRICING

We have shown how buyers can specify value over complex combinations of resources and how sellers can specify their costs for providing these resources. And we have shown that a MIP formulation can be used to clear the resulting market and produce an allocation. But what payment should be asked of the buyer in return for these resources?

The simplest solution would be to have fixed prices associated with each resource type and simply charge the users for the amount of resources they actually use. This is the same pricing model as that used in Amazon EC2 (note that here decisions about quantity and quality of machines to buy is optimized by the system; on EC2, such a decision is left to the user who typically does not have the information needed to make an optimal choice). Fixed prices have the advantage of being simple for buyers to understand; they are often a fixed markup over the seller cost, making them easy for sellers to compute. However, they are not adaptive and thus do not take advantage of the information revelation that occurs in the market. Sellers may set the markup too high and thereby miss out on profitable business when buyers are unwilling to meet the price. Alternatively, they may price too low, forgoing revenue they might otherwise enjoy by leaving much of the surplus from trade with the buyers. However, it is important to emphasize that even when using fixed exogenous prices, a market such as the one described here can still be useful as an internal allocation method that attempts to establish the proper trade-off between performance and consumption.

Another alternative is to still use easily understood linear prices (e.g., a single price per unit of each type of good, as opposed to more complex prices based on *bundles* of goods), but to determine what these prices should be, using the bids submitted to the system. This has the advantage of adaptivity, by choosing prices based on the information revealed by the buyers in the course of interacting with the system. However, because the market is combinatorial in nature, we cannot actually find perfect market clearing prices without using nonlinear bundle prices (which we consider shortly). One way around this is to find and use a set of linear prices that come as close to clearing the market as possible using a linear programming approximation; see Reference 21 for details.

However, such formulations are very complex. While participants will immediately understand the linear prices produced, the same cannot be said for the process used to arrive at them. Consequently, we may instead choose to simply charge the buyers their stated value: the so-called "first-price auction." This is both adaptive in the above sense and simple—it is instantly recognizable to anyone who has seen a single good "English" auction for, for example, art, cars, or cattle. There is, however, a problem. As is common we will assume quasi-linearity in the sequel, that is, a buyer's utility is exactly his/her value minus the payment he/she is charged. Suppose buyers act so as to maximize their expected utility and may, therefore, submit bids to the system inconsistent with their true value, if they believe they will get either a better set of resources or a lower payment. The first price rule provides a large such incentive for buyers to benefit from these manipulations [22].

The classic solution to this problem is to instead use the Vickrey—Clarke—Groves (VCG) mechanism [23]. This mechanism clears at the efficient trade and charges each buyer its marginal impact on the economy, that is, each agent *i* is charged the difference between the optimal trade without *i* and with *i*, as measured by the total reported value of the agents *other than i*. This mechanism has many desirable properties. It is *individually rational*, meaning that agents are never forced to incur a negative profit. It is *budget balanced*, meaning that it will never run at a deficit. And, it is *strategyproof*, meaning that agents will have a dominant strategy to truthfully report their bids to the system, solving the problem identified in the first-price auction.

The VCG payment rule has justifiably received an enormous amount of attention in the literature. However, it too has drawbacks, many of which are listed in an oft-cited paper by Rothkopf [24]. Chief among these is that it can produce unacceptably low revenue. Consider an example from Ausubel et al. [25], with two items for sale, A and B. Bidder 1 wants the AB bundle for $1. Bidder 2 wants just A for $1, and Bidder 3 wants B for just $1. VCG gives the goods to Bidders 2 and 3, but charges $0, the marginal impact of each of these players. For these reasons, adopting VCG may not be appropriate in the data center setting, despite its many desirable properties.

Another consequence of the low revenue of VCG is that it can often be profitable for sets of participants to come together and manipulate their bids so as to subvert the system. Informally, VCG is not "collusionproof"—it is not strategyproof for coalitions of agents working together. Payment rules that have this property are said to choose payments in the "Core," and such rules are now available for combinatorial markets of the type proposed here [26]. However, such prices are typically not strategyproof (i.e., the VCG payments, which are the only ones that are strategyproof in sufficiently complex settings, are not in the core). Because of this, it is now common to choose those payments in the buyer-optimal core that are closest to the VCG payments, in an attempt to mitigate the strategic incentives that must inevitably arise [27]; see also Reference 28 for related ideas in the context of two-sided combinatorial markets [28]. Such pricing represents an attempt to balance the need for both reasonable revenue and reasonable incentive properties and as such have recently been used in the United Kingdom to clear combinatorial bandwidth auctions [18, 19, 29].

While appealing in theory and offering increasing practicality, core payments still retain one major disadvantage. They are exceedingly complex and hard for buyers to understand. This is less of a problem when they can be imposed by governments on large players purchasing exceedingly valuable resources, as in bandwidth auctions. But it is less clear that they can be adopted in more modest settings.

One such setting is the sale of sponsored-search advertising on sites such as Google. Most systems in this space have adopted a simple yet effective payment rule known as the generalized second price (GSP) auction [30]. In these settings, the goods being sold can be easily sorted into a value order that is known to be consistent with all the bids—specifically, ads placed at the top of the page are more valuable to every bidder than those placed further down. The basic rule is then very simple: the market is cleared efficiently at the bids, and players are charged not their bid, but the bid of the player they displaced into the next lower spot. That is, they are charged the price required for them to win their slot holding the other bids constant—or equivalently, a separate second-price auction is effectively applied at each slot in turn from highest value to low, among only those bidders who have not already been satisfied. The GSP mechanism is easily understood, reasonably efficient in practice, results in relatively small amounts of strategic behavior (even though it is not strategyproof), and generates reasonable amounts of revenue for the seller. In short, for settings that meet its requirements, it is an extremely effective mechanism.

A payment rule similar in spirit to the GSP rule can easily be constructed for our setting. First, the efficient allocation at the reported bid is determined, as we have described. We then consider each agent-allocation pair in turn, ordered from highest reported value to lowest. For each such pair, we find the highest value for the buyer's allocation among all buyers further down the list and charge this for the payment. In other words, we charge each buyer the price needed to win the bundle away from the next-most eager bidder. While not strategyproof, or core, there is reason to believe such a rule should work well in practice, and it is certainly simple.

Any of the above payment rules could potentially be adopted for use with our mechanism. As we have discussed, each comes with its own pros and cons. Considering them in balance, we can easily see a role for several of the choices. Clearly fixed prices will continue to be important; we argue that they can indeed be coupled with more complex bidding and allocation systems. We believe core-based rules will continue to garner significant academic interest and, provided that implementations can hide their complexity, may well see practical application in these settings. But we see the fastest immediate impact in GSP-like rules that are simple and should have reasonable properties in practice.

19.6 CONCLUSIONS

We have established that suitably designed combinatorial markets can find practical application to power-managed resource allocation in data centers. Further, it is possible to inject revenue-based utility functions directly into the present data-center business/allocation model without the large changes associated with utility computing; this creates the streamlined migration path required for rapid industry adoption. Such markets obviate the need for operators to divine their customers' value profile, quantify the trade-offs of multiobjective optimization, and facilitate the use of combinatorial optimization in a scalable way, provided carefully designed models are used.

There are many intriguing avenues for future work in this space. First, the results presented here are in simulation; a real-world implementation and trial is thus a clear direction to take. One initial step in this direction might be to evaluate the market approach on a model of a Google compute cluster based on the real-world demand traces they have recently released [31]. More generally, one possibility is to consider richer SLA models, such as models that capture the importance of the timing of failures. For most applications, it is better to suffer an occasional transaction failure over a long period of time than to become completely unresponsive for even a short time period—yet the standard percentile-based SLA model does not distinguish between these cases. The present system is straightforwardly scalable by simply deploying multiple copies of the market. However, this approach makes each market an island. Consequently, very large installations may need a way to transfer loads across these islands, and a higher level market is a reasonable way to perform such coordination. Thus it makes sense to consider a hierarchical allocation paradigm, where a market clears at each level, a compelling generalization of the present approach.

ACKNOWLEDGMENTS

Our profound thanks to Jeff Kephart and Rajarshi Das, our coauthors on much of the work described herein. Any errors are our own. An earlier version of this work appeared in IJCAI-09 [32]. We would also like to thank IBM Research where much of this work took place.

REFERENCES

1. US EPA. *Report to Congress on Server and Data Center Energy Efficiency*, Aug. 2007.
2. US DOE and EPA. *Carbon Dioxide Emissions from the Generation of Electric Power in the United States*, Jul. 2000.
3. US EIA. *Emissions of Greenhouse Gases Report: Carbon Diaoxide Emissions*, Dec. 2008.
4. J. Koomey. *Growth in Data Center Electricity Use 2005 to 2010*. Analytics Press, Oakland, CA. Aug. 1 2010, 2011.
5. J. Glanz. Power, polution and the internet. New York Times, Sept. 22 2012.
6. I. E. Sutherland. "A futures market in computer time," *Communications of the ACM*, 11(6), 1968, 449–451.
7. T. Sandholm. "Expressive commerce and its application to sourcing: how we conducted \$35 billion of generalized combinatorial auctions," *AI Magazine*, 28(3), 2007, 45.
8. J. S. Chase, D. C. Anderson, P. N. Thakar, A. M. Vahdat, and R. P. Doyle. Managing energy and server resources in hosting centers. In *Proceedings of the 18th ACM Symposium on Operating Systems Principles (SOSP-01)*, pp. 103–116. ACM, New York, NY, 2001.
9. Y. Chen, A. Das, W. Qin, A. Sivasubramaniam, Q. Wang, and N. Gautam. Managing server energy and operational costs in hosting centers. In *Proceedings ACM SIGMETRICS Int. Conf. on Measurement and Modeling of Computer Systems (05)*, pp. 303–314. ACM, New York, NY, 2005.
10. C. Low and A. Byde. Market-based approaches to utility computing. Technical Report 23, HP Laboratories, Bristol, Feb. 2006.
11. A. Byde. A comparison between mechanisms for sequential compute resource auctions. In *Proceedings of the 5th International Joint Conference on Autonomous Agents and Multiagent Systems (AAMAS-06)*, 8(12), 1199–1201, 2006.
12. C. S. Yeo and R. Buyya. "A taxonomy of market-based resource management systems for utility-driven cluster computing," *Software, Practice & Experience*, 36(13), 2006, 1381–1419.

13. J. Broberg, S. Venugopal, and R. Buyya. "Market-oriented grids and utility computing: the state-of-the-art and future directions," *Journal of Grid Computing*, 6(3), 2007, 255–276.

14. P. Cramton, Y. Shoham, and R. Steinberg, editors, *Combinatorial Auctions*. MIT Press, Cambridge, MA, Jan. 2006.

15. J. O. Kephart and R. Das. "Achieving self-management via utility functions," *IEEE Internet Computing*, 11(1), 2007, 40–48.

16. M. Steinder, I. Whalley, J. E. Hanson, and J. O. Kephart. Coordinated management of power usage and runtime performance. In *Proceedings of the 9th International Symposium on Integrated Network Management (NOMS-08)*, pp. 387–394, 2008.

17. M. Guevara, B. Lubin, and B. C. Lee. Navigating heterogeneous processors with market mechanisms. In *Proceedings of the 19th International Symposium on High Performance Computer Architecture (HPCA-13)*, 2013. Forthcoming.

18. P. Cramton. A review of the l-band auction. Technical report, Office of Communications, United Kingdom, Aug. 2008.

19. P. Cramton. A review of the 10-40ghz auction. Technical report, Office of Communications, United Kingdom, Sept. 2008.

20. G. Pacifici, W. Segmuller, M. Spreitzer, and A. Tantawi. "CPU demand for web serving: measurement analysis and dynamic estimation," *Performance Evaluation*, 65(6), 2008, 531–553.

21. B. Lubin, D. Parkes, J. Shneidman, S. Lahaie, R. Cavallo, and A. Juda. "Ice: an expressive iterative combinatorial exchange," *Journal of Artificial Intelligence Research*, 33, 2008, 33–77.

22. B. Lubin and D. C. Parkes. Quantifying the strategyproofness of mechanisms via metrics on payoff distributions. In *Proceedings of the 25th Conference on Uncertainty in Artificial Intelligence (UAI-09)*, 2009.

23. V. Krishna. *Auction Theory*. Academic Press, University Park, PA, 2002.

24. M. H. Rothkopf. "Thirteen reasons why the vickrey-clarke-groves process is not practical," *Operations Research*, 55(2), 2007, 191–197.

25. L. M. Ausubel and Paul Milgrom. "Ascending auctions with package bidding," *Frontiers of Theoretical Economics*, 1(1), 2002, 1–42.

26. R. Day and P. Milgrom. Core-selecting package auctions. *International Journal of game Theory*, 36(3), 2008, 393–407.

27. R. W. Day and P. Cramton. The Quadratic Core-Selecting Payment Rule for Combinatorial Auctions. Working Paper, Sept. 2008.

28. D. C. Parkes, J. R. Kalagnanam, and M. Eso. Achieving budget-balance with vickrey-based payment schemes in exchanges. In *Proceedings of the 17th International Joint Conference on Artificial Intelligence*, pp. 4–10, 2001.

29. P. Cramton, E. Kwerel, G. Rosston, and A. Skrzypacz. "Using spectrum auctions to enhance competition in wireless services," *Journal of Law and Economics*, 54(4), 2011, S167–S188.

30. B. Edelman, M. Ostrovsky, and M. Schwarz. "Internet advertising and the generalized second price auction: selling billions of dollars worth of keywords," *American Economic Review*, 97(1), 2007, 242–259.

31. C. Reiss, J. Wilkes, and J. L. Hellerstein. Google cluster-usage traces: format + schema. Technical report, Google Inc., Mountain View, CA, USA, Nov. 2011. Revised

2012.03.20. Posted at URL http://code.google.com/p/googleclusterdata/wiki/Trace Version2.

32. B. Lubin, J. Kephart, R. Das, and D. C. Parkes. Expressive power-based resource allocation for data centers. In *Proceedings of the 21st International Joint Conference on Artificial Intelligence (IJCAI-09)*, 2009.

INDEX

Smart Data Pricing, First Edition. Edited by Soumya Sen, Carlee Joe-Wong, Sangtae Ha, and Mung Chiang.

WILEY SERIES ON INFORMATION AND COMMUNICATION TECHNOLOGY

Series Editors: T. Russell Hsing and Vincent K. N. Lau

The Information and Communication Technology (ICT) book series focuses on creating useful connections between advanced communication theories, practical designs, and end-user applications in various next generation networks and broadband access systems, including fiber, cable, satellite, and wireless. The ICT book series examines the difficulties of applying various advanced communication technologies to practical systems such as WiFi, WiMax, B3G, etc., and considers how technologies are designed in conjunction with standards, theories, and applications.

The ICT book series also addresses application-oriented topics such as service management and creation and end-user devices, as well as the coupling between end devices and infrastructure.

T. Russell Hsing, PhD, is the Executive Director of Emerging Technologies and Services Research at Telcordia Technologies. He manages and leads the applied research and development of information and wireless sensor networking solutions for numerous applications and systems. Email: thsing@telcordia.com

Vincent K.N. Lau, PhD, is Associate Professor in the Department of Electrical Engineering at the Hong Kong University of Science and Technology. His current research interest is on delay-sensitive cross-layer optimization with imperfect system state information. Email: eeknlau@ee.ust.hk

Wireless Internet and Mobile Computing: Interoperability and Performance
Yu-Kwong Ricky Kwok and Vincent K. N. Lau

Digital Signal Processing Techniques and Applications in Radar Image Processing
Bu-Chin Wang

The Fabric of Mobile Services: Software Paradigms and Business Demands
Shoshana Loeb, Benjamin Falchuk, and Euthimios Panagos

Fundamentals of Wireless Communications Engineering Technologies
K. Daniel Wong

RF Circuit Design, Second Edition
Richard Chi-Hsi Li

Networks and Services: Carrier Ethernet, PBT, MPLS-TP, and VPLS
Mehmet Toy

Equitable Resource Allocation: Models, Algorithms, and Applications
Hanan Luss

Vehicle Safety Communications: Protocols, Security, and Privacy
Luca Delgrossi and Tao Zhang

WiFi, WiMAX, and LTE Multi-hop Mesh Networks: Basic Communication Protocols and Application Areas
Hung-Yu Wei, Jarogniew Rykowski, and Sudhir Dixit

Smart Data Pricing
Edited by Soumya Sen, Carlee Joe-Wong, Sangtae Ha, and Mung Chiang